Algal Adaptation to Environmental Stresses

Springer

Berlin
Heidelberg
New York
Barcelona
Hong Kong
London
Milan
Paris
Singapore
Tokyo

L. C. Rai • J. P. Gaur (Eds.)

Algal Adaptation to Environmental Stresses

Physiological, Biochemical and Molecular Mechanisms

With 67 Figures and 23 Tables

Springer

EDITORS:

Professor Dr. L. C. Rai
Professor Dr. J. P. Gaur
Banaras Hindu University
Department of Botany
Varanasi, 221005
India

Springer-Verlag Berlin Heidelberg New York

Library of Congress Cataloging-in-Publication Data
Algal adaptation to environmental stresses: physiological, biochemical and molecular mechanisms / L.C. Rai, J.P. Gaur, (eds.).
p. cm.
Includes bibliographical references (p.). ISBN-13: 978-3-642-63996-8 (hardcover) 1. Algae--Ecophysiology. 2. Algae--Adaptation. I. Rai, L. C. (Lal Chand), 1950- II. Gaur, J. P. (Jai Prakash), 1952-
QK565 .A3748 2001 579.8'17--dc21 2001042016

Springer-Verlag Berlin Heidelberg New York
a member of BertelsmannSpringer Science+Business Media GmbH
http://www.springer.de

ISBN-13: 978-3-642-63996-8 e-ISBN-13: 978-3-642-
DOI: 10.1007/978-3-642-59491-5

Softcover reprint of the hardcover 1st edition 2001

Cover Design: *design & production*, Heidelberg
Typesetting: Camera-ready by the editors

SPIN: 10743684 31/3130xz – 5 4 3 2 1 0 – Printed on acid free paper

Preface

Algae, including cyanobacteria, are primarily aquatic organisms. They display a wide range of morphological diversity and inhabit diverse aquatic systems. Algae often occur in extremely hostile environments, and this is a reflection of their remarkable ability to tolerate various kinds of stresses — natural as well as those resulting from human activities. They have to negotiate with low concentrations of essential nutrients (especially carbon, nitrogen, phosphorus and trace elements) in natural waters and, low availability of light and temperature on one hand. On the other, they have to survive and grow in habitats enriched with salts, toxic metals and pesticides, elevated temperature, UV-B radiation and light intensity. These organisms are also often stressed by extremes of pH. Algae are the principal primary producers of waterbodies — from a small rain puddle to the big oceans. Hence the tolerance of these organisms to diverse stresses assumes tremendous relevance from an ecological standpoint.

The basic metabolic machinery of algae is by and large similar to that of higher plants. This characteristic makes the study of various mechanisms of stress tolerance in algae particularly befitting. It is quite likely that information generated on algae may eventually prove useful for higher plants as well. In fact, many important metabolic pathways of plants were discovered using algae as the experimental organisms; of course, they were subsequently reported from higher plants as well. This fact is best exemplified by the initial researches on photosynthesis that were largely based on information generated using microalgae as the experimental organisms.

The idea to produce this book crystallized during our visits to Germany and Japan, where we had some free time and, could freely exchange ideas with distinguished researchers of our host institutes.

Twelve chapters of the book present a comprehensive account of algal adaptation to major environmental stresses. The introductory chapter has been contributed by Prof. G.E. Fogg, FRS, one of the most distinguished phycologists of our time. Initially, we were hesitant in requesting Prof. Fogg to write the chapter because of his age, but his kind acceptance of our proposal speaks volume of his commitment for Phycology. The other chapters are authored by workers who are widely acclaimed in their respective fields of research. We are grateful to each one of them for graciously complying with various suggestions that we put before them as editors of this book.

Dr. D. Czeschlik and Ms. Christiane Glier, Editorial Office, Springer, rendered valuable help during preparation and editing of the present volume. We were ably assisted by our research students in checking references and other parts of various manuscripts, Subhash Singh, Ajay Singh and Sanjay Singh helped us in word-processing and preparation of camera-ready version of the book. LCR is thankful to the Ministry of Environment and Forests, Govt. of India, for the Pitamber Pant National Environment Fellowship Award, and to NSF, USA and DST, India, for financial support through an Indo-US Project.

The book is dedicated to the memory of the late Prof. Ram Nagina Singh, the legendary phycologist and a great human being.

Varanasi, India, March 2001 L.C. Rai and J.P. Gaur

Contents

Contributors

Axelsson, Lennart
Kristineberg Marine Research Station,
Kristineberg 2130, SE-450 34 Fiskebäckskil, Sweden
l.axelsson@kmf.gu.se

Beer, Sven
Department of Plant Sciences, Tel Aviv University,
Tel Aviv 69978, Israel
svenbeer@post.tau.ac.il

Degenhardt, B.
Julius-von-Sachs Institute of Bioscience
University of Würzburg, Julius-von-Sachs Platz 2,
D-97082 Würzburg, Germany

Erdmann, Norbert
Department of Biosciences,
University of Rostock, 18051 Rostock, Germany
ne@bio4.uni-rostock.de

Falkner, Gernot
Institute of Limnology, Austrian Academy of Sciences,
Gaisberg 116, A-5310 Mondsee, Austria
gernot.falkner@oeaw.ac.at

Fogg, G.E.
School of Ocean Sciences,
University of Wales (Bangor), Marine Science Laboratories,
Menai Bridge, Isle of Anglesey, LL59 5EY, UK

Fujita, Yoshihiko
Department of Marine Bioscience, Fukui Prefectural University,
Obama, Fukui 917, Japan
fujitay@fpu.ac.jp

Gaur, J.P.
Department of Botany,
Banaras Hindu University, Varanasi 221 005, India
jpgaur@banaras.ernet.in

Gimmler, H.
Julius-von-Sachs Institute of Bioscience
University of Würzburg, Julius-von-Sachs Platz 2,
D-97082 Würzburg, Germany
gimmler@botanik.uni-wuerzburg.de

Häder, Donat-P.
Friedrich-Alexander-Universität, Erlangen-Nürnberg,
Institut für Botanik und Pharmazeutische Biologie, Staudtstr. 5,
D-91058 Erlangen, Germany

Hagemann, Martin
Department of Biosciences,
University of Rostock, 18051 Rostock, Germany
mh@bio4.uni-rostock.de

Kleiner, Diethelm
Mikrobiologie, Universität Bayreuth,
Universitatstrasse 30, Postfach 101251,
D-95440 Bayreuth, Germany
diethelm.kleiner@uni-bayreuth.de

Koenig, F.
Fachbereich 2: Biologie/Chemie, Molekulare Pflanzenphysiologie,
Universität Bremen, Postfach 33 04 40, 28334, Bremen, Germany
fkoenig@uni-bremen.de

McKay, R. Michael L.
Department of Biological Sciences,
Bowling Green State University, Bowling Green,
OH 43403, USA
rmmckay@bgnet.bgsu.edu

Murakami, Akio
Kobe University Research Centre for Inland seas,
Iwaya, Awajicho, Hyougo 656, Japan
akiomura@kobe-u.ac.jp

Nalewajko, Czesia
Life Sciences Division, Scarborough College,
University of Toronto, Scarborough, ON, Canada.
nalewajko@scar.utoronto.ca

Ohki, Kaori
School of Marine Sciences and Technology,
Tokai University, Shimizu, Shizuoka 424, Japan
kaorohki@fpu.ac.jp

Peschek, Guenter A.
Molecular Bioenergetics Group, Institute of Physical Chemistry,
University of Vienna, Währingerstrasse 42 and Althanstrasse 14,
A-1090 Wien, Austria
guenter.Peschek@univie.ac.at
http://mailbox.univie.ac.at/Guenter.Peschek

Rai, L.C.
Department of Botany,
Banaras Hindu University, Varanasi 221 005, India.
lcrai@banaras.ernet.in

Shehawy, Rehab M.
Mikrobiologie, Universität Bayreuth,
Universitatstrasse 30, Postfach 101251,
D-95440 Bayreuth, Germany

Twiss, Michael R.
Department of Applied Chemical and Biological Sciences,
Ryerson Polytechnic University, 350 Victoria Street, Toronto,
ON, M5B 2K3, Canada.
m2twiss@acs.ryerson.ca

Wagner, Ferdinand
Institute of Plant Physiology,
University of Salzburg, Hellbrunnerstr. 34,
50202, Salzburg, Austria

Zoder, Roland
Molecular Bioenergetics Group, Institute of Physical Chemistry,
University of Vienna, Währingerstrasse 42 and Althanstrasse 14,
A-1090 Wien, Austria

1 Algal Adaptation to Stress — Some General Remarks

G.E. Fogg*

Contents

*School of Ocean Sciences, University of Wales (Bangor), Marine Science Laboratories, Menai Bridge, Isle of Anglesey, LL59 5EY, UK

1.1 Introduction: Definition of Stress

Stress is a difficult term to define in a biological context. The relevant meaning of the word given by the Shorter Oxford English Dictionary (1933) was "the overpowering pressure of some adverse force or influence". In science it was originally used in a purely mechanical sense, e.g. a "force acting on contiguous surfaces of a body and tending to disarrange its particles" (Beadnell 1942), which still remains an appropriate definition to apply to the operation of water movements on an algal thallus. A reasonable extension is to take "tending to disarrange its particles" as including alterations in physiological processes and metabolic patterns, so making it applicable generally to effects on the activities of living organisms.

However, it is an inherent characteristic of any living organism that it responds to stimulus, that is to say to any change, large or small, in its environment which affects biological processes. The problem is where to draw the line between the almost incessant succession of varying stimuli, to which an organism adapts without dislocation of normal functions, and more powerful influences which seriously disorganize vital activities. As we shall see, response to stimuli and response to stress have it in common that they tend to act to the ultimate benefit of the organism. The two intergrade imperceptibly. This difficulty is evident in the fact that in different books on the same biological topic the word stress figures prominently in the indices of some, but not at all in others.

Among the more recent definitions, Grime (1989) has designated stress as an external constraint "limiting the rates of resource acquisition, growth or reproduction of organisms". Much the same definition was adopted by Fowden et al. (1993) in a book that is to a considerable extent concerned with the effects of stress on the yields of crop plants. However, this definition would not necessarily include mechanical stress and, indeed, Dring (1982), to cover this point, distinguishes between "disturbance", which limits algal growth by partial or total destruction of biomass, and "stress", which limits the amount of dry matter produced. Grimes' definition requires measurements which are not easily made in the field, and is highly dependent on the time factor. The short- and long-term effects of stress may be diametrically opposed (Rees and Lawton 1993). Thus, diatom community growth may be inhibited by short-term exposures to UV radiation but stimulated by similar treatment over a long term (Bothwell et al. 1993). The definition of stress "as an environmental condition that reduces Darwinian fitness when first applied" (Sibly and Calow 1989) is perhaps even more difficult to use at the practical level. Yet another definition —"Any change in the environmental conditions is experienced as a stress that threatens the normal metabolic balance and produces a response designed to counter the effects of the disturbing influence" (Franks et al. 1990) — scarcely distinguishes between stimuli and stresses, "normal" is an ambiguous word, and "designed" introduces a teleological element. Another subjective factor may creep in: we often have the presumption to assume what may be best for an organism. An example is that we instinctively think of as stressful the conditions of intense cold, desiccation, minimal or excessive light and mineral deficiency under which lives the green alga *Hemichloris antarctica*, endemic in the dry valleys of Antarctica. Nevertheless, this species has survived there, successfully and seemingly knowing no better, at the limits of tolerance for a green alga for many thousands of years (Nienow and Friedmann 1993). It now seems fully adapted to the situation and should not be thought of as stressed under present conditions. Algae in hot springs and highly saline waters similarly seem fully adapted to extreme stress.

Perhaps precise definition is not necessary. The chapters in this book have coherence and together make a valuable contribution to important aspects of algal physiology and biochemistry.

1.2 Types of Stress

The stresses to which algae may be subjected are divided by Davison and Pearson (1996) into two kinds — limitation stress, caused by inadequate supply of resources (e.g. low light or nutrient deficiency), and disruptive stress (resulting from damage caused by adverse conditions or by allocation of resources to prevent damage). For present purposes it seems better to distinguish kinds — physical, nutritional and biochemical — according to the nature of the stress itself rather than to its outcome. Grazing by animals, parasitization, disease, and human interference, while undoubtedly stressful, are outside the scope of this book and are not considered.

1.2.1 Mechanical

Besides the disturbance by mechanical agencies, which may result in destruction of algal biomass, there are also stress effects in the sense of Dring (1982). Apart from the damage which algae anywhere may sustain by impact with solid bodies, aquatic forms may be stressed by water flow and turbulence. The biologically important dividing line between turbulent and laminar flows is determined by the Reynolds number, defined as $R_r = \rho\ uL/\eta$, where ρ = density, u = velocity, L = a linear dimension characteristic of the situation under consideration, and η = viscosity. The critical value of R_r for water lies between 500 and 2000. Below 500, flow is laminar and organisms in this domain, ultraplankton less than 20 μm in linear dimensions, are not subject to turbulent stress (Fogg 1991). Larger plankton algae are dependent on turbulence for transport of nutrients but are not stressed mechanically by it to any appreciable extent. However, water movement has a major mechanical impact on benthic macroalgae attached to a substratum. Water currents exert a drag on such algae, the magnitude of which varies according to the pattern of flow and the shape and size of the algal body. Most macroscopic algae are pliant and respond to the tension, shear, bending, and twisting imposed by moving water. This flexibility reduces drag. The hydrodynamic forces can also be reduced by mucilage secretion, which increases viscosity locally and so favours laminar flow rather than turbulence (see the above equation). Nevertheless, the mechanical stresses imposed on the thallus may be powerful enough to tear tissues or dislodge the plant altogether (Koehl 1986). Natural situations in which algal communities are stressed by water movement are usually complicated in both pattern and variations with time. Stream periphyton is subject to shear stress, which for much of the time may not vary to any great extent, but spates may impose damaging stress. Communities differ in their resistance, e.g. non-filamentous diatom assemblages are more resistant than filamentous communities. Spates have important effects on the longer-term composition and dynamics of periphyton growths (Biggs and Thomsen 1995). Algae on wave-swept shores experience accelerating forces as well as drag. It is not easy to measure these forces. Jones and Demetropoulos (1968), using a simple dynamometer found maximum dynamic pressures of over 1 kg cm^{-2} on an exposed shore on Anglesey, N Wales. It is obvious on casual inspection that such forces are frequently sufficient to tear the tissues of seaweeds. The total drag force exerted is proportional to surface area and the reaction force is proportional to volume, so that these become increasingly important as the size of an alga increases. The form and flexibility of large intertidal algae are such as to minimise these stresses. Their cell walls are composed of a variety of different long-chain polysaccharides, the bonding of which may be varied during synthesis to produce gels or fibres, making different degrees of tensile strength or flexibility possible. However, a structural feature which improves resistance to mechanical stress may reduce performance of some other function and there have to be tradeoffs between the two (Koehl 1986). A simple example of this is density of plants per unit area; the higher it is, the greater will be the resistance to the stress of water movement, but the less will be photosynthetic efficiency. The balance struck will depend on the conditions under which the alga has grown. Undoubtedly, resistance to mechanical stress is an important factor determining the location and success of a seaweed on a rocky shore but interactions with other stresses, all compounded by the daily rises and falls of the tide, make it difficult to interpret its role in establishing the observed patterns of the distribution of littoral and sublittoral algae (see Sect. 1.6).

1.2.2 Other Physical Stresses

1.2.2.1 Visible Radiation

Here, one assumes that algae are adapted to the normal diurnal variations in intensity and quality and that the smaller, planktonic, algae have the capacity to adapt to changes in irradiance according to position in the water column, within a cell division cycle. Under natural conditions stress arises because of interposition of shading material or short-term change in the light field caused by emersion or immersion. Excess of photosynthetically active radiation or prolonged dim light or darkness are unequivocally stressful for a photosynthetic plant. Photoperiod may also be important (Dring 1984). Algae have endogenous rhythms in activities such as photosynthesis, bioluminescence and cell division. Thus, when average generation times are equal to or longer than 24 h, cell division may become entrained by the light-dark cycle with bursts of cell division during the dark period (Fogg and Thake 1987). Alteration of the light-dark cycle will upset the balance of rhythmically varying processes and can be regarded as causing stress.

1.2.2.2 Ultraviolet Radiation

Outside the visible spectrum the only stressful radiation likely to be encountered under natural conditions is ultraviolet. UV-B (wavelengths between 280 and 320 nm) is particularly destructive of DNA, causing mutations, and also of proteins and photosynthetic pigments. The intensity of UV-B reaching the Earth's surface is largely controlled by the concentration of ozone in the stratosphere. The discovery of a deficiency of this screening gas, a "hole", over Antarctica in the austral spring has concentrated attention on the factors determining ozone concentrations in the atmosphere and on the stresses imposed by exposure to UV-B (Fogg 1998).

1.2.2.3 Temperature

Many algae are adapted to live at temperatures which are extremely high or low by human standards. These cannot, strictly speaking, be said to suffer from stress. Stress may occur when the temperature is suddenly changed. The multifarious biochemical processes making up the metabolic network in an organism are affected differently by temperature change so that the pattern is distorted to a greater or lesser extent — individual enzymes may be inactivated or increased in activity, metabolites degraded or altered in concentration, and vital structures such as cell membranes become altered in conformation. There is considerable scope for adjustment and the organism may be able to adapt to the new temperature within a short time. If it cannot, then it functions less efficiently, is stressed, and may die if the distortion is extreme. Temperature change may occur with wind direction, sunshine coming or going, the tide retreating or advancing, frost or snow. In monitoring such changes in the natural environment it must always be remembered that the temperature in the microclimate in the alga's immediate vicinity may differ considerably from that of the wider surroundings. In studying the effects of low temperatures it has to be borne in mind that low temperature per se is not necessarily damaging to an organism but the formation of ice crystals may be lethal by causing mechanical damage or by changing concentrations of osmotically active solutes (Fogg 1998). Thus, it is necessary to distinguish between chill and freezing. At high temperatures deficiency of oxygen, which is much less soluble in hot than in cold water, may be the proximal cause of stress (Brock 1978).

1.2.2.4 Osmotic Stress

An algal cell in an aqueous medium has to maintain a balance between intake and loss of water to avoid disruptive change in volume. An abrupt change in the osmotic potential of the medium will impose stress by plasmolysis, leading to inhibition of photosynthesis and other functions, on the one hand, or distension ultimately leading to bursting on the other. Osmoregulation may be achieved and stress ameliorated, usually within an hour or two, by regulated uptake of ions, by synthesis of osmotically active substances compatible with metabolic processes, by expulsion of water via contractile vacuoles or, in species with rigid cell walls, by counterbalancing osmotic pressure by turgor pressure. As with tolerance to extremes of temperature, capacity to grow under hypersaline conditions by virtue of an effective osmoregulatory mechanism is an example of successful genetic adaptation to stress.

1.2.2.5 Desiccation

Terrestrial algae, especially those in frigid environments, freshwater algae in shallow waters and littoral marine algae are often subjected to desiccation. Some species are remarkably resistant to this and may remain dormant when dry for as much as a century, but loss of water always results, sooner rather than later, in reduction, then cessation, of photosynthesis and growth. There can be no doubt that here we are dealing with stress in the proper sense of the word.

1.2.3 Nutritional and Biochemical Stress

1.2.3.1 Nutrient Deficiency

This is an area of particular interest to limnologists and oceanographers. Lack of sufficient amounts of essential nutrients, either inorganic or organic, produces stress in the strict sense, but this does not simplify its detection. A first point, often overlooked, is that the effect of deprivation of a particular nutrient depends on whether this takes place gradually or abruptly. In laboratory experiments a favoured approach has been to grow batch cultures in medium with limiting amounts of the nutrient under consideration and compare the state of the cells when the stationary stage of growth is reached with that in control material supplied with ample nutrient. Another approach is to transfer material growing on a complete medium into one that contains none of the particular nutrient and to compare the state of the cells before and after the transfer. These approaches give entirely different results because the cycle of cell division is stopped at different points in the two treatments. In a nitrogen-deficient batch culture in the stationary phase, for example, cells are halted just before cell division, probably as a result of the accumulation of inhibitory products as well as nutrient depletion. Photosynthesis then becomes directed towards highly reduced "storage" products, such as fat. An actively growing population, on the other hand, consists mostly of young cells, which, when deprived of nitrogen, can still mature and divide to produce a second generation which is then unable to develop normally and has its metabolic pattern fixed in a state favouring synthesis of carbohydrates (Fogg and Thake 1987). Continuous cultures, which are equilibrium systems in which nutrient levels are kept constant, give quite different results again from those just described (Van Liere 1979).

Detection of nutrient stress in natural populations is equally fraught with difficulty. Ratios of average contents of nutrient elements, e.g. the Redfield ratio of C_{106}:N_{16}:P_1 characteristic of oceanic phytoplankton, are no good guide, since determination of such ratios is subject to considerable experimental error and appears to be valid only for populations which are not light-limited (Tett et al. 1985). Tissue concentrations of particular elements or biochemical components are similarly unsatisfactory indices of nutrient status in seaweeds (Davison and Pearson 1996). Nutrient concentrations in natural waters are also a misleading guide, since it is the rate of turnover, not the actual concentration, which determines the supply to the algal cells. A low concentration of ammonia, for example, may suggest nitrogen stress, but if this source is being regenerated rapidly an organism may still receive an ample supply. Furthermore, algae may accumulate reserves of nitrogen or phosphorus which will carry them through several successive divisions before a stress is imposed. Thus, the freshwater diatom *Asterionella formosa* can store sufficient polyphosphate reserve to carry it through nearly seven doublings without further supplies (Mackereth 1953). Cyanobacteria accumulate nitrogen in the form of granules of the polypeptide cyanophycin and may also use their nitrogenous photosynthetic pigments, phycobilins, as reserves (Simon 1987). Enzymic activity, however, can be a good indicator of stress due to lack of certain elements. This works particularly well with phosphorus-deficient algae, which show marked increase in phosphatase activity as compared with similar material with ample phosphate (see Fogg and Thake 1987; Davison and Pearson 1996). Bioassay by supplementation of water with nutrients, singly or in combination, and observation of any resulting growth or activity, is a useful technique for detecting nutrient stress but needs to be conducted with meticulous care and careful interpretation (see Fogg and Thake 1987). A potentially unambiguous method for detecting which nutrients limit the growth of algae depends on the production of specific proteins by stressed organisms. An example of this is the production of siderophores (proteins having a strong affinity for iron) by microorganisms stressed by iron deficiency. There is a possibility that antibodies to such proteins can be used as probes to indicate which nutrient limits photosynthetic carbon fixation in the ocean (La Roche et al. 1993; Davison and Pearson 1996; Graziano et al. 1996).

1.2.3.2 pH

Natural waters may be extremely acid, notably in geothermal and mining areas, whereas saline deserts are usually strongly alkaline. A few algal species are to be found growing at extreme pHs and, again, we may argue as to whether these highly adapted organisms can correctly be described as stressed. Fluctuations in pH which may cause stress in normally intermediate environments may be brought about by circumstances such as influx of ammonia from a bird colony or intense photosynthesis in an already alkaline water removing carbon dioxide. pH affects organisms principally by determining the extents of ionization of metabolites, consequently influencing their uptake and reactivity. Investigation of the effects of pH on aquatic plants is complicated because of its involvement in the carbon dioxide-bicarbonate-carbonate equilibrium (Falkowski and Raven 1997).

1.2.3.3 Toxic Substances

These can occur naturally; algal blooms, for example, often produce powerful toxins and spontaneous oil seepages are to be found in many parts of the world, but the majority are introduced into the environment by human activities and include a variety of chemical types: heavy metals, inorganic substances, hydrocarbons, alcohols, sterols and others. Some are waste products, others used as fuels, pesticides, industrial reagents or drugs. In all, according to the Chemical Abstracts Service Agency, numbering over 5 million different substances. If present in the environment in sufficient concentrations they are regarded as pollutants and we need have no hesitation in describing their effects on algae as stress. A vast amount of work, in which algae have played a notable part as bioassay organisms (Haglund 1997), has been done on the toxicology and biology of pollution but it has to be said that much of this is of poor quality. Scientific guidelines for pollution research and management are to be found in the Report of A Royal Society Study Group (Royal Society 1994).

1.3 Interactions Between Stresses

In their natural environments algae are often subjected simultaneously or within a short space of time to more than one kind of stress. This is very obvious in the intertidal zone, where a seaweed may undergo changes in mechanical stresses, aqueous to aerial environment, desiccation, intensity of both visible and UV radiation, temperature and salinity, within one tidal cycle (Davison and Pearson 1996). Another example is found during ice formation in polar seas when algae incorporated in the ice are subject to mechanical stress, decrease in temperature and increase in salinity as brine is extruded during freezing (Fogg 1998). Resistances to freezing, desiccation and osmotic stress often seem to be associated. This has been investigated in detail in *Fucus distichus* (Pearson and Davison 1994). It may simply be that these stresses tend to occur together in natural environments, but it is striking that resistance to them is particularly well developed in the two groups Chlorophyceae and Cyanobacteria (Fogg 1969). Whether there is some basic cellular feature common to these various kinds of resistance remains to be established. A synergistic effect may be shown when two different stresses occur together, i.e. the effect on the alga is greater than the sum of the effects produced by the two stresses applied separately. An example of this is that phosphorus deficiency sufficient to cause partial inhibition of photosynthesis renders *Scenedesmus acutus* more sensitive to the toxic effect of copper (Twiss and Nalewajko 1992). The brown kelp *Laminaria saccharina*, subjected to a combination of heat and nitrogen deficiency, suffers more devastating effects than when either stress is applied alone (Gerard 1997). There is also the question of the effects of single and multiple applications of the same stress. Experimental data are sparse but the indications are that multiple exposures, if applied only once in a sublethal treatment, may be lethal (Davison and Pearson 1996).

1.4 Reactions to Stress and Its Avoidance

1.4.1 Behaviour

One would expect the immediate reaction to stress by a motile organism would be movement away from the source. This seems to be true with respect to visible radiation. Flagellates and gliding organisms such as diatoms and cyanobacteria generally show positive phototopotaxis, moving towards the light source, when stressed by low light intensity, and negative phototopotaxis when stressed by high light intensity (Nultsch 1974). This behaviour is paralleled by some planktonic Cyanophyceae which do not possess flagella but achieve buoyancy by means of gas vesicles. These are produced at low light intensities and cause the organisms to float up in the water column to more favourable illumination. If they rise to the surface they may be exposed to high irradiance causing photooxidation, but this stress can be avoided if photosynthesis is able to continue sufficiently to produce osmotically active products which increase turgor pressure, collapsing the vesicles. Accumulation of carbohydrates which act as ballast assists sinking to a more favourable level (Walsby 1994). It seems that phytoplankton species able to control position in the water column show no avoiding action to UV-B stress, possibly because they do not possess photoreceptors for these wavelengths. The ecological outcome is that, responding to the visible light field, the cells may move towards high UV-B radiation near the water surface and suffer damage (Häder 1993). Motile microorganisms also exhibit chemotaxis which presumably may enable them to avoid chemical stress but there seem to have been few studies on this and no unequivocal relation between chemotactic activity and chemical structure has been found (Nultsch 1974).

1.4.2 Metabolism

Stress, even if mechanical, dislocates the complicated web of biochemical processes which maintain life. The most obvious effect is usually the impairment of some vital function such as respiration, photosynthesis, or cell division. The cyanobacterial mats in Antarctic freshwater ponds provide a good example. A mat consisting mainly of *Nostoc* dried almost completely under field conditions in three hours, photosynthesis falling to about a fifth of its initial value in this time and ceasing entirely after 24 h. Respiration fell almost to zero after three hours. Rewetting after 48 h brought about almost full recovery of both functions within a few minutes. Mats dominated by *Phormidium* behaved differently on rewetting, showing a massive increase in respiration but with photosynthesis only becoming evident after six days (Hawes et al. 1992). Another example is *Stichococcus bacillaris*, in which protein synthesis is strongly inhibited by salt-shock, although uptake of amino acids shows little change, but recovers remarkably quickly when the alga is returned to lower salinity (Ahmad and Hellebust 1993).

Alteration in the rate of any constituent process is likely to result in loss of balance of the whole metabolic web with diversion of organic material along new pathways. There are numerous examples of stressed algae producing conspicuous amounts of substances, usually liberated extracellularly, which are not normally particularly evident. Phytoplankton commonly releases from 5 to 40% of the total carbon fixed in photosynthesis in extracellular form, glycollic acid often being the principal product, when stressed by high light intensities. The biochemical pathway of this diversion of fixed carbon is well understood and it is reasonable to suppose that it serves a useful purpose by disposing of absorbed light energy which would otherwise fuel photooxidation. Once released, the glycollate is not reabsorbed and utilized by the algae but is used as a substrate by heterotrophs (Fogg 1983). Photooxidation is also countered by active oxygen metabolism. As with plants generally, algae contain antioxidants, such as ascorbic acid, glutathione and α-tocopherol, with high affinity for active oxygen species. Enzymes, such as superoxide dismutase, peroxidase and catalase, promote the conversion of active oxygen into harmless products (Polle and Rennenberg 1993; Davison and Pearson 1996). An increase in catalase activity has been observed in the freshwater dinoflagellate *Peridinium gatunense* during the decline of the spring bloom period when depletion of carbon dioxide promotes photooxidation (Butow et al. 1994). Collén and Davison (1999) have measured the production of and damage caused by reactive oxygen in three species of intertidal brown algae differing in their abilities to withstand freezing, desiccation and high light stress. Production of active oxygen was relatively low in the unstressed seaweeds but increased following stress. In general, the results were consistent with the hypothesis that metabolism of active oxygen is involved in stress tolerance.

Another kind of protective mechanism is that of production by *Phaeocystis* spp. of the osmoregulatory substance dimethylsulphoniopropionate (DMSP). Its interacellular concentration is increased about twofold under high light conditions in iron-deficient cells. The mechanism seems to be that iron deficiency brings about nitrogen deficiency and DMSP acts as a substitute for nitrogen-containing osmoregulatory substances (Stefels and van Leeuwe 1998). DMSP, it may be noted, is of special interest in giving rise to dimethyl sulphide, which contributes substantially to atmospheric sulphate — a matter of climatic consequence. The aqueous phase concentrations of DMSP and DMS are often exceeded by that of a derivative, dimethylsulphoxide (DMSO). Little is known of the origin of DMSO but it is possible that it is produced intracellularly in planktonic algae. This is of interest in the present context because this compound has properties which make it potentially important in stress resistance as a cryoprotectant, a cryoosmoregulator, and a free-radical scavenger (Lee and de Mora 1999).

As in other types of organisms, heat shock causes rapid conformational change or denaturation of previously stable proteins and production of heat-shock proteins, notably ubiquitin. There is strong, but circumstantial, evidence that enhanced thermotolerance is causally linked to the induction of heat-shock proteins (Pollock et al. 1993). In *Skeletonema costatum* a temperature shift from 18 to 30 °C greatly increases the total amount of ubiquitin and, especially, the ubiquitin fraction found conjugated with protein (Penna et al. 1996).

In each of these examples the metabolic dislocation brought about by stress results in processes which circumvent the damage and favour the survival of the organism. This seems to be a general feature which may arise by natural selection. The situation is similar in flowering plants. Much research has been devoted to the stress tolerance of these (Fowden et al. 1993) and it seems likely that the biochemical mechanisms which have been found in them will also be found in algae (Davison and Pearson 1996). The knowledge derived from higher plants should be used as a starting point for further work on algae, always bearing in mind the probability that algae, facing stress under rather different conditions, possess some unique mechanisms.

1.4.3 Structure

If exposed to prolonged sublethal stress algae may show structural changes. Microalgae subjected to drought often develop a lipid pellicle around their cells, produce mucilage, thicken cell walls and accumulate intracellular globules of fat (Fogg 1969). Seaweeds regularly stressed by wave action may show striking change in form. Thus, blades of the giant bull kelp *Nereocystis luetkeana* from plants growing in protected habitats are wider than those exposed to rapid water flow and have undulate margins rather than being flat. The drag on the protected type of blade is about twice that on the other (Koehl 1986). This morphological difference is surely brought about biochemically but the nature of the transponder converting mechanical stimulus into biochemical activity is unknown. Juveniles of *Egregia menziesii* have greater tissue strength, breaking strain, and stiffness when grown under a high force regime than those grown under quieter conditions.

However, this mechanical adaptation could not be related to the content of the main structural component, alginic acid (Kraemer and Chapman 1991).

1.4.4 Symbiosis

A symbiotic association of an algal primary producer with a consumer constitutes a simple, almost self-contained, microcosm which may have properties enabling it to survive stresses which non-symbiotic organisms in the same habitat could not. Coral reefs are evidently able to flourish in oligotrophic waters because the dinoflagellate/coelenterate symbiosis which is their basic feature not only provides physical protection for the flagellate and organic matter for the animal but constitutes a system within which scarce nutrients are recycled and conserved (Muscatine 1973). Lichens are conspicuous in the polar regions and high alpine situations, growing under conditions in which neither symbiont would be likely to survive. However, the capacity to resist stress by species from these situations is not actually correlated with habitat and lichens from temperate or even warm regions show a similar resistance to cold (Ahmadjian 1982). Lichens are to some extent independent of the trophic conditions of their habitat and have adaptations to various kinds of stress. They are able to survive long periods of almost complete desiccation during which the thallus becomes inactivated and safe from heat-induced respiratory loss and stress. Combined with this is a capacity to reactivate metabolism quickly when water becomes available. Mostly the lichens of frigid regions are mechanically robust although not proof against sand- and ice-blast in windswept places. Most Antarctic lichens can withstand gradual or rapid freezing to as low as −196 °C and can photosynthesize perceptibly at temperatures down to −24 °C. Dark pigment provides protection for the phycobiont against strong insolation and enables the lichen to warm up under several centimetres of snow. Lichens which include certain cyanobacterial phycobionts are able to fix atmospheric nitrogen and so avoid stress through deficiency of this element (Kappen 1993).

1.4.5 Phenotypic Adaptation

Examples of this have been given above (Sects. 1.4.1, 1.4.2, 1.4.3). Short-term or fluctuating stress is best met by specific phenotypic response systems (Bradshaw and Hardwick 1989). Thus, lower-shore algae show phenotypic adaptation to the generally prevailing conditions whereas the more stress-tolerant species of the upper shore appear to have this characteristic genetically determined (Davison and Pearson 1996). Acclimation to light intensity and temperature have perhaps received most attention from research workers and the mechanisms appear to be no different in polar algae, which show a broad potential of acclimation and adaptation, from those of algae in less stressing environments (Kirst and Wiencke 1995; Fogg 1998).

1.4.6 Genotypic Adaptation

If appropriate genetic variability is available, classic evolutionary changes should occur in populations subject to consistent stress (Bradshaw and Hardwick 1989). There is considerable evidence for such evolution, producing genetic adaptation to specific stresses, having occurred amongst flowering plants. An algal example is provided by the filamentous brown seaweed, *Ectocarpus siliculosus,* of which both copper-tolerant and copper-sensitive strains have been isolated (Hall 1981). In flowering plants, increases in genome size can occur quickly, sometimes within a generation and sometimes in response to stress. This may be an important factor in stress resistance but does not seem to have been recorded in algae (Bachmann 1993).

1.5 Comparisons of Capacity to Resist Stress

1.5.1 At Different Stages of Development

The impact of stresses in a given habitat will change as an alga grows and develops and at the same time the capacity of the organism to resist may increase or decrease. Such changes will usually be minimal for microalgae, but the snow alga *Chlamydomonas nivalis* is an exception. The green, positively phototactic, zoospores liberated when snow begins to thaw are susceptible to UV-B damage whereas the aplanospore stage contains large amounts of the red carotenoid astaxanthin which provides an effective protection from radiation damage (Fogg 1998). Changes of stress resistance during development are, however, greater in the macroalgae of the seashore. This point has often been ignored in studies of intertidal communities, in which explanation of zonation has been looked for in the resistance to stress of the mature seaweeds whereas recruitment is the crucial determining phase (Davison and Pearson 1996). Successful colonization of an area of shore depends on a "recruitment window" in which emersion combines with favourable non-desiccating conditions. The microscopic stages of seaweeds may find microhabitats in which they can avoid wave action, desiccation and insolation, but they settle over a wider range than is eventually occupied by mature plants and are generally more susceptible to stress than these. Twenty four- to 48-h-old embryos of *Pelvetia fastigiata* are more susceptible during natural emersion than either 6-h-old zygotes or 1-week-old embryos. Changes in acclimation to stress during development may be influenced by growth conditions. Adult plants of *Fucus* spp. grown at 5 °C are more freeze-tolerant than if grown at 15 °C. However, germlings grown at 15 °C are more tolerant than germlings of the same age grown at 5 °C, evidently because freeze tolerance increases rapidly during the first 48 h of development so that germlings acquire tolerance sooner at 15 °C than at the lower temperature (Davison and Pearson 1996).

1.5.2 Differences Between Algal Groups

It appears that some algal groups are more resistant to stress than others, but considering that, although some stresses seem to share common features, others are of entirely different natures, and that precise, objective, assessment of stress resistance is not yet possible, such a conclusion can be only tentative. Amongst the algae (in the broader sense) the Cyanobacteria have an outstanding capacity to resist extreme temperatures, drought and high salt concentrations. In this they resemble the Eubacteria, which is not surprising since both are prokaryotic. Amongst the eukaryotic algae, some unicellular Chlorophyceae have more than ordinary resistance to these particular adverse conditions. It should be noted that only the Cyanobacteria and Chlorophyceae amongst algal forms have achieved success in colonizing terrestrial habitats (Fogg 1969). The resistance to stress of these two groups is illustrated strikingly in studies of the ecology of the McMurdo Dry Valleys in Antarctica, in which they figure prominently, whereas diatoms make only a tentative appearance (see Sect. 1.5.3). Although a few species belonging to other algal groups may show particular resistance to extreme temperature or desiccation, and although these groups show resistance to other stresses, such as, notably in planktonic species, prolonged darkness, they have mostly remained confined to aquatic environments in which extremes of temperature and desiccation are not encountered. Resistance to mechanical stress is evident only in the groups which include the large seaweeds — Chlorophyceae, Rhodophyceae and Phaeophyceae. Size is a measure of resistance here and puts the Phaeophyceae, with *Macrocystis pyrifera*, attaining 60 m or more in length, well above the Rhodophyceae, which are rarely above 50 cm, and Chlorophyceae, which sometimes reach 30 cm in linear dimensions. Whether these differences can be related to the chemical compositions of their respective cell-wall materials is a matter for investigation.

1.5.3 Differences Between Algae and Other Forms of Life

Again, there is difficulty in making other than very general comparisons. An initial point might be that the majority of algae, being small and morphologically simple, must accept the environmental conditions in their immediate vicinity and cannot alter them or exploit the favourable characteristics of spatially separate phases, such as soil and the aerial environment to avoid stress as a higher plant can. It is difficult to compare the effects of a particular stress on two organisms of widely different form. Isolated records are of little use. Data from the Antarctic dry valleys might seem to give some valid comparisons. These valleys can be arranged in a sequence, from extreme desiccation and exposure to comparatively less severe environments, in which organisms appear in the following order: (1) aerobic, heterotrophic, non-pigmented bacteria; (2) microaerophilic, heterotrophic, pigmented bacteria; (3) actinomycetes; (4) coccoid green algae and both coccoid and filamentous cyanobacteria; (5) moulds, yeasts and protozoa; (6) lichens; (7) mosses, with filamentous green algae, diatoms and nitrogen-fixing cyanobacteria (Cameron et al. 1970). Algae seem to be more resistant than any other group except eubacteria. However, dispersal mechanisms must be taken into account. All the classes of organisms listed can be disseminated by wind and there is ample evidence that viable propagules belonging to these classes are brought to Antarctica from elsewhere. The comparatively heavy seeds of flowering plants and eggs of invertebrates have to face greater difficulties in reaching the dry valleys and this probably explains their absence. In the Arctic, with overland dispersal routes, flowering plants establish themselves in polar deserts which correspond approximately to the least hostile of the Antarctic dry valleys (Fogg 1998). The best that can be said is that chlorophycean algae and cyanobacteria are more resistant to stress than the classes of organisms listed in categories 5, 6 and 7 above. Another situation in which the presence or absence of different kinds of organisms can be related to a gradient of stress is that of thermal springs and this has the advantage that in temperate and tropical regions dispersion is not an important factor. Brock (1978) has tabulated the approximate upper temperature limits for various groups, in which eukaryotic algae figure at 55 to 60 °C, just above protozoa at 56 °C and just below fungi at 60 to 62 °C. Cyanobacteria are placed at 70 to 73°C, whereas chemolithotrophic and heterotrophic bacteria survive at more than 90 °C. Animals range from 38 °C for fish to 49 to 50 °C for ostracod crustaceans, and vascular plants and mosses survive to 45 and 50 °C, respectively. This tallies well with the dry-valley sequence and other observations and it seems safe to conclude that eukaryotic algae are distinctly more resistant to stress in general than are other plants, but about the same as fungi and much less resistant than cyanobacteria and eubacteria. The explanation of the relatively high stress resistance of the algae may lie partly in rapid cell division and great flexibility in metabolism.

1.6 The Ecological Role of Stress

The ecological importance of stress tolerance seems evident, but is not universally recognized by higher plant ecologists (Grime 1989). As the preceding discussion will have shown, the difficulties facing precise assessment of its role are great. Ultimately, mathematical models, although they have their drawbacks, seem to provide the most rigorous approach (Rees and Lawton 1993) but have scarcely been employed with respect to algae. Problems are perhaps least in extreme situations — polar habitats, thermal springs and saline water — in which the number of species is low and competition between them often minimal. As we have seen (Sect. 1.5.3), there may be gradients of stress in these along which organisms with different susceptibilities sort themselves out. In the aquatic realm the permanently frozen lakes of the dry valleys, in which stratification of the water columns provides a sequence of distinct layers with their own particular stresses and opportunities for different types of microorganisms (Fogg 1998). For such a system numerical modelling taking account of stresses becomes feasible. An exposed rocky seashore, on the other hand, presents an extreme of complication (Dring 1982). True, there is a definite gradient in stress up the shore but it comprises a number of different kinds of stress, all of which interact to a greater or lesser extent. These stresses not only vary with the tides, but the vagaries of weather impose an element of uncertainty. There is a great deal of difference in ecological impact between the sublethal and lethal stresses which arise. The dominant algae are macrophytes which develop from microscopic juvenile stages having different susceptibilities to stress from the mature plants and being placed on the shore during "windows of opportunity" at levels which may bear no relation to positions on the shore which are most favourable for the adults. The density of the vegetation is often such that competition for substratum, protection and light is severe. At all stages the plants may be grazed by animals which themselves are subject to stress but by reason of their mobility may not be confined to zones. Davison and Pearson (1996) have discussed the various possible approaches to determining the role of stress in such a community. Experiments on plants brought into the laboratory can establish precise relationship between stresses and physiological effects but offer little more than general guidance to what happens in the natural environment with its interacting factors and episodic stresses. In situ physiological experiments should be done where possible and ideally would involve an array of physiological equipment together with instruments for monitoring microclimates such as are used in observations of agricultural crops. It is likely, however, that such a setup on the shore would succumb to stress before the seaweeds. Meanwhile, it seems that stress factors cannot be incorporated in numerical models of the littoral habitat, nor can simple budgets of the costs of stress and stress tolerance in terms of carbon be drawn up, because of lack of data (Davison and Pearson 1996). Production of reactive oxygen appears to provide a measure of the stress associated with vertical zonation (Collén and Davison 1999) but needs to be applied to developmental stages as well as to the mature community if it is to provide a way forward.

It remains to be seen whether the hypothesis favoured by some terrestrial ecologists, that competitiveness is inversely related to stress tolerance (Grime 1993) applies in the aquatic environment.

1.7 The Value of Studies of Stress in Algae

Apart from its intrinsic interest and the essential contribution it can make towards understanding the ecology of many different kinds of community, this subject is not without its practical value. Application of stress is a much-used experimental approach. For example, the remarkable toughness of *Chlorella* under all sorts of physical and chemical stress once made it a favoured experimental material for study of the metabolism of green plants. Warburg and Negelein (1920) were able to isolate the process of nitrate reduction from photosynthesis and respiration by incubation of this green alga in a medium of pH 2 to promote the uptake of undissociated nitrate and swamp these other processes. Another green alga, *Dunaliella salina*, produces enough glycerol when salt-stressed to make it worth consideration as a commercial source of this chemical (Ben-Amotz and Avron 1989). Research on cold stress in algae is valuable because it may contribute to better storage of food during refrigeration or of biological material at low temperatures and in identifying strains which might be exploited in biotechnology to avoid the expense of providing elevated temperatures.

References

Ahmad I, Hellebust JA (1993) Protein biosynthesis in salt-shocked cells of *Stichococcus bacillaris* (Chlorophyceae). J Phycol 29:294-300

Ahmadjian V (1992) Algal/fungal symbioses. In: Round FE, Chapman DJ (eds) Progress in phycological research. Elsevier, Amsterdam, pp 179-233

Bachmann K (1993) Effects of stress on the genome. In: Fowden L, Mansfield T, Stoddart J (eds) Plant adaptation to environmental stress. Chapman & Hall, London, pp 275-292

Beadnell CM (1942) Dictionary of scientific terms. Watts, London

Ben-Amotz A, Avron M (1989) The biotechnology of mass culturing *Dunaliella* for products of commercial interest. In: Cresswell RC, Rees TAV, Shah N (eds) Algal and cyanobacterial biotechnology. Longman Scientific & Technical, Harlow, Essex, pp 91-114

Biggs BJF, Thomsen HA (1995) Disturbance of stream periphyton by perturbations in shear stress: time to structural failure and differences in community resistance. J Phycol 31:233-241

Bothwell ML, Sherbot D, Roberge AC, Daley RJ (1993) Influence of natural ultraviolet radiation on lotic periphytic diatom community growth, biomass accrual, and species composition: short-term versus long-term effects. J Phycol 29:24-35

Bradshaw AD, Hardwick K (1989) Evolution and stress — genotype and phenotype components. Biol J Linn Soc 37:137-155

Brock TD (1978) Thermophilic microorganisms and life at high temperatures. Springer-Verlag, New York Heidelberg Berlin

Butow B, Wynne D, Tel-Or E (1994) Response of catalase activity to environmental stress in the freshwater dinoflagellate *Peridinium gatunense*. J Phycol 30:17-22

Cameron RE, King J, David CN (1970) Microbiology, ecology and microclimatology of soil sites in dry valleys of southern Victoria Land, Antarctica. In: Holdgate MW (ed) Antarctic ecology, vol 2. Academic Press, London, pp 702-716

Collén J, Davison IR (1999) Reactive oxygen production and damage in intertidal *Fucus* spp. (Phaeophyceae). J Phycol 35:54-61

Davison IR, Pearson GA (1996) Stress tolerance in intertidal seaweeds. J Phycol 32:197-211

Dring MJ (1982) The biology of marine plants. Edward Arnold, London

Dring M.J (1984) Photoperiodism and phycology. In: Round FE, Chapman DJ (eds) Progress in phycological research. Biopress, Bristol, pp 159-192

Falkowski PG, Raven JA (1997) Aquatic photosynthesis. Blackwell, Oxford

Fogg GE (1969) Survival of algae under adverse conditions. In: Woolhouse HW (ed) Dormancy and survival. 23rd symp of the society of experimental biology. Cambridge University Press, Cambridge, pp 123-142

Fogg GE (1983) The ecological significance of extracellular products of phytoplankton photosynthesis. Bot Mar 26:3-14

Fogg GE (1991) The phytoplanktonic ways of life. New Phytol 118:191-232

Fogg GE (1998) The biology of polar habitats. Oxford University Press, Oxford

Fogg GE, Thake B (1987) Algal cultures and phytoplankton ecology. University of Wisconsin Press, Madison

Fowden L, Mansfield T, Stoddart J (eds) (1993) Plant adaptation to environmental stress. Chapman and Hall, London

Franks F, Mathias SF, Hatley RHM (1990) Water, temperature and life. Philos Trans R Soc Lond B 326:517-533

Gerard VA (1997) The role of nitrogen nutrition in high-temperature tolerance of the kelp, *Laminaria saccharina* (Chromophyta). J Phycol 33:800-810

Graziano LM, La Roche J, Gieder RJ (1996) Physiological responses to phosphorus limitation in batch and steady-state cultures of *Dunaliella tertiolecta* (Chlorophyta): a unique stress protein as an indicator of phosphate deficiency. J Phycol 32:825-838

Grime JP (1989) The stress debate: symptom of impending synthesis? Biol J Linn Soc 37:3-17

Grime JP (1993) Stress, competition, resource dynamics and vegetation processes. In: Fowden L, Mansfield T, Stoddart J (eds) Plant adaptation to environmental stress. Chapman & Hall, London, pp 45-63

Häder DP (1993) Risks of enhanced solar ultraviolet radiation for aquatic ecosystems. In: Round FE, Chapman DJ (eds) Progress in phycological research, 9. Biopress, Bristol, pp 1-45

Haglund K (1997) The use of algae in aquatic toxicity assessment. In: Round FE, Chapman DJ (eds) Progress in phycological research, 13. Biopress, Bristol, pp 181-212

Hall A (1981) Copper accumulation in copper-tolerant and non-tolerant populations of the marine fouling alga, *Ectocarpus siliculosus* (Dillw.) Lyngbye. Bot Mar 24:223-228

Hawes I, Howard-Williams C, Vincent WF (1992) Desiccation and recovery of antarctic cyanobacterial mats. Polar Biol 12:587-594

Jones WE, Demetropoulos A (1968) Exposure to wave action: measurements of an important ecological parameter on rocky shores on Anglesey. J Exp Mar Biol Ecol 2:46-63

Kappen L (1993) Lichens in the Antarctic region. In: Friedmann EI (ed) Antarctic microbiology. Wiley-Liss, New York, pp 433-490

Kirst GO, Wiencke C (1995) Ecophysiology of polar algae. J Phycol 31:181-199

Koehl MAR (1986) Seaweeds in moving water: form and mechanical function. In: Givnish TJ (ed) On the economy of plant form and function. Cambridge University Press, Cambridge, pp 603-634

Kraemer GP, Chapman DJ (1991) Biomechanics and alginic acid composition during hydrodynamic adaptation by *Egregia menziesii* (Phaeophyta) juveniles. J Phycol 27:47-53

La Roche J, Geider RJ, Garziano LM, Murray H, Lewis K (1993) Induction of specific proteins in eukaryotic algae grown under iron-, phosphorus-, or nitrogen-deficient conditions. J Phycol 29:767-777

Lee PA, de Mora SJ (1999) Intracellular dimethylsulfoxide (DMSO) in unicellular marine algae: speculations on its origin and possible biological role. J Phycol 35:8-18

Mackereth FJ (1953) Phosphorus utilization by *Asterionella formosa* Hass. J Exp Bot 4:296-313

Muscatine L (1973) Nutrition of corals. In: Jones OA, Endean R (eds) Biology and geology of coral reefs vol II: Biology 1. Academic Press, New York, pp 77-115

Nienow JA, Friedmann EI (1993) Terrestrial lithophytic (rock) communities. In: Friedmann EI (ed) Antarctic microbiology. Wiley-Liss, New York, pp 343-412

Nultsch W (1974) Movements. In: Stewart WDP (ed) Algal physiology and biochemistry. Blackwell, Oxford, pp 864-893

Pearson GA, Davison IR (1994) Freezing stress and osmotic dehydration in *Fucus distichus* (Phaeophyta): evidence for physiological similarity. J Phycol 30: 257-267

Penna A, Crinelli R, Magnani M (1996) Modulation of the heat shock ubiquitin pool in *Skeletonema costatum* (Bacillariophyceae). J Phycol 32:409-415

Polle A, Rennenberg H (1993) Significance of antioxidants in plant adaptation to environmental stress. In: Fowden L, Mansfield T, Stoddart J (eds) Plant adaptation to environmental stress. Chapman & Hall, London, pp 263-273

Pollock CJ, Eagles CF, Howarth CJ, Schünmann PHD, Stoddart JL (1993) Temperature stress. In: Fowden L, Mansfield T, Stoddart J (eds) Plant adaptation to environmental stress. Chapman & Hall, London, pp 109-132

Rees M, Lawton JH (1993) What can models tell us? In: Fowden L, Mansfield T, Stoddart J (eds) Plant adaptation to environmental stress. Chapman & Hall, London, pp 65-83

Royal Society (1994) Pollutant control priorities in the aquatic environment: scientific guidelines for management. Royal Society, London

Sibly RM, Calow P (1989) A life-cycle theory of responses to stress. Biol J Linn Soc 37:101-116

Simon RD (1987) Inclusion bodies in the cyanobacteria: cyanophycin, polyphosphate, polyhedral bodies. In: Fay P, Van Baalen C (eds) The cyanobacteria. Elsevier, Amsterdam, pp 199-225

Stefels J, van Leeuwe MA (1998) Effects of iron and light stress on the biochemical composition of Antarctic *Phaeocystis* sp. (Prymnesiophyceae), I Intracellular DMSP concentrations. J Phycol 34:486-495

Tett P, Heaney SI, Droop MR (1985) The Redfield ratio and phytoplankton growth rate. J Mar Biol Assoc UK 65:487-504

Twiss MR, Nalewajko C (1992) Influence of phosphorus nutrition on copper toxicity to three strains of *Scenedesmus acutus* (Chlorophyceae). J Phycol 28:291-298

Van Liere L (1979) On *Oscillatoria agardhii* Gomont, experimental ecology and physiology of a nuisance bloom-forming cyanobacterium. PhD Thesis, University of Amsterdam, Amsterdam

Walsby AE (1994) Gas vesicles. Microbiol Rev 58: 94-144

Warburg O, Negelein E (1920) Über die Reduktion der Salpetersäure in grünen Zellen. Biochem Z 110:66-115

2 Carbon Limitation

Lennart Axelsson[1] and Sven Beer[2]

Contents

[1] Kristineberg Marine Research Station, Kristineberg 2130, SE-450 34 Fiskebäckskil, Sweden
[2] Department of Plant Sciences, Tel Aviv University, Tel Aviv 69978, Israel

2.1 Introduction

This chapter deals with aspects of uptake and photosynthetic reduction of inorganic carbon (Ci) in algae. The treatise is restricted to marine macroalgae, although some of the principles described may certainly be valid also for freshwater species and both marine and freshwater microalgae. Whether or not the marine Ci concentration and/or composition is limiting for macroalgae, and whether they constitute a "stress factor", will depend on a variety of factors. A priori, nearly all Ci assimilated is used in photosynthesis. In this process, captured light energy is used to synthesise carbohydrates (and produce O_2) from CO_2 and water. While water, as a substrate for photosynthesis, is never limiting, the light and CO_2 supplies will vary considerably and independently of each other on both short- and long-term scales. Neither captured light energy (i.e. ATP and NADPH) nor CO_2 can be stored for longer periods of time (or to any great extent). Since excess of captured light energy might be very harmful (Demmig-Adams and Adams 1992, 2000), the plants in the sea apparently have to compromise between maximum productivity and safe light utilisation (avoidance of over-excitation; cf. Long et al. 1994). While low productivity is perhaps more related to competition (rather than to stress sensu stricto), overexcitation is by far the most important stress effect by low CO_2 supply to the plants. According to the specific role of irradiance in carbon stress, certain parts of Chapter 6 are also relevant to the present text.

A low CO_2 supply could have several reasons. Firstly, since seawater may contain various amounts of CO_2, it will depend on the specific CO_2 concentration of a certain habitat. Secondly, since normal, air-equilibrated seawater contains some 180 times more ionic Ci (bicarbonate, HCO_3^- and carbonate, CO_3^{2-}) than CO_2, it will depend on to which degree the algae can utilise these ionic carbon forms. Thirdly, it has recently become clear that certain algae may change the preferred way of utilising the ionic carbon forms according to their surroundings, and transient Ci stresses may thus be present in cases where the algae have not yet fully adapted to one way of utilising a certain carbon form or another. Finally, as with most other environmental components, the degree of Ci limitation depends on the status of other factors. Besides light, which has a general importance, factors like nutrient availability, temperature and water movements should also be considered.

Inorganic carbon acquisition in marine macroalgae has been surveyed rather recently, and the reader is thus also referred to the following reviews: Raven et al. (1990); Axelsson et al. (1991); Johnston (1991); Beer (1994); Raven (1997a,b). Except for some special cases dealing directly with measurements on Ci limitations, only works from the last ca. 10 years have been used as references in the following; for older works, the reader is, again, referred to the reviews. Since much research on carbon acquisition and CO_2-concentrating mechanisms (CCMs) on microalgae is relevant also for macroalgae, the reader is referred also to the recent reviews by Kaplan and Reinhold (1999); Moroney and Somanchi (1999) and references therein, as well as to relevant chapters in this Volume.

2.2 The Inorganic Carbon Composition and Carbon Equilibria of Seawater

CO_2 is close to "ideal" gas, and the atmospheric CO_2 dissolves in water according to Henry's law, with only minor corrections for non-ideality (Weiss 1974). This law prescribes that the concentration of dissolved CO_2 is proportional to its partial pressure in the air phase. The proportionality is given by the solubility constant (concentration in water/concentration in air), which for this gas is close to 1, and dissolved CO_2 concentrations are thus within the same order of magnitude as atmospheric ones (per volume). Temperature affects the solubility constant in that the water contains less CO_2 in equilibrium with the gas phase at increasing temperatures. In seawater, the second most important factor affecting the solubility of CO_2 is the salinity. Here, increasing salinity lowers the dissolution of CO_2. In oceanic seawater (with a salinity of 3.9% or 39 PSU) equilibrated with air containing 350 parts per million (ppm) CO_2 (i.e. ca. 16 μM), the dissolved CO_2 concentration is around 13 μM (20 °C and normal air pressure).

As CO_2 dissolves in water, some of the molecules also react with the water (hydrate) so as to form carbonic acid (H_2CO_3): $CO_2 + H_2O <> H_2CO_3$. The concentration of H_2CO_3 is assumed to be considerably lower than that of CO_2, and is usually included in the values given for CO_2 concentrations. This weak acid then dissociates to form HCO_3^- and CO_3^{2-} according to the first (for HCO_3^-) and second (for CO_3^{2-}) dissociation constants of H_2CO_3: $H_2CO_3 <> HCO_3^- + H^+ <> CO_3^{2-} + 2H^+$. Due to the high pH of seawater, most of the protons generated react with OH^- to form water, thus driving the equilibrium to the right. While the CO_2 concentration of seawater is fixed (in relation to that of the atmosphere, assuming full equilibration), the concentrations of HCO_3^- and CO_3^{2-} vary with the H^+ concentration (i.e. the pH); the higher the pH, the more HCO_3^- and CO_3^{2-}. Just like the solubility constant for CO_2, the dissociation constants for H_2CO_3 depend on temperature and salinity. In the above-mentioned oceanic seawater, the air-equilibrium concentrations of HCO_3^- and CO_3^{2-} under an atmosphere containing 350 ppm CO_2 (and, thus, 13 μM dissolved CO_2) are ca. 2000 and 400 μM, respectively.

Perhaps the best way to envision the carbonic acid system of seawater is to regard the water as being derived from a carbon-free solution of "sea salts", which contains a certain amount of (total) alkalinity (more or less a surplus of OH^- ions, or a titratable deficit in H^+ ions; cf. Uusitalo 1996). If such a solution is brought in contact with air, this alkalinity will result in the formation of HCO_3^- and CO_3^{2-} as CO_2 gradually dissolves, eventually resulting in normal seawater. In the example given above, the total alkalinity will be close to the carbonate alkalinity, which is ca. 2.8 equivalents OH^- m^{-3} (i.e. $[HCO_3^-] + 2[CO_3^{2-}]$). While the (total) alkalinity in most cases can be regarded as constant, the concentrations of dissolved CO_2 (usually including H_2CO_3), HCO_3^- and CO_3^{2-}, as well as the pH of the seawater, depend on the CO_2 partial pressure of the air and the above mentioned dissociation constants. CO_2 can be removed or added to the seawater either as a consequence of exchanges with the atmosphere, or due to respiration or photosynthetic activities, and this occurs usually without any changes in (total) alkalinity (although the pH of the seawater will change, cf. Uusitalo 1996). Inherent to these considerations is the fact that alkalinity, pH, $[CO_2]$, $[HCO_3^-]$, $[CO_3^{2-}]$ and [total Ci] depend on one another, and if two of these are measured, the rest may be calculated (provided temperature and salinity are known). For example, if the seawater is fully equilibrated with the atmosphere, it is possible to predict the concentration of CO_2 in the water. The concentration of ionic carbon can thus be calculated from the air concentration of CO_2 (again, at normal air pressure) and the pH (while taking into account also the more subtle effects of temperature and salinity). Such calculations are often termed open systems calculations. If, on the other hand, the alkalinity is constant (and known), and no CO_2 exchange with the atmosphere is taking place (a so-called closed system), photosynthetic Ci uptake can be calculated with a high resolution from changes in pH of the seawater (Haglund et al. 1987; Axelsson 1988; Axelsson et al. 1989a,b). Closed-systems calculations of concentrations of the various Ci forms are useful in artificially closed experimental setups in which the concentrations of various Ci forms can be made to vary on account of one another as a function of pH (by modifying the alkalinity via additions of H^+ or OH^-). Examples of using pH as a factor determining the distribution of various Ci forms in both open systems (where the CO_2 concentration per se is kept stable by exchange with the atmosphere while the concentrations of ionic carbon forms change) and closed systems (where the concentrations of the ionic Ci forms change on account of the dissolved CO_2) can be found in Beer and Eshel (1983).

While the dissociation of H_2CO_3, (to form HCO_3^- and CO_3^{2-}) is an instantaneous process, the spontaneous hydration of CO_2 to form H_2CO_3, and the reverse process, i.e. the dehydration of H_2CO_3 to form CO_2 and H_2O, is a slow process with half times of up to 1 min. There is, however, an enzyme, carbonic anhydrase (CA), which catalyses the hydration of CO_2 and the dehydration of H_2CO_3 so as to make the process virtually instantaneous (the reaction catalysed by CA is actually $CO_2 + H_2O <> HCO_3^- + H^+$; Lindskog 1997). In algae that lack the activity of CA at the cellular surfaces that face the seawater, it is likely that CO_2 cannot be formed from extracellular HCO_3^- fast enough to meet the intracellular photosynthetic potential of the plant and, as a consequence, they may be Ci-limited (if no other system for HCO_3^- utilisation is present).

While most open-water conditions can be expected to be open in terms of CO_2 fluxes between the water and the air, some are clearly not. For example, if the photosynthetic or respiratory activities of densely populated, and/or slow water-exchange habitats are faster than the CO_2 exchange rate across the water surface, then these habitats will become sub- (in the case of excessive photosynthesis, e.g. during the day) or super- (in the case of excessive respiration at night) saturated with respect to aerial CO_2. An extreme habitat to be considered is represented by rock pools densely populated by algae where, during the day, the pH becomes very high (because of excessive uptake of CO_2 and utilisation of HCO_3^-) and, consequently, the CO_2 and, eventually, the entire Ci, concentration become very low (see e.g. Larsson et al. 1997).

2.3 Inorganic Carbon Acquisition in Marine Macroalgae

One basic consideration regarding potential Ci limitation in marine macroalgae (and many other photosynthetic organisms as well) is related to the function of the enzyme Rubisco for CO_2 fixation (cf. Raven 1997b). This enzyme evolved its fundamental properties some 3-4 billion years ago in environments quite different from those of the present days, with a much (ca. 100 times) higher CO_2 concentration in the atmosphere (and, according to Henry's law, in the seawater as well; Raven 1997b,d) and a complete absence of oxygen. As photosynthetic performance improved, the CO_2 concentration gradually decreased and the oxygen concentration increased to the present-day values. In response to these changes, some aquatic plants developed means of creating a microenvironment within their cells in order to maintain suitable conditions for the proper operation of Rubisco. It was recently suggested that, in fact, a coevolution of CO_2-concentrating mechanisms (CCMs) and the kinetic properties of Rubisco occurred (Badger et al. 1998; cf. Kaplan and Reinhold 1999). During the past 10 years, our knowledge of the past atmospheric conditions has increased. It appears that the concentration of CO_2 has been subjected to considerable variations on both long-term and (relatively) short-term scales (Falkowski and Raven 1997; Kuypers et al. 1999; Norris and Röhl 1999; Petit et al. 1999), thus greatly affecting the prerequisites for both HCO_3^- utilisation and intracellular CO_2 concentration mechanisms. This would especially be true for non-green algae having the type ID Rubisco (red type present, e.g. in Rhodophyta, as well as Bacillariophyta), which has a high affinity for CO_2 and a (comparatively) low sensitivity to O_2 (cf. Raven 1997b; Badger et al. 1998). This may, in turn, explain the great multitude of HCO_3^--utilising and CO_2-concentrating mechanisms so far described (and suggested) with little taxonomic significance. It is, however, not surprising that there appear also to be some general, taxonomically related, differences in the occurrence of mechanisms for HCO_3^- utilisation such as between green and non-green macroalgae (Larsson and Axelsson 1999, cf. below).

Rubisco is often regarded as a non-efficient enzyme due to its low catalytic turnover rate and its tendency to react with O_2 at normal O_2 concentrations (i.e. in an environment where the CO_2 and O_2 activities correspond to those of the present atmosphere). This depends partly on its high half-saturation concentration (K_m)

for CO_2. The $K_m(CO_2)$ of Rubisco is especially high in marine macroalgae (varying between 30 and 90 µM depending on species, cf. Beer 1994; Badger et al. 1998) as compared with ca. 13 µM CO_2 in seawater. Therefore, photosynthetic CO_2 fixation via Rubisco (which is the ultimate way of carbon fixation in all plants) would be far from even half-saturated if it were based on a diffusional supply of CO_2 from the seawater. This would thus limit the photosynthetic rates of these algae. Further, one marked difference between marine and terrestrial environments regarding carbon acquisition of plants is the more than 10 000 times higher molecular diffusion resistance (or 1000 times higher resistance to turbulent diffusion) in water than in air, and this would further augment a potential Ci-limitation. These two factors, i.e. the high $K_m(CO_2)$ of macroalgal Rubisco relative to the seawater CO_2 concentration and the high diffusion resistance of solutes in aquatic media, would render photosynthetic performances of marine macroalgae inefficient to the point where growth would often be severely limited. When comparing with those terrestrial C_3 plants that are CO_2-limited in today's atmosphere, but which still have higher Rubisco affinities for CO_2 than macroalgae, one is also intuitively prone to think of the latter as, at least potentially, Ci-limited. Despite the potential Ci limitation in algae that use only CO_2 as their external Ci source for photosynthesis, some such algae do exist. These algae include the red species *Lomentaria articulata* and *Delesseria sanguinea* (Johnston et al. 1992) and *Durvillaea potatorum* (Raven et al. 1989).

2.3.1 Mechanisms Facilitating Ci Entry

All macroalgae can probably use CO_2 from the seawater to some extent. The CO_2 could either come directly from the dissolved seawater CO_2 pool, or from HCO_3^- via non-catalysed dehydration (HCO_3^- utilising mechanism type 0, cf. Fig. 2.1). Despite the potential limitation of Ci utilisation based on such a CO_2 uptake, the type 0 mechanism might be quite sufficient for some species, e.g. sublittoral algae with high surface to volume ratios (Larsson and Axelsson 1999) or with low growth rates (Johnston et al. 1992). Unfortunately, the frequent (and inappropriate, see below) use of buffers (e.g. TRIS) in studies of Ci affinity (cf. Axelsson et al. 2000, Fig. 2.2 and below) and lack of accurate methods for the determination of extracellular CA activity (Mercado et al. 1998) may invalidate some earlier conclusions on Ci acquisition mechanisms. Despite this, there are indications that a few species of macroalgae (preferentially sublittoral red forms) may depend more or less completely on this means of Ci utilisation, at least in their natural habitat (Johnston et al. 1992; Maberly et al. 1992).

There are two principally different ways in which algae may reduce Ci limitations. One is to improve their Ci uptake into the photosynthesising cells, e.g. by using Ci forms other than CO_2 as their external Ci source. The other way is to improve the efficiency of Ci transport within the cells. As stated above, seawater contains, beside CO_2, the Ci forms H_2CO_3, HCO_3^- and CO_3^{2-} also. While the H_2CO_3

concentration is negligible, the concentrations of the ionic carbon forms are substantially higher than that of CO_2. Especially HCO_3^- is present at a concentration some 150 times that of CO_2. Given the above CO_2 restrictions, it is not surprising that a large number of marine macroalgae have been found able to utilise HCO_3^- as their external Ci source for intracellular CO_2 fixation (e.g. Maberly 1990; cf. Beer 1994).

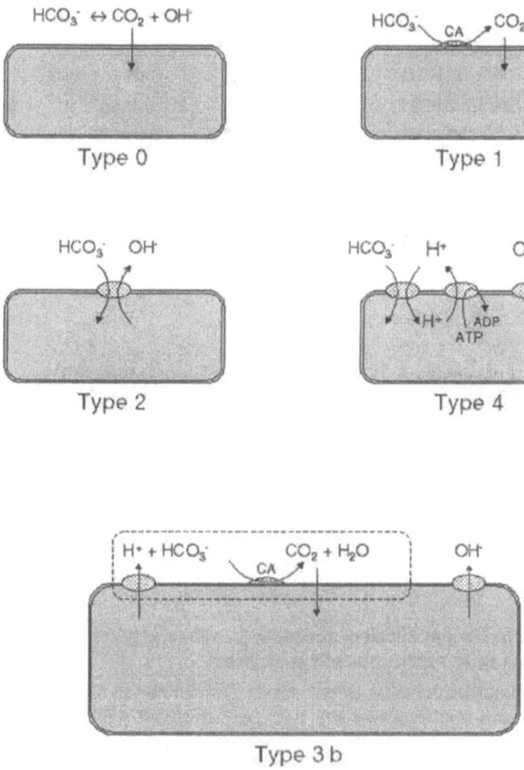

Fig. 2.1. Schematic presentation of various HCO_3^- utilisation methods in marine macroalgae. *Type 0* Bicarbonate use via non-catalysed dehydration; *type 1* bicarbonate use via CA-catalysed dehydration; *type 2* direct HCO_3^- uptake via an anion exchanger (facilitated transport); *type 3b* proton-enhanced HCO_3^- dehydration, catalysed by periplasmic CA activity; *type 4* ATP-ase driven active transport of HCO_3^- (via H^+/HCO_3^- symport). The *bordered part* in type 3b comprises net HCO_3^- uptake, illustrating why type 3 mechanisms could be termed indirect HCO_3^- uptake

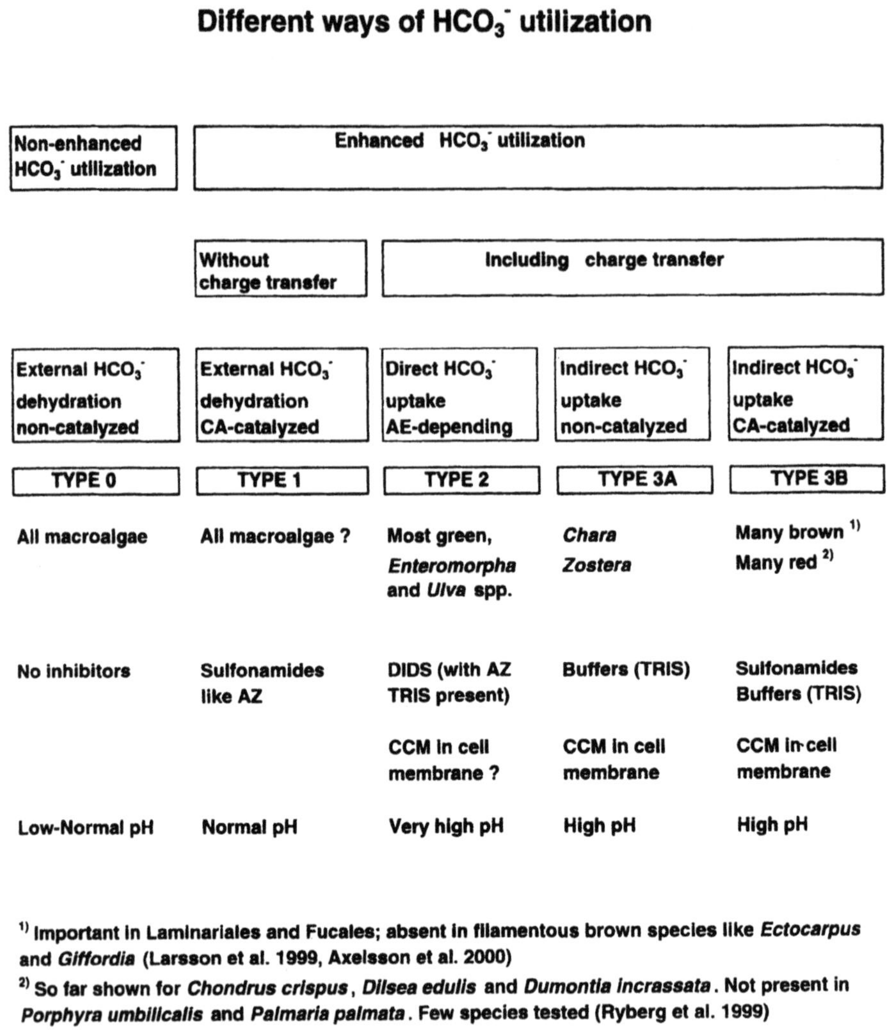

Fig. 2.2. Characterisation and occurrence of the HCO_3^--utilising mechanisms described in Fig. 2.1

The different ways by which marine macroalgae can improve their HCO_3^- utilisation (above that according to type 0) are listed in the following (and are also schematically depicted in Figs. 2.1 and 2.2):

1. Bicarbonate can constitute the extracellular Ci form from which CO_2 is derived via CA catalysed dehydration exterior to the cellular plasma membrane (type 1). The CA may either be associated with the plasma membrane (periplasmic CA) or occur free in the cell wall.
2. Bicarbonate can be taken up directly through the plasma membrane. In this case, since ions cannot easily penetrate such membranes by diffusion, an ion-

transporting protein (anion exchanger) is present as an integral part of the plasma membrane. This does not constitute active, but rather facilitated, transport along an electrochemical gradient (type 2).

3. An active efflux of H^+ may facilitate extracellular HCO_3^- dehydration. Due to the low pH, a higher than ambient CO_2 concentration will cause diffusion of CO_2 into the cells (type 3a). This means of HCO_3^- utilisation can be further catalysed by periplasmic CA activity (type 3b).
4. Bicarbonate can be transported into the cells against an electrochemical gradient (type 4). This could be done in several different ways and would require a direct energy input at the plasma membrane level. One possibility (HCO_3^-/H^+ symport) is exemplified in Fig. 2.1, where the energy is provided via a plasmalemma associated (P-type) ATP-ase.

While the type 1-3 mechanisms have been shown to be present in various algal species, type 4 still needs to be experimentally verified. It should be noted that the type 4 mechanism exemplified in Fig. 2.1 may be very difficult to separate experimentally from type 3a.

Many marine macroalgae have been shown to feature type 1 HCO_3^- utilisation as an integral part of their Ci acquisition system. These include green algal genera such as *Enteromorpha, Cladophora, Monostroma* (Maberly 1990) and various species of *Ulva* (Beer 1994 and references therein, Axelsson et al. 1999), the brown algae *Ascophyllum nodosum, Fucus serratus* (Giordano and Maberly 1989) and *Laminaria saccharina* (Haglund et al. 1992b; Axelsson et al. 2000), and the red algae *Chondrus crispus* (Smith and Bidwell 1989) and *Gracilaria tenuistipitata* (Haglund et al. 1992a). This mechanism is indicated by the presence of external CA activity. Recent investigations using a more sensitive method for the determination of external CA activity (Mercado et al. 1997) and inhibitors of external CA activity (e.g. Mercado et al. 1998; Larsson and Axelsson 1999), have suggested this way of HCO_3^- utilisation to be almost universal in macroalgae (although in many cases it is accompanied by additional mechanisms, see below).

The type 2 HCO_3^- utilisation system appears to be confined to green macroalgae, where it has been shown for several taxonomic groups (Larsson and Axelsson 1999). It has been thoroughly investigated for *Ulva fasciata* (reviewed in Beer 1994), *Ulva lactuca* (Axelsson et al. 1995, 1999) and *Enteromorpha intestinalis* (Larsson et al. 1997). These algae typically rely on the type 1 utilisation when this mechanism is sufficient to supply enough CO_2 for photosynthesis (Axelsson et al. 1995). Under conditions of carbon stress (e.g. high pH or high light intensities, cf. Larsson et al. 1997; Larsson and Axelsson 1999), these algae are capable of inducing the type 2 direct HCO_3^- uptake within 6 to 12 h. This induction is very temperature-dependent, with a Q_{10} value close to 4 (Carlberg et al. 1990), and appears to fail completely at temperatures close to 0 °C (L. Axelsson, unpubl.).

A number of other algae have also been suggested to feature properties of HCO_3^- uptake (without defining how such an uptake occurs), e.g. the brown algae *Fucus distichus*, *Pelvetiopsis litata* (Cook et al. 1986), and the red algae

Gracilaria conferta (Israel and Beer 1992), *Palmaria palmata* (Cook and Colman 1987), *Laurencia pinnatifida* (Johnston et al. 1992), *Chondrus crispus* (Brechignac and Andre 1985) and species of *Ceramium*, *Iridea*, *Gigartina*, *Porphyra*, *Gelidium* and *Pterocladia* (Cook et al. 1986, 1988). With the exception of *Palmaria palmata*, these algae probably depend on external CA activity for their HCO_3^- utilisation (cf. Mercado et al. 1997; Larsson and Axelsson 1999), suggesting instead the type 1 and possibly also the type 3b HCO_3^- utilisation to be operating.

Type 3a has so far been shown only for submerged higher plants (Prins et al. 1982; Rascio et al. 1999) and some Characean algae (Price and Badger 1985). Recent results have shown that this mechanism may be important also in seagrasses (Fernández et al. 1999; Hellblom et al. 1999, 2001). Type 3b was recently described for *Laminaria saccharina* (Axelsson et al. 2000). A combined sensitivity to buffers (interfering with extracellularly accumulating H^+, Axelsson et al. 2000) and inhibitors of extracellularly operating CA (Larsson and Axelsson 1999) suggests this mechanism to be quite common among brown macroalgae (Fig. 2.2, cf. Schmid and Dring 1993,1996; Schmid et al. 1996). Ongoing investigations suggest a similar situation for many species of red macroalgae (Fig 2.2; L. Axelsson, P. Moulin and J. Andria unpubl.; Ryberg et al. 1999).

2.3.2 Mechanisms Facilitating Intracellular Ci Transport

Activity of CA may also be important (and seems to be generally present) inside the cells. For example, if HCO_3^- is transported into the photosynthesising cells (and/or chloroplasts), it needs to be transformed to CO_2 since Rubisco recognises only CO_2 for the carboxylation reaction. In general, internal CA would facilitate CO_2 transport to Rubisco as transport of both CO_2 and HCO_3^- can take place (Raven 1997a). On the other hand, internal CA in a "wrong" location may interfere with CCM, as was shown for the cyanobacterium *Synechococcus* (Price and Badger 1989). Some sulphonamides (inhibitors of CA) like ethoxyzolamide (EZ), which rapidly penetrate cell membranes, are assumed to demonstrate the importance of intracellular CA (Moroney et al. 1985). If added after a prior inhibition of external CA activity (using the sulphonamide acetazolamide, AZ), EZ usually causes an additional inhibition for red and green algae (Beer 1994; Axelsson et al. 1995; Mercado et al. 1998; Badger et al. 1998), thus supporting the role of internal CA in Ci transport. This is, however, not the case for some brown algae, notably those of the families Laminariaceae and Fucaceae, where additions of AZ and EZ caused almost the same response (L. Axelsson, J. Uusitalo and J.M. Mercado, unpubl.). As these algae are assumed to have a comparatively higher internal CA activity (Giordano and Maberly 1989; Surif and Raven 1989), this behaviour is not understood. It might be that internal CA is not present primarily for Ci transport, and that CO_2 used by Rubisco is formed in the thylakoid lumen (cf. Raven 1997c), possibly without the aid of CA (i.e. in a way similar to the type

3a mechanism). These algae also possess a type 3b mechanism, which includes a certain CCM capability located in the plasma membrane (Fig. 2.3). Alternative functions for internal CA could be in a H^+ buffer system (in cooperation with the CO_2/HCO_3^- pool of the cell). There are several metabolic reactions involving H^+ exchanges in the cell, and these might benefit from the presence of such a buffer system. Such a system may even help to maintain a Ci buffer (such as HCO_3^-), without being directly involved in Ci transport (cf. Johnston 1991). It may also prevent the CO_2 formed in the chloroplast from escaping out through the cell (see below).

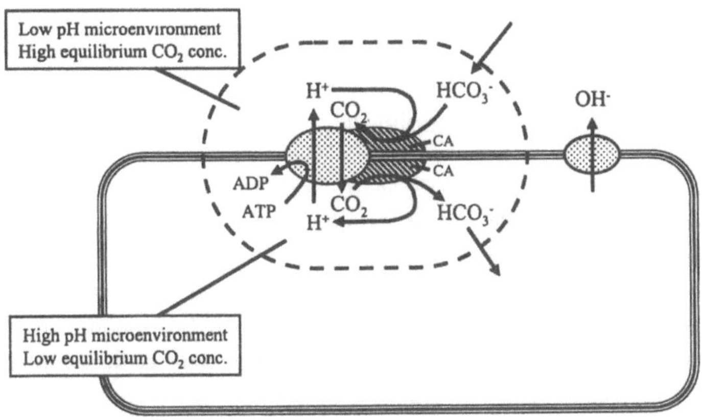

Fig. 2.3. Schematic presentation of type 3b HCO_3^- utilisation, combined with CA activity on the inside of the cell membrane

One arrangement that has been suggested for green microalgae (and which would facilitate both CO_2 entry and Ci transport) is the location of CA just inside the plasma membrane. This CA is suggested to be associated with alkaline pockets, thus achieving an almost complete transformation of entering CO_2 into HCO_3^- (cf. Kaplan and Reinhold 1999). Such an arrangement would comprise a CA activity localised to the cytoplasm compartment of the cell without interfering with CCM associated HCO_3^- utilisation due to outleakage of CO_2. Besides, it would also act as a trap for any outdiffusion of CO_2. It is not unlikely that also (at least some) green macroalgae (such as *Ulva* and *Enteromorpha*) have this possible mechanism. The arrangement would, in effect, behave like a pump for active CO_2 transport through the cell membrane. In combination with type 3b, it would constitute an extremely efficient HCO_3^--utilising mechanism, operating at moderately increased pH of the surrounding water. The proton-transporting system (either an ATP-ase, as illustrated in Fig. 2.3, or an electron transport chain, cf. Nimer et al. 1999) could contribute both to the low pH outside and the high pH inside the cell membrane, and could also provide a proteinaceous pore for CO_2

entrance through a lipid membrane with low CO_2 permeability (Fig. 2.3, cf. below).

The transport of Ci to Rubisco involves the passage through a number of membranes. The different mechanisms described above for HCO_3^- utilisation (Figs. 2.1 and 2.2) can also potentially be used to facilitate the passage through such inner membranes. The chloroplasts of all plants appear to stem from the same blue-green ancestor (Raven 1997b; Palmer 1999) and Ci acquisition of cyanobacteria has been fairly well investigated (Kaplan and Reinhold 1999). These organisms appear to be able to actively transport both HCO_3^- and CO_2 into the cell, and the carboxysome of cyanobacteria and the pyrenoid of many (eukaryotic) algae appear to function in a similar way (McKay and Gibbs 1991), being largely responsible for the CCM of these organisms.

2.3.3 CO_2-Concentrating Mechanisms

A CO_2-concentrating mechanism (CCM) is one that allows an alga to create a higher CO_2 concentration at the site of Rubisco than can be explained by diffusion of Ci (even when facilitated by CA) from the external medium. While the extracellular activity of CA can instantaneously reinstate the HCO_3^-/CO_2 ratio as CO_2 diffuses into the photosynthesising cells, it must be pointed out that this system per se cannot cause a higher CO_2 concentration close to the cellular plasma membrane (or inside the cell) than that of the surrounding seawater medium.

If algae relying on type 1 HCO_3^- utilisation have a CCM, thus alleviating possible Ci limitation by concentrating CO_2 around Rubisco in the chloroplasts, a second system must also be present which would actively concentrate the indiffusing CO_2 towards the chloroplast. While many macroalgae relying only on type 1 HCO_3^- utilisation possess features which points to the presence of a CCM (e.g. O_2-insensitive photosynthetic rates), the actual Ci-concentrating step has not yet been elucidated. Possible candidates for such a step comprise: CA in alkaline pockets just inside the plasma membrane, HCO_3^- utilisation mechanisms similar to types 3-4, but located at the chloroplast membranes, and the generation of CO_2 from HCO_3^- at any acid compartment within the cell (e.g. thylakoid lumen or acid vacuoles, cf. Raven 1997b,c). In addition, at least some types of HCO_3^- utilisation (notably types 3 and 4) imply a CCM at the plasma membrane (cf. Fig. 2.2). Thus, while many red and brown macroalgae appear to have a CCM at the plasma membrane level, little is known of CCMs inside the cells. This is true also for the type 2 mechanism because the HCO_3^- transporter is a proposed anion exchange protein that only facilitates this transport along an electrochemical gradient but cannot transport HCO_3^- against it (Drechsler et al. 1993, 1994; Axelsson et al. 1999).

Unlike in terrestrial plants, C_4 metabolism has not been reported to be a common CCM in algae. In fact, the only marine macroalga that has convincingly been attributed C_4 metabolism is a species of the green genus *Udotea* (Reiskind et al. 1989). CAM-like metabolism, although reported for some brown algae, seems not to have a large quantitative importance among the marine macroalgae (e.g. Johnston 1991, although a photoprotective role may be important, see below). Thus, the CCM in marine algae seems largely to be based on the Ci-utilisation systems.

2.4 Algal Responses to Carbon Limitation

2.4.1 Responses to Carbon Stress

A low supply of CO_2 to Rubisco during photosynthesis may have two consequences. Firstly, a priori there will be a lower rate of photosynthesis and growth. Since the different types of improved HCO_3^- utilisation depicted above will increase the CO_2 supply, they may thus be of competitive advantage in increasing photosynthetic rates. The second consequence is potential overexcitation of the photosynthetic apparatus, eventually leading to a decreased photosynthetic performance over longer periods (primarily through photoinhibition, Hanelt et al. 1992, 1993, 1994) or even to irreversible damage of the photosynthetic apparatus (Long et al. 1994; Demmig-Adams and Adams 2000). To avoid this situation, photosynthetic organisms have developed a number of regulatory mechanisms ranging from adjustment of light trapping and dissipation of light energy to specific metabolic adaptations allowing re-use of ADP and NADP without Ci reduction (e.g. photorespiration, Demmig-Adams and Adams 1992, 2000). The great number and variety of such mechanisms is perhaps the best indication of the importance of avoiding stress caused by a low CO_2 supply. These mechanisms may be initiated in periods from seconds (light dissipation, e.g. via the xanthophyll cycle, Demmig-Adams and Adams 1992) to several hours (induction of HCO_3^- uptake in green macroalgae; Axelsson et al. 1991, 1995, 1999) or even days (e.g. pigment regulations), and have been thoroughly studied in higher terrestrial plants (Long et al. 1994).

Photorespiration, which is an important protection mechanism during CO_2 shortage (and excess light energy) in higher plants, would be quite unsuitable in aquatic organisms like algae. This should be evident from a simple comparison between physical conditions (O_2 solubility and concentration, diffusion rates, Ci availability) of the atmosphere and the sea (Axelsson et al. 1991). It has also been shown in PAM fluorometry experiments that photorespiration is largely missing in macroalgae (Franklin and Badger 1999; Beer et al. 2000).

2.4.2 Regulation of Carbon Acquisition

Gas-exchange studies using mass spectrometric technologies have shown that algae tend to be much more leaky to Ci than hitherto assumed (Kaplan and

Reinhold 1999). Since energy is often required for an HCO_3^- utilisation process, this method of recycling inorganic carbon via CCMs points to an additional way of getting rid of excess energy (Kaplan and Reinhold 1999). Although little is known about Ci leakage in macroalgae, such a recycling has the potential of functionally replacing photorespiration of higher plants (or, from an evolutionary point of view, photorespiration of terrestrial plants replaces the Ci recycling mechanism of algae, which cannot function in air). Algae contain several kinds of CA, the occurrence and activities of which are regulated by low CO_2 conditions in a way that is not quite understood. Although it has been generally assumed that CA activity always increases at low CO_2 concentrations, *Chlamydomonas reinhardtii* (perhaps the best examined green microalga in this sense) contains two extracellular CAs, one of which is expressed at high and one at low CO_2 conditions (cf. Sültemeyer 1998). It might be that these regulations constitute a fine-tuning of both Ci acquisition mechanisms and light energy dissipation via Ci recycling. It has, for example, been shown that *Ulva lactuca*, acclimatised to direct CO_2 uptake (type 2) and photosynthesising at high pH, will have a higher CO_2 leakage if external CA activity is present (as the addition of AZ will cause an increase in photosynthesis; L. Axelsson and S. Beer unpubl.; shown also for *Enteromorpha*, Larsson et al. 1997). It has also been shown that external CA activity was downregulated when *Ulva* acclimatised to photosynthesis at high pH (Axelsson et al. 1995). There are strong indications that (the lipid layers of) membranes are far less permeable to CO_2 than previously assumed (Baier et al. 1990; Gimmler et al. 1990). This may imply that CO_2 diffusion is more or less restricted to proteinaceous pores in the membranes (e.g. aquaporines, Raven 1997a). Such a situation would, of course, greatly improve the potential of fine-tuning both Ci acquisition and Ci recycling via the above-mentioned mechanisms.

The different mechanisms which algae can use to improve their HCO_3^- utilisation probably require different environmental conditions for functioning optimally. *Ulva lactuca*, for example, can be treated to feature HCO_3^- utilisation via both the type 1 and type 2 mechanisms. The type 1 mechanism is more efficient in seawater of normal pH under conditions of good stirring, while type 2 is more efficient in seawater of increased pH, or at low concentrations of Ci, especially CO_2 (Axelsson et al. 1995, 1999). As there is also a cost involved in running these mechanisms (especially those comprising a CCM, i.e. types 3 and 4) one should expect unnecessary mechanisms to be downregulated. Another reason for downregulation would be interference between the two mechanisms (for example between direct uptake and extracellular CA activity, as in the example given above).

There are a few examples of regulation of Ci acquisition by macroalgae in the literature. Almost all green algae appear to have a capability of inducing HCO_3^- utilisation at high pH, probably via direct uptake (type 2). This possibility was not found to be present in any brown or red macroalgae investigated (Larsson and Axelsson 1999). There are several reports on decrease in HCO_3^- utilisation via the type 1 mechanism (i.e. extracellular CA activity) as a consequence of growth in seawater with high CO_2 concentrations. However, due to the profound effect this treatment has also on nitrate metabolism, some of these results are questionable (Axelsson et al. 1991; Magnusson et al. 1996).

Many macroalgae appear to possess several ways of HCO_3^- utilisation, which might be regulated according to the external conditions (Fig. 2.2). Such a regulation has so far been demonstrated most convincingly for *Ulva lactuca.* In this alga, direct uptake (type 2) can take over from CA-catalysed extracellular HCO_3^- dehydration (type 1) under conditions where the extracellular pH is increased experimentally (Axelsson et al. 1995), and a possible trigger is the low CO_2 concentration adjacent to the cellular plasma membrane. In nature, it is likely that such conditions can occur when plants are photosynthesising at high rates, i.e. when the temperature, irradiance and nutrient supply are high. Thus, these algae could alleviate a possible Ci limitation under such favourable growth conditions by making the plasma membrane more permeable for HCO_3^-. Although this ability to switch between HCO_3^- utilisation systems and, thus, also optimise the energetic cost of these systems, must be related to competition between species, it may also be important in avoidance of over-excitation (i.e. carbon stress in a strict sense).

2.4.3 Molecular Responses

Little is known about the actual molecular sensor responding to carbon stress in algae. Due to the high potential impact of the stress, it is likely to be a general stress sensor. Grossman et al. (1999) have isolated a mutant altered in nblS, which encodes a sensor kinase protein that plays a role in regulating both nutrient stress and high light responses in cyanobacteria. This protein probably binds to a heme or flavin group, and appears to integrate a variety of stress conditions by sensing the redox status of the cell. There are indications of a tight coupling between photosynthesis and CCMs via redox potential (for example plasma membrane based HCO_3^- utilisation in a number of marine phytoplankton, Nimer et al. 1999). This, together with the likely importance of CCMs in light energy dissipation (perhaps the primary function in their evolution, Kaplan and Reinhold 1999), is in support of the above suggestion. The dual function of CCMs probably requires more than one sensor, and a sensor of the partial pressure of CO_2 (functioning also in darkness!) might also be of general importance (Sültemeyer 1998).

2.4.4 Situations and Places Where Ci Limitation May Occur

As mentioned above, organic buffers (e.g. TRIS) inhibit an important way of HCO_3^- acquisition (Axelsson et al. 2000; Fig. 2.2). Many studies made to evaluate Ci limitations in the sea (e.g. Ci affinity studies) have involved the use of such buffers with the intent to vary the relative concentrations of various Ci forms according to the pH, but these buffers can also per se cause both a pronounced decrease (for example in *Chondrus crispus*, Fig. 2.2) and a pronounced increase (for example *Porphyra umbilicalis*, Ryberg et al. 1999) in the Ci affinity. Thus, there are few reliable data on Ci affinities, and there is a need for re-evaluating many of the available data. With this precaution in mind, the present views based on the literature will be presented in the following.

2.4.5 Light Gradients

Since both Ci and light are required in photosynthesis, and since both may be considered as potential limiting factors for the process, it follows that one is more prone to be limiting when the other is not. For example, it is unlikely that Ci is a limiting factor at depths where light is scarcely available because photosynthetic generation of ATP and reducing power, both needed for CO_2 fixation, would be limited. On the other hand, since various algae can feature more or less efficient HCO_3^- utilisation properties, one may expect the more efficient ways to have evolved under high light conditions. This plasticity in Ci utilisation can thus yield to the question: is Ci generally limiting in a habitat of a certain light regime? a rather complex answer.

Along a depth gradient, the brown algae growing highest up (mainly the Fucales) featured more efficient Ci utilisation systems than did members of the Laminariales growing lower down (Surif and Raven 1989, 1990). This confirms that Ci utilisation would adapt to the average surrounding irradiance, thus minimising potential Ci limitations. In the intertidal, adaptation to high light could also be seen against the background that ca. 350 ppm CO_2 is available as the only bulk Ci source during emergence, and that efficient Ci utilisation systems would be available to transport this Ci efficiently into the photosynthesising cells (probably as HCO_3^- formed by CA catalysed CO_2 hydration within a wet boundary layer in the cell wall, just outside the plasma membrane). Incidentally, a pattern similar to this was observed in marine angiosperms (seagrasses); it was found that shallow-growing species featured much more efficient HCO_3^- utilisation systems than did deeper-growing ones (Björk et al. 1997).

Some adaptational features of Ci acquisition to the surrounding irradiance can also be found among the red algae. *Palmaria palmata* and *Laurencia pinnatifida* from high-light intertidal habitats showed a high degree of HCO_3^- utilisation while *Delesseria sanguinea* and *Lomentaria articulata* from more shaded habitats were restricted to CO_2 use only (Johnston et al. 1992).

In addition to these adaptations, algae may also acclimatise to changing conditions on a shorter time scale. Under daily or weekly fluctuating light, it may be that an alga acclimatises rather sluggishly so that there is a transient limitation in Ci utilisation as the light increases, but that Ci is not limiting should the irradiance decrease. Superimposed on this is also the likelihood that the light-capturing and -transducing systems themselves adapt to variations in irradiance, and so the question becomes, which system adapts first?

2.4.6 Emersion, Tidal Pools and Rock Pools

Most algae contain an internal pool of HCO_3^- which, together with internal CA, can constitute a buffer allowing CO_2 supply over short-term increases in light intensity or decreases in CO_2 availability. Some littoral algae, especially of the family Fucaceae, have developed more elaborate photosynthetic buffers, i.e. stores of CO_2 (and probably nitrate as well) which can supply the photosynthetic apparatus with reducible substances over several hours (Axelsson et al. 1989a,b, 1991). The time course for operation of these buffers suggests that they are important for algae growing under the periodically changing conditions in the intertidal zone. There are indications that different species have adapted to both the buffer (its function and size) and their degree of HCO_3^- utilisation (via the type 3b mechanism, Fig. 2.2) for functioning in a certain habitat with respect to both wave exposure and the duration of emersion (Axelsson et al. 1989b). These algae show a number of adaptational features such as storage of nitrate, rapid nitrate uptake, efficient CO_2 uptake from air and intracellular storage of CO_2, and these features allow for improved productivity (by allowing photosynthesis in air) that protect the plants from photoinhibition under adverse conditions. The question: which mechanism is most important, and for which function? is probably not relevant in such a complex system of integrated functions. Also in this case, the possibility of a CCM acting in light energy dissipation has to be considered. The specific ultrastuctural organisation, with mitochondria covering the inner side of the cell membrane (Axelsson et al. 1989b) might be perfect for such a function, and could explain both the (so far missing, cf. above) role of internal CA (localised in the mitochondria?) and the (so far not altogether understood) possible function of C_4 like metabolism (cf. Axelsson et al. 1991; Raven 1997a).

The photosynthetic buffer system of littoral brown algae also functions well in tidal pools. Here, the algae are trapped in still water instead of being emersed. As pH increases, these algae have (for a period of time) the ability to excrete protons. Thus, a considerable part of the DIC can be assimilated without the pH increasing to the high levels where only the direct uptake of green macroalgae can function (Axelsson and Uusitalo 1988). Also in this case, a specific function in light stress avoidance should be considered.

Usually, rock pools that have been isolated from natural seawater for up to days are occupied by green algae of the genera *Ulva* and/or *Enteromorpha*. If so, the water of such pools usually has a high pH and O_2 concentration during the day, and Ci levels are declining. While *Ulva* has the possibility to change its preferred way of HCO_3^- utilisation (from type 1 to type 2) in response to high pH (Axelsson et al. 1995), it has been shown that *Enteromorpha* growing in rock pools at high pH values actually does feature the type 2 HCO_3^- uptake system as compared with specimens growing nearby in natural seawater which featured type 1 HCO_3^- utilisation (Larsson et al. 1997).

In general, we view direct HCO_3^- uptake as the most efficient way of utilising this Ci source, at least when the water exchange rate is limited. This is because high photosynthetic rates usually lead to high pH values near the surface of the algae (or in a larger enclosed water volume such as a rock pool or shallow bay). Under such conditions, the Ci-equilibrium CO_2 concentration is very low (even if it is readily resupplied by CA-catalysed HCO_3^- dehydration), and the HCO_3^- concentration is high. There would thus be a great advantage if HCO_3^- could be transported through the plasma membrane via a transport protein.

2.4.7 Adaptation to Terrestrial Life

It is well known that only the green algae have been able to adapt to life on land (with the photosynthetic function intact). This adaptation must have been very much dependent on the evolution of systems for eliminating light stress. It is becoming increasingly clear that one of the most important systems in marine photosynthesis could be the CCMs coupled with HCO_3^- utilisation. We are only beginning to realise this new function of CCMs, but maybe the differences between green and non-green algae in this respect is the key to better understanding of important questions regarding the evolution of adaptational features that made plant life on land possible.

2.5 Inorganic Carbon Limitation: How Can It Be Measured?

It follows from the above considerations on the availability of CO_2 in seawater that Ci limitations are possible. The various ways of HCO_3^- utilisation confer, however, a sufficient supply of CO_2 and, in most cases, even a CCM to the algae, and these systems often alleviate such potential limitations. In all, it seems that the green algae have developed the most efficient Ci-acquisition systems, and this has rendered some of the more common species such as *Ulva fasciata* and *Ulva lactuca* Ci saturation, at least under laboratory conditions, for both photosynthesis (cf. Beer 1994) and growth (Frost-Christensen and Sand Jensen 1990). Actual measurements on whether marine macroalgae in nature could be Ci limited or not are scarce, and none of them were done in situ, thus questioning their validity to some degree. Also, buffers have been used in many cases, and they may per se have affected some of the HCO_3^- utilisation systems adversely (see above). Within these limitations, it seems that not all green algae are Ci saturated under close to natural conditions. For example, *Ulva rotundata* (Levavasseur et al. 1991), *Enteromorpha compressa* (Beer and Shragge 1987) and *Cladophoropsis membranacea* (Holbrook et al. 1988) were at times found to be Ci-limited. Regarding algae other than the green, *Fucus vesiculosus* (Raven and Osmond 1992) was found to be Ci-saturated while *Durvilaea potatorum* (Raven et al. 1989) was Ci-limited in normal seawater.

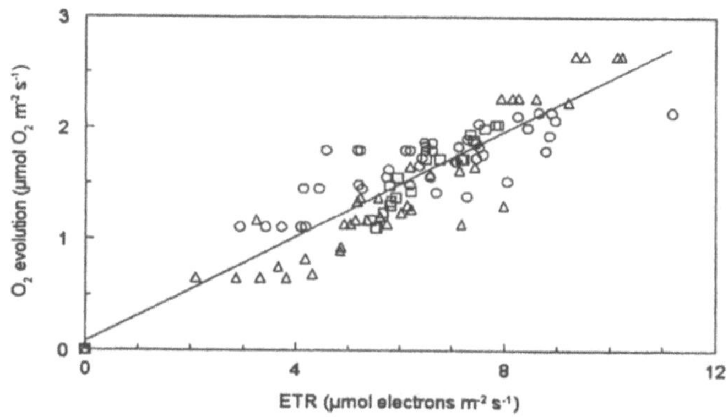

Fig. 2.4. Relationships between simultaneously measured rates of photosynthetic O_2 evolution (net O_2 exchange corrected for dark respiration) and electron transport (ETR) for *Ulva lactuca* at various concentrations of Ci (up to 1.6 mM, generally increasing towards the *upper right* of the figure). The *data points* were taken from three different experiments (depicted with *different symbols*). For details, see Beer et al. (2000)

In all the above cited works, photosynthetic (and also growth when measured) responses to Ci have been measured in the laboratory, usually as gas (O_2 or CO_2) exchange. Since these methods require the enclosure of the algae while CO_2 is taken up from, or O_2 is released to, the medium, they become cumbersome to use outside of the laboratory and are therefore usually not done in situ. Enclosures as such may also yield feed-back effects on the results and, thus, may not reflect the true responses under field conditions. Recently, pulse amplitude-modulated (PAM) chlorophyll fluorometry has been introduced as a measure of photosynthetic activity also for in situ measurements under submerged conditions. This was done with the development of a submersible instrument (the Diving-PAM fluorometer; Walz, Germany), and there is little doubt that this instrument will be of great importance when measuring in situ photosynthetic rate responses of marine macroalgae. It was recently shown that responses to Ci could also be measured in this way (Beer et al. 2000), and that those measurements were just as good in determining photosynthetic rates as a function of Ci as were O_2 electrode measurements (Fig. 2.4). This method will thus provide for quick, non-invasive, in situ estimates of photosynthetic performances under natural conditions, including those of variations in the Ci environment.

References

Axelsson L (1988) Changes in pH as a measure of photosynthesis by marine macroalgae. Mar Biol 97:287-294

Axelsson L, Uusitalo J (1988) Carbon acquisition strategies for marine macroalgae. 1. Utilisation of proton exchanges visualized during photosynthesis in a closed system. Mar Biol 97:295-300

Axelsson L, Carlberg S, Ryberg H (1989a) Adaptations by macroalgae to low carbon availability. I. A buffer system in *Ascophyllum nodosum*, associated with photosynthesis. Plant Cell Environ 12:765-770

Axelsson L, Carlberg S, Ryberg H (1989b) Adaptations by macroalgae to low carbon availability. II. Ultrstructural specializations, related to the function of a photosynthetic buffer system in the Fucaceae. Plant Cell Environ 12:771-778

Axelsson L, Uusitalo J, Ryberg H (1991) Mechanisms for concentrating and storage of inorganic carbon in marine macroalgae. In: Garcia-Reina G, Pedersén M (eds) Seaweed cellular biotechnology. Universidad de Las Palmas de Gran Canaria, Spain, pp 185-198

Axelsson L, Ryberg H, Beer S (1995) Two modes of bicarbonate acquisition in *Ulva lactuca*. Plant Cell Environ 18:439-445

Axelsson L, Larsson C, Ryberg H (1999) Affinity, capacity and oxygen sensitivity of two different mechanisms for bicarbonate utilization in *Ulva lactuca* L. (Chlorophyta). Plant Cell Environ 22:969-978

Axelsson L, Mercado J, Figueroa F (2000) Utilization of HCO_3^- at high pH by the brown macroalga *Laminaria saccharina*. Eur J Phycol 35:53-59

Badger MR, Andrews TJ, Whitney SM, Ludwig M, Yellowlees DC, Leggat W, Price D (1998) The diversity and coevolution of Rubisco, plastids, pyrenoids, and chloroplast-based CO_2-concentrating mechanisms in algae. Can J Bot 76:1057-1071

Baier M, Gimmler H, Hartung W (1990) The permeability of the guard cell plasma membrane and tonoplast. J Exp Bot 41:351-358

Beer S (1994) Mechanism of inorganic carbon acquisition in marine macroalgae (with special reference to the Chlorophyta). In: Round FE, Chapman VJ (eds) Progress in phycological research. Biopress, Bristol, pp 179-207

Beer S, Eshel A (1983) Photosynthesis of *Ulva* sp. II. Utilization of CO_2 and HCO_3^- when submerged. J Exp Mar Biol Ecol 70:99-106

Beer S, Shragge B (1987) Photosynthetic carbon metabolism in *Enteromorpha compressa* (Chlorophyta). J Phycol 23:580-584

Beer S, Larsson C, Poryan O, Axelsson L (2000) Measuring photosynthetic rates of the marine macroalga *Ulva* by pulse amplitude modulated (PAM) fluorometry. Eur J Phycol 35:69-74

Björk M, Weil A, Semesi S, Beer S (1997) The photosynthetic responses of eight seagrass species from Zanzibar to increased inorganic carbon concentrations. Mar Biol 129:363-366

Brechignac F, Andre M (1985) Continuous measurements of the free dissolved CO_2 concentration during photosynthesis of marine plants: evidence for HCO_3^- use in *Chondrus crispus*. Plant Physiol 78:551-554

Carlberg S, Axelsson L, Larsson C, Ryberg H, Uusitalo J (1990) Inducible CO_2 concentrating mechanisms in green seaweeds. I. Taxonomical and physiological aspects. In: Baltscheffsky M (ed) Current research in photosynthesis, vol IV. Kluwer, Dordrecht, pp 529-532

Cook CM, Colman B (1987) Some characteristics of photosynthetic inorganic carbon uptake of a marine macrophytic red alga. Plant Cell Environ 10:275-278

Cook CM, Lanaras T, Colman B (1986) Evidence for bicarbonate transport in species of red and brown marine macrophytic marine algae. J Exp Bot 37:977-984

Cook CM, Lanaras T, Roubelakis-Angelakis KA (1988) Bicarbonate transport and alkalinization of the medium by four species of Rhodophyta. J Exp Bot 39:1185-1198

Demmig-Adams B, Adams WW III (1992) Photoprotection and other responses of plants to high light stress. Annu Rev Plant Physiol Plant Mol Biol 43:599-626

Demmig-Adams B, Adams WW III (2000) Harvesting sunlight safely. Nature 403:371-374

Drechsler Z, Sharkia R, Cabantchik ZI, Beer S (1993) Bicarbonate uptake in the marine macroalga *Ulva* sp. is inhibited by classical probes of anion exchange by red blood cells. Planta 191:34-40

Drechsler Z, Sharkia R, Cabantchik ZI, Beer S (1994) The relationship of arginine groups to photosynthetic HCO_3^- uptake in *Ulva* sp. mediated by a putative anion exchanger. Planta 194:250-255

Falkowski PG, Raven JA (1997) Aquatic photosynthesis. Blackwell, Malden, Massachusetts

Fernández, JA, García-Sánchez MJ, Felle HH (1999) Physiological evidence for a proton pump and sodium exclusion mechanisms at the plasma membrane of the marine angiosperm *Zostera marina* L. J Exp Bot 50:1763-1768

Franklin L, Badger M (1999) Can fluorometric measurements of PSII charge separation be used to infer photosynthetic electron transport in macroalgae? Conf on carbon acquisition in marine macroalgae, COST 49, Uppsala University, Sweden, p 15

Frost-Christensen H, Sand-Jensen K (1990) Growth rate and carbon affinity of *Ulva lactuca* under controlled levels of carbon, pH and oxygen. Mar Biol 104:497-501

Gimmler H, Weiss C, Baier M, Hartung W (1990) The conductance of the plasmalemma for CO_2. J Exp Bot 41:785-795

Giordano M, Maberly SC (1989) Distribution of carbonic anhydrase in British marine macroalgae. Oecologia 81:534-539

Grossman AR, van Waasbergen L, Schwartz R, Dolganov N (1999) Acclimation of photosynthetic organisms to adverse environmental conditions. 2nd Eur Phycol Congr, September 20-26, Montecatini Terme, Italy, p 4

Haglund C, Axelsson L, Pedersén M (1987) Photosynthesis and respiration in the alga *Ahnfeltia plicata* in a flow-through system. Mar Biol 96:409-412

Haglund K, Björk M, Ramazanov Z, Garcia-Reina G, Pedersén M (1992a) Role of carbonic anhydrase in photosynthesis and inorganic-carbon assimilation in the red alga *Gracilaria tenuistipitata*. Planta 187:275-281

Haglund K, Ramazanov Z, Mtolera M, Pedersén M (1992b) Role of external carbonic anhydrase in light-dependent alkalinization by *Fucus serratus* L. and *Laminaria saccharina* (L.). Lamour (Phaeophyta). Planta 188:1-6

Hanelt D, Hyppertz K, Nultsch W (1992) Photoinhibition of photosynthesis and its recovery in red algae. Bot Acta 105:278-284

Hanelt D, Hyppertz K, Nultsch W (1993) Daily course of photosynthesis and photoinhibition in marine macroalgae investigated in the laboratory and field. Mar Ecol Prog Ser 97:31-37

Hanelt D, Li J, Nultsch W (1994) Tidal dependence of photoinhibition of photosynthesis in marine macrophytes of the South China Sea. Bot Acta 107:66-72

Hellblom F, Beer S, Björk M, Axelsson L (1999) Buffer sensisivity in *Zostera marina* — inhibition of a proton-mediated HCO_3^- utilisation? Conf on carbon acquisition in marine macroalgae, COST 49, November 10-13, Uppsala University, Sweden, p 20

Hellblom F, Beer S, Björk M, Axelsson L (2001) A buffer-sensitive inorganic carbon utilisation system in the seagrass *Zostera marina*. Aquat Bot (in press)

Holbrook G, Beer S, Spencer WE, Reiskind JB, Davis JS, Bowes G (1988) Photosynthesis in marine macroalgae: evidence for carbon limitation. Can J Bot 66:577-582

Israel A, Beer S (1992) Photosynthetic carbon acquisition in the red alga *Gracilaria conferta*. 2. Rubisco carboxylase kinetics, carbonic anhydrase and HCO_3^- uptake. Mar Biol 112:697-700

Johnston AM (1991) The acquisition of inorganic carbon by marine macroalgae. Can J Bot 69:1123-1132

Johnston AM, Maberly SC, Raven JA (1992) The acquisition of inorganic carbon by four red macroalgae. Oecologia 92:317-326

Kaplan A, Reinhold L (1999) CO_2 concentrating mechanisms in photosynthetic microorganisms. Annu Rev Plant Physiol Plant Mol Biol 50:539-570

Kuypers MMM, Pancost RD, Sinninghe Damst JS (1999) A large and abrupt fall in atmospheric CO_2 concentration during Cretaceous times. Nature 399:342-345

Larsson C, Axelsson L (1999) Bicarbonate uptake and utilization in macroalgae. Eur J Phycol 34:79-86

Larsson C, Axelsson L, Ryberg H, Beer S (1997) Photosynthetic carbon utilization in *Enteromorpha intestinalis* from a rockpool. Eur J Phycol 32:49-54

Larsson C, Schmid R, Axelsson L (1999) Blue light stimulation of photosynthesis and buffer sensitive inorganic carbon assimilation in Phaeophyceae — a comparison. Conf on carbon acquisition in marine macroalgae, COST 49, November 10-13, Uppsala University, Sweden, p 22

Levavasseur G, Edwards GE, Osmond CB, Ramus J (1991) Inorganic carbon limitation of photosynthesis in *Ulva rotundata* (Chlorophyta). J Phycol 27:667-672

Lindskog S (1997) Structure and mechanism of carbonic anhydrase. Pharmacol Ther 74:1-20

Long SP, Humphries S, Falkowski PG (1994) Photoinhibition of photosynthesis in nature. Annu Rev Plant Physiol Plant Mol Biol 45:633-662

Maberly SC (1990) Exogenous sources of inorganic carbon for photosynthesis by marine macroalgae. J Phycol 26:439-449

Maberly SC, Raven JA, Johnston AM (1992) Discrimination between ^{12}C and ^{13}C by marine plants. Oecologia 91:481-492

Magnusson G, Larsson C, Axelsson L (1996) Effects of high CO_2 treatment on nitrate and ammonium uptake by *Ulva lactuca* grown in different nutrient regimes. Sci Mar 60 Suppl 1:179-189

McKay RML, Gibbs SP (1991) Composition and function of pyrenoids: cytochemical and immunocytochemical approaches. Can J Bot 69:1040-1052

Mercado JM, Figueroa FL, Niell FX, Axelsson L (1997) A new method for estimating carbonic anhydrase activity in macroalgae. J Phycol 33:999-1006

Mercado JM, Javier F, Gordillo F, Figueroa FL, Niell FX (1998) External carbonic anhydrase and affinity for inorganic carbon in intertidal macroalgae. J Exp Mar Biol Ecol 221:209-220

Moroney JV, Somanchi A (1999) How do algae concentrate CO_2 to increase the efficiency of photosynthetic carbon fixation. Plant Physiol 119:9-16

Moroney JV, Husic HD, Tolbert NE (1985) Effect of carbonic anhydrase inhibitors on inorganic carbon accumulation by *Chlamydomonas reinhardtii*. Plant Physiol 79:177-183

Nimer NA, Xiao Ling M, Brownlee C, Merrett MJ (1999) Inorganic carbon limitation, exofacial carbonic anhydrase activity and plasma membrane redox activity in marine phytoplankton species. J Phycol 35:1200-1205

Norris RD, Röhl U (1999) Carbon cycling and chronology of climate warming during the Palaeocene/Eocene transition. Nature 401:775-778

Palmer JD (1999) The origin of plastids and their spread by secondary symbiosis. 2nd Eur Phycol Congr, September 20-26, Montecatini Terme, Italy, p 5

Petit JR, Jouzel J, Raynaud,D, Barkov NI, Barnola J-M, Basile I, Bender M, Chappellaz J, Davis M, Delaygue G, Delmotte M, Kotlyakov VM, Legrand M, Lipenkov VY, Lorius C, Pépin L, Ritz C, Saltzman E, Stievenard M (1999) Climate and atmospheric history of the past 420,000 years from the Vostok ice core, Antarctica. Nature 399:429-436

Price GD, Badger MR (1985) Photosynthetic uitilization of bicarbonate in *Chara corallina*. Aust J Plant Physiol 12:257-267

Price GD, Badger MR (1989) Expression of human carbonic anhydrase in the cyanobacterium *Synechococcus* PCC7942 creates a high carbon dioxide-requiring phenotype: evidence for a central role for carboxysomes in the carbon dioxide-concentrating mechanism. Plant Physiol 91:505-513

Prins HBA, Snel JFH, Zanstra PE, Helder RJ (1982) The mechanism of bicarbonate assimilation by the polar leaves of *Potamogeton* and *Elodea*. CO_2 concentrations at the leaf surface. Plant Cell Environ 5:207-214

Rascio N, Cuccato F, Vecchia FD, La Rocca N, Larcher W (1999) Structural and functional features of the leaves of *Ranunculus trichophyllus* Chaix., a freshwater submerged macrophyte. Plant Cell Environ 22:205-212

Raven JA (1997a) Inorganic carbon acquisition by marine autotrophs. Adv Bot Res 27:85-209

Raven JA (1997b) Putting the C in phycology. Eur J Phycol 32:319-333

Raven JA (1997c) CO_2 concentrating mechanisms: a direct role for thylakoid lumen acidification? Plant Cell Environ 20:147-154

Raven JA (1997d) The role of marine biota in the evolution of terrestrial biota: gasses and genes. Biogeochemistry 39:139-164

Raven JA, Osmond CB (1992) Inorganic carbon acquisition processes and their ecological significance in inter- and sub-tidal macroalgae of North Carolina. Funct Ecol 6:41-47

Raven JA, Beardall J, Roberts S (1989) The ecophysiology of inorganic carbon assimilation by *Durvillaea potatorum* (Durvillaeales, Phaeophyta). Phycologia 28:429-437

Raven JA, Johnston AM, McFarlane JJ (1990) Carbon metabolism. In: Cole KM, Sheath RG (eds) Biology of the red algae. Cambridge University Press, Cambridge, pp 171-219

Reiskind JB, Seamon PT, Bowes G (1989) Photosynthetic responses and anatomical features of two marine macroalgae with different CO_2 compensation points. Aquat Bot 33:71-86

Ryberg H, Larsson C, Axelsson L (1999) Three red macroalgae with different mechanisms for HCO_3^- utilization. Conf on carbon acquisition in marine macroalgae, COST 49, November 10-13, Uppsala University, Sweden, p 36

Schmid R, Dring MJ (1993) Rapid blue-light-induced acidifications at the surface of *Ectocarpus* and other marine macroalgae. Plant Physiol 101:907–913

Schmid R, Dring MJ (1996) Blue light and carbon acquisition in brown algae: an overview and recent developments. Sci Mar 60 Suppl 1 :115–124

Schmid R, Mills JA, Dring MJ (1996) Influence of carbon supply on the stimulation of light-saturated photosynthesis by blue light in *Laminaria saccharina*: implications for the mechanism of carbon acquisition in higher brown algae. Plant Cell Environ 19:383-391

Smith RG, Bidwell RGS (1989) Mechanisms of photosynthetic carbon dioxide uptake by the red macroalga, *Chondrus crispus*. Plant Physiol 89:93-99

Sültemeyer D (1998) Carbonic anhydrase in eukaryotic algae: characterization, regulation and possible function during photosynthesis. Can J Bot 76:962-972

Surif MB, Raven JA (1989) Exogenous inorganic carbon sources for photosynthesis in water by members of the Fucales and the Laminariales (Phaeophyta): ecological and taxonomical implications. Oecologia 78:97-105

Surif MB, Raven JA (1990) Photosynthetic gas exchange under emeresed conditions in eulittoral and normally submersed members of the Fucales and the Laminariales: interpretation in relation to C isotope ratio and N and water use efficiency. Oecologia 82:68-80

Uusitalo J (1996) Algal carbon uptake and the difference between alkalinity and hıgh pH ("alkalinization") exemplified with a pH-drift experiment. Sci Mar 60 Suppl 1:129-134

Weiss RF (1974) Carbon dioxide in water and seawater: the solubility of a non-ideal gas. Mar Chem 2:203-215

3 Nitrogen Limitation

Rehab M. Shehawy and Diethelm Kleiner[1]

Contents

[1] Mikrobiologie, Universität Bayreuth, Universitatstrasse 30, Postfach 101251, 95440 Bayreuth, Germany

3.1 Introduction

Nitrogen (N) is an indispensable element incorporated in the most important structural and functional macromolecules (proteins, nucleic acids, and to a lesser extent polysaccharides). On the average, proteins contain about 15% and nucleic acids about 13% N. Thus, depending on the polysaccharide content, cyanobacteria and algae contain between 3 and 10% N in their dry weight. In general, both classes of organisms can use a wide variety of nitrogenous compounds to fulfil their N requirements: ammonia, nitrate, nitrite, urea, amino acids, nucleosides or bases, and (in the case of some cyanobacteria only) N_2.

When confronted with different N sources, microorganisms including algae and cyanobacteria are mostly able to choose the "best" one before assimilating the "less tasty" ones. Although belonging to different kingdoms of the living world, both classes have evolved a preference for the ubiquitous N source ammonia. Its presence elicits a special kind of catabolite repression, the so-called ammonia effect: assimilatory enzymes for other N sources are repressed, until ammonia is depleted from the environment. This depletion is felt as a mild N limitation. For most organisms the next-best N sources are urea or nitrate, followed by different organic compounds. The worst N sources are assimilated only when there is nothing else to derive N from. This is the case of severe N limitation.

In order to give an appropriate answer, the organism must be able to sense its intracellular N nutritional status. This signal then must be transduced to enzymes in charge of the adaptive response, or the genome to induce or repress the synthesis of appropriate enzymes.Therefore, sensing and transmission of N status signals will be an important part of this chapter.

3.2 General Strategies to Cope with Nitrogen Limitation

Frequently, uptake and assimilation of nitrogenous compounds are achieved by parallel pathways involving different sets of enzymes for the same net reaction. In general, one pathway is used when the N source is abundant, the alternative route is chosen under N-limiting conditions. Under N-replete conditions the physiological goal is high turnover, the enzymes involved have a low affinity (high K_m value) for the substrate and a high capacity (high V_{max}) for the reaction. This pathway is termed low-affinity/high-capacity route. Alternatively, limiting conditions may lead to induction of enzymes with a high affinity (low K_m) and a low capacity (low V_{max}); they aim at efficient use of low substrate concentrations. This high-affinity/ low capacity route may be viewed as a powerful scavenging system.

The well-known examples are the alternative routes for ammonia assimilation in many microorganisms. These are of prime importance, since ammonia is not only the preferred N source, but also a key metabolite, situated at the junction of catabolism and anabolism.

N sufficiency in many microorganisms induces the low-affinity enzyme glutamate dehydrogenase (GluDH, K_m for NH_4^+ generally > 1 mM), which catalyses the reductive amination of 2-oxo glutarate into glutamate without energy expenditure:

$$\text{2-oxo glutarate} + NH_3 + NAD(P)H \rightarrow \text{glutamate} + NAD(P) + H_2O. \qquad (1)$$

N limitation, on the other hand, induces the scavenging enzyme glutamine synthetase (GS), which catalyses the energy-dependent conversion of ammonia into an amide group:

$$\text{glutamate} + NH_3 + ATP \rightarrow \text{glutamine} + ADP + P_i. \qquad (2)$$

The amide group can be regarded as activated ammonia, its hydrolysis energy ΔG amounts to -14 kJ mol^{-1}. This activation adds thermodynamic pressure to the subsequent transfer reaction, which is catalysed by the enzyme glutamate synthase (officially glutamine:oxo glutarate amido transferase, acronym GOGAT)

$$\text{2-oxo glutarate} + \text{glutamine} + NAD(P)H \rightarrow 2\ \text{glutamate} + NAD(P). \qquad (3)$$

The combination of Eqs. (2) and (3) is termed glutamate cycle (or GS/GOGAT cycle), because one of the glutamate molecules produced by Eq. (3) again enters the cycle via Eq. (2).

Thus, both pathways result in glutamate synthesis from NH_3, NAD(P)H and 2-oxo glutarate.

The glutamate cycle as the high-affinity/low-capacity route is marked by a high affinity of GS for NH_4^+ (K_m around 0.1 mM or less). The ATP hydrolysis renders glutamate formation essentially irreversible. This cycle reduces intracellular ammonia concentrations to values probably below 0.1 mM. In contrast, the GluDH-catalysed reaction is reversible, and is employed when high ammonia levels are present in the cell.

A special case of the high-affinity/low-capacity (scavenging) systems are transport proteins involved in a process called cyclic retention. Although most metabolites produced intracellularly are rather hydrophilic, their solubility in the membrane lipid phase frequently is not negligible. This implies that some molecules will pass the membrane by unspecific diffusion and constantly leak to the outside (or periplasmic space). In the case of N metabolism it is significant for NH_3 (but not NH_4^+, (Kleiner 1985) and amino acids (Antonucci and Oxender 1988; Fig. 3.1). Under N-limiting conditions this leakage is a severe challenge, since it curtails the growth of the organism. As a counteraction many organisms induce synthesis of specific high affinity carriers to retrieve the fugitive molecules in an energy-dependent uptake reaction. Abundant ammonia frequently leads to repression of the ammonium carriers. Under these conditions the inward diffusion of NH_3 apparently is fast enough to satisfy the cell's demand for N (Kleiner 1985, 1993).

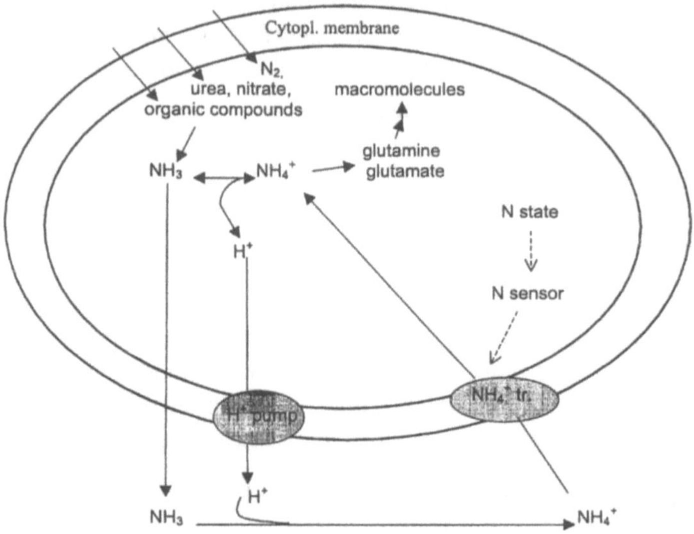

Fig. 3.1. Schematic representation of the NH_3/NH_4^+ retention cycle in microorganisms. ——▶ reactions including transport; ········> regulatory interactions; NH_4^+ tr. = ammonium transport system. (After Kleiner 1985; For details see text)

3.3 Cyanobacteria

Cyanobacteria mainly use ammonia and nitrate as N sources, but urea and amino acids can also be assimilated by some species. Specialised groups are able to fix N_2.

3.3.1 Assimilation of Ammonia, Nitrate and Nitrite

When supplied in low, limiting concentrations, ammonia is generally taken up by specific NH_4^+ transport systems (Boussiba and Gibson 1991), which drive concentrative uptake with the membrane potential as the energy source (Kleiner 1993). Some organisms express multiple NH_4^+ carriers for yet unknown physiological reasons (Montesinos et al. 1998). At least some of these transport systems are repressed when the organism is cultivated with abundant ammonia (Boussiba and Gibson 1991; Flores and Herrero 1994). Under these conditions, passive diffusion of NH_3 into the cell is probably sufficient to fulfil the N demand.

The uptake of nitrate takes place through different kinds of high-affinity transport systems, which either are H^+/NO_3^- symporters (Sakamoto et al. 1999) or belong to the widespread ABC-type transporters (Flores and Herrero 1994; Forde and Clarkson 1999). These membrane-bound proteins share a conserved ATP-binding domain, and are able to set up large concentration gradients at the expense of ATP hydrolysis. Thus, cyanobacteria are able to grow with very low nitrate levels (below 1 μM). Nitrite in at least some strains is taken up via the same carriers. Inside the cell both nitrate and nitrite are ultimately reduced to ammonia by the (combined) actions of nitrate and nitrite reductase with reduced ferredoxin as the reductant (Flores and Herero 1994). The genes coding for the ABC-type nitrate and nitrite transport proteins (*nrtABCD*) in some organisms are found in a cluster together with the genes for nitrite (*nir*) and nitrate (*nar*) reductase, possibly forming an operon. All genes are induced by nitrate and/or nitrite, and repressed in the presence of ammonia (Flores and Herero 1994). The H^+/NO_3^- symporter (encoded by *nrtP*) is additionally repressed by urea (Sakamoto et al. 1999).

3.3.2 Assimilation of Urea

Urea is a major compound in the global N cycle. It is the excretion product of ureotelic animals, and thus wide-spread (0.1-5 μM in seawater). Urea is taken up by cyanobacteria via specific, energy-dependent transport systems and then hydrolytically degraded by urease to NH_3 and CO_2 (Flores and Herero 1994). Both the urea transporter and urease seem to be subject to repression by ammonia (Singh 1991; Collier et al. 1999).

3.3.3 Assimilation of Amino Acids

Several amino acids have been reported to serve as N sources for cyanobacteria, notably arginine, glutamine, asparagine and proline (Singh et al.1991; Flores and Herrero 1994). The general route seems to be energy-dependent uptake followed by breakdown to ammonia/and or glutamate. No report seems to deal with regulation, notably with an ammonia effect.

3.3.4 Nitrogen (N_2) Fixation

This ability is distributed among numerous cyanobacterial strains, both unicellular and filamentous. Like in other prokaryotes, fixation of N_2 in cyanobacteria is catalysed by the enzyme complex nitrogenase, which is irreversibly inactivated by O_2. Therefore obligate aerobes have evolved mechanisms for protection of their N_2 fixing machinery from being damaged by either external or intracellularly generated O_2 (Fay 1992).

As in other bacteria, N_2 fixation in cyanobacteria requires a high energy input (about 16 ATP molecules have to be split for the reduction of one N_2 molecule).

Therefore the organisms resort to this reaction only when other N sources are exhausted — we are dealing with severe N limitation.

3.3.4.1 Heterocystous Cyanobacteria

Many filamentous cyanobacteria (notably the genera *Anabaena, Nostoc, Fischerella* and *Calothrix*) have developed a complex mechanism for performing aerobic N_2 fixation by restricting expression of nitrogenase to specialized cells, the heterocysts. These thick-walled cells upon N deprivation appear at regular intervals in the filaments and undergo a series of modifications devoted to expression of the nitrogenase proteins and the protection of nitrogenase from O_2 (see reviews Adams and Carr 1989; Buikema and Haselkorn 1993; Wolk et al. 1994; Wolk 1996).

The formation of heterocysts in the filamentous cyanobacteria is a collective process, in which a sequential change in expression of around 1000 genes is observed (Buikema and Haselkorn 1993). The primary decision, which cells are to develop into heterocysts, is not yet understood. The spacing in *Anabaena* is achieved by the excretion of a pentapeptide (Arg-Gly-Ser-Gly-Arg) by emerging heterocysts, which diffuses along the filament in the periplasmic space (i.e. between cytoplasmic membrane and cell wall) and suppresses the heterocyst formation in about five adjacent cells (Yoon and Golden 1998). Thus cyanobacteria can count to 10 (Haselkorn 1998).The subsequent differentiation of heterocysts includes two major stages (see Fig. 3.2), which can be followed by

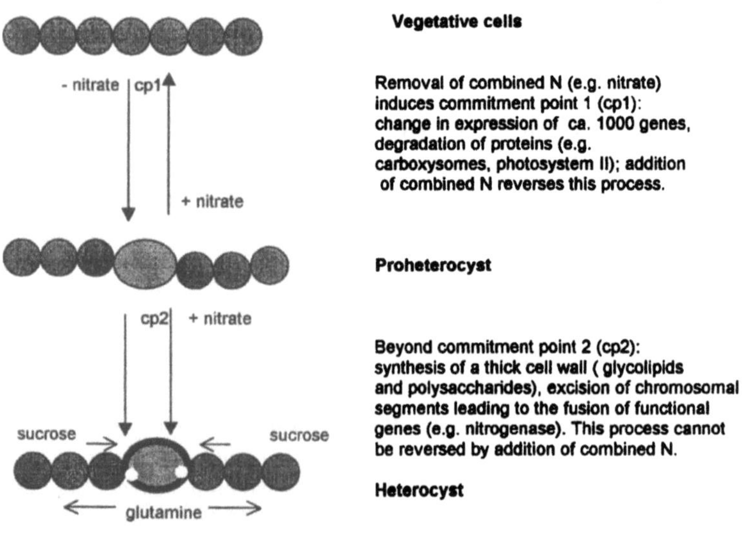

Fig. 3.2. Characteristic morphological, physiological and molecular biological events during heterocyst formation

simple light microscopy. The first stage (taking about 10-20 h, depending on the strain) comprises the formation of proheterocysts, which appear less granulated than vegetative cells, but still lack the thick wall. This process, starting at the so-called commitment point 1, involves specific degradation of proteins. Of special importance is the partial destruction of photosystem II, by which O_2 evolution is suppressed, and of carboxysomes, rendering heterocysts incapable of CO_2 fixation. Formation of proheterocysts is reversible: upon addition of fixed nitrogen (ammonia, nitrate) the proheterocyst reverts to a vegetative cell.

The second stage comprises the development of the proheterocyst into the mature heterocyst with formation of a thick cell wall containing specific glycolipids and polysaccharides (Wolk 1996, 2000). This wall is thought to limit entrance of O_2, thus enabling stable expression of nitrogenase as a consequence of specific signalling mechanisms (see below). Stage 1 and 2 are separated by commitment point 2, beyond which further development can no longer be reversed by addition of nitrate or ammonium salts (point of no return). This irreversibility, which is unique among prokaryotic differentiation processes, shows interesting parallels to senescence of eukaryotic cells (Shehawy and Kleiner 1999a). In the final stage, heterocysts and vegetative cells exchange N (in the form of glutamine) or C (as sucrose).

3.3.4.2 Nonheterocystous Strains

A number of strains avoid the O_2 problem by fixing N_2 only under anaerobic condition. Some unicellular organisms are able to temporally separate N_2 fixation from photosynthesis. For example, several strains restrict N_2 fixation to the night (Fay 1992).

3.3.5 Nitrogen Sensing and Nitrogen Regulation

Bacteria sense and respond to environmental changes via widely distributed and well-conserved signalling systems (the two-component signal-transducing systems), which generally consist of a sensor that receives the environmental stimulus and phosphorylates a response regulator protein, that directly regulates transcription of target genes (Stock et al. 1989).

In enterobacteria, sensing of the intracellular N state and resulting N regulated transcriptions are mediated by the two-component Ntr (nitrogen regulation) system. This system ultimately responds to the intracellular levels of 2-oxo glutarate and glutamine, and transmits this signal in a cascade via phosphorylation or dephosphorylation of the transcriptional activator protein NtrC to the genetic level where specific genes, mainly associated with N catabolism are switched on or off. N (especially ammonia) abundance is sensed as a rise in the intracellular glutamine level, which ultimately leads to the repression of, e.g. nitrogenase, nitrate reductase or ammonium transport proteins (Merrick and Edwards 1995; Magasanik 1996).

As outlined above, a similar phenomenon of N control operates in cyanobacteria. Analysis of mutants pleiotropically impaired in N catabolism led to the discovery of a global N-regulatory gene, *ntcA* (nitrogen control: Vega-Palas et al. 1990). This gene is widespread among unicellular and filamentous cyanobacteria (Frias et al. 1993) and encodes a regulatory protein which is required for the expression of genes known to be subject to repression by ammonium. The NtcA protein binds to specific DNA sequences upstream of the promoters and generally enhances transcription (Frias et al. 1994; Wei et al. 1994; Ramasubramanian et al. 1996). Transcription of *ntcA* is induced during N starvation (Lindell et al. 1998) in an unknown way. Discussions centre around the question whether the primary signal is intracellular ammonia or a derivative of it (e.g. glutamine). NtcA synthesis additionally is subject to autoregulation (Ramasubramanian et al. 1994). Further signal transmission involves different signal-transducing proteins, apparently depending on the target (Aichi and Omata 1997; Lee et al. 1999).

In heterocystous forms special targets are genes that regulate development of heterocysts and expression of nitrogenase, especially *hetR, xisA* and *hetC* (Buikema and Haselkorn 1991, 1993; Black et al. 1993; Muro-Pastor et al. 1999). A tentative summary is proposed in Fig. 3.3 (see the reviews of Adams and Carr 1989; Buikema and Haselkorn 1993; Wolk et al. 1994; Wolk 1996 for references and details).

NtcA directly or indirectly induces expression of *hetR*. The HetR protein is considered as master switch in heterocyst development, influencing expression of many genes in an unknown manner. HetR is a positive autoregulator (Black et al. 1993), but lacks any known DNA-binding ability. The protein is rapidly degraded after the readdition of a combined nitrogen source, probably by autodigestion, since HetR was shown to be a serine-type proteinase (Zhou et al. 1998a). The protein escapes this fate, when nitrogen deprivation is suggested to result in its modification, possibly by phosphorylation, in an unknown way (Zhou et al. 1998b). Stable expression of the HetR protein in this cascade-like manner leads to formation of enzymes for glycolipid and polysaccharide synthesis, the products of which seem to be transported to the outside by specific ABC-type transport proteins and deposited in the nascent wall (Cai and Wolk 1997; Wolk 2000).

In vegetative cells one of the nitrogenase structural genes (*nifD*) is rendered inactive by interruption with an 11-kb-long stretch of DNA. During heterocyst formation this part is excised by the site-specific recombinase XisA (Brusca et al. 1990). The *xisA* promoter region contains three binding sites for NtcA, which apparently can be bound as monomer or as multimer (Chastain et al. 1990). NtcA seems to be involved in a complex regulation of *xisA* expression in an unknown manner.

After fusion the *nif* genes are activated by NtcA (Frias et al. 1994), resulting in expression of nitrogenase, apparently only after the O_2 level has been reduced as a consequence of cell wall synthesis. The entire regulatory sequence thus is temporarily coordinated in a way which ensures nitrogenase appearance in the late stage of heterocyst development (see also Fig. 3.2).

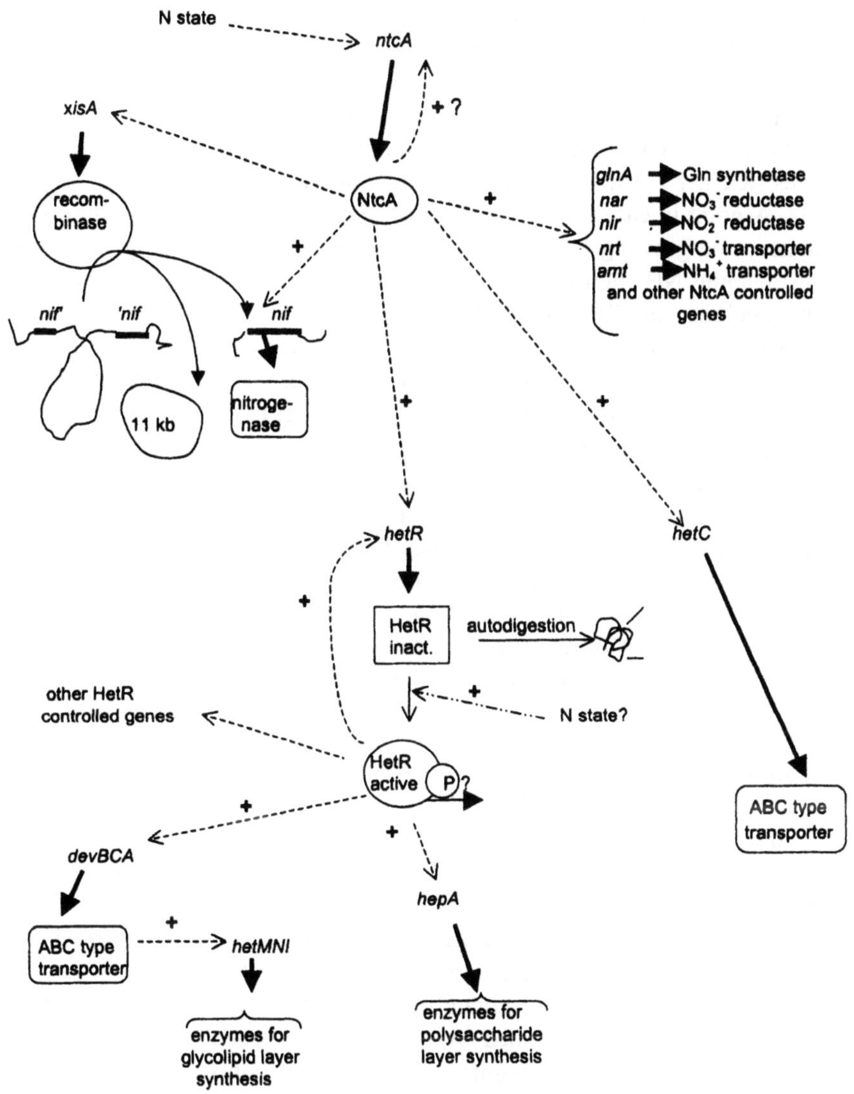

Fig. 3.3. Regulatory events during heterocyst development in *Anabaena*. –> Allosteric interaction; ⟶ modification or conformation change of protein; ·········> regulation of transcription (+ indication, - repression); ⟶▶ gene expression (transcription plus translation)

3.3.6 Metabolism of Intracellular Ammonia

From the preceding paragraphs it is evident that catabolism of inorganic and organic N sources mainly results in production of intracellular ammonia and glutamate. Under N-limiting conditions (absence of external ammonia) this leads to an NH_3 gradient formation resulting in leakage of NH_3 to the outside, where it is protonated to NH_4^+ (Fig. 3.1). Although not proven for all N sources, it is likely that catabolism of N sources — other than abundant ammonia — leads to induction of NH_4^+ carriers serving in cyclic retention (Kleiner 1985). In *Synechocystis*, expression of ammonium transport (*amt*) genes increases with the degree of N starvation; this signal is transferred via NtcA (Montesinos et al. 1998). Therefore, we may assume that energy-dependent cyclic retention of NH_3/NH_4^+, as in other prokaryotes (Kleiner 1985, 1993), operates under all kinds of N limitation in cyanobacteria. A point in case is the discovery of ammonium transport by the fully differentiated, non-growing heterocysts, which makes sense only in view of retrieval of escaping NH_3 molecules which had been produced by nitrogenase in these cells (Shehawy and Kleiner 1999b).

Assimilation of intracellular ammonia, at least under N-limiting conditions, takes place via the high-affinity/low-capacity route, i.e. via the glutamate cycle (Flores and Herero 1994) [(Eqs. (2) and (3)]. The GluDH catalysed reaction [(Eq. (1)] seems to be realised only in a limited number of cyanobacterial strains, where it may play an auxiliary role (Chavez et al. 1999), even in the presence of abundant anmmonia. At least some cyanobacteria have multiple GS enzymes, encoded by different genes (*gln*). Expression of some of these *gln* genes is N regulated via NtcA in such a way that N limitation increases GS synthesis (Flores and Herrero 1994; Reyes et al. 1997). Another regulatory mechanism is a fast, reversible inactivation of GS by binding of regulatory peptides (Garcia-Dominguez et al. 1999), which ensures a rapid answer to changes in the environmental N status.

3.3.7 Protein Degradation

As outlined in Section 3.3.4.1 (Fig. 3.2), degradation of specific proteins is of importance in heterocyst development. A different type of protein degradation in response to severe N deprivation has been found in the non-N_2-fixing species *Synechococcus*. This organism showed proteolytic degradation of the phycobiliproteins, its major light-harvesting pigments, and of chlorophyll a, a process known as chlorosis. After the addition of nitrate to the culture, the cells regenerate pigmentation. Chlorosis is accompanied by extensive differential gene expression, suggesting the operation of tight regulatory mechanisms (Gorl et al. 1998; Sauer et al. 1999). Chlorosis can also be induced by limitation of nitrate uptake due to a shift to low temperaratures (Sakamoto and Bryant 1999).

3.4 Eukaryotic Algae

Eukaryotic algae are able to assimilate a wide range of N sources, but not N_2. Investigations, especially on the molecular biological level, have mainly concentrated on species of the unicellular genera *Chlamydomonas* (Chlorophyta, Volvocales) and *Chlorella* (Chlorophyta, Chlorellales). Therefore, main emphasis will be put on these species.

3.4.1 Assimilation of Ammonia, Nitrate and Nitrite

Both *Chlamydomonas* and *Chlorella* express ammonium transport systems. In *Chlamydomonas* two systems have been detected, one high-affinity/low-capacity, one low- affinity/high-capacity (Franco et al. 1988). The former is repressed by ammonia, and thus seems to serve in cyclic retention; the second system is constitutive. A single, NH_4^+ repressible, but high-affinity ammonium transporter, probably serving in cyclic retention, has been found in *Chlorella* (Schlee and Komor 1986). Interestingly, its synthesis is considerably stimulated by glucose.

Both organisms express several transport sytems for nitrate and nitrite as well as the nitrate/nitrite reductase enzymes. *Chlamydomonas* expresses three high affinity transport systems, one for nitrate, one for nitrite and one for both (Galvan et al. 1996). All transport systems as well as the reducing enzymes are subject to repression by ammonia and induction by nitrate (Quesada and Fernandez 1994; Quesada et al. 1998; Rexach et al. 1999). As to *Chlorella*, several reports support the existence of nitrate transport systems (Schlee et al. 1985; Tischner et al. 1989). Some evidence points towards involvement of nitrate reductase in this process (Tischner et al. 1989). As in *Chlamydomonas*, the synthesis of nitrate reductase is repressed by ammonia and induced by nitrate (Sherman and Funkhouser 1989; Zeiler and Solomonson 1989; Dawson et al. 1996).

3.4.2 Assimilation of Urea

Urea is a good N source for both *Chlorella* and *Chlamydomonas*. Carrier-mediated uptake of urea and its repression by ammonia has been reported for *Chlamydomonas* only (Williams and Hodson 1977; Kirk and Kirk 1978). Further metabolism in plants can proceed either by urease-mediated hydrolysis yielding ammonia and CO_2, or by the sequence

$$\text{urea} + CO_2 + \text{ATP} \rightarrow H_2N\text{-CO-NH-COO}^- \text{ (allophanate)} + \text{ADP} + P_i, \quad (4)$$

$$\text{allophanate} + H_2O \rightarrow 2\,NH_3 + 2\,CO_2, \quad (5)$$

catalysed by the enzymes urea carboxylase and allophanate lyase (urea amidolyase pathway).

In contrast to all other algae, Chlorophyta exclusively express the amidolyase pathway (Leftely and Syrett 1973), which may also serve as a device for CO_2 accumulation. While both carboxylase and allophanase are subject to induction by urea and repression by ammonia in *Chlamydomonas*, only carboxylase is regulated in *Chlorella* (Hodson et al. 1975).

3.4.3 Assimilation of Amino Acids

Both *Chlamydomonas* and *Chlorella* are able to assimilate all amino acids, albeit along different pathways. Under heterotrophic conditions, *Chlamydomonas* extracellularly converts all amino acids except arginine by a broad range L-amino acid oxidase to oxo acids and ammonia, which subsequently is taken up (Piedras et al. 1992). This enzyme is repressed by ammonia but not by nitrate (Munoz-Blanco et al. 1990). In contrast, *Chlorella* expresses multiple-transport systems for all amino acids, the synthesis of which is not repressed by ammonia, but mostly stimulated by glucose (Cho and Komor 1985). At least the high-affinity systems supposedly function in cyclic retention of the cognate amino acids like in bacteria (Antonucci and Oxender 1988).

3.4.4 Assimilation of Other N Sources

The purine bases adenine, guanine, hypoxanthine, xanthine and urate are good nitrogen sources for the growth of *Chlamydomonas*. They are transported against a concentration gradient by two different high-affinity systems, one for adenine, guanine and hypoxanthine and another one for xanthine and urate. Both transport systems are repressed by ammonia (Pineda and Cardenas 1996). Further assimilation proceeds via established routes (Vogels and van der Drift 1976):

$$\begin{array}{l} \text{Adenine} \rightarrow \text{hypoxanthine} \\ \qquad\qquad\qquad \downarrow \\ \text{Guanine} \rightarrow \text{xanthine} \rightarrow \text{uric acid} \rightarrow \text{allantoin} \rightarrow \text{allantoic acid} \rightarrow\rightarrow\rightarrow \text{urea.} \end{array} \quad (6)$$

Chlamydomonas also uses allantoin and allantoate as sole N sources. The uptake system(s) for allantoin and allantoate are induced by both ureides. However, the enzymes involved in their degradation (allantoinase and allantoicase) are regulated differently. Allantoinase seems to be constitutive irrespective of the nitrogen sources, while allantoicase was inducible, being present only in cells cultured in ureides or any of their precursors. Neither allantoinase nor allantoicase activities were repressed by ammonium in the presence of ureides (Piedras et al. 1998).

For *Chlorella*, uptake and assimilation of xanthine, uric acid and allantoin were reported (Devi-Prasad 1983), but no data are available concerning N regulation.

3.4.5 Nitrogen Sensing and Nitrogen Regulation

The preceding sections demonstrate the existence of N regulation (ammonia effect) in eukaryotic algae. In comparison to cyanobacteria, less is known about sensing of the cell's N state and about transmission of this signal to the genome or to adaptive machineries. Again, as with cyanobacteria, discussions centre around the question whether the primary signal is intracellular ammonia or a derivative of it (e.g. glutamine). Although the field is in rapid progress, no unifying theory has yet emerged. In higher plants a central role in signalling is thought to be played by cytokinins, the levels of which are N-regulated (Beck 1999; Gaudinova 1999). Possibly this signal is transferred to the target genes via phosphorelay signal transduction systems similar to the ones found in bacteria (D'Agostino and Kieber 1999). Interesting examples for interaction of cytokinin with novel phosphorelays, elicited by N starvation, were recently discovered in maize (Sakakibara et al. 1998, 1999) and *Arabidopsis* (Taniguchi et al. 1998).

Presence of a limited number of cytokinins has been reported in green algae (Zhang et al. 1989; Auer 1997), including *Chlamydomonas* (Ivanova et al. 1992). Although different aspects of signal transduction are under intensive investigation in both *Chlorella* (Maetschke et al. 1997; Matsuda et al. 1998) and *Chlamydomonas* (Bloodgood 1992; Quarmby 1994; Quarmby and Hartzell 1994; Rochaix 1995; Beck et al. 1996; Im et al. 1996; Hegemann 1997; Matsuda et al. 1998; Rodriguez et al. 1999), it was so far not possible to form a hypothesis on N sensing and signalling.

3.4.6 Metabolism of Intracellular Ammonia

Evidence for cyclic retention of ammonia has been summarized in Section 3.4.1 (but see Rees 1995). Further metabolism, as in cyanobacteria (Sect. 3.3.6), can proceed via the high- or low-affinity route (Sect. 3.2). The regulation is complex, since, like in higher plants, ammonia assimilation in algae is catalysed by multiple forms (isoenzymes) of both glutamine synthetase (GS) and glutamate dehydrogenase (GluDH). Nevertheless, strong indications suggest that high- and low-affinity routes are regulated by the N state of the cell, as outlined in Section 3.2.

In *Chlamydomonas* two GS forms occur. The level of GS-1 transcripts was shown to be repressed by ammonium and induced by nitrate. The level of GS-2 transcripts was not affected by either (Chen and Silflow 1996). The occurrence of GOGAT in this organism (Munoz-Blanco and Cardenas 1989) completes the glutamate cycle. The organism was reported to express three GluDH isoenzymes. When the extracellular ammonium concentration was high, only GluDH-1 activity increased with growth whereas the other activities remained unchanged. This indicates a role of GluDH-1 in ammonia assimilation, while GluDH-2 and GluDH-3 are probably involved in the production of 2-oxoglutarate to fuel the tricarboxylic acid cycle (Moyano et al. 1995).

Chlorella autotrophica expresses two GS isoenzymes. A significant decline in the levels of GS, and a marked increase in the concentration of NADPH-dependent GDH was observed when nitrate was replaced by ammonium as the sole nitrogen source (Ahmad and Hellebust 1993). Other reports confirm induction of GluDH upon ammonia addition in *Chlorella* (Beudeker and Tabita 1984; Tischner 1984).

3.4.7 Effect of Nitrogen Limitation on Photosynthesis and Pigment Patterns

A widespread effect of N limitation in eukaryotic algae seems to be a significant reduction in photosynthetic activity concomitant with considerable changes in pigmentation, generally a decline in chlorophyll content and a rise in carotenoid levels (Czygan 1970; Pyliotis and Goodchild 1975; Vechtel et al. 1992; Grung et al. 1994; Bar et al. 1995; Skoda 1997; Sayed 1998; Hanagata and Dubinsky 1999; Boussiba et al. 1999). Under severe N starvation photosynthetic activity can even decline to zero (Petersen-Mahrt et al. 1994). While the step-down in photosynthesis may be a response to a reduced demand of carbon precursors for nitrogenous building blocks, the enhanced pigment production is thought to protect the cells from irradiation damage.

3.4.8 Nitrogen Starvation and Gametogenesis

This well-investigated differentiation process is the subject of several comprehensive recent reviews (Musgrave 1993; Quarmby 1994; Beck et al. 1996).

In *Chlamydomonas reinhardtii,* deprivation of nitrogen, but no other factor, is an essential signal for the first stage of differentiation into gametes. The stimulus for the second stage is light. It is likely, though not proven, that gametogenesis is induced by the same mechanism sensing N deprivation that governs induction of N-responding enzymes (see Sect. 3.4.5). In the following steps of differentiation extensive degradation of soluble proteins and chloroplast ribosomes concomitant with synthesis of new proteins has been observed. The subsequent process, apparently governed by a tightly regulated sequential activation of genes, finally results in the conversion of vegetative cells into gametes. Pairs of sexually different gametes (called *mt+* and *mt-*) after specific alignment of their flagella fuse to form a zygote. Mature zygotes are resistant to desiccation and freezing, and are able to survive starvation periods. Thus, they provide a special answer to severe N limitation.

3.5 Resume and Outlook

Although cyanobacteria and green algae are phylogenetically unrelated, both have evolved congruent strategies to solve the N problem. Since the N demand can be satisfied by numerous compounds, both groups have developed discriminatory mechanisms to set up a nutritional hierarchy ranging from very tasty compounds to last choices (e.g. from ammonia to N_2 in the case of N_2-fixing cyanobacteria), culminating in the ammonia effect. Stepdown in this hierarchy is felt as a gradual increase in N limitation. This limitation is apparently reflected by the internal N state of the organism, which, in turn, elicits various adaptive responses. Prominent features are the switchon of the cyclic NH_3/NH_4^+ retention mechanism and the change from the low-affinity to the high-affinity route for ammonia assimilation. For both groups the detailed mechanisms of N sensing and signalling of the N state are yet to be elucidated, with the most advanced knowledge apparently existing for the formation of cyanobacterial heterocysts.

Since intracellular N will be ultimately fed into organic compounds, mostly amino acids and nucleic acid bases, metabolism of N compounds probably is tightly connected with the C state of the cell. Therefore, coordinating control mechanisms are expected to exist which function like the bacterial Ntr system. An interesting example for the existence of such systems may be the stimulation of uptake of nitrogenous compounds by glucose in *Chlorella* (Cho and Komor 1985; Schlee et al. 1985; Schlee and Komor 1986). Another, extreme example is represented by the chlorosis effects (stepdown in energy and C metabolism) brought about by severe N starvation in both cyanobacteria and algae.

Last, but not least, another striking parallel may be seen in the observation that N limitation can be an inducer of morphogenic phenomena, e.g. heterocyst formation in filamentous cyanobacteria or gametogenesis in *Chlamydomonas*.

References

Adams DG, Carr NG (1989) Control of heterocyst development in the cyanobacterium *Anabaena cylindrica*. J Gen Microbiol 89:839-849

Ahmad I, Hellebust JA (1993) Partial characterization of enzymes of nitrogen metabolism in *Chlorella autotrophica* Shihira and Krauss. New Phytol 123:685-692

Aichi M, Omata T (1997) Involvement of NtcB, a LysR family transcription factor, in nitrite activation of the nitrate assimilation operon in the cyanobacterium *Synechococcus* sp. strain PCC 7942. J Bacteriol 179:4671-4675

Antonucci TK, Oxender DL (1988) The molecular biology of amino-acid transport in bacteria. Adv Microb Physiol 28:145-180

Auer CA (1997) Cytokinin conjugation: recent advances and patterns in plant evolution. Plant Growth Regul 23:17-32

Bar E, Rise M, Vishkautsan M, Arad S (1995) Pigment and structural changes in *Chlorella zofingiensis* upon light and nitrogen stress. J Plant Physiol 146:527-534

Beck CF, Haring MA, Jeon KW (1996) Gametic differentiation of *Chlamydomonas*. Int Rev Cytol 168:259-302

Beck E (1999) Towards an understanding of plant growth regulation: cytokinins as major signals for biomass distribution. In: Strnad M, Pec P, Beck E (eds) Advances in regulation of plant growth and development. Peres, Praha, pp 97-110

Beudeker RF, Tabita FR (1984) Glycolate metabolism is under nitrogen control in *Chlorella sorokiniana*. Plant Physiol 75:516-520

Black TA, Cai Y, Wolk CP (1993) Spatial expression and autoregulation of *hetR*, a gene involved in the control of heterocyst development in *Anabaena*. Mol Microbiol 9:77-84

Bloodgood RA (1992) Directed movements of ciliary and flagellar membrane components: a review. Biol Cell 76:291-301

Boussiba S, Gibson J (1991) Ammonia translocation in cyanobacteria. FEMS Microbiol Rev 88:1-14

Boussiba S, Bing W, Yuan JP, Zarka A, Chen F (1999) Changes in pigments profile in the green alga *Haematococcus pluvialis* exposed to environmental stresses. Biotechnol Lett 21:601-604

Brusca JS, Chastain CJ, Golden JW (1990) Expression of the *Anabaena* sp. strain PCC 7120 *xisA* gene from a heterologous promoter results in excision of the *nifD* element. J Bacteriol 172:3925-3931

Buikema W, Haselkorn R (1991) Characterization of a gene controlling heterocyst differention in the cyanobacetrium *Anabaena* 7120. Genes Dev 5:321-330

Buikema W, Haselkorn R (1993) Molecular genetics of cyanobacterial development. Annu Rev Plant Physiol 44:33-52

Cai Y, Wolk CP (1997) *Anabaena* sp. strain PCC 7120 responds to nitrogen deprivation with a cascade-like sequence of transcriptional activations. J Bacteriol 179:267-271

Chastain CJ, Brusca JS, Ramasubramanian TS, Wei TF, Golden JW (1990) A sequence-specific DNA-binding factor (VF1) from *Anabaena* sp. strain PCC 7120 vegetative cells binds to three adjacent sites in the xisA upstream region. J Bacteriol 172:5044-5051

Chavez S, Lucena JM, Reyes JC, Florencio FJ, Candau P (1999) The presence of glutamate dehydrogenase is a selective advantage for the cyanobacterium *Synechocystis* sp. strain PCC 6803 under nonexponential growth conditions. J Bacteriol 181:808-813

Chen Q, Silflow CD (1996) Isolation and characterization of glutamine synthetase genes in *Chlamydomonas reinhardtii*. Plant Physiol 112:987-996

Cho B-H, Komor E (1985) The amino acid transport systems of the autotrophically grown green alga *Chlorella*. Biochim Biophys Acta 821:384-392

Collier JL, Brahamsha B, Palenik B (1999) The marine cyanobacterium *Synechococcus* sp. WH7805 requires urease (urea amidohydrolase, EC 3.5.1.5) to utilize urea as a nitrogen source: molecular-genetic and biochemical analysis of the enzyme. Microbiology 145:447-459

Czygan F-C (1970) Blutregen und Blutschnee: Stickstoffmangel-Zellen von *Haematococcus pluvialis* und *Chlamydomonas nivalis*. Arch Mikrobiol 74:69-76

D'Agostino IB, Kieber J (1999) Phosphorelay signal transduction: the emerging family of plant response regulators. Trends Biochem Sci 24:452-456

Dawson HN, Pendleton LC, Solomonson LP, Cannons AC (1996) Cloning and characterization of the nitrate reductase-encoding gene from *Chlorella vulgaris*: structure and identification of transcription start points and initiator sequences. Gene 171:139-145

Devi-Prasad PV (1983) Hypoxanthine and allantoin as nitrogen sources for the growth of some freshwater green algae. New Phytol 93:575-580

Fay P (1992) Oxygen relations of nitrogen fixation in cyanobacteria. Microbiol Rev 56:340-373

Flores E, Herrero A (1994) Assimilatory nitrogen metabolism and its regulation. In: Bryant DA (ed) The molecular biology of cyanobacteria. Kluwer, Amsterdam, pp 487-517

Forde BG, Clarkson DT (1999) Nitrate and ammonium nutrition of plants: physiological and molecular perspectives. Adv Bot Res 30:1-90

Franco AR, Cardenas J, Fernandez E (1988) Two different carriers transport both ammonium and methylammonium in *Chlamydomonas reinhardtii*. J Biol Chem 263:14039-14043

Frias JE, Merida A, Herrero A, Martin-Nieto J, Flores E (1993) General distribution of the nitrogen control gene *ntcA* in cyanobacteria. J Bacteriol 175:5710-5713

Frias JE, Flores E, Herrero A (1994) Requirement of the regulatory protein NtcA for the expression of nitrogen assimilation and heterocyst development genes in the cyanobacterium *Anabaena* sp. PCC 7120. Mol Microbiol 14:823-832

Galvan A, Quesada A, Fernandez E (1996) Nitrate and nitrite are transported by different specific transport systems and by a bispecific transporter in *Chlamydomonas reinhardtii*. J Biol Chem 271:2088-2092

Garcia-Dominguez M, Reyes JC, Florencio FJ (1999) Glutamine synthetase inactivation by protein-protein interaction. Proc Natl Acad Sci USA 96:7161-7166

Gaudinova E (1999) Effects of phytohormones on nitrate reductase. In: Strnad M, Pec P, Beck E (eds) Advances in regulation of plant growth and development. Peres, Praha, pp 59-66

Gorl M, Sauer J, Baier T, Forchhammer K (1998) Nitrogen-starvation-induced chlorosis in *Synechococcus* PCC 7942: adaptation to long-term survival. Microbiology 144:2449-2458

Grung M, Metzger P, Berkaloff C, Liaaen-Jensen S (1994) Studies on the formation and localization of primary and secondary carotenoids in the green alga *Botryococcus braunii*, including the regreening process. Comp Biochem Physiol B 107:265-272

Hanagata N, Dubinsky Z (1999) Secondary carotenoid accumulation in *Scenedesmus komarekii* (Chlorophyceae, Chlorophyta). J Phycol 35:960-966

Haselkorn R (1998) How cyanobacteria count to 10 ? Science 282:891-892

Hegemann P (1997) Vision in microalgae. Planta 203:265-274

Hodson RC, Williams SK, Davidson WR (1975) Metabolic control of urea catabolism in *Chlamydomonas reinhardtii* and *Chlorella pyrenoidosa*. J Bacteriol 121:1022-1035

Im CS, Matters GL, Beale SI (1996) Calcium and calmodulin are involved in blue light induction of the *gsa* gene for an early chlorophyll biosynthetic step in *Chlamydomonas*. Plant Cell 8:2245-2253

Ivanova M, Todorov I, Pashankov P, Kostova L, Kaminek M (1992) Estimation of cytokinins in the unicellular green algae *Chlamydomonas reinhardtii* Dang. In: Kaminek, M, Mok DWS, Zazimalova E (eds) Physiology and biochemistry of cytokinins in plants. SFB Academic, Den Haag, pp 483-485

Kirk DL, Kirk MM (1978) Carrier-mediated uptake of arginine and urea by *Chlamydomonas reinhardtii*. Plant Physiol 61:556-560

Kleiner D (1985) Bacterial ammonium transport. FEMS Microbiol Rev 32:87-100

Kleiner D (1993) Ammonium transport systems — an overview. In: Bakker EP (ed) Alkali cation transport systems in prokaryotes. CRC Press, Boca Raton, pp 379-396

Lee HM, Vazquez-Bermudez MF, de Marsac NT (1999) The global nitrogen regulator NtcA regulates transcription of the signal transducer PII (GlnB) and influences its phosphorylation level in response to nitrogen and carbon supplies in the cyanobacterium *Synechococcus* sp. strain PCC 7942. J Bacteriol 181:2697-2702

Leftley JW, Syrett PJ (1973) Urease (EC 3.5.1.5) and ATP urea amido lyase activity in unicellular algae. J Gen Microbiol 77:109-115

Lindell D, Padan E, Post AF (1998) Regulation of *ntcA* expression and nitrite uptake in the marine *Synechococcus* sp. strain WH 7803. J Bacteriol 180:1878-1886

Maetschke M, Riedel J, Tischner R (1997) Evidence for signal transduction in the stimulation of nitrate uptake by blue light in *Chlorella saccharophila*. Photochem Photobiol 66:128-132

Magasanik B (1996) Regulation of nitrogen utilization. In: Neidhardt FC (ed) *Escherichia coli* and *Salmonella* , 2nd edn. ASM Press, Washington, DC, pp 1344-1356

Matsuda Y, Bozzo-Gale G, Colman B (1998) Regulation of dissolved inorganic carbon transport in green algae. Can J Bot 76:1072-1083

Merrick MJ, Edwards RA (1995) Nitrogen control in bacteria. Microbiol Rev 59:604-622

Montesinos ML, Muro-Pastor AM, Herrero A, Flores E (1998) Ammonium/ methylammonium permeases of a cyanobacterium. J Biol Chem 273:31463-31470

Moyano E, Cardenas J, Munoz-Blanco J (1995) Involvement of NAD(P)+glutamate dehydrogenase isoenzymes in carbon and nitrogen metabolism in *Chlamydomonas reinhardtii.* Physiol Plant 94:553-559

Munoz-Blanco J, Cardenas J (1989) Changes in glutamate dehydrogenase activity of *Chlamydomonas reinhardtii* under different trophic and stress conditions. Plant Cell Environ 12:173-182

Munoz-Blanco J, Hidalgo-Martinez J, Cardenas J (1990) Extracellular deamination of L-amino acids by *Chlamydomonas reinhardtii* cells. Planta 182:194-198

Muro-Pastor AM, Valladares A, Flores E, Herrero A (1999) The *hetC* gene is a direct target of the NtcA transcriptional regulator in cyanobacterial heterocyst development. J Bacteriol 181:6664-6669

Musgrave A (1993) Mating in *Chlamydomonas.* Prog Phycol Res 9:193-237

Petersen-Mahrt SK, Ekelund NGA, Widell S (1994) Influence of UV-B radiation and nitrogen starvation on daily rhythms in phototaxis and cell shape of *Euglena gracilis.* Physiol Plant 92:501-505

Piedras P, Aguilar M, Pineda M (1998) Uptake and metabolism of allantoin and allantoate by cells of *Chlamydomonas reinhardtii* (Chlorophyceae). Eur J Phycol 33:57-64

Piedras P, Pineda M, Munoz J, Cardenas J (1992) Purification and characterization of an L-amino acid oxidase from *Chlamydomonas reinhardtii.* Planta 188:13-18

Pineda M, Cardenas J (1996) Transport and assimilation of purines in *Chlamydomonas reinhardtii.* Sci Mar 60 Suppl 1:195-201

Pyliotis NA, Goodchild DJ (1975) The regreening of nitrogen-deficient *Chlorella fusca* II. Structural changes during synchronous regreening. Arch Microbiol 103:259-270

Quarmby LM (1994) Signal transduction in the sexual life of *Chlamydomonas.* Plant Mol Biol 26:1271-1287

Quarmby LM, Hartzell HC (1994) Dissection of eukaryotic transmembrane signalling using *Chlamydomonas.* Trends Pharmacol Sci 15:343-349

Quesada A, Fernandez E (1994) Expression of nitrate assimilation-related genes in *Chlamydomonas reinhardtii.* Plant Mol Biol 24:185-194

Quesada A, Hidalgo J, Fernandez E (1998) Three *Nrt2* genes are differentially regulated in *Chlamydomonas reinhardtii.* Mol Gen Genet 258:373-377

Ramasubramanian TS, Wei TF, Golden JW (1994) Two *Anabaena* sp. strain PCC 7120 DNA-binding factors interact with vegetative cell- and heterocyst-specific genes. J Bacteriol 176:1214-1223

Ramasubramanian TS, Wei TF, Oldham AK, Golden JW (1996) Transcription of the *Anabaena* sp. strain PCC 7120 *ntcA* gene: multiple transcripts and NtcA binding. J Bacteriol 178:922-926

Rees TAV (1995) On ammonia futile cycling in a marine unicellular alga. Biochim Biophys Acta 1228:254-260

Rexach J, Montero B, Fernandez E, Galvan A (1999) Differential regulation of the high affinity nitrite transport systems III and IV in *Chlamydomonas reinhardtii.* J Biol Chem 274:27801-27806

Reyes JC, Muro-Pastor MI, Florencio FJ (1997) Transcription of glutamine synthetase genes (*glnA* and *glnN*) from the cyanobacterium *Synechocystis* sp. strain PCC 6803 is differently regulated in response to nitrogen availability. J Bacteriol 179:2678-2689

Rochaix JD (1995) *Chlamydomonas reinhardtii* as the photosynthetic yeast. Annu Rev Genet 29:209-230

Rodriguez H, Haring MA, Beck CF (1999) Molecular characterization of two light-induced, gamete-specific genes from *Chlamydomonas reinhardtii* that encode hydroxyproline-rich proteins. Mol Gen Genet 261:267-274

Sakakibara H, Suzuki M, Takei K, Deji A, Taniguchi M, Sugiyama T (1998) A response-regulator homologue possibly involved in nitrogen signal transduction mediated by cytokinin in maize. Plant J 14:337-344

Sakakibara H, Hayakawa A, Deji A, Gawronski SW, Sugiyama T (1999) His-Asp phosphotransfer possibly involved in the nitrogen signal transduction mediated by cytokinin in maize: molecular cloning of cDNAs for two-component regulatory factors and demonstration of phosphotransfer activity in vitro. Plant Mol Biol 41:563-573

Sakamoto T, Bryant DA (1999) Nitrate transport and not photoinhibition limits growth of the freshwater cyanobacterium *Synechococcus* species PCC 6301 at low temperature. Plant Physiol 119:785-794

Sakamoto T, Inoue-Sakamoto K, Bryant DA (1999) A novel nitrate/nitrite permease in the marine cyanobacterium *Synechococcus* sp. strain PCC 7002. J Bacteriol 181:7363-7372

Sauer J, Gorl M, Forchhammer K (1999) Nitrogen starvation in *Synechococcus* PCC 7942: involvement of glutamine synthetase and NtcA in phycobiliprotein degradation and survival. Arch Microbiol 172:247-255

Sayed OH (1998) Analysis of photosynthetic responses and adaptation to nitrogen starvation in *Chlorella* using in vivo chlorophyll fluorescence. Photosynthetica 35:611-619

Schlee J, Komor E (1986) Ammonium uptake by *Chlorella vulgaris*. Planta 168:232-238

Schlee J, Cho B-H, Komor E (1985) Regulation of nitrate uptake by glucose in *Chlorella vulgaris*. Plant Sci 39:25-30

Shehawy RM, Kleiner D (1999a) Heterocyst formation as a process of senescence. Abstr 4th Eur Workshop Mol Biol Cyanobact, Berlin, 103 pp

Shehawy RM, Kleiner D (1999b) Ammonium (methylammonium) transport by heterocysts and vegetative cells of *Anabaena variabilis*. FEMS Microbiol Lett 181:303-306

Sherman TD, Funkhouser EA (1989) Induction and synthesis of nitrate reductase in *Chlorella vulgaris*. Arch Biochem Biophys 274:525-531

Singh AK, Singh HN, Rai AN (1991) Evidence for a role of glutamine synthetase in assimilation of amino acids as nitrogen source in the cyanobacterium *Nostoc muscorum*. Biochem Int 25:887-894

Singh S (1991) Urea metabolism in the cyanobacterium *Anabaena cycadeae*: regulation of urea uptake and urease by ammonia. FEMS Microbiol Lett 84:177-182

Skoda B (1997) Contributions to the biochemical taxonomy of the genus *Chlorella* Beijerinck s.l.-pigment composition. II. Biochemotaxonomical differences in pigment composition of the strains growing under nitrogen deficient nutritional conditions. Arch Hydrobiol Suppl 122:109-136

Stock JB, Ninfa AJ, Stock AM (1989) Protein phosphorylation and regulation of adaptive responses in bacteria. Microbiol Rev 53:450-490

Taniguchi M, Kiba T, Sakakibara H, Ueguchi C, Mizuno T, Sugiyama T (1998) Expression of *Arabidopsis* response regulator homologs is induced by cytokinins and nitrate. FEBS Lett 429:259-262

Tischner R (1984) Evidence for the participation of NADP glutamate dehydrogenase in the ammonium assimilation of *Chlorella sorokiniana*. Plant Sci Lett 34:73-80

Tischner R, Ward MR, Huffaker RC (1989) Evidence for a plasma-membrane-bound nitrate reductase involved in nitrate uptake of *Chlorella sorokiniana*. Planta 178:19-24

Vechtel B, Kahmann U, Ruppel HG (1992) Secondary carotenoids of *Eremosphaera viridis* de Bary (Chlorophyceae) under nitrogen deficiency. Bot Acta 105:219-222

Vega-Palas MA, Madueno F, Herrero A, Flores E (1990) Identification and cloning of a regulatory gene for nitrogen assimilation in the cyanobacterium *Synechococcus* sp. strain PCC 7942. J Bacteriol 172:643-647

Vogels GD, van der Drift C (1976) Degradation of purines and pyrimidines by microorganisms. Bacteriol Rev 40:403-468

Wei TF, Ramasubramanian TS, Golden JW (1994) *Anabaena* sp. strain PCC 7120 *ntcA* gene required for growth on nitrate and heterocyst development. J Bacteriol 176:4473-4482

Williams SK, Hodson RC (1977) Transport of urea at low concentrations in *Chlamydomonas reinhardtii* . J Bacteriol 130:266-273

Wolk CP (1996) Heterocyst formation. Annu Rev Genet 30:59-78

Wolk CP (2000) Heterocyst formation in *Anabaena*. In: Brun YV, Shimkets LJ (eds) Prokaryotic development, 1st edn. ASM Press, Washington, pp 83-104

Wolk CP, Ernst A, Elhai J (1994) Heterocyst metabolism and development. In: Bryant DA (ed) The molecular biology of cyanobacteria. Kluwer, Dordrecht, pp 769-823

Yoon HS, Golden JW (1998) Heterocyst pattern formation controlled by a diffusible peptide. Science 282:935-938

Zeiler KG, Solomonson LP (1989) Regulation of *Chlorella* nitrate reductase: control of enzyme activity and immunoreactive protein levels by ammonia. Arch Biochem Biophys 269:46-54n

Zhang HW, Yamane H, Takahashi M, Chapman DJ, Phinney BO (1989) Identification of a cytokinin in the green alga *Chara globularis*. Phytochemistry 28:337-338

Zhou R, Wei X, Jiang N, Li H, Dong Y, Hsi K-L, Zhao J (1998a) Evidence that HetR protein is an unusual serine-type protease. Proc Natl Acad Sci USA 95:4959-4963

Zhou R, Cao Z, Zhao J (1998b) Characterization of HetR protein turnover in *Anabaena* sp. PCC 7120. Arch Microbiol 169:417-423

4 Phosphate Limitation

Ferdinand Wagner[1] and Gernot Falkner[2]

Contents

---

[1] Institute of Plant Physiology, University of Salzburg, Hellbrunnerstr. 34, 50202 Salzburg, Austria
[2] Institute of Limnology, Austrian Academy of Sciences, Gaisberg 116, 5310 Mondsee, Austria

4.1 Introduction

4.1.1 The Evolution of Phosphorus Limitation in Lakes and the Essential Role of a Phosphorus Store

Phosphorus is an essential constituent of all living cells. This element is part of fundamental building blocks that constitute nucleic acids, phospholipids and complex carbohydrates. Phosphorus compounds play a central role in anabolic and catabolic pathways and in the energy conversion of the cell via transfer of energy-rich phosphoanhydride bonds, a reaction which is also involved in posttranslational regulation of enzyme activities.

In nature, phosphorus is usually present in the fully oxidized state, either as inorganic phosphate or polyphosphate, or as a component of organic matter originating from organismic activities. Any phosphate cycling in biotic and abiotic processes in the biosphere is ultimately originating from the weathering of phosphate rocks. The important natural source of phosphorus is the mineral apatite that consists of impure calcium-phosphate compounds. Although phosphorus is actually widely distributed in the Earth's crust (about 0.1% by weight), its availability for organisms is frequently restricted (Bieleski and Ferguson 1983), and the main portion of phosphate in natural soils is bound in poorly soluble mineral phosphates (e.g. calcium-phosphate), which can be utilized solely by highly specialized microorganisms (Goldstein 1994). Therefore, loss of phosphate from untreated soils due to rainfall is usually scanty. The enormous use of fertilizers in agriculture, however, has often led to oversaturation of soils with nutrients. Any phosphate exceeding the immediate demand of soil organisms and plants causes the retention capacity of the soil to be washed out from the upper layers. The dissolved nutrients are finally swept from the catchment areas into lakes where massive inflow of phosphate may cause severe algal bloom formation. An incredibly high proportion (90%) of total phosphate use in the world is due to the application of fertilizers (Bieleski and Ferguson 1983) and, together with direct disposal of phosphate-containing materials like detergents and sewage, continuous human phosphate input seriously threatens the stability of characteristic ecosystems.

In oligotrophic lakes, however, the free concentration of orthophosphate is very low, and complex precipitation and binding processes in which iron and calcium compounds are involved usually prevent an increase in the concentration of this nutrient above nanomolar levels. During overturn periods, when the water stratification breaks down, considerable amounts of iron and phosphate, as well as organophosphates originating from previous sedimentation of debris and dead organisms, may be transported from the sediment to the waterbody, and may subsequently cause transient algal blooms. However, usually most of the released phosphate rapidly adsorbs again to iron hydroxide and sediment particles or is precipitated again, predominantly as $FePO_4$ under aerobic conditions (Schwoerbel 1984). Despite these transient phases of relatively high phosphate supply that may occur, but which are more decisive for shallow lakes, the production of phytoplankton biomass in oligotrophic lakes can be expected to be limited by the amount of available phosphate.

Although analysis of nutrient ratios between carbon, nitrogen and phosphorus of lake samples in some cases suggested potential limitation of phytoplankton productivity by the first two elements, Schindler (1977) confirmed that the growth-limiting nutrient in lakes is usually phosphate. He fertilized some smaller Canadian lakes with different N/P ratios for several years, and observed a proportional dependency of phytoplankton growth on the total phosphate concentration. He explained his results by the impact of external and biological mechanisms that compensate for potential deficiencies in carbon and nitrogen. Turbulences in the waterbody enable the gas exchange between atmosphere and upper water layers. Thus, an equilibrium distribution of carbon dioxide between the two phases is supported, and any decrease in the content of inorganic carbonate species caused by the photosynthetic organisms of a lake will be balanced by redissolution of atmospheric CO_2 in the water. For this reason there is usually sufficient carbon for maintenance and growth of phytoplankton communities. A similar importance is attributed to gaseous N_2, a huge potential nitrogen reservoir in the air: deficiency of this nutrient in the water will favour nitrogen-fixing microorganisms, resulting in shifts in the composition of the community, for example, towards filamentous cyanobacteria which can utilize N_2 and are consequently not nitrogen-limited.

However, since no comparable gaseous atmospheric cycle exists for phosphate, and since more or less all organisms need the same form of phosphorus, namely soluble orthophosphate (Reynolds 1993), there is no compensatory mechanism for losses of this nutrient, for example, due to sedimentation. Accordingly, in most lakes the supply of phosphate to microorganisms and algae is restricted to essentially two sources: external nutrient input, mainly inflowing water from the catchment area that may contain appreciable amounts of fertilizers, and internal release of phosphate in biotic and abiotic phosphate cycles that occur within the waterbody and sediment (Schwoerbel 1984). It is generally acknowledged that most of the inorganic and organic phosphate cycles in the epilimnion occur via an uptake by phytoplankton and bacteria and a subsequent release due to microbial activities and predation and excretion by zooplankton (Schwoerbel 1984; Currie and Kalff 1984; Vadstein et al. 1993; Salonen et al. 1994; Sommer 1994). While inorganic phosphate can be immediately utilized by phytoplankton, organic phosphates usually have to be hydrolysed before the released phosphate is incorporated by the respective inorganic phosphate transport systems. Cleavage of organic phosphate compounds is mediated by phosphatases that are excreted by the cells into the extracellular space. Among these, the most prominent are the rather unspecific alkaline phosphatases which are capable of utilizing polyphosphates and a wide range of organic phosphates as substrate. Due to their massive biosynthesis during phosphate starvation, the activity of these enzymes in lake samples was long considered as an indicator for assessing the phosphate status of phytoplankton (see Jenssen et al. 1988; Hernandez 1996; Ruiz et al. 1997; Huang and Hong 1999).

In most temperate lakes phosphate deficiency is a common phenomenon, and earlier limnological studies confirmed this proposition. A classical experiment was carried out by Einsele (1941), who added 94 kg of phosphate in two portions to Lake Schleinsee. He observed that the concentration of this nutrient in the water decreased within a few days below the limits of chemical detection. Concomitantly, the ratio of phosphorus to nitrogen in the algae changed from about 1:20 to 1:3, indicating that the phytoplankton community was able to store a multitude of its original cellular phosphorus content. Further studies confirmed that phosphate-deficient organisms are capable of accumulating this nutrient in severalfold excess of their immediate demand when exposed to considerable amounts of phosphate. The tendency of phosphate-starved cells to store phosphate, predominantly in the form of polyphosphates (Kulaev and Vagabov 1983), was later termed Phosphat-Überkompensation by Liss and Langen (1962) and is now referred to as the polyphosphate overplus (or surplus) phenomenon.

Since limnological studies with phytoplankton have demonstrated that under phosphate limitation the uptake system for this nutrient is highly activated and that an abrupt increase in the external concentration results in rapid incorporation and storage in the form of polyphosphates, interest in this compound has soon intensified. Inorganic polyphosphates are linear polymers in which two to several thousand orthophosphate residues are linked by energy-rich phosphoanhydride bonds (Yoshida 1955), and short- to long-chain forms of these polymers can be found in all cell regions or compartments, for example, in nuclei, or vacuoles of algae (Kulaev and Vagabov 1983). It is now assumed that virtually all prokaryotic and eukaryotic organisms are capable of synthesizing polyphosphates (Kornberg 1994; Kornberg et al. 1999). The terminal phosphate group of ATP is thereby transferred to polyphosphates in a reversible reaction catalyzed by polyphosphate kinases which can only elongate preexisting polymers. Besides polyphosphate kinase itself, several specific proteins are involved in the degradation of polyphosphates (exo-/endopolyphosphatases, phosphotransferases), resulting in different phosphate compounds which can then be utilized in assimilatory or regulatory pathways. A number of functions, besides that of a phosphate store, have recently been attributed to the polyphosphates: energy store and ATP substitute, as buffer, chelation of metal ions (see Chap. 12.5), initiating competence for bacterial transformation, mRNA processing and degradation, as well as "a regulatory role in the physiological adjustments to growth, development, stress, and deprivation" (Kornberg et al. 1999). The latter function seems to be confirmed by the so-called phosphate luxury consumption behaviour that has been described in organisms deficient in nutrients other than phosphate. These cells also accumulate polyphosphates in huge amounts, for example, sulfate-deficient *Synechococcus* sp. (Lawry and Jensen 1979) or *Synechocystis* sp. (Lawerence et al. 1998). Similar observations have been made in *E. coli* when exposed to nitrogen starvation or salt stress (Ault-Riché et al. 1998), and have led to the development of a model for stress-induced polyphosphate formation (Kornberg et al. 1999), which awaits further empirical evidence.

The main physiological role of polyphosphate formation in phosphate-deficient microorganisms and algae, is however, obvious if we consider the natural growth conditions to which these organisms are exposed in an oligotrophic environment. As mentioned above, increases in the external phosphate supply are usually transient, for example, after excretion by organisms of the aquatic fauna or by diffuse runoff of phosphate following heavy rainfall. Since most of the time phosphate is not available, phytoplankton cells have to incorporate this nutrient with a high velocity during these short periods of elevated phosphate supply, otherwise the limiting nutrient will be taken up by surrounding organisms or is lost by inorganic binding or precipitation processes. The high incorporation rates, however, exceed the immediate demand to sustain the ongoing growth rate, so the organisms must be capable to store this nutrient in a physiologically favourable form, i.e. the osmotically inert polyphosphates, which can be later used for growth. The possibility of accumulating considerable amounts of inorganic phosphate ions in the cytoplasm can be excluded, since in deficient cells the influx rates of this nutrient are usually so high that the internal phosphate concentration would immediately increase to levels that would disturb the homeostasis of cells (phosphorylation potential, osmotic pressure, etc.), and terminate the incorporation process within a few minutes: a typical phosphate uptake rate for the unicellular freshwater cyanobacterium (blue-green alga) *Anacystis nidulans* (*Synechococcus* sp.) at higher external phosphate concentrations, for example, 100 nM, is 50 μmol P_i (mg Chl a h)$^{-1}$. If we assume that 1 mg chlorophyll corresponds to a cell volume of 80 μl (Allen 1968), we can readily calculate that the cytoplasmic phosphate concentration of this organism would increase by about 100 mmol l^{-1} after only 10 min of phosphate incorporation. During that time, however, only about two thirds of the phosphate necessary for a single cell doubling is incorporated (1 mg chlorophyll equals about 12.5 μM total phosphate in non-growing *Anacystis* cells).

Similar considerations may apply to higher algae which can potentially accumulate considerable amounts of inorganic phosphate in their vacuoles. If these cells are challenged with high external pulses and store a multitude of their initial phosphate content, again the major portion of the incorporated phosphate has to be deposited in the polyphosphates.

Hence, the capability of synthesizing polyphosphates is a prerequisite to sustain high phosphate incorporation rates. Inside the cell the incorporated phosphate is immediately converted with ADP to ATP, in the light via photophosphorylation which operates at sufficiently high rates under normal physiological conditions to keep the cytoplasmic phosphate concentration fairly constant. In an energy-independent reaction the terminal phosphate group of ATP is finally condensed to polyphosphates.

The decisive physiological importance of polyphosphates is further revealed by their mediator role between phosphate incorporation that proceeds in a discontinuous mode due to the variable external phosphate supply, and the continuous growth process itself. For obvious reasons, growth of phytoplankton must be independent of irregular fluctuations of the external phosphate that cause irregular fluxes of phosphate into the cell. Otherwise the growth process would be initiated each time the cell acquires phosphate, and stop again as soon as the external phosphate source is depleted. Under such conditions synthesis of virtually all components which are necessary for the production of biomass could not be performed in a coordinated way, and continuous growth would be restricted to periods when the external phosphate supply is constantly high. The ability to build up a polyphosphate pool, however, enables the cells to maintain continuous growth, even if the external concentration remains low for a prolonged period of time during which no uptake occurs. This feature will be outlined in more detail in the following section.

4.1.2 The Interrelationship Between the Activity of the Uptake System, the Size of the Phosphorus Store and the Growth Rate

Another important classical field study with respect to phosphate limitation of phytoplankton growth was performed by Rigler (1956), who uniformly labelled an entire lake with ^{32}P-phosphate. Using radioactive phosphate of the same specific activity, he was able to follow in lake samples the net uptake behaviour of the phytoplankton. After having offered radioactive phosphate in the nanomolar range, he noticed a rapid decrease in the concentration of external orthophosphate, until a quasiequilibrium state was attained at which the external concentration of phosphate in the lake water remained constant. After a second addition of phosphate, the nutrient was again rapidly utilized by the algae and ceased at the same external phosphate concentration that had been observed before nutrient addition. Apparently, the phytoplankton community could not deplete the ambient phosphate concentration to zero, and Rigler concluded that a distinct external equilibrium concentration exists at which the rate of incorporation of phosphate by the organisms is balanced by the rate of loss of phosphate from the phytoplankton. Further increases in the phosphate content of phytoplankton cells would then only be possible, if the external concentration rises above this equilibrium level due to an input of phosphate from the in-flowing water or a release of phosphorus from within the lake.

In his experiments Rigler proved the existence of a limiting or threshold concentration in phytoplankton communities, a result that was confirmed in additional studies which demonstrated that phosphate uptake in algal cultures or phytoplankton ceases at a distinct critical threshold value, usually in the nanomolar range in phosphate-deficient organisms (e.g. Kuenzler and Ketchum 1962; Mueller 1972; Brown and Button 1979; Falkner et al. 1984a,b, 1995; Wagner et al. 1995).

Since the threshold value of organisms or populations is a decisive ecological parameter that has been (and is still) neglected in most models of phosphate incorporation and growth kinetics, it deserves a further description. For kinetic reasons the cytoplasmic phosphate concentration necessary to sustain cytosolic reactions — such as ATP formation by photophosphorylation or oxidative phosphorylation — at a physiologically reasonable rate is usually much higher than the external phosphate level in oligotrophic lakes (an intracellular phosphate concentration of 1 nM would leave only less than 1000 molecules in the cytosol of small organisms). Determination of cytoplasmic orthophosphate concentrations by direct extraction or NMR analysis yielded high values for *Synechococcus* (*Anacystis nidulans*) and *Escherichia coli* between 1 and 10 mM (Bornefeld and Weis 1981; Kallas and Dahlquist 1981; Rao et al. 1993). Although these values seem somewhat overestimated for free-living organisms (e.g. in *Anacystis nidulans* the cytoplasmic phosphate concentration is about 100 µM; Wagner and Falkner 1992), we may safely assume that a concentration gradient of 4–5 orders of magnitude exists at external levels in the low nanomolar range. Such a steep concentration gradient of phosphate across the cell membrane can only be maintained if the uptake process is coupled to an energy source. It is obvious that the energy necessary for incorporation increases with decreasing external phosphate concentrations, until a distinct external phosphate concentration is attained at which the energy expended for phosphate translocation is no longer sufficient to drive further accumulation of this ion. As in almost any chemical reaction, in this equilibrium state (in the simplest case of complete coupling, Sect. 4.4.1) the rate of the forward reaction, i.e. influx of phosphate into the cell, is the same as the rate of the backward reaction, as was suggested by Rigler (1956). This results in a constant equilibrium distribution between phosphate outside and inside the cell, at least as long as no modifications of the uptake system or subsequent reactions occur (Sect. 4.4). Further incorporation of the nutrient is then only possible if the external concentration exceeds the respective equilibrium value that therefore represents a threshold for phophate uptake. For this reason the threshold value characterizes the lowest possible concentration of ambient phosphate that can be utilized by organisms, while the kinetic properties of the transport system are decisive for the rate at which elevated phosphate concentrations are utilized. Accordingly, the interplay between the energetic and kinetic properties of the phosphate uptake system finally determines how much phosphate an organism can acquire under conditions of discontinuous nutrient supply.

As has been outlined above, in oligotrophic lakes the external phosphate concentration usually fluctuates around a threshold value in the nanomolar range (Rigler 1956, 1968; Falkner et al. 1984a,b), and growth occurs at the expense of stored polyphosphates. Thus, in a growth situation typical for a nutrient-poor environment, phosphate incorporation and growth are not directly coupled. However, an indirect linkage between these two processes exists, since the growth rate somehow conforms to the amount of stored phosphate (Droop 1973, 1974; Rhee 1973). In order to coordinate the growth rate with the internal polyphosphate pool under conditions of pulse-wise phosphate supply, the phosphate uptake system must attain properties which assure that this balanced adjustment is not disturbed by subsequent phosphate pulses. If this were not the case, the phosphate flux into the cell during each nutrient pulse would interfere with the once established balance inside the cell.

It was previously shown that *Anacystis nidulans* solves this complex regulatory constraint by developing some kind of information-processing capacity that allows information about the previous fluctuation pattern to be stored in the properties of the uptake system via an adaptive response to elevated phosphate concentrations. These adapted properties are maintained even during the periods where no uptake occurs (Wagner et al. 1995). In a fluctuating environment fluxes of energy and matter are permanently changing on a short scale, so we may speculate that the survival of a population depends on its adaptive flexibility to respond adequately to these alterations (cf. Reynolds 1998). The adapted properties of organisms have to be stable enough thereby to assure a continuous growth behaviour over a distinct period of time but, on the other hand, cells must be capable of responding quickly to major changes in the environment to readjust their cellular metabolism to the new growth situation.

The different biological disciplines investigate the problem of phosphate limitation with different approaches and aims. Molecular-genetic work on phosphate transport systems in microorganisms is mainly focused on the *E. coli* paradigm (Sect. 4.2), and though much valuable information has been gained in these studies, adaptation to phosphate limitation is not an intrinsic topic in this discipline. In ecological studies nutrient incorporation and growth of phosphate-limited microorganisms and algae are usually analyzed by kinetic models, nowadays mainly with respect to the competitive capability of distinct species. These studies exclude short-term adaptation processes, and assume that their subjects are in an invariant state under defined growth conditions (Sect. 4.3). The last section of this chapter describes the physiological adaptation of microorganisms to fluctuating phosphate supply, and tries to give some insight into the regulation of this complex phenomenon (cf. Sect. 4.4).

4.2 Molecular Composition of the Phosphate Uptake Systems of Bacteria and Cyanobacteria

4.2.1 The Molecular Basis of Inorganic Phosphate Uptake Systems in the Model Organism *Escherichia coli*

In an environment in which phosphate supply is variable and intermittent, cyanobacteria and algae have to be capable of reacting to changes in the ambient nutrient status by rearranging the phosphate uptake system so that it becomes adjusted to the prevailing growth conditions. It is well known that the onset of phosphate limitation activates phosphate uptake in such a way that the maximum velocity of the incorporation rate increases by several orders of magnitude. Thus, lower external concentrations of the nutrient can be utilized. The observed kinetic and energetic alterations in the uptake behaviour are the result of biosynthesis of distinct proteins that constitute a high-affinity phosphate uptake system. Thus, alterations in the molecular composition of this system are the outcome of (at least long-term) adaptation to phosphate limitation. For this reason, it is important to understand the underlying molecular mechanisms that enable and regulate phosphate uptake at low external phosphate concentrations.

Most of our knowledge about the molecular basis of incorporation and utilization of phosphorus sources is derived from molecular, biochemical and physiological studies with the Gram-negative bacterium *Escherichia coli*, which has been the subject of intensive research for many decades. Usually, this bacterium is not exposed to phosphate-deficient growth conditions in its natural environment so, although a good model system, it is possibly not representative of free-living organisms that have to cope with phosphate limitation frequently or even permanently. Nevertheless, it still serves as the best-studied organism with regard to phosphate transport on the molecular level.

For this reason, an overview of the molecular basis of phosphate-transport systems has to start with *E. coli*, which has developed several systems for the transport of inorganic phosphate, phosphonates and organic phosphates as potential phosphorus sources. Here we will consider only the transport systems for inorganic phosphate [originating from free inorganic (poly)phosphate or organic phosphates]. For further information the reader is referred to the excellent and very detailed reviews on the molecular basis of transport and utilization of individual phosphorus sources in *E. coli* (e.g. Rosenberg 1987; Silver and Walderhaug 1992; Wanner 1996).

Incorporation of phosphate requires not only translocation of this nutrient per se, but also distinct ion circuits in order to keep the overall balance of ion movements neutral. For example, in *E. coli*, phosphate incorporation is dependent on the presence of potassium ions in the incubation medium, and a model of ion circulation was established in which phosphate, protons and potassium are involved (Russell and Rosenberg 1979, 1980). Originally, it was assumed that HPO_4^{2-} is the proper substrate that is incorporated under deficient as well as non-deficient conditions, and that phosphate enters the cell in cotransport with protons. In order to revoke the resulting acidification of the cytosol, an immediate (direct or indirect) counterexchange between protons and potassium occurs which leads to the extrusion of protons and the concomitant accumulation of potassium.

Additionally, incorporation and storage of phosphate inside the cell is accompanied by coordinated ion fluxes that are involved in the basic reactions of energy metabolism or, more specifically, those that deliver the necessary counterions (e.g. Mg^{2+}, K^+) to compensate for the negative charges of the phosphate residues in polyphosphates (Kulaev and Vagabov 1983). If we consider that in the cell numerous reactions of ion acquisition and release occur simultaneously, a rather complicated and complex image emerges of a living cell which has to coordinate the individual ongoing fluxes permanently in order to respond adequately to the prevailing external conditions.

In *E. coli* two uptake systems for inorganic phosphate have been identified (Bennett and Malamy 1970; Medveczky and Rosenberg 1971; Rosenberg et al. 1977; Willsky and Malamy 1980a): the Pit system (P_i (inorganic phosphate transport system), which is assumed to be the main uptake system under phosphate rich growth conditions, and the highly specific Pst system (phosphate-specific transport system), which is part of the so-called Pho regulon, an assembly of genes and operons that are under common molecular control and which are regulated according to the availability of external phosphate.

The Pit system in *E. coli* is a constitutive transport system with only low affinity and specificity for phosphate: its K_T for phosphate is about 25–40 µM (Rosenberg et al. 1977; Willsky and Malamy 1980a,b) and it uses arsenate as a substrate with similar affinity. Synthesis of Pit is not under the control of the Pho regulon and this transport system consists probably of a single-membrane component with a molecular weight of around 39 kDa. Under phosphate-rich conditions the cytoplasmic phosphate concentration of *E. coli* was estimated to be around 8–10 mM (Rao et al. 1993), well above natural external phosphate concentrations. Early studies revealed that Pit is energized by the protonmotive force (Rosenberg et al. 1977), which can be established by primary proton translocation associated with the respiratory electron transfer in the presence of oxygen or by a proton-pumping ATPase under anaerobic conditions (Rosenberg et al. 1979).

The Pit system has been considered as the major phosphate uptake system in *E. coli* through which phosphate enters the cell under conditions of excess phosphate. More recent studies, however, question this view and interpret the Pit as a transporter for neutral metal-phosphate complexes (Van Veen et al. 1994 a,b; Van Veen 1997). Using membrane vesicles and proteoliposomes reconstituted with the carrier protein of *Acinetobacter johnsonii* and *E. coli*, these authors showed that phosphate transport is strictly dependent on the presence of divalent cations (Mg^{2+}, Ca^{2+}, Mn^{2+}, Co^{2+}). From a series of refined experiments it was deduced that the actual substrate for Pit is not inorganic orthophosphate but a $MeHPO_4$ complex (K_M of 8.8 µM for the reconstituted *E. coli* Pit) which is translocated in symport with a proton. The proposed kinetic model for the operation mode of the Pit system (derived from efflux studies) describes a translocation loop involving six discrete states of the carrier that reflect the ordered mechanism of binding and release of a proton and the neutral $MeHPO_4$ complex at both sides of the membrane (Van Veen et al. 1994b; Van Veen 1997). In the proposed model the primary function of the Pit system could actually be that of a transporter of divalent metal ions for which phosphate serves as a suitable counterion. In this context it is noteworthy to mention that the lack of Pit does not inhibit growth of the bacterium on phosphate (Wanner 1996). Although this puzzling phenomenon may be attributed to the activity of compensatory phosphate uptake sytems, possibly Pst, the question about the physiological function of Pit is open again.

Onset of phosphate starvation induces the enhanced synthesis of more than 80 proteins in *E. coli.* About 30 of these are products of Pho regulon genes, and can be considered to be involved in the specific molecular response of this organism to phosphate limitation. The genes of the Pho regulon are transciptionally activated by a gene product designated PhoB which itself is a component of the regulon. The Pho regulon comprises the following genes and operons:

- *phoE*, encoding a porin protein, located in the *E. coli* outer membrane.
- *phoA*, encoding an alkaline phosphatase, a non-specific phosphomonoesterase that hydrolyzes organic phosphates.
- The gene products of the *pst-phoU* operon are PstS,A,C,B and PhoU. The Pst proteins constitute the transport complex *per se*, the function of PhoU is still unclear.
- *phoB* and *phoR* are the two main regulatory genes and constitute a single operon.
- The *phn* operon comprises 14 genes. The respective proteins are involved in the coordinated transport and utilization of phosphonates (C-P bonds).
- The *ugp* operon encodes components involved in transport and utilization of *sn*-glycerol-3-phosphate which is suitable as the only P- (but not C-) source for *E. coli* (rev. in Brzoska et al. 1994). The transport system for this substrate is a periplasmic-binding protein-dependent transport system, and thus similar in its mode of operation to that for inorganic phosphate: a G3P-binding protein (UgpB) captures the molecule and directs it to two membrane proteins (UgpA,E) that are supposed to form a channel through the membrane. The complex is probably energized by ATP hydrolysis occurring at the UgpC protein, which is tightly associated with the transmembrane complex. UgpQ, a protein which is not essential for transport of the substrate, reveals catalytic properties; its function, however, is not completely clear.
- The *psiE* and the *phoH* genes are also under regulatory control of *phoB*, their function, however, is still unknown.

The conception of how these individual components interact and regulate the uptake of inorganic phosphate in *E. coli* is summarized in Fig. 4.1 (Torriani 1990; Silver and Walderhaug 1992; Wanner 1993, 1994). In Gram-negative bacteria and cyanobacteria the outer membrane is the outermost layer that separates the cell structurally from its environment. This barrier enables the retention of excreted enzymes that should operate in the vicinity of the cytoplasmic membrane (e.g. periplasmic binding proteins) and whose presence potentially opens new phosphorus sources for the cell (e.g. phosphatases). On the other hand, the synthesis of porins, more or less specific proteins that mediate passage of substrates into the periplasm, assures that this additional layer is not a major barrier for essential nutrients that have to reach the cytoplasmic membrane. In *E. coli* the porin protein of the Pho regulon is encoded by *phoE*. This protein contains 330 amino acid residues and exhibits specificity for negatively charged molecules (Overbeeke et al. 1983). Although its synthesis is enhanced by phosphate starvation, PhoE does not show preference for phosphate ions (Overbeeke and Lugtenberg 1980). The functional complex through which phosphate and organic phosphates pass the outer membrane is formed by three identical PhoE subunits. Despite its trimeric association in the membrane (De Cock et al. 1992), each active monomer contains its own pore which operates independently of the other channels. PhoE does not reveal hydrophobic features like integral membrane proteins of the cell membrane; it is, however, assumed that integration of the trimer into the outer membrane is partly stabilized by a hydrophobic boundary at the outer surface of the complex.

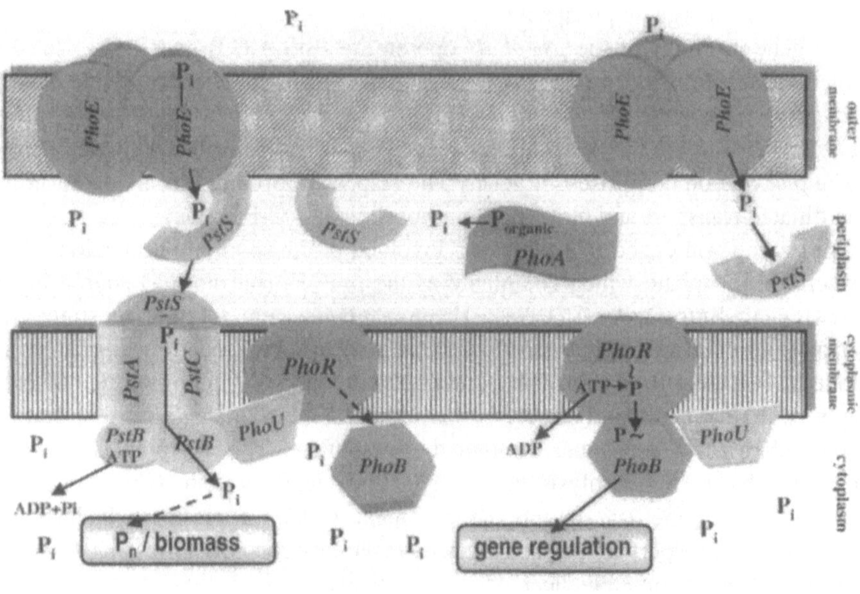

Fig. 4.1. Pathway of external phosphate from the extracellular space into the cytoplasm in *E. coli* (Torriani 1990; Silver and Walderhaug 1992; Wanner 1993,1994). For details on mechanisms and proteins involved see text. The *symbols* do not reflect actual size proportions or shapes of the proteins. P_i denotes inorganic phosphate, P_n designates polyphosphates

Only a few organic phosphates that have passed the outer membrane can be utilized by the cells in their intact form. Most organic phosphorus compounds must be cleaved by periplasmic phosphatases into the organic component and inorganic phosphate, which is then incorporated. *E. coli* synthesizes at least seven phosphatases with more or less specific mode of activity (Wanner 1996). However, the phosphatase that becomes dominant during phosphate starvation is alkaline phosphatase (Coleman 1992), the gene product of *phoA*. This enzyme shows the highest activity at a pH range between 8 and 10, and can be present in multiple isozymes of the active 94.4-kDa dimer (Kantrowitz 1994). Alkaline phosphatase is a phosphomonoesterase of low specificity. Therefore, many organic phosphates are only poor substrates for this enzyme, but this disadvantage is balanced by the broad range of potential phosphorus sources and by the huge amount of synthesized enzyme (up to 6% of total cell protein; Wanner 1996).

The inducible high affinity phosphate transport complex per se is constituted by the gene products of the *pst-phoU* operon. The Pst system has a Michaelis constant of transport of about 0.2 – 0.4 μM phosphate (Rosenberg et al. 1977; Wanner 1996), and the complex is energized by phosphate-bond energy. It exhibits high specificity for phosphate and transports arsenate or sulfate only poorly (Medveczky and Rosenberg 1971), a feature which is now attributed to the specificity of the periplasmic-binding protein. The Pst system is a typical exponent of the so-called ABC (ATP binding cassette) transporter family (reviewed in Boos and Lucht 1996). It is characterized by the interplay of a periplasmic binding protein (PstS) and a membrane-associated complex that is composed of two integral membrane proteins (PstA,C) and two proteins located at the inner surface of the cytoplasmic membrane (dimeric PstB). The latter protein contains an ATP-binding domain and is probably the place of energization of the transport complex.

The periplasmic phosphate binding protein PstS is the first specific component of the high-affinity uptake system that is encountered by the inorganic phosphate ion on its way from the external medium into the cell (Medveczky and Rosenberg 1970). PstS can be easily released from the cells by cold osmotic shock treatment and is mostly considered as a freely diffusible periplasmic protein that "harvests" phosphate and mediates diffusion of this ion through the periplasm to the translocation complex in the cytoplasmic membrane. At first sight, however, this operation mode seems disadvantageous: the substrate has to be bound to a protein first, and, more decisive, subsequent diffusion of the protein-substrate complex should be slower in an aqueous phase than diffusion of the free ion, which is considerably smaller than the complex. One possible explanation for the diffusion problem of binding proteins was seen in the potential reduction in dimensionality in the diffusion process: if the protein had an increased probability for associating with the cell membrane, then the three-dimensional diffusion problem of the protein-substrate complex would be reduced to the two-dimensional one of scanning the cell membrane surface for the translocating complex (Brass et al. 1986; Higgins 1992).

Boos and Lucht (1996) assume an additional, important role of the periplasmic binding proteins in functioning as a trap for the substrate: due to the high concentration of binding proteins, substrate is held back in the periplasm and prevented from rediffusion into the medium in which the concentration of the substrate may be extremely low.

The observation that the presence of PstS enhanced the passage of phosphate through PhoE (Korteland et al. 1982) could be interpreted in the way that these two proteins are somehow associated, not only spatially but potentially also in their function. Continuing these considerations, one could imagine some kind of "channeling" pathway (cf. channeling models which have been proposed for several cellular reactions, e.g. Welch and Easterby 1994) in which the components of the outer membrane and the uptake system are arranged in a way that enables the directed transfer of the substrate from the external medium to the cytoplasmic membrane.

The mode of operation of binding proteins is still unclear, and even their function as proteins that mediate the diffusion of the substrate to the translocation complex in the cell membrane is questionable again since the discovery that binding protein-dependent transport systems are not unique to Gram-negative bacteria which synthesize an outer membrane as a retention barrier for periplasmic proteins. Several of these ABC transporters have now been characterized in Gram-positive bacteria (review in Boos and Lucht 1996); the binding proteins in these systems, however, are anchored in the cell membrane, a molecular organization which excludes the diffusion catalyst function. Other observations inconsistent with the original idea of diffusible binding proteins further complicated the interpretation of the function of these proteins: in fluorescence bleaching studies it was shown that the lateral diffusion of proteins in the periplasm of *E.coli* was about 3 orders of magnitude slower than could be expected in an aqueous system. These and similar results suggest that the periplasm is actually not comparable to an aqueous phase but reveals gel-like properties which favour the movement of proteins.

Intensive biochemical studies elucidated the structure and properties of the periplasmic phosphate-binding protein (Kubena et al. 1986; Luecke and Quiocho 1990; Wang et al. 1994). This 34.4-kDa protein has one binding site with a K_D for phosphate of about 1 μM. The high-resolution crystal structure of PstS revealed that the anhydrous monovalent or (preferably) divalent phosphate ion is bound without counterion in a deep cleft between two globular domains of the protein. Binding is achieved by coordination via 12 hydrogen bonds to 3 essential amino acid residues. The specificity for phosphate is attributed to the distinct acidic residue D56, the presence of which excludes binding of the similar sulfate ion.

The two integral membrane proteins, PstA and PstC (Surin et al. 1985; Amemura et al. 1985; Webb et al. 1992), constitute the transmembrane translocator system that allows passage of phosphate through the cell membrane. It is probably energized by ATP hydrolysis occurring at the peripheral PstB subunits. Since PstA and PstC show more homology in their amino acid sequences with each other than with comparable membrane transporters of the ABC family, it is possible that they are the product of gene duplication (Ames 1986). PstA and PstC consist of six hydrophobic helices each (A1-6 and C1-6, respectively) that span the membrane and are joined by five hydrophilic loops at the periplasmic or cytoplasmic side of the cell membrane. A proposed mechanism for phosphate transport involves the tight cooperation of the domains of the protein pair (cf. Webb and Cox 1994): three charged residues located on A5 and C6 that resemble in their electrostatic properties the three essential charged residues of the PstS-binding site form the putative binding site that receives the phosphate ion from PstS. Release of PstS and further movement of phosphate through the membrane is attributed to a conformational change initiated by peptidyl-prolyl isomerization on C3 and C4 (or A3 and A4). The respective coordinated pair of helices is in close connection with a PstB subunit which catalyzes the isomerization step at the expense of ATP. Due to the resulting conformational change membrane-binding site 1 comes into contact with a membrane-binding site 2 that accepts the phosphate ion. Release of phosphate into the cytosol and reversal of the conformational change requires the ATP-hydrolyzing activity of the second PstB protein that interacts with helices A3 and A4 (or C3 and C4).

Successful adaptation of organisms to alterations in the environmental nutrient status necessitates the capability of cells to develop structures and mechanisms to sense these changes and respond to them accordingly. It was suggested that phosphate depletion in the medium leads to subsequent dramatic decreases in the cytoplasmic phosphate concentration, which, in turn, would activate expression of phosphate starvation-inducible genes. This explanation, however, has been excluded for *E. coli* since no regulatory effect of the cytoplasmic phosphate concentration – which remained relatively constant between 9 and 13 mM – on *pho* genes could be observed in wild-type and mutant cells grown on different phosphorus sources (Rao et al. 1993,1994). Probably the extracellular phosphate concentration is the proper signal that regulates expression of Pho regulon genes, although the identity of the actual sensor is still unclear, since PhoR is no longer considered as the proper candidate. It is now assumed that the intact Pst system is required in sensing the external phosphate status and transferring the phosphate signal (see below).

Transcriptional regulation of the Pho regulon is mainly attributed to the operation of a two-component signal transduction mechanism that comprises PhoR as the putative (but probably only indirect) sensor for external phosphate and PhoB as the DNA-binding response regulator. PhoR is an integral membrane protein containing 431 amino acid residues, while PhoB is a soluble cytoplasmic protein which consists of 229 amino acids (Makino et al. 1986a,b). PhoR is involved in activation and repression of genes of the Pho regulon. If the external phosphate concentration decreases below a critical value, the sensor kinase PhoR is autophosphorylated on a histidinyl residue, and subsequently transphosphorylates an aspartyl residue of the PhoB protein. Phospho-PhoB binds then, probably in dimeric form, to a 18-nucleotide consensus sequence (*pho* box; cf. Makino et al. 1994) upstream of the Pho regulon promoters, and transcriptionally activates expression of the phosphate starvation inducible proteins.

Repression of Pho regulon genes at high external phosphate concentrations is less transparent and requires the presence of PhoR, PhoU and the intact, but not necessarily functional, Pst system, probably as a sensor. In an attempt to explain the interactions of these constituents, Wanner proposed a model based on a repressor complex the formation of which is initiated if the phosphate-binding sites of PstS are nearly saturated due to high ambient phosphate concentrations (Wanner 1994,1996). The phosphate-loaded PstS induces then a conformational change in the Pst system, and via protein-protein interactions this conformational change is transmitted to PhoU and finally to PhoR. In this reaction sequence PhoR is transformed to a special repressor form that potentially inactivates the phospho-PhoB by dephosphorylation.

The understanding of the regulation of the Pho regulon has been further complicated by the discovery of CreC, a homologue of PhoR, and the fact that acetylphosphate is able to activate PhoB by phosphorylation in the absence of PhoR. Astonishingly, these components are members of control mechanisms that operate independently of the external phosphate status. Since their potential effect on regulation of the Pho regulon has been shown in PhoR mutants solely, the actual importance in wild-type cells is unclear (reviewed in Wanner 1996).

4.2.2 Cyanobacterial Phosphate Transport Systems

Although cyanobacteria are widely used objects of molecular genetic work, specific interest in the molecular composition and regulation of the phosphate uptake system in these organisms is sparse. Only few working groups deal with this topic (cf. Scanlan and Wilson 1999), probably because it is mainly assumed that the high-affinity transport system of cyanobacteria essentially resembles that of *E. coli*. However, as stated before, this view is not necessarily correct if we consider that *E. coli* is not a typical free-living organism that had to cope regularly with phosphate-deficient growth conditions during its evolution. Although the listing below confirms that genes found in *E. coli* are actually present in many cyanobacteria, one has to keep in mind that cyanobacterial components of the high-affinity phosphate transport system are usually identified by homology comparison with the *E. coli* genome. The putative function of the cyanobacterial protein is then assumed to be equivalent to the function of the respective protein in *E. coli*. However, the detected levels of homologies between genes of *E. coli* and of cyanobacteria are frequently not high enough to exclude the possibility that the respective protein(s) may exhibit different properties or mode of operation in the assembled transport system of cyanobacteria. Furthermore, for obvious reasons, homology searches are not suitable to detect additional components — or the interplay between individual components — which may be necessary for the coordinated phosphate incorporation process. Additionally, in higher algae the molecular composition of high-affinity phosphate transport systems is almost completely obscure. Hence, in cyanobacteria and higher algae more direct information about the molecular basis of phosphate utilization and the concomitant physiological characterization of the mode of operation of transport systems is needed for a sound approach to ecophysiological problems, such as the adaptation of organisms to limiting phosphate supply.

The freshwater cyanobacterium *Synechocystis* sp. PCC6803 is one of the few blue-green algae of which the whole genome has been sequenced (http:/www.kazusa.or.jp/cyano/cyano.html). As a quasi-by-product of these sequencing efforts, molecular information about the putative composition of the high-affinity phosphate uptake system is available; however, specific biochemical and physiological characterization of phosphate utilization in this organism is still missing. According to homology searches, the genome of *Synechocystis* contains two separate operons that carry *pst* genes (D. Scanlan, pers. comm.): one operon comprising the genes encoding *pstA,B,C,S,* and a second one with two *pstB* genes, one gene for *pstA,C,S,* and an additional *sphX* gene which encodes another putative phosphate binding protein (see below). Due to this duplicate occurrence of *pst* system genes this organism may be especially interesting for future research on phosphate transport in cyanobacteria.

Another whole genome sequence is now available (http://spider.jgi-psf.org/JGI_microbial/html/prochlorococcus_homepage.html), namely that of the marine alga *Prochlorococcus* sp. Homology searches reveal that the genome contains an operon with all the necessary genes to constitute a functional Pst system. This unicellular organism is an abundant component of phytoplankton assemblages in marine ecosystems. Due to its importance in primary production in these ecosystems it has started to become a preferred subject matter of investigation.

In the marine cyanobacterium *Synechococcus* sp., WH7803 phoR and phoB homologues have been identified (Watson et al. 1996), as well as the *ptrA* gene encoding an additional putative transcriptional activator protein (Scanlan et al. 1997). The *ptrA* which is located downstream of *pstS* shows similarity to the global nitrogen regulator of cyanobacteria (*ntcA*) (see Chap. 3.4.2), and indicates a more complicated regulatory mechanism with regard to controlling the high-affinity phosphate uptake system existing in this organism. The presence of a *pstS* homologue in *Synechococcus* sp. WH7803 (Scanlan et al. 1993) suggests that the Pst system constitutes the functional phosphate transport system, however, other *pst* genes have not yet been identified in this organism. Synthesis of PstS is greatly enhanced in *Synechococcus* sp. WH7803 after transfer from phosphate-rich to phosphate-deplete medium. Fractional separation of the cell components revealed that this protein was predominantly present in the cell wall fraction. Recently a *pstS* mutant of *Synechococcus* sp. WH7803 has been constructed (D.J. Scanlan, unpubl.), and preliminary results suggest that PstS of *Synechococcus* sp. WH7803 operates like a "classical" binding protein that catalyzes and thereby accelerates the transfer of phosphate through the periplasm to the translocator complex (F. Wagner and D. Scanlan, unpubl.).

In the closely related freshwater cyanobacterium *Synechococcus* sp. PCC7942 (*Anacystis nidulans*), two alkaline phosphatases have been identified: PhoA is a phosphate-repressible cell wall-associated protein of unusually large size (145 kDa; Ray et al. 1991), and PhoV a phosphate-non-repressible membrane-associated protein with a molecular mass of about 61 kDa (Wagner et al. 1995).

Furthermore, this organism contains a two-component regulatory system comparable to the PhoR/PhoB system: the sensor kinase SphS and the response regulator SphR (Aiba et al. 1993). To our knowledge, there is no indication for the presence of constituents of the Pst system in this cyanobacterium so far. In *Synechococcus* sp. PCC7942 (*Anacystis nidulans*) phosphate starvation induces the abundant synthesis of a protein which could be identified as a component of the high-affinity phosphate uptake system (Wagner et al. 1994). This protein has a molecular mass of about 33 kDa, and was designated SphX by Aiba and Mizuno (1994), who also constructed a *sphX* mutant of this organism. The SphX polypeptide reveals highest homology to the phosphate-binding protein of *Mycobacterium tuberculosis*, but, interestingly, little similarity to the PstS of the marine *Synechococcus* species (Mann and Scanlan 1994). SphX is probably a peripheral component of the cytoplasmic membrane, and exhibited a high phosphate-binding capacity in reconstituted proteoliposomes with a dissociation constant of the high-affinity binding site of about 0.3 μM. Since this protein did not exhibit transport activity, it was assumed that SphX is a phosphate-binding protein, though somewhat atypical (Wagner et al. 1994). Further studies on a SphX mutant yielded first indications that the proposed function of this protein was questionable (Falkner et al. 1998). Growth of the mutant under permanent phosphate deficient conditions was not evidently impaired and the lowest detectable threshold values of phosphate uptake were in the low nanomolar range, as well. However, deletion of *sphX* obviously led to the loss of the adaptive capability of the mutant to adjust the kinetic and energetic properties of the uptake system to elevations in the external phosphate concentration (this characteristic feature of wild-type *Synechococcus* sp. will be described in detail in Sect. 4.4.).

4.3 Conventional Kinetic Models of Phosphate Uptake and Growth

4.3.1 The Applicability of the Michaelis-Menten Equation for Analysing Phosphate Uptake

In oligotrophic lakes the production of biomass is usually limited by the availability of phosphate. Since practically all aquatic microorganisms and phytoplankton meet their phosphorus demand from the same source, organisms have to compete for slightest increases in external phosphate. Therefore selection of the species with the most efficient phosphate uptake system could be expected. Astonishingly, however, the biodiversity within oligotrophic lakes is rather high, and a number of species coexist under these growth conditions, while during eutrophication – when the supply of phosphate is constantly high for a prolonged period of time – usually one species becomes dominant and outgrows its competitors. The cooccurrence of a multitude of phytoplankton species under oligotrophic growth conditions has been referred to as the paradox of the plankton

(Hutchinson 1961), and is explained by the existence of ecological niches that emerge from an ever-changing environment. A major constraint, however, for phytoplankton to survive under oligotrophic growth conditions is that the phosphate uptake system is regulated in a way that meets the requirements imposed on the organisms by a low and variable phosphate supply.

Virtually all models that analyze phosphate acquisition and the subsequent growth behaviour of microorganisms and algae under conditions of permanent or fluctuating nutrient supply suppose that incorporation of the nutrient follows Michaelis-Menten kinetics. According to this model, the reaction sequence comprises formation of an intermediate complex (ES) from the free enzyme (E) and substrate (S). Due to the catalytic activity of the enzyme the substrate is converted to product (P) which is released by decay of the complex:

$$E + S \xleftrightarrow{k1/k-1} ES \xrightarrow{k2} E + P$$

For a simple mathematical description of the concentration dependence of the rate of this catalyzed process, the assumption is made that the decay of the complex ES is irreversible so that no backward reaction occurs in which ES is formed from E+P. If this requirement is fulfilled, the familiar Michaelis-Menten equation can be derived for a steady state situation in which dES/dt = 0 (cf. Segel 1975):

$$v = \frac{v_{max}[S]}{K_M + [S]} . \qquad (1)$$

Here v is the velocity of substrate conversion, attaining a maximum value (v_{max}) when the substrate concentration is so high that all enzyme is present as ES. The Michaelis constant K_M is numerically equal to the free substrate concentration at which $v = v_{max}/2$. A plot of the substrate concentrations vs. the corresponding rates yields a rectangular hyperbolic curve as shown in Fig. 4.2. At very low substrate levels [$S \ll K_M$) Eq. (1)] practically follows the first-order kinetics, while at saturating substrate concentrations the reaction is of zero order and the velocity of the reaction is no longer concentration-dependent ($v=v_{max}$).

The K_M of an enzyme is a consequence of the physical properties of the active site, and a low value indicates that the enzyme attains already at low substrate concentrations a saturation state that results in a kinetically efficient substrate conversion relative to the catalytic capability of the enzyme. Analogously, a transport system with a lower Michaelis constant is capable of utilizing lower ambient nutrient concentrations kinetically more efficiently. When applied to the phosphate uptake process, the substrate corresponds to the external phosphate, and the product is the incorporated nutrient. Since the Michaelis-Menten model is based on the irreversible conversion of substrate to product, no equilibrium state exists for the overall reaction. Accordingly, incorporation of phosphate should proceed until the external concentration is zero, and no threshold value can be

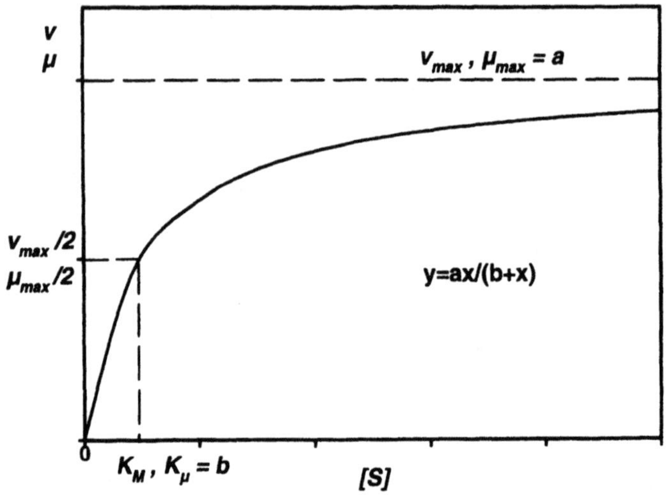

Fig. 4.2. The Michaelis-Menten and Monod models: dependence of the velocity of nutrient incorporation v or growth rate μ, respectively, on the external substrate concentration [S]

attained. As mentioned earlier, it was confirmed in numerous studies with natural phytoplankton and pure laboratory cultures that incorporation of phosphate actually ceases at a distinct external concentration below which phosphate uptake is impossible for energetic reasons (cf. Sect. 4.1.2). The discrepancy between the model and the empirical observations did not evoke, however, rejection of the Michaelis-Menten kinetics for analysis of phosphate uptake, an obvious thing to do, but resulted in a simple modification of the equation by which the threshold concentration is subtracted from the external phosphate concentration (Droop 1974; Button 1978; Perez-Llorens and Niell 1995) — if the threshold value is not neglected at all. Despite the lack of a theoretical foundation (Droop 1974) in the case of phosphate incorporation, Michaelis-Menten kinetics are generally accepted (cf. Reynolds 1993; Sommer 1994) and applied for the description of phytoplankton and microbial phosphate accumulation (e.g. Rhee 1973; Ritchie et al. 1997; Ikeya et al. 1997; Hameed et al. 1999).

The properties of the phosphate uptake system of organisms can vary within a wide range and are usually tuned to the prevailing growth conditions. Accordingly, the observed Michaelis constants vary not only between individual bacterial and algal species, ranging from low nanomolar to micromolar K_M values (Nalewajko and Lean 1980; Healey 1982), but depend also on the state of phosphate starvation of the respective population — although some authors consider the Michaelis constant to be independent of the prevailing growth state for most species (e.g. Rhee 1982; Morel 1987). The dependency of the activity of the transport system on the growth rate is most clearly revealed by the considerable variation in its maximum velocity: v_{max} usually rises with increasing phosphate starvation, and decreases if the algae are supplied with higher amounts of the limiting nutrient. Rhee (1973) investigated the interdependence between phosphate incorporation, the amount of stored phosphate and the growth rate, and observed that the algae with the highest proliferation rate exhibited the lowest maximum velocity of incorporation and concomitantly the highest polyphosphate pool. From the inverse relationship between the activity of the uptake system and the cellular content of the limiting nutrient, he concluded that the pool of internally stored nutrient has an inhibitory effect on the incorporation process (Rhee 1973, 1974, 1982). However, the simplest, and biologically more comprehensible, explanation for the interdependence between activity of the uptake system and the degree of phosphate limitation is that under different growth conditions the number of phosphate transport systems per average cell is not constant. Accordingly, a maximum value is attained in the non-growing cell, while fast-growing cells synthesize a considerably lower number of transport systems. Due to the observed variability of the kinetic constants, Healey (1980) proposed characterizing transport systems by the ratio v_{max}/K_M, and considered this affinity to be a suitable indicator of the competitive capability of individual species with respect to phosphate acquisition. However, the maximum velocity is only meaningful for assessing the capacity of a transport system if it can be referred to the respective protein (complex). In studies with living organisms the numerical value of v_{max} is strongly dependent on the chosen reference system (chlorophyll, total protein content, cell number) which varies according to the growth conditions in a non-uniform manner in individual species, rendering interspecific comparisons problematic.

4.3.2 Kinetic Basis of Conventional Growth Models

Applied and theoretical interests in the growth behaviour of populations have led to an abundance of literature on this topic. Microbiological studies are often focused on the (optimization of) yield of cultures, which is the quantity of biomass that can be produced by the cells from the amount of consumed substrate, while aquatic ecologists are interested in the factors that govern the productivity as well as the composition of phytoplankton assemblages. Therefore, it is important to understand the interplay between acquisition of the limiting nutrient and its subsequent utilization in the growth process. This crucial issue was investigated in discontinuous (batch) and continuous (chemostat) cultures to a great extent. In a batch culture an inoculum of organisms is diluted into growth medium which contains all necessary nutrients in excess. After a certain lag phase during which the cells become acclimatized to the new growth conditions, a state of exponential growth is attained which is indicated by a linear increase in biomass if the logarithmic cell number is plotted against time. The slope of the line corresponds to the specific growth rate μ, since during exponential growth the increase in biomass (d X/dt) is proportional to the amount of biomass(X) present:

$$\frac{dX}{dt} = \mu\ X\text{, and therefore } \mu = \frac{dX}{dt}\frac{1}{X}.$$

After adequate transformation, the relation between the specific growth rate and the doubling time (t_d) is derived:

$$\mu = \frac{\ln 2}{t_d}.$$

The exponential growth phase of the cells is only maintained as long as all nutrients are present in excess, or if the concentration of a limiting nutrient is supplied at a constant level. The latter condition is usually accomplished by growing the microorganisms or algae in chemostat cultures (Monod 1950) in which a constant inflow of fresh growth medium, balanced by an outflow of cell suspension at the same rate, supplies the cells with the desired amount of the limiting nutrient in an otherwise unmodified medium. If a steady-state situation is attained, cell proliferation follows the exponential growth law, and by fixing a distinct inflow rate of nutrient, a defined specific growth rate is adjusted. Under these growth conditions the chemical environment of the cells is assumed to be constant. Monod (1942) considered also the chemical composition of the organisms with respect to the limiting nutrient to be constant, and proposed a yield coefficient that is equivalent to the ratio between the growth rate (dX/dt) and the steady-state uptake rate (-dS/dt). This yield coefficient is the reciprocal of the cell quota Q (Droop 1973), which represents the total amount of internal nutrient per average cell under the respective growth conditions. Accordingly, a steady-state relationship between uptake rate, growth rate and internal nutrient store can be established:

$$v = \mu \cdot Q. \tag{2}$$

In his studies on carbon nutrition of bacterial chemostat cultures, Monod (1942) observed that the relationship between the specific growth rate and the external substrate concentration followed a hyperbolic function. In order to describe his empirical data mathematically, he suggested an equation that is derived from the formalism of the Michaelis-Menten kinetics. Accordingly, the dependence of the specific growth rate on the external steady state concentration of the limiting substrate (S) attains the general form:

$$\mu = \frac{\mu_{max} \cdot [S]}{K_\mu + [S]}. \tag{3}$$

μ_{max} is the maximum specific growth rate, and K_μ is a half-saturation constant which is numerically equal to the external concentration at which $\mu=\mu_{max}/2$ (cf. Fig. 4.2). The Monod equation has since become one of the most influential and widespread models to describe the growth behaviour of nutrient-limited cultures under steady state — and sometimes even non-steady state — conditions (e.g. Mueller 1970; Tilman 1977; Sommer 1989; Grover 1989, 1991a). K_μ and μ_{max} are considered to be species-specific constants that allow interspecific comparisons of the ecological capability of populations: a species with a high maximum proliferation rate has a potential ecological advantage and may outgrow its competitors under favourable conditions, and a low half-saturation constant indicates an enhanced capability to utilize lower external phosphate concentrations for growth. Accordingly, attempts were made to refer the dominance or inferiority of distinct species in natural environments to the interplay between the half-saturation constant and the maximum growth rate (e.g. Titman 1976; Sommer 1983; affinity ratio μ_{max}/K_μ Healey 1980).

However, the Monod model is based on several assumptions that soon provoked criticism about its applicability to nutrient-limited growth (e.g. Droop 1973). Empirical data revealed that the Monod model does not adequately describe growth behaviour in the low concentration range (e.g. Mueller 1970; Brown and Button 1979; Olsen 1989), since growth of organisms ceases at a finite external phosphate concentration, a simple consequence of the fact that the incorporation process is characterized by a threshold value. However, in analogy to the Michaelis-Menten kinetics of nutrient uptake, this dilemma was sometimes removed by subtracting the threshold value from the external nutrient concentration, so that S is replaced by $(S\text{-}S_0)$ in Eq. (3) (cf. Sommer 1994).

The Monod model seems suitable for defining bacterial growth in carbon nutrition (cf. Schlegel 1985), but contradicts the phosphate uptake behaviour of microorganisms and algae in many respects, especially its inadequacy to explain the existence of a threshold concentration for uptake and growth. Since the Monod model assumes the immediate utilization of the incorporated nutrient for the ongoing growth process, it is based on a constant yield factor – with respect to the limiting nutrient – that is independent of the growth rate. Accordingly, it does not cover variations in the internal concentration of the limiting nutrient, and no growth should occur if the external concentration is zero. However, if growing cells are transferred to phosphate-free medium they usually continue to grow for some time due to their internal stores of the nutrient. Consequently, growth of organisms is not solely a function of the acquisition of the limiting nutrient, but also of the utilization of the incorporated nutrient in the ongoing growth process.

Several authors observed a linkage between internal quantities of the limiting nutrient and the proliferation rate of the population (e.g. Caperon 1968; Fuhs 1969). Droop (1968) originally studied the relation between growth rate and the respective cell quota using chemostat cultures of vitamin B12-deficient *Monochrysis lutheri*, but his results were later confirmed to be valid for phosphate, too, in many species (e.g. Droop 1973; Rhee 1973; Ducobu et al. 1998; Spijkerman and Coesel 1998). According to the Droop model, the growth rate is a hyperbolic function of the total cell content of the limiting nutrient, Q, which comprises in the case of phosphate the internal orthophosphate, the polyphosphates and organic phosphorus compounds:

$$\mu = \mu'_{max} \cdot \left(1 - \frac{Q_0}{Q}\right). \tag{4}$$

A graphical presentation of the hyperbolic relation between Q and μ is given in Fig. 4.3. The intercept on the abscissa corresponds to Q_0 which is the minimum cell quota or subsistence quota of the non-growing cell. At this critical internal threshold of the limiting nutrient no further growth occurs, obviously because at this level the intracellular nutrient is utilized in the most efficient way that still guarantees coordinated maintenance of basic cellular functions. The cell quota of phosphorus can vary by more than 1 order of magnitude in slow- and fast-growing cultures even under steady-state conditions of continuous nutrient supply (Healey 1982) — mainly due to the variable polyphosphate content established under different growth conditions — not to mention the surplus storage capability of many microorganisms and algae when exposed to high nutrient pulses. μ'_{max} corresponds to the maximum specific growth rate at infinite Q levels, a hypothetical value that cannot be attained since the storage capability of organisms is limited. According to Eq. (4), the ratio between μ_{max}, actually observed when the maximum amount Q_{max} of phosphate has been stored, and μ'_{max} is determined by the term $(Q_{max} - Q_0)/ Q_{max}$. The Droop equation was applied by countless authors for analyzing the growth behaviour of microorganisms and algae and is still the most accepted model in this field (e.g. Borchardt et al. 1994; Spijkerman and Coesel 1998; Ducobu et al. 1998).

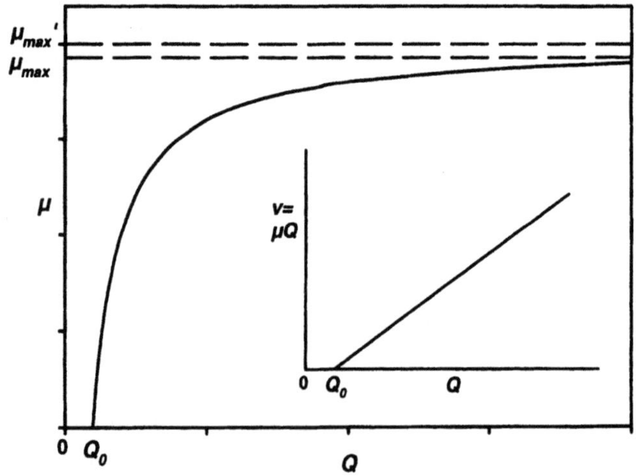

Fig. 4.3. Dependence of the specific growth rate μ on the cell quota Q according to the Droop model. Q_0 designates the minimum cell quota. *Inset* Graphic representation of the steady state incorporation rate μQ as a function of the cell quota Q

Simple rearrangement of Eq. (4) reveals the interdependence between the steady-state incorporation rate μ Q [cf. Eq. (2)] and the cell quota Q:

$$\mu \cdot Q = \mu'_{max} \cdot (Q - Q_0)\,. \tag{5}$$

Droop (1973) used this linear equation for confirming the validity of the relationship between μ and Q for vitamin B12, iron, nitrogen and phosphorus as the limiting nutrients. It can be seen from Eq. (5) that a plot of the steady-state uptake rate vs. the cell quota — determined at varying specific growth rates — should yield a straight line that intercepts the Q-axis at Q_0 (Fig. 4.3 inset). Although the Droop equation does not describe the relation between the specific growth rate and the cellular phosphate quota satisfactorily in all the cases (cf. Healey 1982), his and other studies revealed an important feature of nutrient-limited growth by demonstrating that in many species the prevailing growth rate is somehow adjusted to the internal phosphorus content. Though perhaps intuitively comprehensible, it is not evident from his or similar models (e.g. Fuhs 1969) why this interdependence exists, and how it is regulated, or, in other words, how the organisms monitor their polyphosphate content, and adjust the uptake system accordingly to maintain an apparently favourable internal level of the limiting nutrient at a distinct growth rate.

The empirically derived Monod and Droop equations represent steady-state models which assume a direct coupling between the steady-state incorporation rate of the nutrient and the growth process. While Monod proposed a dependence of the specific growth rate on the external phosphate concentration, Droop established a simple internal stores model that relates the cellular phosphate content and the growth process. However, neither of the two models considers the inverse interdependence between the short-term phosphate uptake behaviour and the growth rate which is frequently observed in microorganisms and algae: usually the activity of the phosphate uptake system increases with reduced specific growth rate, and attains a maximum in the non-growing cell. Hence, if a phosphate-limited population is exposed to a sudden increase in the external nutrient concentration, the initial incorporation rates exceed by far the rates of phosphate utilization for biosynthesis at the prevailing growth rate. Therefore, phosphate-deficient organisms potentially accumulate considerable amounts of the nutrient, and are capable of growing for several generations at the expense of the internal store even if no further uptake of the nutrient is possible. It is obvious that under non-steady-state conditions in which the supply of phosphate occurs not constantly but pulse-wise, the incorporation of the nutrient and the growth of the population are uncoupled processes. More complex internal-stores models (e.g. Tilman 1977; Kilham 1978) try to account for non-steady-state growth conditions by distinguishing the kinetics of nutrient incorporation from the kinetics of the growth process: the uptake step is assumed to follow Michaelis-Menten kinetics, and the subsequent growth step is characterized by an interdependence between internal cell quota of the limiting nutrient and the specific growth rate, usually modelled by means of the Droop equation (e.g. Grover 1991a,b; Spijkerman and Coesel 1998; Ducobu et al. 1998). Accordingly, several-state equations derived from steady-state assumptions are combined to explain the growth behaviour of populations under non-steady-state conditions. Probably the most elaborate and systematic study in this field was given by Morel (1987). His formal treatment of the interrelationship between short-term nutrient incorporation and the long-term demands of the growing cell seems convincing due to its consequently logical structure: he reparameterizes familiar basic steady-state models, but considers only the kinetic properties of the uptake system, and not a threshold value for nutrient incorporation.

Outgoing from Eqs. (1) – (4) he demonstrates that the half-saturation constant of steady-state incorporation ($K_{\mu Q}$) is necessarily higher than the half-saturation constant of growth (K_μ) due to the variable cell quota. The ratio between the two constants is dependent on the ratio between the maximum and minimum cell nutrient content, Q_{max} and Q_{min}, respectively.

Morel assumes that short-term nutrient incorporation follows Michaelis-Menten kinetics, and that the half-saturation constant (K_ρ) of the transport system is invariant under different growth conditions. However, he accounts for the empirical observation that the maximum velocity of instantaneous nutrient incorporation (ρ_{max}) generally increases with increasing phosphatelimitation of the population which is accompanied by decreases in the cell quota. Accordingly, the maximum velocity of short-term incorporation can vary between a maximum value in non-growing cells ($\rho_{max}{}^{hi}$), and a minimum value in non-limited growing cells ($\rho_{max}{}^{lo}$), resulting in a family of incorporation hyperbolae that characterize any intermediate growth state with the respective cell quota. Following his line of argument, the saturation constant of instantaneous nutrient incorporation is higher than the half-saturation constant of steady-state incorporation by a factor that equals $\rho_{max}{}^{hi}/\rho_{max}{}^{lo}$. Combining his equations obtained from steady-state and instantaneous nutrient incorporation considerations, he establishes a coupling between short-term uptake, the cell quota and the specific growth rate, which is expressed in the interrelationship between the half-saturation constants for growth and short-term nutrient uptake:

$$K_\mu / K_\rho = \rho^{lo}_{max} / \rho^{hi}_{max} \cdot Q_{min} / Q_{max} \, .$$

Since the maximum specific growth rate is given by the quotient $\rho_{max}{}^{lo}/ Q_{max}$ [cf. Eq. (2)], the equation above attains the form:

$$K_\mu / K_\rho = Q_{min} \cdot \mu_{max} / \rho^{hi}_{max} \, .$$

If the external nutrient concentration is within the boundaries given by the values of these two half-saturation constants, the growth rate can be maintained fairly constant due to concomitant alterations in the properties of the transport system and the cell quota. Morel denotes this concentration range as a range of acclimation, since here major physiological changes occur without substantially influencing the growth behaviour of the organism. He identifies two physiological strategies within the range of K_μ and K_ρ, strategies which are decisive for maintaining as high a growth rate as possible at the prevailing nutrient status: if the external nutrient concentration is low, and thus between K_μ and $K_{\mu Q}$, predominantly alterations in the internal nutrient content should occur. At higher external concentrations, between $K_{\mu Q}$ and K_ρ, the major response of the organism is assumed to comprise modifications of the activity of the uptake system.

As mentioned above, in Morel's considerations the instantaneous incorporation rate is inversely related to the specific growth rate of the organisms, which, in turn, is adjusted to the cell quota. He derives a linear relationship between the activity of the uptake system and the cell nutrient content by making Q a function of the maximum short-term incorporation rate:

$$\rho_{max} = \rho^{hi}_{max} - (\rho^{hi}_{max} - \rho^{lo}_{max}) \frac{(Q - Q_{min})}{(Q_{max} - Q_{min})} .$$

The applicability of this function is apparently restricted to species in which ρ_{max} varies, while K_ρ remains constant at different growth rates, and, consequently, deviations from the above equation in the uptake behaviour of algae have been reported (Grover 1991a).

In countless studies the Monod, Droop or complex internal stores models are used to analyze the growth behaviour of microorganisms and algae with respect to the prevailing phosphate supply. A general consensus exists about the applicability of the cited models to obtain species-specific parameters that define the physiological range of adaptive responses of a population to distinct modes of phosphate supply. The major proportion of the recent ecological literature on phosphate limitation is based on comparative studies in single- and multispecies systems with respect to the competitive capability of populations (e.g. Tilman 1977; Sommer 1983, 1985, 1989, 1993; Olsen et al. 1989; Egge 1998; Vadstein 1998; Nicklisch 1999). Combinations of kinetic constants are used to interpret the dominance of distinct species under defined growth conditions, and elaborate models exist for explaining the composition of natural algal assemblages on the basis of competitive exclusion (e.g. Armstrong and McGhee 1980; Tilman et al. 1982; Tilman and Sterner 1984). The vast majority follows the main ideas of these hypotheses that if only one nutrient is limiting, only one species will survive, and if several species coexist, there must be several ecological niches (e.g. limitation by a second nutrient). Some authors, however, have called these statements a tautology, and not sufficient to explain the coexistence of a multitude of species under oligotrophic growth conditions (cf. Ghilarov 1984).

4.4 Adaptation as a Process: the Transition from Adaptive to Adapted States in Cyanobacteria

4.4.1 Theoretical Treatment of the Adaptive Faculty of the Phosphate Uptake System on a Phenomenological Basis

Conventional models of phosphate uptake and growth kinetics suffer from the decisive shortcoming that they ignore the above described threshold concentration, a parameter of major ecophysiological significance that determines whether or not external phosphate can be utilized by the microorganisms. If at all introduced in contemporary models, this parameter remains a somewhat artificial component that lacks theoretical justification. In consequence, these models do not provide substantial insight into the interplay between the high- and the low-affinity transport systems and portray a rather static picture of the interaction between microorganisms and their environment. Originating from experiments with organisms in a defined physiological state, i.e. grown under distinct phosphate-limiting conditions, the phosphate uptake system degenerates to an invariant machinery that determines the growth behaviour in a definable way. Thus, the properties of the respective transport system are assumed to be constant for a given growth rate, and only long-term adaptation processes are assumed to occur when growth conditions are altered, leading to modification of the maximum velocity of phosphate incorporation. This static concept can neither account for short-term adaptive responses to phosphate fluctuations, nor does it consider the influence of the growth history on a distinct adaptive behaviour. Therefore, contemporary models do not allow us to analyze the adaptive potential of the phosphate uptake system by which cyanobacteria are able to respond within minutes to alterations in the ambient phosphate supply, a period of time not sufficient for substantial changes in the protein composition of the cells.

In order to be able to analyze the adaptive dynamics of the uptake behaviour of microorganisms and algae, an alternative, energetic approach was suggested by Falkner (1984), who derived a linear flow-force relationship on the basis of non-equilibrium thermodynamics (Falkner et al. 1989,1994). The relevant features of this bioenergetic model will be briefly outlined in this section.

For the simplest case of a cyanobacterial cell the assimilation of external phosphate to polyphosphates inside the cell comprises three steps in the light: (1) translocation of external phosphate (P_e) across the cell membrane, presumably electroneutral in cotransport with protons, (2) conversion of internal phosphate (P_i) and ADP to ATP via photophosphorylation, (3) transfer of the terminal phosphate group of ATP to the polyphosphates (P_n, P_{n+1}):

$$P_e \leftrightarrow P_i$$

$$P_i + ADP \leftrightarrow ATP$$

$$ATP + P_n \leftrightarrow ADP + P_{n+1} \qquad (I)$$

$$sum: P_e + P_n \leftrightarrow P_{n+1}.$$

Incorporation of phosphate against a concentration gradient and the subsequent synthesis of ATP are endergonic processes that require an energy source. Since in the light the input energy is provided by photophosphorylation (Simonis and Urbach 1973), the pH gradient across the thylakoid membrane is the driving force for the overall incorporation process. Thereby the energy consuming reactions are either indirectly (via ATP consumption for transport) or directly linked to the flow of a distinct number (n_p) of protons from the thylakoid (H_T^+) to the cytoplasmic space (H_C^+):

$$n_p H_T^+ \leftrightarrow n_p H_C^+. \qquad (II)$$

Condensation of the terminal phosphate group from ATP to the polyphosphates occurs by an equilibrium reaction that does not require an additional energy source. The overall reaction sequence for the formation of polyphosphates in the case of complete coupling comprises the energy-consuming phosphate assimilation (I) and the energy-delivering proton flux (II):

$$P_e + P_n + n_p H_T^+ \leftrightarrow P_{n+1} + n_p H_C^+. \qquad (III)$$

According to non-equilibrium thermodynamics, close to equilibrium the flux of phosphate into the cell (J_p) can be assumed to be proportional to the affinity -ΔG of the overall reaction (III) {$-\Delta G = 2.3RT(\log(K[P_e]) + n_p \Delta pH_T$}, where $K = 10^{-5}\ M^{-1}$ is the overall equilibrium constant for polyphosphate formation under the respective environmental conditions, and ΔpH_T is the pH gradient across the thylakoid membrane; P_n and P_{n+1} have been cancelled out, since they have the same thermodynamic activity). Summarizing the constant factors and the proportionality factor into a conductivity coefficient L_p, leads for an ideal case of complete coupling between the proton flux and the phosphate flux into the polyphosphate pool to the following flow-force equation:

$$J_p = L_p \log(K'[P_e]) + n_p \Delta pH_T.$$

In reality, complete coupling cannot be expected since additional leakage fluxes (i.e. unspecific losses of phosphate through the cell membrane), slippage phenomena (e.g. Pietrobon et al. 1982, 1983), futile cycles etc., occur in the cell and thus lead to a decrease in the degree of coupling between the energy-delivering proton flux and the energy-consuming reactions described above. Using the linear energy converter model to analyze the dependence of the uptake behaviour on the degree of coupling (Kedem and Caplan 1965), it was shown that for incomplete coupling the flow-force equation given above attains the general form:

$$J_P = L_P\left(\log\left(K'[P_e]\right) + q^2 n_P \Delta pH_T\right). \tag{6}$$

In this equation q is a coupling factor that can attain any value between 1 (complete coupling) and 0 (no coupling).

At the threshold concentration the overall affinity of the coupled reaction is zero, and net incorporation of phosphate ceases. Solving Eq. (6) for the external concentration where J_P is zero yields for the logarithmic threshold concentration $[P_e]_A$:

$$\log[P_e]_A = -\log K' - q^2 n_P \Delta pH_T). \tag{7}$$

Thus, for phosphate-deficient *Anacystis nidulans* with a pH gradient of about 1.8 (F. Wagner unpubl.) and a H^+/ATP-stoichiometry of 4 (Wagner and Falkner 1992), n_P is 8, since one ATP is invested in the transport across the membrane and a second ATP must be formed for the conversion of incorporated phosphate into the subsequent polyphosphate pool. This yields, for example, for a coupling coefficient of 0.99 a threshold value of about 1 nM. If the coupling coefficient is lowered, the threshold value will increase correspondingly. Equation (7) also predicts that the threshold value of microorganisms is altered when the input energy, i.e. the pH gradient or the stoichiometric factor n_P or both are changed.

The linear energy converter model (Kedem and Caplan 1965) allows us to derive a second characteristic parameter that also depends on the degree of coupling, namely the external phosphate concentration $[P_e]_{opt}$ at which uptake proceeds with optimal efficiency. It was shown (Falkner et al. 1989, 1994) that this concentration (in logarithmic form) can be expressed as:

$$\log[P_e]_{opt} = [(1-q^2)^{1/2} - 1]n_P\Delta pH_T - \log K. \tag{8}$$

Thus, for $\Delta pH_T = 1.8$, $n_p = 8$ and a coupling coefficient of 0.99, as in the example above, $[P_e]_{opt} = 43$ nM. The two equations reveal a basic property of linear energy converters: the higher the degree of coupling, the smaller becomes the difference between $[P_e]_A$ and $[P_e]_{opt}$, and the nearer to thermodynamic equlibrium does the system operate with optimal efficiency.

Combining Eqs. (6) and (7) leads to the following relationship:

$$J_P = L_P\left(\log[P_e] - \log[P_e]_A\right). \tag{9}$$

It can be seen that a plot of the incorporation rate J_P vs. the logarithm of the external phosphate concentration $\log[P_e]$ should yield a straight line with the slope representing the conductivity coefficient L_P which is a measure for the activity of the uptake system (cf. Falkner et al. 1993). The intercept on the abscissa corresponds to the logarithm of the external phosphate concentration where net incorporation stops, i.e. the threshold value ($\log[P_e]_A$) (cf. Fig. 4.6B). This function allowed the validity range of the flow-force relationship to be tested and it was shown that this equation is also obeyed under stationary growth conditions (Falkner et al. 1994).

Analysis of uptake systems by this semilogarithmic plot was originally suggested by Thellier (1970), and consequently this type of replot has been termed Thellier plot. If the external phosphate concentration drops below the threshold concentration, an efflux of phosphate from the cell to the medium should occur. Such an efflux, however, would be very disadvantageous for the cells from a biological point of view, and was never observed in *Anacystis nidulans*. Therefore, it must be postulated that below the threshold value the cells are able to adopt distinct properties that prevent an efflux, and, consequently, that the equations given above are valid only if the external phosphate concentration exceeds the threshold value.

Although this model was originally derived to describe the phosphate uptake behaviour of cyanobacterial cells, it was successfully applied to the analysis of phosphate uptake of higher algae and lake communities (Falkner et al. 1984a,b; Falkner and Falkner 1989; Istvanovics et al. 1994; Istvanovics and Herodek 1995; Aubriot et al. 2000).

4.4.2 The Transition from an Adaptive to an Adapted State

Due to its structural simplicity and its energy metabolism (photosynthesis and respiration) resembling that of higher plants, the unicellular cyanobacterium (blue-green alga) *Synechococcus* sp. (*Anacystis nidulans*) has been widely used in physiological and genetical studies for many decades. A long tradition in investigating the phosphate uptake behaviour of this organism exists (e.g. Simonis et al. 1974; Falkner et al. 1974), and *Anacystis nidulans* has proved to be also a suitable object for analyzing adaptive processes in algae.

Anacystis nidulans grown under phosphate deficiency has an extremely active high-affinity phosphate uptake system and can store appreciable amounts of phosphate. Accordingly, if a phosphate-deficient bacterial suspension contains such a high cell number that a short phosphate pulse in the nanomolar range can be utilized within a few minutes, the external phosphate concentration decreases rapidly to the threshold value which is under this condition usually between 1 and 3 nmoles. Addition of a second or third pulse of phosphate initiates the incorporation process again, and after a few minutes the same low threshold value is attained (Fig. 4.4). Under these experimental conditions the algae can incorporate several times their initial phosphorus content, and no changes in the kinetic or energetic properties of the phosphate uptake system can be observed.

Due to the high phosphate storage capacity, prolonged exposure of cells to high phosphate concentrations is possible and the adaptive alterations provoked by this pretreatment can be studied. When *Anacystis nidulans* is preincubated with micromolar phosphate concentrations for at least 10 – 20 min and the phosphate uptake behaviour of the preincubated cells is then compared to that of control cells

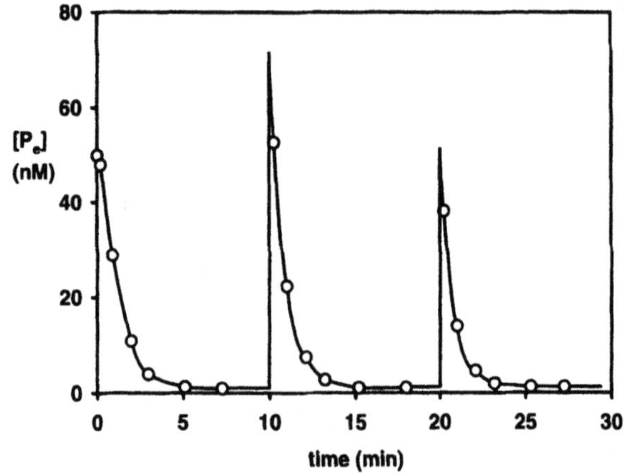

Fig. 4.4. Time course of phosphate removal by *Anacystis nidulans* (*Synechococcus* sp.). The population containing 20 µg Chl a l^{-1} was exposed to subsequent pulses of phosphate (50 nM at $t = 0$; ~ 70 nM at $t = 10$ min; ~ 50 nM at $t = 20$ min); for details see Falkner et al. (1998)

Net incorporation of phosphate (Figs. 4.4, 4.5 and 4.8) was determined as follows: the cyanobacteria had been cultivated in a phosphate-limited mode on radioactively labelled phosphate of the same specific activity as was later used in the experiments. Uptake rates were determined by the decrease of phosphate in the external medium: after addition of phosphate pulses aliquots of the bacterial suspensions were filtered at the indicated time intervals, and the external phosphate concentration was calculated from the radioactivity in the filtrate. Evaluation of the time courses was performed with the Mlab computer programme and data were replotted (Fig. 4.6)

that had not been exposed to a high phosphate supply prior to the uptake experiment, remarkable changes in the phosphate uptake behaviour can be observed. While the control cells incorporate the ambient phosphate rapidly and attain a low threshold value (Fig. 4.5A), comparable to the previous figure, the preincubated cells utilize the external phosphate at a considerably lower rate, and phosphate incorporation usually stops at a higher external concentration (Fig. 4.5B). Apparently, incubation with a high phosphate concentration caused an adaptive response in the organism by which a slower uptake rate and an increased threshold value was established. Addition of a second pulse to the preincubated cells resulted in essentially the same uptake kinetics and threshold value. Thus, the observed alterations in the kinetic and energetic properties of the high-affinity uptake system could not be attributed to an exhausted capacity to store polyphosphates. Furthermore, the identical time courses of phosphate removal after the two pulses reveal that the adapted phosphate uptake system represents a stable physiological state with distinct properties.

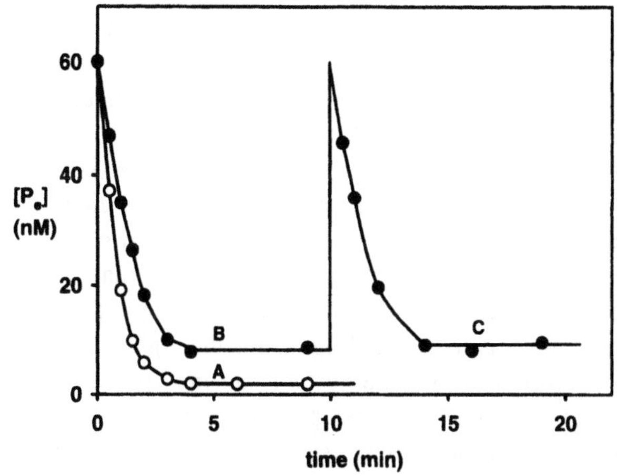

Fig. 4.5. Effect of preincubation on net phosphate uptake in *Anacystis nidulans*. *A* non preincubated population, *B* and *C* preincubated population (preincubation of cells corresponding to 67 μg Chl a l^{-1} with 4 μM phosphate), first and second addition of 60 nM phosphate. The chlorophyll content in all cell suspensions was 20 μg Chl a l^{-1} (cf. Wagner et al. 1995)

If the phosphate uptake behaviour before and after exposure to high phosphate pulses is analyzed in semilogarithmic Thellier plots, characteristic features of the transition from an adaptive to an adapted state are revealed. In the non-adapted state the dependence of the incorporation rate on the logarithm of the external phosphate concentration follows a non-linear, upward-curved function (Fig. 4.6A), and Eq. (9) is valid only in the vicinity of the threshold concentration, in

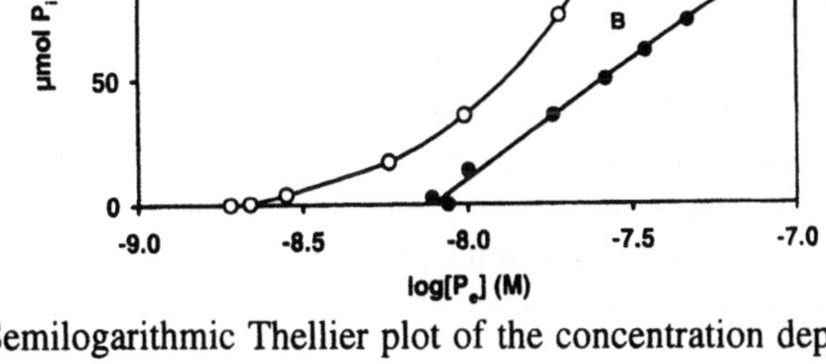

Fig. 4.6. Semilogarithmic Thellier plot of the concentration dependence of the incorporation rates obtained from the time-courses of Fig. 4.5 *A* and *B*

accordance with expectations of non-linear thermodynamics. In the adapted state, however, a linear dependency of the incorporation rate on the driving force of this process is established (Fig. 4.6B), which is valid over a wide concentration range.

The tendency of the phosphate uptake system to attain a linear operation mode was further analyzed on the basis of the theory of linear energy converters (Falkner et al. 1989, 1994), using the formalism described in Section 4.4.1. It was shown that the regulation of the phosphate uptake system is under strict energetic control: when the organisms are cultivated for a prolonged period of time under phosphate-limited conditions, under which the external concentration is maintained close to the threshold value, the uptake system will only operate with optimal efficiency if the degree of coupling attains the highest possible value close to 1. Under this condition the uptake system can incorporate phosphate with a sufficiently high rate only when the transport machinery is highly activated [see Eq. (9)]. If in this situation the cells are suddenly exposed to elevated phosphate concentrations to which the transport system is not adjusted, the uptake system operates in an energetically inefficient and non-linear mode (Stucki 1980; Stucki et al. 1983; cf. Fig. 4.6A). In order to attain the efficient state of minimal entropy production again, a definite conductivity coefficient must be adjusted such that the resulting steady-state uptake rate matches the phosphorus needs of the growing cell when the system operates with optimal efficiency at the prevailing external concentration. This can only be accomplished by lowering the degree of coupling, leading to an increase in the threshold value [cf. Eqs. (7) and (8)], and resulting in a new, adapted state.

Following these considerations, one can say that the actual stress for the preadapted cells is not low ambient phosphate concentrations near the threshold value, but the sudden availability of higher amounts of phosphate to which their uptake system is not adjusted. During adaptation, however, the transport system attains new kinetic and energetic properties which are tuned to the altered growth conditions.

The adapted state is then stable (cf. Fig. 4.5B and C), at least as long as no dramatic changes in the external phosphate supply necessitate a readaptation of the uptake system. It was previously shown that the linear operation mode of the transport system is maintained for several hours, even if no uptake occurred during that time (Wagner et al. 1995). Astonishingly, the kinetic and energetic properties of the adapted phosphate transport system remained invariant even during cell division. Two important features of the uptake system can be derived from its tendency towards linearization: firstly, the new kinetic and energetic properties necessarily reflect the previous phosphate supply to which they have been adjusted, and, secondly, due to the stability of the linearized uptake behaviour, the transport system reveals some kind of memory function, since it is capable of storing information about the previous phosphate status in its properties for a prolonged period of time, even if during that time no further phosphate uptake occurred.

Consequently, analysis of the phosphate uptake behaviour using the linear flow-force relationship may serve as an indicator of prevailing phosphate supply in lakes where other methods are insufficient to detect the phosphate status of organisms: chemical analysis of soluble reactive phosphorus cannot detect the actual availability of phosphate in lakes, since increases in the nutrient are frequently very low, and the available phosphate is utilized with high velocity by the phytoplankton community so that the external concentration returns rapidly to the threshold value.

Field experiments have shown that the linear flow-force relationship could indeed be usefully applied for analyzing the phosphate uptake behaviour of lake phytoplankton (Falkner et al. 1984a,b; Falkner and Falkner 1989; Istvanovics et al. 1994; Istvanovics and Herodek 1995), and the studies indicated that the properties of the phosphate uptake behaviour reflected the prevailing nutrient status in natural communities.

Thellier plots of the concentration dependence of uptake rates have revealed that after adaptation to elevated phosphate concentrations the proportional flow-force relation is valid far from equilibrium. A kinetic explanation of the linearization of the phosphate uptake behaviour was previously derived (Falkner et al. 1995; Wagner et al. 1995) on the basis of the simple pore model (Segel 1975; Stein 1981) that describes the reversible reaction sequence:

$$C + P_e \leftrightarrow CP \leftrightarrow C + P_i,$$

in which C denotes the free carrier, CP the carrier-phosphate complex and P_e and P_i the external and internal phosphate, respectively. In a situation in which translocation of phosphate through the cell membrane is the rate limiting step for the overall reaction sequence, the steady-state equation for incorporation of the nutrient can be expressed by the equation:

$$J_P = \frac{([P_e]-[P_e]_A)V_e}{[P_e]+K_e+\frac{V_e}{V_i}[P_e]_A},$$

with V_e and V_i representing the maximum velocities of forward (transport into the cell) and backward reactions, respectively, and K_e the kinetic constant for the forward reaction. In this kinetic model $[P_e]_A$ corresponds to the true equilibrium concentration of external phosphate (as can be seen, this equation reduces to the Michaelis-Menten equation if $[P_e]_A$ becomes zero).

Figure 4.7 shows a semilogarithmic Thellier plot of the above function for the simplest case of a symmetrical carrier for which V_e=V_i. If K_e exceeds by far the respective equilibrium concentration, the resulting graph reveals an upward curved shape, and linearity is confined to the vicinity of the equilibrium concentration (B in Fig. 4.7). If, however, K_e is of the same order of magnitude as $[P_e]_A$, and the uptake system operates efficiently close to its equilibrium value, a wide linear dependence of the incorporation rates on the logarithm of the external phosphate concentration is achieved (A in Fig. 4.7). It has previously been shown (Wagner et al. 1995; Falkner et al. 1995) that the simultaneous operation of several transport systems with different uptake properties results in appreciable extensions of the width of the linear range, and potentially leads to elevations in the threshold concentration.

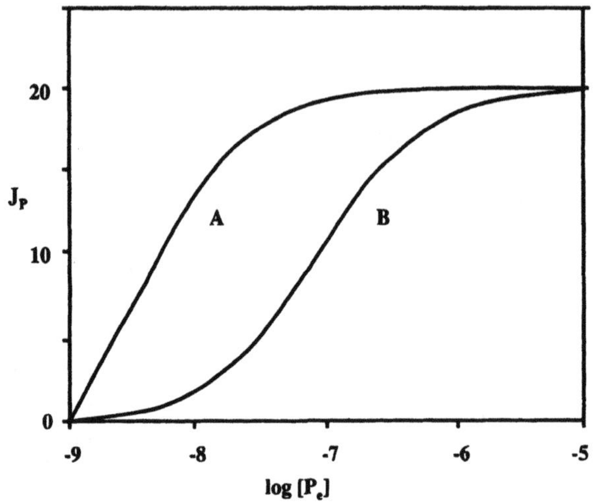

Fig. 4.7. Transition of the phosphate uptake system from a non-linear to a linear operation mode as explained with the simple pore model. Semilogarithmic plot of the concentration dependence of incorporation rates, using Eq. (9) and the following values: *A* threshold $[P_e]_A$ = 1nM, K_e = 2 nM, V_e = 20 (arbitrary units); *B* $[P_e]_A$ = 1nM, K_e = 80 nM, V_e = 20

Accordingly, we may assume that the phosphate uptake system can attain a linear operation mode if the cell is capable of an adaptive reponse in which the K_e of phosphate influx is conformed to the respective threshold value. As could be seen, this kinetic strategy is usually accompanied by a concomitant energetic one in which the threshold value is adjusted to the prevailing phosphate status, and any interplay between these two adaptive mechanisms seems possible, resulting in a highly variable uptake behaviour of the same population.

4.4.3 The Adaptive Response of Cyanobacterial Populations to Relative Changes in Their Environment

In preincubation experiments like the one presented in Fig. 4.5, the cells are exposed to rather high phosphate pulses to cause transition to an adapted state. However, the adaptive event is not a simple response to high, micromolar external phosphate concentrations, but depends in a complex manner on the pulse height and the density of the population.

Due to the phosphate storage capacity, *Anacystis nidulans* can incorporate considerable amounts of this nutrient during subsequent pulses, provided that each pulse is utilized within a few minutes (cf. Fig. 4.4). Under these experimental conditions, the uptake system is not deactivated, and a low threshold value is attained after each pulse. In this type of experiment the cell density determines the time course of phosphate removal, and consequently, how long the algae are exposed to external concentrations exceeding the threshold value. In the experiment of Fig. 4.8 two cell suspensions, originating from the same batch of

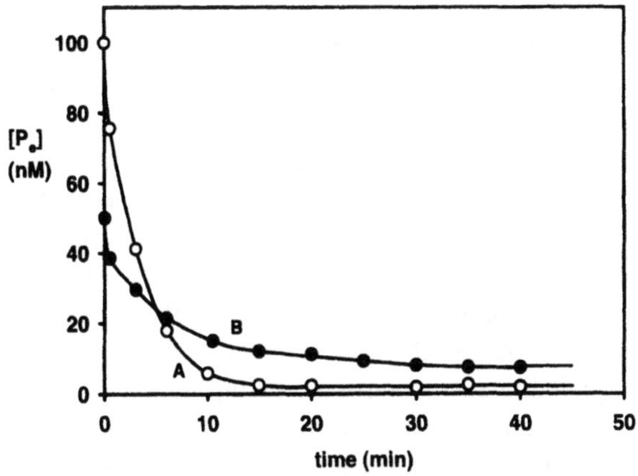

Fig. 4.8. Time course of phosphate incorporation followed by the decrease in the external phosphate concentration in two differently diluted suspensions of *Anacystis nidulans*. The cells originated from the same batch of cyanobacteria, however, suspension *A* contained 8 μg Chl a l^{-1} and was exposed to an initial phosphate concentration of 100 nM, while suspension *B* contained 4 μg Chl a l^{-1} and received 50 nM phosphate (cf. Wagner et al. 1995)

algae, were diluted to different cell densities, and challenged with phosphate pulses in a way that the amount of phosphate supplied was the same per average cell in both the suspensions. It can be seen that the external phosphate concentration decreased rapidly in the less diluted suspension (A in Fig. 4.8), and within a few minutes a low threshold value was established. In the more diluted suspension (B in Fig. 4.8), however, removal of external phosphate proceeded slowly, and the cells were exposed to elevated phosphate concentrations for a prolonged period of time. In this suspension incorporation of phosphate ceased at a significantly higher external phosphate concentration. In numerous experiments we have observed that suspensions with a cell density resembling the natural condition in oligotrophic lakes underwent an adaptive response, provided the decrease in the external phosphate concentration was so slow that the external concentration remained above the (initial) threshold value sufficiently long for an adaptation to occur.

An explanation for the alteration of the threshold value on energetic grounds has been given before. Here, it is interesting to look at some features that cause transition from the adaptive to the adapted state. It is noteworthy that the adaptive response is independent of the amount of stored phosphate and the absolute external concentration, which was rather low in suspension B in Fig. 4.8. The relevant information for the organism to respond is obviously the time course of the decrease in ambient phosphate which, in turn, depends on the cell density. Hence, the cells adapt to changes in the environment which they themselves have caused (cf. discussion about phosphate uptake behaviour as a self-referential system, Falkner et al. 1996).

These examples showed that even during the course of an uptake experiment which potentially takes less than 1 hour, adaptive changes in the phosphate uptake system may occur, and that the organisms respond rather to relative changes in the ambient phosphate supply than to the absolute external concentration (cf. Bray 1995). It has been repeatedly observed in culture suspensions with very low cell density resembling natural conditions that after exposure to a phosphate pulse the incorporation process ceased at unexpectedly high values due to the adaptive response (for example after addition of an initial phosphate concentration of 100 nM, a threshold value of about 50 nM was attained, Wagner et al. 1999). In many experiments with highly diluted suspensions, incorporation of phosphate ceased even before the cells had stored the amount of phosphate that would be necessary for only one cell doubling. It must be stressed, however, that these very high threshold values are transient states, and after a certain period of time the phosphate uptake system is re-arranged and the available phosphate is further utilized. Thus, when the uptake behaviour of a population or a community is studied, it depends on the experimental protocol as to what extent an adaptation may or may not occur and, consequently, to what extent phosphate is stored during the course of the experiment. For this reason the kinetic and energetic properties of the uptake system cannot be expressed by fixed, determinate parameters that are representative for a species in a given growth state, but vary significantly as a function of the prevailing measurement conditions. If this is ignored, erroneous conclusions about the threshold value of an organism may be drawn, and the same cells may be classified as highly or poorly capable of storing phosphate, depending on the experimental protocol.

The adaptive variability of *Anacystis* has an interesting ecophysiological aspect concerning the behaviour of the population as a unity. The observed increase in the threshold value in the adapted state indicates that the whole population joins in the adaptive response: let us assume that during an uptake experiment only a part of the population changed the properties of the uptake system by adopting a new, higher threshold value, and the rest of the population maintained the original, lower value. In this case, the external phosphate would nevertheless return to the initial, lower threshold concentration —though at a slower rate — since the algae that had not adapted would continue to utilize phosphate at concentrations below the higher threshold value of the adapted cells. Hence, the observation that a higher threshold value is established and maintained shows that the transition from an adaptive to an adapted state is accomplished by the whole population in a coherent manner (Wagner et al. 2000; cf. Shapiro 1998). It is noteworthy that the adaptive response of phosphate-deficient cells that are supposed to compete for slightest amounts of phosphate results in a physiological state in which they incorporate, at least transiently, less of the limiting nutrient than they potentially could. However, we have to assume that this behaviour reflects a biologically sensible strategy of the organisms to survive phosphate-deficient growth conditions. Recalling that the adaptive response is characterized by a transition of the uptake system from an energetically unfavourable to an energetically favourable one, we may conclude, that regulation of the phosphate uptake system is not primarily dominated by the necessity to incorporate and store as much phosphate as possible within a short time, but by the tendency to utilize the available phosphate in a most efficient way to assure a coordinated growth process.

Acknowledgement. The authors are grateful to Dr. David Scanlan, University of Warwick, UK, for carefully reading and improving the manuscript.

References

Aiba H, Mizuno T (1994) A novel gene whose expression is regulated by the response-regulator, SphR, in response to phosphate limitation in *Synechococcus* species PCC7942. Mol Microbiol 13:25-34

Aiba H, Nagaya M, Mizuno T (1993) Sensor and regulator proteins from the cyanobacterium *Synechococcus* species PCC7942 that belong to the bacterial signal-transduction protein families: implication in the adaptive response to phosphate limitation. Mol Microbiol 8:81-91

Allen MM (1968) Photosynthetic membrane system in *Anacystis nidulans*. J Bacteriol 96:836-841

Amemura M, Makino K, Shinagawa H, Kobayashi A, Nakata A (1985) Nucleotide sequence of the genes involved in phosphate transport and regulation of the phosphate regulon in *Escherichia coli*. J Mol Biol 184:241-259

Ames GF (1986) Bacterial periplasmic transport systems: structure, mechanism, and evolution. Annu Rev Biochem 55:397-425

Armstrong RA, McGhee R (1980) Competitive exclusion. Am Nat 115:151-170

Aubriot L, Wagner F, Falkner G (2000) The phosphate uptake behaviour of phytoplankton communities in eutrophic lakes reflects alterations in the phosphate supply. Eur J Phycol 35:255-262

Ault-Riché D, Fraley CD, Tzeng CM, Kornberg A (1998) Novel assay reveals multiple pathways regulating stress-induced accumulations of inorganic polyphosphate in *Escherichia coli*. J Bacteriol 180:1841-1847

Bennett RL, Malamy MH (1970) Arsenate resistant mutants of *Escherichia coli* and phosphate transport. Biochem Biophys Res Commun 40:496-503

Bieleski RL, Ferguson IB (1983) Physiology and metabolism of phosphate and its compounds. In: Pirson A, Zimmermann MH (eds) Encyclopedia of plant physiology vol 15A. Laeuchli

A. Bieleski RL (eds) Inorganic plant nutrition. Springer, Berlin Heidelberg New York, pp 422-449

Boos W, Lucht JM (1996) Periplasmic protein-dependent ABC transporters. In: Neidhardt FC (ed) *Escherichia coli* and *Salmonella*. ASM Press, Washington, pp 1175-1209

Borchardt MA, Hoffmann JP, Cook PW (1994) Phosphorus uptake kinetics of *Spirogyra fluviatilis* (Charophyceae) in flowing water. J Phycol 30:403-417

Bornefeld T, Weis U (1981) Adenylate energy charge and phosphorylation potential in blue-green bacterium *Anacystis nidulans*. Biochem Physiol Pflanz 176:71-82

Brass JM, Higgins CF, Foley M, Rugman PA, Birmingham J, Garland PB (1986) Lateral diffusion of proteins in the periplasm of *Escherichia coli*. J Bacteriol 165:787-794

Bray D (1995) Protein molecules as computational elements in living cells. Nature 376:307-312. Errata: (1995) Nature 378:419

Brown EJ, Button DK (1979) Phosphate-limited growth kinetics of *Selenastrum capricornutum* (Chlorophyceae). J Phycol 15:305-311

Brzoska P, Rimmele M, Brzostek K, Boos W (1994) The Ugp paradox: the phenomenon that glycerol-3-phosphate, exclusively transported by the *Escherichia coli* Ugp system, can serve as a sole source of phosphate but not as a sole source of carbon is due to trans inhibition of Ugp-mediated transport by phosphate. In: Torriani-Gorini A, Yagil E, Silver S (eds) Phosphate in microorganisms. American Society for Microbiology, Washington, DC, pp 30-36

Button DK (1978) On the theory of control of microbial growth kinetics by limiting nutrient concentrations Deep Sea Res 25:1163-1177

Caperon J (1968) Population growth response of *Isochrysis galbana* to variable nitrate environment. Ecology 49:866-872

Coleman JE (1992) Structure and mechanism of alkaline phosphatase. Annu Rev Biophys Chem 21:441-483

Currie DJ, Kalff J (1984) The relative importance of bacterioplankton and phytoplankton in phosphorus uptake in freshwater. Limnol Oceanogr 29:311-321

De Cock H, Overeem W, Tommassen J (1992) Biogenesis of outer membrane protein PhoE of *Escherichia coli*. J Mol Biol 224:369-379

Droop MR (1968) Vitamin B12 and marine ecology. IV: The kinetics of uptake, growth and inhibition in *Monochrysis lutheri*. J Mar Biol Assoc UK 48:689-733

Droop MR (1973) Some thoughts on nutrient limitation in algae. J Phycol 9:264-272

Droop MR (1974) The nutrient status of algal cells in continuous culture. J Mar Biol Assoc UK 54:825-855

Ducobu H, Huisman J, Jonker RR, Mur LR (1998) Competition between a prochlorophyte and a cyanobacterium under various phosphorus regimes: comparison with the Droop model. J Phycol 34:467-476

Egge JK (1998) Are diatoms poor competitors at low phosphate concentrations? J Mar Syst 16:191-198

Einsele W (1941) Die Umsetzung von zugefuehrtem, anorganischem Phosphat im eutrophen See und ihre Rückwirkungen auf seinen Gesamthaushalt. Z Fisch Hilfswissensch 39:407-488

Falkner G (1984) Die Phosphataufnahme bei Algen unter besonderer Berücksichtigung der Cyanophyceen. Sitzungsber Oesterr Akad Wissensch 193:33-88

Falkner G, Werdan H, Horner F, Heldt HW (1974) Energieabhaengige Phosphataufnahme der Blaualge *Anacystis nidulans*. Ber Dtsch Bot Ges 87:263-266

Falkner G, Strasser P, Graffius D (1984a) Phosphate uptake by blue-green algae in vitro and in a lake during an algal bloom: useful application of a force-flow relationship. Hydrobiologia 108:265-271

Falkner G, Strasser P, Graffius D (1984b) Phosphate uptake by blue-green algae during an algal bloom. Verh Int Ver Limnol 22:195-199

Falkner G, Falkner R, Schwab AJ (1989) Bioenergetic characterization of transient state phosphate uptake by the cyanobacterium *Anacystis nidulans*. Theoretical and experimental basis for a sensory mechanism adapting to varying environmental phosphate levels. Arch Microbiol 152:353-361

Falkner G, Falkner R, Wagner F (1993) Adaptive phosphate uptake behaviour of the cyanobacterium *Anacystis nidulans*: analysis by a proportional flow-force relation. C R Acad Sci Paris Life Sci 316:784-787

Falkner G, Wagner F, Falkner R (1994) On the relation between phosphate uptake and growth of the cyanobacterium *Anacystis nidulans*. C R Acad Sci Paris Life Sci 317:535-541

Falkner G, Wagner F, Small JV, Falkner R (1995) Influence of fluctuating phosphate supply on the regulation of phosphate uptake by the blue-green alga *Anacystis nidulans*. J Phycol 31:745-753

Falkner G, Wagner F, Falkner R (1996) The bioenergetic coordination of a complex biological system is revealed by its adaptation to changing environmental conditions. Acta Biotheor 44:283-299

Falkner R, Falkner G (1989) Phosphate uptake by eukaryotic algae in cultures and by a mixed phytoplankton population in a lake: analysis by a force-flow relationship. Bot Acta 102:283-286

Falkner R, Wagner F, Aiba H, Falkner G (1998) Phosphate-uptake behaviour of a mutant of *Synechococcus* sp. PCC7942 lacking one protein of the high-affinity phosphate-uptake system. Planta 206:461-465

Fuhs GW (1969) Phosphorus content and rate of growth in the diatoms *Cyclotella nana* and *Thalassiosira fluviatilis*. J Phycol 5:312-321

Ghilarov AM (1984) The paradox of the plankton reconsidered; or, why do species coexist? Oikos 43:46-52

Goldstein AH (1994) Involvement of the quinoprotein glucose dehydrogenase in the solubilization of exogenous mineral phosphates by Gram-negative bacteria. In: Torriani-Gorini A, Yagil E, Silver S (eds) Phosphate in microorganisms. American Society for Microbiology, Washington, DC, pp 197-203

Grover JP (1989) Phosphorus-dependent growth kinetics of 11 species of freshwater algae. Limnol Oceanogr 34:341-348

Grover JP (1991a) Non-steady state dynamics of algal population growth: experiments with two chlorophytes. J Phycol 27:70-79

Grover JP (1991b) Resource competition in a variable environment: phytoplankton growing according to the variable-internal-stores model. Am Nat 138:811-835

Hameed HA, Kilinc S, McGowan S, Moss B (1999) Physiological tests and bioassays: aids or superfluities to the diagnosis of phytoplankton nutrient limitation? A comparative study in the Broads and the Meres of England. Eur J Phycol 34:253-269

Healey FP (1980) Slope of the Monod equation as an indicator of advantage in nutrient competition. Microb Ecol 5:281-286

Healey FP (1982) Phosphate. In: Carr NG, Whitton BA (eds) The biology of cyanobacteria. Blackwell, Oxford, pp 105-124

Hernandez I (1996) Analysis of the expression of alkaline phosphatase activity as a measure of phosphorus status in the red alga *Porphyra umbilicalis* (L) Kützing. Bot Mar 39:255-262

Higgins CF (1992) ABC transporters: from microorganisms to man. Annu Rev Cell Biol 8:67-113

Huang BQ, Hong HS (1999) Alkaline phosphatase activity and utilization of dissolved organic phosphorus by algae in subtropical coastal waters. Mar Pollut Bull 39:205-211

Hutchinson GE (1961) The paradox of the plankton. Am Nat 95:137-145

Ikeya T, Ohki K, Takahashi M, Fujita Y (1997) Study on phosphate uptake of the marine cyanophyte *Synechococcus* sp. NIBB1071 in relation to oligotrophic environments in the open ocean. Mar Biol 129:195-202

Istvanovics V, Herodek S (1995) Estimation of net uptake and leakage rates of orthophosphate from ^{32}P-uptake kinetics by a linear force-flow model. Limnol Oceanogr 40:17-32

Istvanovics V, Padisak J, Pettersson K, Pierson DC (1994) Growth and phosphorus uptake of summer phytoplankton in Lake Erken (Sweden). J Plankton Res 16:1167-1196

Jenssen M, Olsson H, Pettersson K (1988) Phosphatases: origin, characterization, and function in lakes. Hydrobiologia 170:157-175

Kallas T, Dahlquist FW (1981) Phosphorus-31 nuclear magnetic resonance analysis of internal pH during photosynthesis in the cyanobacterium *Synechococcus*. Biochemistry 20:5900-5907

Kantrowitz ER (1994) Introduction: sructure and function of *Escherichia coli* alkaline phosphatase. In: Torriani-Gorini A, Yagil E, Silver S (eds) Phosphate in microorganisms. American Society for Microbiology, Washington, DC, pp 319-328

Kedem O, Caplan SR (1965) Degree of coupling and its relation to efficiency of energy conversion. Trans Faraday Soc 61:1897-1911

Kilham SS (1978) Nutrient kinetics of freshwater planktonic algae using batch and semicontinuous methods. Mitt Int Ver Limnol 21:147-157

Kornberg A (1994) Inorganic polyphosphate: a molecular fossil comes to life. In: Torriani-Gorini A, Yagil E, Silver S (eds) Phosphate in microorganisms. American Society for Microbiology, Washington, DC, pp 204-208

Kornberg A, Rao NN, Ault-Riche D (1999) Inorganic polyphosphate: a molecule of many functions. Annu Rev Biochem 68:89-125

Korteland J, Tommassen J, Lugtenberg B (1982) PhoE protein pore of the outer membrane of *Escherichia coli* K12 is a particularly efficient channel for organic and inorganic phosphate. Biochim Biophys Acta 690:282-289

Kubena BD, Luecke H, Rosenberg H, Quiocho FA (1986) Crystallization and X-ray diffraction studies of a phosphate-binding protein involved in active transport in *Escherichia coli*. J Biol Chem 261:7995-7996

Kuenzler EJ, Ketchum BH (1962) Rate of phosphorus uptake by *Phaeodactylum tricornutum*. Biol Bull Woods Hole 123: 134-145

Kulaev IS, Vagabov VM (1983) Polyphosphate metabolism in micro-organisms. Adv Microb Physiol 24:83-171

Lawerence BA, Suarez C, Depina A, Click E, Kolodny NH, Allen MM (1998) Two internal pools of soluble polyphosphate in the cyanobacterium *Synechocystis* sp. strain PCC6308 — an in vivo P-31 NMR spectroscopic study. Arch Microbiol 169:195-200

Lawry N, Jensen TH. (1979) Deposition of condensed phosphate as an effect of varying sulfur deficiency in the cyanobacterium *Synechococcus* sp. (*Anacystis nidulans*). Arch Microbiol 120:1-7

Liss E, Langen P (1962) Versuche zur Polyphosphat-Überkompensation in Hefezellen nach Phosphatverarmung. Arch Mikrobiol 41:383-392

Luecke H, Quiocho FA (1990) High specificity of a phosphate transport protein determined by hydrogen bonds. Nature 347:402-406

Makino K, Shinagawa H, Amemura M, Nakata A (1986a) Nucleotide sequence of the phoB gene, the positive regulatory gene for the phosphate regulon of *Escherichia coli* K-12. J Mol Biol 190:37-44

Makino K, Shinagawa H, Amemura M, Nakata A (1986b) Nucleotide sequence of the phoR gene, a regulatory gene for the phosphate regulon of *Escherichia coli*. J Mol Biol 192:549-556

Makino K, Amemura M, Kim S-K, Nakata A, Shinagawa H (1994) Mechanism of transcriptional activation of the phosphate regulon in *Escherichia coli*. In: Torriani-Gorini A, Yagil E, Silver S (eds) Phosphate in microorganisms. American Society for Microbiology, Washington, DC, pp 5-12

Mann NH, Scanlan DJ (1994) The SphX protein of *Synechococcus* species PCC7942 belongs to a family of phosphate-binding proteins. Mol Microbiol 14:595-596

Medveczky N, Rosenberg H (1970) The phosphate-binding protein of *Escherichia coli*. Biochim Biophys Acta 211:158-168
Medveczky N, Rosenberg H (1971) Phosphate transport in *Escherichia coli*. Biochim Biophys Acta 241:404-506
Monod J (1942) Recherches sur la croissance des cultures bacteriennes. Hermann, Paris, 211 pp
Monod J (1950) La technique de la culture continue: theorie et applications. Ann Inst Pasteur Lille 79:390-410
Morel FMM (1987) Kinetics of nutrient uptake and growth in phytoplankton. J Phycol 23:137-150
Mueller H (1970) Das Wachstum von *Nitzschia actinastroides* (Lemm.) v. Goor im Chemostaten bei limitierender Phosphatkonzentration. Ber Dtsch Bot Ges 83:537-544
Mueller H (1972) Growth and phosphate requirements of *Nitzschia actinastroides* (Lemm.) v. Goor in batch and chemostat cultures under phosphate limitation. Arch Hydrobiol Suppl 38:399-484
Nalewajko C, Lean DRS (1980) Phosphorus. In: Morris I (ed) The physiological ecology of phytoplankton. Blackwell, Oxford, pp 325-358
Nicklisch A (1999) Competition between the cyanobacterium *Limnothrix redekei* and some spring species of diatoms under P limitation. Int Rev Hydrobiol 84:233-241
Olsen Y (1989) Evaluation of competitive ability of *Staurastrum luetkemuellerii* (Chlorophyceae) and *Microcytis aeruginosa* (Cyanophyceae) under P limitation. J Phycol 25:486-499
Olsen Y, Vadstein O, Andersen T, Jensen A (1989) Competition between *Staurastrum luetkemuellerii* (Chlorophyceae) and *Microcytis aeruginosa* (Cyanophyceae) under varying modes of phosphate supply. J Phycol 25:499-508
Overbeeke N, Lugtenberg B (1980) Expression of outer membrane protein e of *Escherichia coli* K12 by phosphate limitation. FEBS Lett 112:229-232
Overbeeke N, Bergmans H, Mansfeld FV, Lugtenberg B (1983) Complete nucleotide sequence of phoE, the structural gene for the phosphate limitation inducible outer membrane pore protein of *Escherichia coli* K12. J Mol Biol 163:513-532
Perez-Llorens JL, Niell FX (1995) Short-term phosphate uptake kinetics in *Zostera noltii* Hornem: a comparison between excised leaves and sediment-rooted plants. Hydrobiologia 297:17-27
Pietrobon D, Zoratti M, Azzone GF, Stucki JW, Walz D (1982) Non-equilibrium thermodynamic assessment of redox-driven H^+ pumps in mitochondria. Eur J Biochem 127:483-494
Pietrobon D, Zoratti M, Azzone GF (1983) Molecular slipping in redox and ATPase H^+ pumps. Biochim Biophys Acta 723:317-321
Rao NN, Roberts MF, Torriani A, Yashphe J (1993) Effect of glpT and glpD mutations on expression of the phoA gene in *Escherichia coli*. J Bacteriol 175:74-79
Rao NN, Kar A, Roberts MF, Yashphe J, Torriani-Gorini A (1994) Phosphate, phosphorylated metabolites, and the Pho regulon of *Escherichia coli*. In: Torriani-Gorini A, Yagil E, Silver S (eds) Phosphate in microorganisms. American Society for Microbiology, Washington DC, pp 22-29
Ray JM, Bhaya D, Block MA, Grossman AR (1991) Isolation, transcription, and inactivation of the gene for an atypical alkaline phosphatase of *Synechococcus* sp. strain PCC7942. J Bacteriol 173:4297-4309
Reynolds CS (1993) The ecology of freshwater phytoplankton. Cambridge University Press, Cambridge, 384 pp
Reynolds CS (1998) The state of freshwater ecology. Freshwater Biol 39:741-753
Rhee GY (1973) A continuous culture study of phosphate uptake, growth rate and polyphosphate in *Scenedesmus* sp.. J Phycol 9:495-506
Rhee GY (1974) Phosphate uptake under nitrate limitation by *Scenedesmus* sp. and its ecological implications. J Phycol 10:470-475

Rhee GY (1982) Effects of environmental factors and their interactions on phytoplankton growth. In: Marshall KC (ed) Advances in microbial ecology, vol 6. Plenum Press, New York, pp 33-74

Rigler FH (1956) A tracer study of phosphorus cycle in lake water. Ecology 37:550-562

Rigler FH (1968) Further observations inconsistent with the hypothesis that the molybdenium blue method measures orthophosphate in lake water. Limnol Oceanogr 13:7-13

Ritchie RJ, Trautman DA, Larkum AWD (1997) Phosphate uptake in the cyanobacterium *Synechococcus* R-2 PCC7942. Plant Cell Physiol 38:1232-1241

Rosenberg H (1987) Phosphate transport in prokaryotes. In: Rosen BP, Silver S (eds) Ion transport in prokaryotes. Academic Press, New York, pp 205-248

Rosenberg H, Gerdes RG, Chegwidden K (1977) Two systems for the uptake of phosphate in *Escherichia coli*. J Bacteriol 131:505-511

Rosenberg H, Gerdes RG, Harold FM (1979) Energy coupling to the transport of inorganic phosphate in *Escherichia coli* K-12. Biochem J 178:133-137

Ruiz RG, Hernandez I, Lucena J, Niell FX (1997) Preliminary studies on the significance of alkaline phosphatase activity in the diatom *Phaeodactylum tricornutum* Bohlin. Sci Mar 61:517-525

Russell LM, Rosenberg H (1979) Linked transport of phosphate, potassium ions and protons in *Escherichia coli*. Biochem J 184:13-21

Russell LM, Rosenberg H (1980) The nature of the link between potassium transport and phosphate transport in *Escherichia coli*. Biochem J 188:715-723

Salonen K, Jones RI, de Haan H, James M (1994) Radiotracer study of phosphorus uptake by plankton and redistribution in the water column of a small humic lake. Limnol Oceanogr 39:69-83

Scanlan DJ, Wilson WH (1999) Application of molecular techniques to addressing the role of P as a key effector in marine ecosystems. Hydrobiologia 401:149-175

Scanlan DJ, Mann NH, Carr NG (1993) The response of the picoplanktonic marine cyanobacterium *Synechococcus* species WH7803 to phosphate starvation involves a protein homologous to the periplasmic phosphate-binding protein of *Escherichia coli*. Mol Microbiol 10:181-191

Scanlan DJ, Bourne JA, Mann NH (1997) A putative transcriptional activator of the Crp/Fnr family from the marine cyanobacterium *Synechococcus* sp. WH7803. J Appl Phycol 8:565-567

Schindler DW (1977) Evolution of phosphorus limitation in lakes. Science 195:260-262

Schlegel HG (1985) Allgemeine Mikrobiologie. Thieme, Stuttgart, 571 pp

Schwoerbel J (1984) Einfuehrung in die Limnologie. Fischer, Stuttgart, 233 pp

Segel I (1975) Enzyme kinetics. Behavior and analysis of rapid equilibrium and steady-state enzyme systems. Wiley, New York, 957 pp

Shapiro JA (1998) Thinking about bacterial populations as multicellular organisms. Annu Rev Microbiol 52:81-104

Silver S, Walderhaug M (1992) Gene regulation of plasmid- and chromosome-determined inorganic ion transport in bacteria. Microbiol Rev 56:195-228

Simonis W, Urbach W (1973) Photophosphorylation in vivo. Annu Rev Plant Physiol 24:89-114

Simonis W, Bornefeld T, Lee-Kaden J, Majumdar K (1974) Phosphate uptake and photophosphorylation in the blue-green alga *Anacystis nidulans*. In: Zimmermann U, Dainty J (eds) International workshop on membrane transport in plants and plant organelles. Springer, Berlin Heidelberg New York, pp 220-225

Sommer U (1983) Nutrient competition between phytoplankton species in multispecies chemostat experiments. Arch Hydrobiol 96:399-416

Sommer U (1985) Comparison between steady state and non-steady state competition: experiments with natural phytoplankton. Limnol Oceanogr 30:335-346

Sommer U (1989) Nutrient status and nutrient competition of phytoplankton in a shallow, hypertrophic lake. Limnol Oceanogr 34:1162-1173

Sommer U (1993) Phytoplankton competition in Plußsee: a field test of the resource-ratio hypothesis. Limnol Oceanogr 38:838-845

Sommer U (1994) Planktologie. Springer, Berlin Heidelberg New York, 274 pp

Spijkerman E, Coesel PFM (1998) Different response mechanisms of two planktonic desmid species (Chlorophyceae) to a single saturating addition of phosphate. J Phycol 34:438-445

Stein WD (1981) Concepts of mediated transport. In: Bonting SL, De Pont JJ (eds) Membrane transport. Elsevier, Amsterdam, pp 123-157

Stucki JW (1980) The optimal efficiency and the economic degrees of coupling of oxidative phosphorylation. Eur J Biochem 109:269-283

Stucki JW, Compiani M, Caplan SR (1983) Efficiency of energy conversion in model biological pumps. Optimization by linear nonequilibrium thermodynamic relations. Biophys Chem 18:101-109

Surin BP, Rosenberg H, Cox GB (1985) Phosphate-specific transport system of *Escherichia coli*: nucleotide sequence and gene-polypeptide relationships. J Bacteriol 161:189-198

Thellier M (1970) An electrokinetic interpretation of the functioning of biological systems and its application to the study of mineral salts absorption. Ann Bot 34:983-1009

Tilman D (1977) Resource competition between planktonic algae: an experimental and theoretical approach. Ecology 58:338-348

Tilman D, Sterner RW (1984) Invasions of equilibria: tests of resource competition using two species of algae. Oecologia 61:197-200

Tilman D, Kilham SS, Kilham P (1982) Phytoplankton community ecology: the role of limiting nutrients. Annu Rev Ecol Syst 13:349-372

Titman D (1976) Ecological competition between algae: experimental confirmation of resource-based competition theory. Science 192:463-465

Torriani A (1990) From cell membrane to nucleotides: the phosphate regulon in *Escherichia coli*. Bioessays 12:371-376

Vadstein O (1998) Evaluation of competitive ability of two heterotrophic planktonic bacteria under phosphorus limitation. Aquat Microb Ecol 14:119-127

Vadstein O, Olsen Y, Reinertsen H, Jensen A (1993) The role of planktonic bacteria in phosphorus cycling in lakes — sink and link. Limnol Oceanogr 38:1539-1544

Van Veen HW (1997) Phosphate transport in prokaryotes: molecules, mediators and mechanisms. Antonie van Leeuwenhoek 72:299-315

Van Veen HW, Abee T, Kortstee GJJ, Konings WN, Zehnder AJB (1994a) Translocation of metal phosphate via the phosphate inorganic transport (Pit) system of *Escherichia coli*. Biochemistry 33:1766-1770

Van Veen HW, Abee T, Kortstee GJJ, Konings WN, Zehnder AJB (1994b) Phosphate inorganic transport (Pit) system in *Escherichia coli* and *Acinetobacter johnsonii*. In: Torriani-Gorını A, Yagil E, Silver S (eds) Phosphate in microorganisms. American Society for Microbiology, Washington, DC, pp 43-49

Wagner F, Falkner G (1992) Concomitant changes in phosphate uptake and photophosphorylation in the blue-green alga *Anacystis nidulans* during adaptation to phosphate deficiency. J Plant Physiol 140:163-167

Wagner F, Gimona M, Ahorn H, Peschek GA, Falkner G (1994) Isolation and functional reconstitution of a phosphate binding protein of the cyanobacterium *Anacystis nidulans* induced during phosphate-limited growth. J Biol Chem 269:5509-5511

Wagner F, Falkner R, Falkner G (1995) Information about previous phosphate fluctuations is stored via an adaptive response of the high-affinity phosphate uptake system of the cyanobacterium *Anacystis nidulans*. Planta 197:147-155

Wagner F, Sahan E, Falkner G (1999) The complex relation between phosphate uptake and photosynthetic CO_2 fixation in the cyanobacterium *Anacystis nidulans*. In: Peschek GA,

Loefffelhardt W, Schmetterer G (eds) The phototrophic prokaryotes. Kluwer Academic/Plenum Publishers, New York, pp 739-744

Wagner F, Sahan E, Falkner G (2000) The establishment of coherent phosphate uptake behaviour by the cyanobacterium *Anacystis nidulans*. Eur J Phycol 35:243-253

Wagner KU, Masepohl B, Pistorius EK (1995) The cyanobacterium *Synechococcus* sp. strain PCC7942 contains a second alkaline phosphatase encoded by phoV. Microbiology 141:3049-3058

Wang Z, Choudhary A, Ledvina PS, Quiocho FA (1994) Fine tuning the specificity of the periplasmic phosphate transport receptor. Site-directed mutagenesis, ligand binding, and crystallographic studies. J Biol Chem 269:25091-25094

Wanner BL (1993) Gene regulation by phosphate in enteric bacteria. J Cell Biochem 51:47-54

Wanner BL (1994) Multiple controls of the *Escherichia coli* Pho regulon by the P_i sensor PhoR, the catabolite regulatory sensor CreC, and acetyl phosphate. In: Torriani-Gorini A, Yagil E, Silver S (eds) Phosphate in microorganisms. American Society for Microbiology, Washington, DC, pp 13-21

Wanner BL (1996) Phosphorus assimilation and control of the phosphate regulon. In: Neidhardt FC (ed) *Escherichia coli* and *Salmonella*. ASM Press, Washington, pp 1357-1381

Watson GMF, Scanlan DJ, Mann NH (1996) Characterization of the genes encoding a phosphate-regulated two-component sensory system in the marine cyanobacterium *Synechococcus* sp. WH7803. FEMS Microbiol Lett 142:105-109

Webb DC, Cox GB (1994) Proposed mechanism for phosphate translocation by the phosphate-specific transport (Pst) system and role of the Pst system in phosphate regulation. In: Torriani-Gorini A, Yagil E, Silver S (eds) Phosphate in microorganisms. American Society for Microbiology, Washington, DC, pp 37-42

Webb DC, Rosenberg H, Cox GB (1992) Mutational analysis of the *Escherichia coli* phosphate-specific transport system, a member of the traffic ATPase (or ABC) family of membrane transporters. J Biol Chem 267:24661-24668

Welch GR, Easterby JS (1994) Metabolic channeling versus free diffusion: transition-time analysis. Trends Biochem Sci 19:193-197

Willsky GR, Malamy MH (1980a) Characterization of two genetically separable inorganic phosphate transport systems in *Escherichia coli*. J Bacteriol 144:356-365

Willsky GR, Malamy MH (1980b) Effect of arsenate on inorganic phosphate transport in *Escherichia coli*. J Bacteriol 144:366-374

Yoshida A (1955) Studies on metaphosphate. II. Heat of hydrolysis of metaphosphate extracted from yeast cells. J Biochem Tokyo 42:163-168

5 Trace Metal Constraints on Carbon, Nitrogen and Phosphorus Acquisition and Assimilation by Phytoplankton

R. Michael L. McKay[1], Michael R. Twiss[2] and Czesia Nalewajko[3]

Contents

[1] Department of Biological Sciences, Bowling Green State University, Bowling Green, OH 43403, USA

[2] Department of Applied Chemical and Biological Sciences, Ryerson Polytechnic University, 350 Victoria Street, Toronto, ON, M5B 2K3, Canada

[3] Life Sciences Division, Scarborough College, University of Toronto, Scarborough, ON, Canada

5.1 Introduction

Reviews of trace metal interactions with phytoplankton by Sunda (1994, 2000) and Bruland et al. (1991) have critically analyzed the importance of the bioactive trace metals: Mn, Fe, Co, Ni, Cu, and Zn in controlling biological activity of primary producers in the marine environment, and conversely, of the importance of biological processes in the regulation of trace metal chemistry. The availability of bioactive trace metals, so called because of their involvement in photosynthetic electron transport (Raven et al. 1999) and other metabolic processes, can affect the acquisition and assimilation of macronutrients such as C, N and P. Our purpose here is not to provide another discussion of the information presented by these authors, but rather, following the theme of this compendium, we will highlight the aspects of phytoplankton living in stressed environments. These environments are stressed from the point of view of essential nutrient availability including bioactive trace metals. We seek to highlight the instances where potential to achieve maximum productivity based on macronutrient availability is limited by the availability of essential trace metals. Of course, our evaluation of "potential" is frequently based on the provision of adequate nutrient levels or optimal physical growth conditions, as often occur during enrichment bottle assays or during the culture of phytoplankton in the laboratory. Therefore, it is quite possible that in the natural environment, phytoplankton frequently exist under suboptimal conditions, and that in any given natural environment, there exists a range of nutrient limitations based on each organism's current nutrient requirements.

Our thesis is to show the linkages between micronutrient availability and concomitant macronutrient acquisition and assimilation. In this chapter, where possible, we place emphasis on freshwater environments, in particular large lakes such as the Laurentian Great Lakes of North America. Here, we focus primarily on three trace metals (Zn, Fe and Ni) and discuss their role in macronutrient acquisition and assimilation.

5.1.1 Historical Considerations

Five trace metals have been recognized for several decades as essential for growth of all algae and cyanobacteria: iron, manganese, copper, zinc and molybdenum (Eyster 1964; Healey 1973; O'Kelley 1974). Other essential trace nutrients apparently required by only some algae and cyanobacteria include cobalt and boron (Wiessner 1962; Nicholas 1963; Healey 1973) nickel (Antia et al. 1991; Morel et al. 1991) and vanadium (Meisch and Bieling 1975). Most of the early work to determine the roles of trace metals was carried out with laboratory culture material, and focused on concentrations required for normal physiology and morphology, and on symptoms of deficiency (Healey 1973; O'Kelley 1974). In defined growth media for freshwater microorganisms, trace metals are typically added at micromolar levels (Nichols 1973), yet the concentrations added vary considerably among media and among trace metals, and some media include no additions other than iron. Marine media (McLachlan 1973) typically also contain micromolar concentrations of trace metals which in some media include metals whose functions have not yet been established, e.g. selenium, strontium, aluminum, rubidium and lithium. To ensure that trace metals in the culture media remain in solution and are available for the microrganisms, chelating agents such as EDTA are typically supplied. Note that many phycological growth media are prepared with dilutions of reagent-grade salts. Trace metals from these non-purified salts typically contribute a nominal concentration of essential trace nutrients to the growth media. Any attempt to establish trace metal nutrition of phytoplankton in defined chemical growth media requires the use of media treated with an ion-exchange resin to remove trace metal impurities (Morel et al. 1975; Price et al. 1989) and a subsequent thermodynamic calculation or analytical determination of trace metal chemical speciation (see below).

The search for trace nutrient limitation of phytoplankton in natural freshwaters followed the laboratory studies with algal cultures and it soon became apparent that trace metals, although required in extremely low concentrations, may nevertheless constrain productivity in lakes. This was demonstrated through several whole-system Fe fertilization efforts in small marl lakes (Schelske 1962; Schelske et al. 1962; Wetzel 1966) as well as through enrichment bioassays in some large oligotrophic lakes (Goldman 1972; Schelske et al. 1972, 1978; Stoermer et al. 1978; Lin and Schelske 1981). Goldman (1972) discussed four methods used to assess trace metal availability to biota: concentration of the metal in the water, cellular metal quota, trace nutrient enrichment followed by measurement of growth of a bioassay organism and, lastly, nutrient enrichment followed by measuring phytoplankton productivity using the ^{14}C bottle method. Using the latter approach, Goldman (1965) demonstrated periodic molybdenum deficiency in Castle Lake and several other lakes in California, as well as in some New Zealand lakes. Other trace metals which Goldman found to stimulate primary productivity in some lakes included Co, Mn, Zn and Fe. Goldman correctly pointed out that the ^{14}C method, which measures the community response, may obscure differences in individual requirements for trace metals of phytoplankton species present in the community. For example, there was a marked increase in numbers of *Dinobryon sertularia*, a smaller increase in *Synedra radians* and a reduction in numbers of *Cyclotella meneghiniana* following Mo addition to natural communities from Castle Lake, California. Because different trace metals are required by different enzymes and participate in different metabolic pathways, and because species differ in their requirements, multiple trace element deficiencies are possible in a community. The likely differences among species in levels of micronutrient requirement limit the usefulness of single-species bioassays (Gerloff and Skoog 1957) for the evaluation of trace metal limitations in natural waters.

The early work on micronutrient limitation of natural waters focused mainly on short-term responses (on the order of hours) of photosynthesis and, less frequently, growth to nutrient enrichment. The short exposure times were thought advisable in order to minimize the chemical and physical changes that occur when water samples are confined in containers, the so-called bottle effects. In spite of these limitations, the bottle or container assays have continued to be employed, possibly because the better alternatives such as whole-ecosystem manipulations are not often possible or desirable.

5.2 Consideration of Trace Metal Chemical Speciation

Trace metals in natural aquatic systems are found in either particulate or dissolved forms. Dissolved metal (operationally defined as metal that passes through a 0.2-0.45-μm pore-size membrane filter) is present as free metal ions (for cationic forms of trace metals) and complexed metal. Trace metals are complexed by a range of inorganic and organic ligands of varying affinity, both in solution and on particle surfaces (including internalization by organisms).

The free-ion model of trace metal interactions with aquatic organisms (see Campbell 1995) states that biological effects, both toxic and nutritive, of a cationic trace metal towards an aquatic organism are proportional to the concentration of the free metal ion [i.e. the uncomplexed aquo-ion, $M(H_2O)_x^{Z+}$], not the total dissolved concentration of the particular trace metal. With few exceptions, the free-ion model has a wide applicability for all nutritive trace metals, the notable exception being iron, which can have diffusion limitation due to the greater rate of iron uptake by some phytoplankton cells compared to the rate of iron dissociation from the ligands that control [Fe^{3+}] (Hudson and Morel 1990).

Understandably, metals associated with biota or biologically produced ligands are subject to the biogeochemical cycles characteristic of that element. Plankton have been shown to be important agents that extensively influence the geochemical fate of trace metals in both marine (Whitfield and Turner 1987; Bruland et al. 1991) and freshwater (Sigg 1994; Twiss and Campbell 1998b) systems. A review of trace metals in the water column of Lake Erie (Twiss 2001) is relevant to the current discussion since it details the generally low levels of trace metals in the surface waters of this large lake and the current level of knowledge regarding the geochemical fate of these metals, which is limited in part by our understanding of phytoplankton — trace nutrient interactions in the natural environment.

5.3 Trace Metal Effects on Carbon Acquisition

The inorganic speciation of carbon in freshwater from pH 7-9 is dominated by the bicarbonate anion. As such, many phytoplankton in natural waters rely on the enzyme carbonic anhydrase (EC 4.2.1.1) to convert $HCO_3^- \rightarrow CO_2$, the substrate for the primary carboxylating enzyme Rubisco. The demands of photosynthesis readily consume CO_2 converted from the bicarbonate anion, thereby reducing the amount of CO_2 that is lost by diffusion from the cell. Additional mechanisms exist to further reduce CO_2 diffusion from the cell, including the colocalization of Rubisco (McKay et al. 1993) and CA (So and Espie 1998; Kaplan and Reinhold 1999) into carboxysomes of cyanobacteria as well as scavenging mechanisms employed to reacquire leaked CO_2 (Miller et al. 1990). In a recent commentary, Tortell (2000) provides an insightful phylogenetic analysis of the capacity of phytoplankton to accumulate inorganic carbon. It is hypothesized that phytoplankton that evolved in the geological past during periods of high atmospheric CO_2 have since evolved efficient carbon-concentrating mechanisms in order to respond to the lower present-day inorganic carbon availability and a low-affinity Rubisco enzyme characteristic of these photoautotrophs.

Carbonic anhydrase is a zinc-requiring metalloenzyme notable for its extremely rapid rate of enzymatic turnover (Silverman 1991; Badger and Price 1994; Sültemeyer 1997). It has been demonstrated that elemental Zn analogues such as Cd and Co can substitute for Zn within this enzyme (Silverman 1991). In the case of Cd, this trace metal, that was long considered to have only a toxicological function, is now accepted as a trace nutrient (see Lee et al. 1995). The first evidence of the nutritive capacity of Cd came from oceanic depth profiles plotting total dissolved Cd, which revealed a nutrient-type profile for this element with a characteristic surface water depletion (Boyle et al. 1976). Typically, those elements that are associated with biotic activity are readily removed from the dissolved phase in surface waters and enter into the particulate phase, viz. living cells. However, the ecological fate of most plankton in surface waters is loss by grazing or some lytic mechanism such as viral attack (Fuhrman 1999; Wilhelm and Suttle 1999). Thus, these elements eventually weave their way through the food web and are ultimately regenerated back into surface waters through a variety of mechanisms (Lee and Fisher 1993; Barbeau et al. 1996; Twiss et al. 1996; Wang et al. 1996).

The concentration depth profile of Cd matches closely that of P. It was this constancy of the total dissolved Cd:P ratios in seawater that led to the hypothesis that cadmium might serve as a trace nutrient (Price and Morel 1990). These workers showed that cadmium at a free ion concentration of $10^{-10.9}$ M, supports 90% of the growth rate of *Thalassirosira weissflogii* relative to the growth of this marine diatom at a free Zn^{+} concentration of $10^{-10.9}$ M. Likewise, cobalt at a similar free ion concentration, $[Co^{2+}] = 10^{-10.9}$ M, supports 60% of the growth rate afforded at the comparable level of free zinc ion. Since the $[Zn^{2+}]$ is approximately $10^{-11.5}$ M in oceanic surface waters, it is concluded that both Cd and Co can effectively substitute for Zn requirements in some phytoplankton. Moreover, it appears that metal substitution is predominantly targeted towards carbonic anhydrase (Morel et al. 1994; Yee and Morel 1996; Cullen et al. 1999). Morel et al. (1994) demonstrated that carbonic anhydrase activity was restored in low Zn cultures of *T. weissflogii* by addition of either Cd or Co, whereas Yee and Morel (1996) showed actual incorporation of radiolabelled ^{57}Co into the dominant CA isoform of this diatom. More recent data indicate that *T. weissflogii* produces both Cd- and Zn-specific isoforms of CA that are accumulated in response to the concentrations of free available Cd^{2+} and Zn^{2+} as well as p_{CO2} (Cullen et al. 1999). Sunda and Huntsman (1995a) demonstrated that Co can substitute for Zn in eukaryotic marine phytoplankton whereas the cyanobacterium *Synechococcus* has no Zn requirement but, rather, an obligate, non-cyanocobalamin, Co requirement. Again, the chemical condition of the geological past is invoked to explain the phylogenetic differences. Since the cyanobacteria evolved during the period of an anoxic ocean with high sulfide levels, it follows that Zn^{2+} availability to organisms would be exceedingly low due to ZnS precipitation. However, Co(II), which is more soluble in the presence of sulfide, would be more available and, hence, it is postulated that biochemical pathways evolved utilizing the available trace metal cofactor, Co. Cobalt activates carbonic anhydrase, yet in the modern oxygenated ocean, dissolved Co(II) concentrations are very low. As a result, Cd may be expected to substitute for Zn (or Co) in situ.

If this is the case, then we expect to observe metal substitution in the surface waters of the lower Great Lakes (Lake Erie, Lake Ontario) during the summer months, mainly due to low Co availability (~150 pM; Saito and M.R. Twiss, unpubl. data). Dissolved CO_2 in surface waters during the summer is low due to calcite precipitation, elevated pH (~8.5), and elevated temperature (~20 °C). Moreover, cyanobacterial biomass, as assessed by size-fractionated Chl a, is appreciable, accounting for on an average 30% of the total Chl a (~ 2 $\mu g \cdot l^{-1}$; Twiss and Campbell 1998a). In fact, the picoplankton size fraction (0.2–2 µm) consistently scavenged more Cd than Zn (Twiss and Campbell 1998a), as might be expected if cyanobacteria required Cd (to substitute for Co) but not Zn. In addition, Cd regenerated into the dissolved phase by protozoan grazing activity on cyanobacteria was more effectively recycled through the plankton than was Zn (Twiss et al. 1996). This observation may be due to the chemical form of the regenerated Zn compared to that of the regenerated Cd or, alternatively, it might reflect the biological demand for each of these elements in the surface water. Further research is warranted to investigate the interactions between Cd, Co and Zn in relation to carbon acquisition in freshwater environments.

5.4 Synergistic Interactions Between Trace Metals and N

5.4.1 Trace Metals and Diazotrophic Activity

Geochemical considerations predict that aquatic environments should be phosphorus-limited (Redfield 1958). Unlike nitrogen, there is no appreciable atmospheric source of P, and that delivered to aquatic systems originates mainly from weathering of rock or through fluvial input (Froelich 1988). This is certainly applicable to freshwater systems where P limitation is a central tenet of modern limnology (Schindler 1977; Hecky and Kilham 1988). Yet, despite the rationale behind a geochemical argument for P limitation, with few exceptions marine systems have been shown to be N-limited (Ryther and Dunstan 1971; Carpenter and Capone 1983).

Arguably, a major factor contributing to P limitation of freshwater systems is the biological process of N_2 fixation (Postgate 1998). Phytoplankton community response to low fixed N (low N:P ratios; Flett et al. 1980; Howarth et al. 1988a,b) is characterized by the appearance of diazotrophic, heterocystous cyanobacteria capable of reducing N_2 into the usable, combined form of ammonium. Diazotrophic activity is also reported in some oceanic environments. Notably, the colonial, non-heterocystous cyanophyte *Trichodesmium* spp. fulfills this role in marine systems; however, its distribution is restricted to tropical and subtropical latitudes (Capone et al. 1997; Zehr et al. 2000). Moreover, within these latitudinal boundaries, its distribution is patchy. Apart from *Trichodesmium*, marine diazotrophic activity is rare, with notable exceptions including the brackish waters of the Baltic Sea (e.g. Kononen et al. 1996; Bianchi et al. 2000) and the Harvey-Peel Inlet of southwestern Australia (Huber 1986).

The inability of diazotrophic cyanobacteria to more fully exploit the marine niche has long been a mystery to oceanographers. Whereas sources of combined N (nitrate and ammonium) are frequently below detection limits in oceanic surface waters, dissolved N_2 is abundant (800 μg atoms N l^{-1}; Sharp 1983). Physical factors such as light intensity and turbulence (Carpenter and Price 1976) have been postulated as factors contributing to the paucity of N_2-fixing cyanobacteria in marine environments. The latter factor can be related to the negative effect of turbulence on maintaining anoxic microzones that would favor diazotrophic activity of a non-heterocystous cyanophyte (Carpenter and Price 1976; Paerl et al. 1987; Paulsen et al. 1991). Low diazotrophic activity may also relate to the inherent metal content of the cellular proteins active in N_2 fixation. The reduction of $N_2 \rightarrow NH_3$ is catalyzed by nitrogenase (EC 1.18.6.1), an enzyme complex comprised of an Fe-containing dinitrogenase reductase subunit encoded by *nifH* and a Mo- and Fe-containing dinitrogenase subunit encoded by *nifDK* (Postgate 1998). Thus, it is feasible that low metal ion availability is linked to N limitation in aquatic environments.

To demonstrate an effect of low metal availability on N_2 fixation, initial efforts were directed toward the potential for Mo limitation of nitrogenase activity

(Howarth and Cole 1985). Despite the fact that levels of Mo in seawater are considered high (~ 115 nM; Howarth et al. 1988b), especially for a trace constituent, acquisition of Mo by phytoplankton is subject to competitive inhibition by sulphate (Howarth and Cole 1985; Cole et al. 1993). The basis for this rests in the fact that molybdate, the thermodynamically stable form of Mo in oxic seawater, is a structural analogue to sulphate, which is present in seawater at levels several orders of magnitude greater than molybdate (Howarth et al. 1988b). The stereochemical similarity between the two anions is postulated to result in competitive inhibition of molybdate acquisition so as to impair nitrogenase activity. Although there has been moderate support for this hypothesis, Mo limitation as a single factor constraint is frequently rejected by available bioassay data (Paerl et al. 1987, 1994; Paulsen et al. 1991; Evans and Prepas 1997; Stal et al. 1999). Moreover, it appears that nitrogenase does not possess an absolute requirement for Mo. Forms of nitrogenase having no discernible requirement for Mo have been characterized (Bishop et al. 1986; Robson et al. 1986), including several from coastal salt marsh environments (Loveless et al. 1999). A recent modelling effort, however, indicates that low Mo availability may play a role in a multifactor control process over diazotrophic activity that includes both top-down and bottom-up regulatory components (Howarth et al. 1999).

Low Fe availability is also postulated to constrain diazotrophic capacity in marine waters (Rueter 1982; Howarth et al. 1988b; Michaels et al. 1996; Falkowski 1997; Falkowski et al. 1998). Nitrogenase possesses a high Fe requirement with ~30 atoms of Fe per 2 atoms of Mo in the dinitrogenase component alone (Shah et al. 1984). Reflecting this, it is calculated that the Fe requirement for diazotrophic growth of a photolithotrophic organism is up to 101 times greater than that for growth on ammonium (Raven 1988). The high Fe requirement is a product of the subunit requirements for Fe as well as the slow reaction rate characteristic of nitrogenase, a feature which necessitates accumulation of a large pool of enzyme (Raven 1988; Postgate 1998). Also contributing to the large Fe burden is the extra energy and reductant requirement associated with N_2 fixation (Raven 1988). Unlike many lacustrine systems, which presumably contain adequate amounts of available Fe to support N_2 fixation, it is now apparent that bioavailable Fe is generally low in marine environments. Arguably, among the most notable developments in the field of ocean sciences during the past decade has been the demonstration that low availability of Fe constrains phytoplankton growth in the vast high-nutrient, low-chlorophyll (HNLC) regions of the world oceans. Support for the iron hypothesis has come from on-deck bottle assays (de Baar 1994; Geider and La Roche 1994; Hutchins 1995) as well as from several mesoscale open-ocean fertilization efforts (Martin et al. 1994; Coale et al. 1996; Boyd et al. 2000). Moreover, it is becoming increasingly apparent that consideration of Fe deficiency is not limited to the open ocean. Fe limitation has been described in terms of a mosaic throughout the Californian coastal upwelling region (Hutchins and Bruland 1998; Hutchins et al. 1998). Coastal Fe limitation has also been demonstrated for regions of the northern Japan Sea (Suzuki et al. 1995) and in a coral reef environment (Entsch et al. 1983). Even in the Baltic Sea, where cyanobacterial blooms recur annually (e.g. Kononen et al. 1996; Bianchi et al. 2000), diazotrophic activity appears to be constrained by low availability of Fe. Amendment with Fe but not Mo, resulted in stimulation of growth and nitrogenase activity of *Nodularia* sp., a heterocystous cyanobacterium endemic to this region (Stal et al. 1999).

What, then, promotes the diazotrophic activity and ecological success of *Trichodesmium*? After all, the Fe quota for *Trichodesmium* is extraordinarily high with N:Fe ratios of field samples ranging between 200 and 700 mol: µmol (Rueter et al. 1992). The high Fe quota is partly attributed to requirements of nitrogenase, although it is presumably also related to the unusually high photosystem I:photosystem II ratio (~ 24:1) reported for *Trichodesmium* (Subramaniam et al. 1999). Photosystem I typically possesses a severalfold higher Fe requirement compared to photosystem II (Raven et al. 1999). Moreover, unlike other cyanophytes (Murphy et al. 1976; Wilhelm 1995), siderophores apparently play no role in Fe acquisition and ecological competitiveness of *Trichodesmium* (Rueter et al. 1992). Rather, the ability of *Trichodesmium* to compete is related primarily to two factors. First, *Trichodesmium* is routinely detected only in oceanic regions receiving an elevated flux of aeolian dust. For much of the open ocean, dust derived from continental weathering represents the most significant allochthonous source of Fe to surface waters (Duce and Tindale 1991). Second, *Trichodesmium* may be uniquely adapted to directly exploit aeolian dust as an Fe source (Rueter et al. 1992). In addition to its colonial nature, *Trichodesmium* is buoyant. Together, these characteristics facilitate trapping of dust by colonies whereby discrete particulates are enveloped by individual trichomes and the associated nutrients, including Fe, are presumably solubilized and accessed (Rueter et al. 1992). This is not to suggest that the diazotrophic capacity of *Trichodesmium* is unimpaired by endemic levels of Fe encountered in these regions. Indeed, addition of Fe to natural samples of *Trichodesmium* spp. collected near Barbados (Rueter 1988) and off the coast of North Carolina (Paerl et al. 1994) resulted in significant increases of nitrogenase activity as measured using acetylene reduction assay. No stimulation of nitrogenase activity was reported following addition of Mo or P (Paerl et al. 1994). Likewise, nitrogenase activity of *Trichodesmium* is stimulated by Fe addition to laboratory cultures (Rueter et al. 1990; Paerl et al. 1994). Nitrogenase and rates of photosynthetic O_2 evolution were stimulated by Fe amendment to cultures of *Trichodesmium* NIBB 1067 (Rueter et al. 1990). Similarly, both nitrogenase activity and yield of cultured *T. thiebautii* maintained in oligotrophic Gulf Stream water were significantly increased following Fe amendment (Paerl et al. 1994).

Perhaps because diazotrophic activity appears to be commonplace in freshwater systems, particularly in response to downward shifts in the N:P ratio below Redfield stoichiometry (Flett et al. 1980; Howarth et al. 1988a,b), little effort has been focused on elucidating potential constraints in these systems imposed as a result of low trace-element availability. Circumstantial evidence for such controls comes from studies demonstrating that diazotrophic activity does not always compensate for low N:P loading in lakes (Flett et al. 1980; Ashton 1981). That is, the activity of nitrogen-fixing cyanophytes in these systems was unable to raise the N:P ratio to Redfield levels. Low availability of trace metals in these lakes could provide an explanation. More direct evidence supporting trace-element limitation of N_2 fixation in lakes is derived from studies that demonstrate that additions of Mo stimulate nitrogenase activity. Bioassays conducted on water collected from Castle Lake, California, showed that amendment with Mo resulted in enhanced rates of nitrogen fixation (Axler et al. 1980). Compared to marine systems, lakes typically possess much lower levels of Mo (Howarth et al. 1988b), with Castle Lake as no exception. Fe deficiency has similarly been invoked to explain low diazotrophic activity in a lacustrine system (Wurtsbaugh and Horne 1983). Fe amendment assessed using bioassays resulted in substantially increased rates of N_2 fixation up to 500% over controls. Freshwater systems are traditionally not believed to be metal-deficient, mainly due to their proximity to terrestrial influence. However, the implementation of metal clean analytical techniques in recent years have demonstrated that levels of dissolved metals in freshwater environments, particularly the Great Lakes, are low and at times approach levels observed in marine systems (Coale and Flegal 1989; Nriagu et al. 1996; Twiss et al. 2000). As a result, future studies of diazotrophic activity in lakes will have to note the potential constraints imposed by low metal availability.

5.4.2 Trace Metal Constraints on Assimilation of Nitrate

As with N_2 fixation, the assimilation of nitrate and nitrite are both Fe-requiring processes. Not only do the enzymes that catalyze these assimilatory processes possess a structural requirement for Fe, the assimilation of NO_3^- and NO_2^- also requires reductant which is derived from Fe-dependent photosynthetic reactions.

Nitrate reductase (NR; EC 1.6.6.1-3), the first enzyme in the process, functions in reducing $NO_3^- \rightarrow NO_2^-$ in a reaction requiring 2 mol electrons mol N^{-1}. Of the three forms of NR associated with eukaryotes, two exist in algae and higher plants and differ only in their requirements for reductant with one form specific for NADH and the other form using either NADH or NADPH. Photosynthetic prokaryotes possess still different forms of NR. For a more thorough consideration of NR, readers are referred to reviews on this topic by Solomonson and Barber (1990) and, more recently, by Campbell (1999). In addition, a consideration of non-chlorophyte, algal NR is provided by Berges (1997).

Although NR is structurally divergent across taxonomic lines, invariant between both prokaryotic and eukaryotic forms is a requirement for both Fe and Mo. In eukaryotic algae, NR is generally viewed as a homodimer, or occasionally a homotetramer (Howard and Solomonson 1982), possessing subunit polypeptides of Mr ~100 kDa and with several functional domains including a cytochrome b domain to which Fe is bound. Among algae, the enzyme has been most extensively characterized from chlorophyte sources (Solomonson and Barber 1990), although NR from several non-chlorophyte algae has also been described (Gao et al. 1993; Ramalho et al. 1995; Vergara et al. 1998). Among cyanobacteria, NR is smaller with Mr ~75 kDa and with Fe present in the form of an Fe-S cluster. Eukaryotic and prokaryotic forms of NR also differ in their source of reductant. Whereas electrons are donated by NAD(P)H in the eukaryotic reaction, reduced ferredoxin, an Fe-S containing electron transfer catalyst, functions in this capacity among prokaryotes.

Nitrite reductase (NiR; EC 1.7.7.1) completes the reaction, further reducing $NO_2^- \rightarrow NH_4^+$ in a process requiring 6 mol electrons mol N^{-1}. In contrast to NR, NiRs from higher plant and cyanobacterial sources are structurally similar, sharing ~50% identity at the amino acid level (Luque et al. 1993). Consistent with this structural relatedness, NiR from both sources contains Fe in Fe-S clusters and siroheme centres. Additionally, NiR from higher plant, algal and cyanobacterial sources requires reduced ferredoxin in the role of electron donor.

As evident from the preceding description, assimilation of NO_3^- and NO_2^- is expected to exert an appreciable Fe burden on cells. Reflecting this, theoretical calculations based on Fe use efficiencies and cellular metabolic demands predict that phytoplankton growing on NO_3^- require 1.6x additional Fe compared to cells provided with NH_4^+ (Raven 1988). Compared to the estimated Fe demands of diazotrophic activity, this is modest. However, under Fe-deplete conditions, the additional Fe burden associated with NO_3^- assimilation is likely to be a factor in determining phytoplankton community dynamics and ecological outcomes.

Several studies have provided experimental data consistent with the predicted Fe burden associated with growth on NO_3^-. Neritic and oceanic congeners of *Thalassiosira* provided with growth-limiting levels of Fe (pFe 21) with NO_3^- as their sole source of N, possessed an average Fe content (inferred from Fe:C ratios) of 1.6 – 1.8 x that of cells provided with NH_4^+ (Maldonado and Price 1996). In contrast, NO_3^--grown cells provided with non-limiting levels of Fe (pFe 18) did not possess a higher Fe content than did their NH_4^+ -grown counterparts. Consistent with this is the observation that Fe-replete diatoms engage in luxury uptake of Fe, accumulating this trace nutrient to levels severalfold higher than required to sustain metabolic activity (Sunda and Huntsman 1995b; Maldonado and Price 1996). Similar results were reported using a marine *Synechococcus* (Kudo and Harrison 1997). Significantly lower growth rates were obtained for high-light (110 μmol $m^{-2}s^{-1}$) NO_3^--grown cells compared to cells provided with NH_4^+ only when Fe was present at a concentration lower than 1 nM. When cells were cultured under low irradiance (26 μmol m^{-2} s^{-1}), the limiting Fe threshold was even higher for NO_3^--grown cells.

Considering the intimate structural association between Fe and both enzymes involved in assimilatory nitrate reduction, it follows that studies have typically assessed the predicted inhibitory effect of Fe deprivation on both NR and NiR activity. Clearly, Fe deficiency results in depressed rates of enzyme activity (Timmermans et al. 1994; Milligan and Harrison 2000). In *Emiliania huxleyi*, NR activity was depressed by 50% in cells cultured with no added Fe (Timmermans et al. 1994). Similarly, rates of NR activity were depressed by 60% in Fe-deficient *T. weissflogii* whereas rates of NiR activity declined by 50-fold in the same cells (Milligan and Harrison 2000). The more pronounced effect of Fe deficiency on NiR activity presumably reflects the higher Fe content of this enzyme. Despite these reductions, the activities of both NR and NiR were deemed sufficient to meet cellular N demands of *T. weissflogii* under the low Fe conditions employed in this study. In fact, Milligan and Harrison (2000) suggest that NiR may serve as a pool of cellular Fe with depressed NiR activity under Fe deficiency, resulting in release of between 11and 22% of total cellular Fe to be used for other purposes. Rather, the factor that appears to limit nitrate assimilation under low Fe is the supply of reductant to NiR. Under Fe deficiency, rates of photosynthetic electron transfer are markedly reduced as a result of lesions located in the electron transfer chain (Falkowski et al. 1992; Greene et al. 1992; Vassiliev et al. 1995). This, in turn, results in a decrease in production of reductant for NiR. Not only is the quantity of reductant decreased under Fe-limited growth, the situation may be exacerbated by the exchange of ferredoxin for flavodoxin under these conditions (McKay et al. 1999). Notably, NiR interacts with ferredoxin with a much higher affinity compared with flavodoxin resulting in lower efficiency electron transfer when the latter catalyst is used (Vigara et al. 1998).

5.4.3 Trace Metal Constraints on the Assimilation of Organic Nitrogen

The above considerations may hold particular relevance for oceanic upwelling and high-nutrient, low-chlorophyll (HNLC) regions of the world oceans where nitrate is abundant (Martin et al. 1994); however, the broad applicability of nitrate assimilation to aquatic systems in general is not so apparent. Even in waters with abundant nitrate, other, more readily assimilated forms of nitrogen such as NH_4^+ and urea are often preferred by the endemic phytoplankton (e.g. Varela and Harrison 1999). Moreover, one should consider that nitrate is frequently depleted in surface waters of the oligotrophic open ocean (Eppley et al. 1990). In these waters, nitrogen is derived mainly from autochthonous sources and exists as NH_4^+, urea, amino acids and various other forms of dissolved organic nitrogen (Antia et al. 1991; McCarthy et al. 1998). Reflecting this, much of the primary production in marine systems is characterized as regenerated as opposed to new production, the latter of which is supported predominantly by allochthonously derived NO_3^- (Dugdale and Goering 1967) and N_2 (Capone et al. 1997; Karl et al. 1997). In fact, several oceanic species are unable to assimilate nitrate, a characteristic most likely reflecting the paucity of nitrate in some waters. For example, a search of the newly sequenced genome of *Prochlorococcus marinus* MED4 using gene sequences encoding components of nitrate assimilation and uptake in marine cyanobacteria scored negative results (A. Post, pers. comm.).

As demonstrated in our consideration of "new" production, trace metals also play an important role in supporting "regenerated" production. Ammonium, whether acquired directly or indirectly through metabolism of other nitrogenous compounds, is assimilated predominantly via the enzyme system glutamine synthetase (EC 6.3.1.2) and glutamate synthase (GOGAT), which function in amino acid biosynthesis. The latter enzyme, GOGAT, occurs in two distinct forms, NADH-dependent GOGAT (EC 1.4.1.14) and ferredoxin-dependent GOGAT (EC 1.4.7.1). Both forms are metalloenzymes, possessing Fe-S clusters. Additionally, the latter form, which is present in all oxygenic photoautotrophs examined to date, as implied by its name, requires ferredoxin as an electron donor. Similarly to NiR, this activity can be replaced by flavodoxin in some green algae; however, substitution comes at the expense of lower GOGAT affinity (Vigara et al. 1995, 1998).

It is becoming increasingly clear that urea is an abundant and important source of nitrogen supporting "regenerated" production in marine systems. Indeed, current estimates indicate that upwards of 80% of the cellular nitrogen quota can be attributed to urea acquisition in some marine environments (McCarthy 1972; Kristiansen 1983; Harrison et al. 1985; Sahlsten 1987; Sahlsten et al. 1988; Probyn 1988; Hansell and Goering 1990; Glibert et al. 1991). Its role in this capacity in freshwater systems is not well studied; however, uptake and assimilation of urea have been documented for many species of algae and cyanobacteria, both freshwater and marine. (Antia et al. 1991). Of particular relevance to this chapter is the fact that hydrolysis of urea is a metal-ion-dependent process catalyzed by the enzyme urease (EC 3.5.1.5), which possesses an obligate Ni(II) requirement (Mobley and Hausinger 1989). At physiological pH, the products of hydrolysis are ammonium and bicarbonate, which are used to support amino acid biosynthesis and photosynthesis, respectively (Price and Harrison 1988).

Although uptake and assimilation of urea are well documented, relatively little attention has been given to the Ni(II) requirement for this activity. Presumably, Ni ions present as contaminants of major salts used in media preparation have served to fulfill the Ni(II) requirement in many previous investigations that have documented urea assimilation (Antia et al. 1991). Consistent with this, Ni is a component of few cellular constituents. The sole requirement for Ni among eukaryotic phytoplankton is accounted for by urease activity, whereas prokaryotes appear to have an additional hydrogenase activity that requires Ni(II) (Papen et al. 1986). Despite the low cellular quota predicted for Ni, actual Ni requirements vary considerably among phytoplankton. As mentioned previously, the ability to assimilate urea has been reported for numerous phytoplankton cultured in "Ni-free" medium. These plankton apparently have the ability to concentrate low endogenous levels of Ni, as demonstrated by the inability of these same species to grow in medium containing the strong Ni chelator, citric acid (Oliveira and Antia 1986; Oliveira and Huynh 1990). In contrast, other species (*Cyclotella cryptica*, Oliveira and Antia 1984; *Thalassiosira nordenskioldii*, Oliveira and Antia 1986;

Amphidinium carterae and *Olisthodiscus luteus*, Oliveira and Huynh 1990; *Emiliania huxleyi*, Palenik and Henson 1997) possess demonstrable Ni requirements with exogenously provided Ni (upwards of 100 nM) required to support growth on urea. Price and Morel (1991) calculated an Ni cell quota for *T. weissflogii* cultured on urea of 18.8 amol cell^{-1}, within the range reported for Zn (66 amol cell^{-1}; Price and Morel 1990) and Fe (80 amol cell^{-1}; Harrison and Morel 1986), both of which are required by multiple cellular components (Vallee and Auld 1990; Raven et al. 1999). Based on the amount of Ni accumulated by *T. weissflogii*, it is calculated that potential urease activitcy would provide in excess of 4000 times the cellular demand for nitrogen in this alga. Clearly, this is not the case, and a substantial amount of cellular Ni is not associated with urease, but rather with uncharacterized, low M_r compounds in *T. weissflogii* (Price and Morel 1991). This storage of excess Ni is thought to offset the slow reaction kinetics characteristic of this trace metal.

By extension, Ni(II) is required for growth on substrates from which urea is derived as a product of catabolism. For example, urea is a product of purine oxidation in a variety of organisms (Mobley and Hausinger 1989; Antia et al. 1991). Consistent with this, a Ni(II) requirement is reported for growth on the purine, hypoxanthine (Oliveira and Huynh 1990; Palenik and Henson 1997). In Ni(II)-replete culture, cells of *Emiliania huxleyi* provided hypoxanthine grew at ~85% of the growth rate achieved using NO_3^-, whereas cells grew poorly in hypoxanthine-containing medium to which no Ni(II) was added (Palenik and Henson 1997). A strict requirement for Ni(II) was also documented for growth on hypoxanthine by the dinoflagellate *Amphidinium carterae* (Oliveira and Huynh 1990). Although purine use is widely documented among phytoplankton (Antia et al. 1991), an Ni(II) requirement for growth on purines other than hypoxanthine is predicted but has yet to be addressed. As with studies of urea assimilation, Ni ion contamination has most likely served to fulfill the Ni(II) requirement in many investigations that have documented purine use. An exception to this line of reasoning comes from the finding that purine degradation can proceed via an alternative pathway which does not result in the formation of urea. Although this alternative pathway was originally described for higher plants, its presence in some phytoplankton may explain the observed Ni(II)-independent metabolism of hypoxanthine in the prymnesiophyte *Pavlova lutheri* (Oliveira and Huynh 1990).

Likewise, a number of green algae are able to metabolize urea without a concomitant requirement for Ni (Rees and Bekheet 1982; Oliveira and Huynh 1990; Antia et al. 1991). This appears to be the result of an alternative, Ni(II)-independent pathway for the catabolism of urea via the enzyme complex, ATP-dependent urea amidolyase (urea carboxylase; EC 6.3.4.6 and allophanate hydrolase; EC 3.5.1.13). It is likely that this mechanism is also behind the Ni(II)-independent metabolism of hypoxanthine in the chlorophyte *Dunaliella tertiolecta* (Oliveira and Huynh 1990). Apart from the lack of an Ni(II) requirement, the most striking difference between urease and urea amidolyase rests in the inducible nature of the latter. To date, ATP-urea amidolyase activity has been reported only among chlorophytes; urea metabolism in all other classes of algae, including several distinct orders of Chlorophyceae and the cyanobacteria, appears to be via Ni(II)-dependent urease activity (Oliveira and Huynh 1990; Antia et al. 1991).

Our preceding consideration of phytoplankton Ni(II) requirements leads to the inevitable question: can phytoplankton growth be colimited by Ni and N? This question has been posed by several groups relating calculated Ni requirements to levels of Ni measured in marine environments (Oliveira and Antia 1986; Price and Morel 1991). Vertical profiles of dissolved Ni resemble those for other algal nutrients and are characterized by a pronounced near-surface depletion (Sclater et al. 1976; Bruland 1980). Upon consideration of concentrations of dissolved Ni alone, it is predicted that observed levels are barely sufficient to sustain maximum growth rates of phytoplankton during urea assimilation (Price and Morel 1991). Note, however, that Ni(II), similarly to other transition metals, generally exists in a complexed state in seawater (van den Berg and Nimmo 1987; Nimmo et al. 1989; Bruland et al. 1991). Upon further consideration of the potentially limited availability of inorganic and organic Ni complexes as well as the slow reaction kinetics characteristic of Ni, colimitation by Ni and N is a distinct possibility.

5.5 Trace Metal Constraints on P Acquisition

Apart from its effect on carbon acquisition, the availability of Zn may also dictate the efficiency with which phytoplankton access phosphate. Inorganic oxyanion phosphate (P_i) , the biologically relevant form of the nutrient, is rare in aquatic environments. This is particularly true of freshwater systems, which are typically P-limited (Schindler 1977; Hecky and Kilham 1988). To counter the paucity of naturally occurring P_i, aquatic organisms frequently rely on phosphatase activity to access P from organic complexes. Of particular relevance to lacustrine environments is alkaline phosphatase (APase ; EC 3.1.3.1), an enzyme catalyzing the hydrolysis of phosphate monoesters and releasing P_i in the process. As a group, APases possess a broad pH optimum in neutral to alkaline environments and also exhibit broad substrate specificity for a range of phosphate monoesters.

Alkaline phosphatase is also notable in being a metalloenzyme (Vincent et al. 1992). In this context, it is similar to carbonic anhydrase, typically possessing a Zn(II) ion requirement. The enzymology has been most extensively studied using protein purified from the enteric bacterium *E. coli* where APase is localized to the periplasmic space forming a homodimeric complex having M_r ~94 kDa. A high-resolution crystal structure elucidated for the enzyme-phosphate complex details the coordination of two Zn(II) ions and a single Mg(II) ion within the active site of each subunit (Wyckoff 1987). Here, the Zn(II) ions most likely possess a catalytic role in rendering phosphate monoesters susceptible to enzymatic hydrolysis (Vincent et al. 1992).

The need to acquire phosphate transcends taxonomic barriers, so it is not surprising that APase is widely distributed among organisms ranging from enteric bacteria to humans. Although the enzymes characterized from *E. coli* and various human cell types possess only moderate (~30%) homology, active site motifs are highly constrained and enzymes share additional features of three-dimensional structure (Wyckoff 1987). Among these "conserved" features is a catalytic role for Zn(II) ions. Consistent with this feature, APase activity is used as an index of human Zn nutritional status (Gibson 1990; Cousins 1996).

Although APase assays are a routine component of aquatic field studies (Tarapchak and Nalewajko 1986; Lean 1987), the enzyme has been characterized from only a few algal and cyanobacterial sources. Consistent with their bacterial and metazoan counterparts, APases from photoautotrophs are generally phosphate-repressible; but it is curious that in the handful of studies in which metal ion requirements have been characterized, there appears to be no consensus for a Zn(II) ion requirement. Arguably the best-characterized APase included in this group is that purified from the freshwater cyanophyte *Synechococcus* sp. PCCC 7942 (Ihlenfeldt and Gibson 1975; Block and Grossman 1988; Ray et al. 1991). The *Synechococcus* APase is a periplasmic enzyme having a single subunit of M_r ~145 kDa. The gene encoding this APase (*phoA*) displays no homology to previously characterized APases and the deduced amino acid sequence provides no basis for a conventional Zn(II) ion binding motif (Ray et al. 1991). Moreover, this APase is irreversibly inhibited by Zn (Ihlenfeldt and Gibson 1975; Block and Grossman 1988) and appears to have only an Mg(II) ion requirement. Likewise, the few algal APases that have been characterized show no evidence for a Zn(II) ion requirement. Rather, the chlorophytes *Chlamydomonas reinhardtii* (Quisel et al. 1996) and *Volvox carteri* (Hallmann 1999) both appear to possess phosphate-repressible, Ca(II) ion-dependent forms of APase.

Recently, a Zn(II)-requiring, constitutively expressed APase was reported for *Synechococcus* sp. PCCC 7942 (Wagner et al. 1995). Encoded by *phoV*, the deduced amino acid sequence displays 34% identity to bacterial APase and predicts a protein of M_r ~61.3 kDa. Similarly to the *phoA* gene product, APase encoded by *phoV* is localized in the periplasmic space in *Synechococcus*.

Given the conserved nature of the metal ion requirement noted for APases across taxonomically distinct groups, the apparent divergence of many of the algal and cyanobacterial forms in this respect is puzzling. This may be a response to the paucity of dissolved Zn(II) characteristic of marine (Bruland 1989; Bruland et al. 1994) and freshwater environments (Coale and Flegal 1989; Nriagu et al. 1996). However, in this context, one might expect a mechanism such as metal replacement, as invoked for carbonic anhydrase, to be more widely adopted. Studies conducted in vitro using APase purified from *E. coli* support metal replacement as a plausible mechanism (Szajn and Csopak 1977; Coleman et al. 1983; Cioni et al. 1989). Of course, as more aquatic APases are characterized, the Zn(II)-requiring form may turn out to be more widespread than presently apparent. In this event, cellular acquisition and allocation of Zn(II) will be of high priority to algae and cyanobacteria under P-limiting conditions. This is particularly relevant given that APase can represent 6% of newly synthesized protein in P-deficient *E. coli* (Yoda et al. 1987).

References

Antia NJ, Harrison PJ, Oliveira L (1991) The role of dissolved organic nitrogen in phytoplankton nutrition, cell biology and ecology. Phycologia 30:1-89

Ashton PJ (1981) Nitrogen fixation and the nitrogen budget of a eutrophic impoundment. Water Res 15:823-833

Axler RP, Gersberg RM, Goldman CR (1980) Stimulation of nitrate uptake and photosynthesis by molybdenum in Castle Lake, California. Can J Fish Aquat Sci 37:707-712

Badger MR, Price GD (1994) The role of carbonic anhydrase in photosynthesis. Annu Rev Plant Physiol Plant Mol Biol 45:369-392

Barbeau K, Moffett JW, Caron DA, Croot PL, Erdner DL (1996) Role of protozoan grazing in relieving iron limitation of phytoplankton. Nature 380:61-64

Berges JA (1997) Algal nitrate reductases. Eur J Phycol 32:3-8

Bianchi TS, Engelhaupt E, Westman P, Andren T, Rolff C, Elmgren R (2000) Cyanobacterial blooms in the Baltic Sea: natural or human-induced? Limnol Oceanogr 45:716-726

Bishop PE, Premakumar R, Dean DR, Jacobson MR, Chisnell JR, Rizzo TM, Kopczynski J (1986) Nitrogen fixation by *Azotobacter vinelandii* strains having deletions in structural genes for nitrogenase. Science 232:92-94

Block MA, Grossman AR (1988) Identification and purification of a derepressible alkaline phosphatase from *Anacystis nidulans* R2. Plant Physiol 86:1179-1184

Boyd PW, Watson A, Law CS, Abraham E, Trull T, Murdoch R, Bakker DCE, Bowie AR, Charette M, Croot P, Downing K, Frew R, Gall M, Hadfield R, Hall J, Harvey M, Jameson G, La Roche J, Liddicoat M, Ling R, Maldonado M, McKay RML, Nodder S, Pickmere S, Pridmore R, Rintoull S, Safi K, Sutton P, Strzepek R, Tannenberger K, Turner S, Waite A, Zeldis J (2000) Mesoscale iron fertilization elevates phytoplankton stocks in the polar Southern Ocean. Nature 407:695-702

Boyle EA, Sclater F, Edmond JM (1976) On the marine geochemistry of cadmium. Nature 263:42-44

Bruland KW (1980) Oceanographic distributions of cadmium, zinc, nickel, and copper in the North Pacific. Earth Planet Sci Lett 47:176-198

Bruland KW (1989) Complexation of zinc by natural organic ligands in the central North Pacific. Limnol Oceanogr 34:269-285

Bruland KW, Donat JR, Hutchins DA (1991) Interactive influences of bioactive trace metals on biological production in the sea. Limnol Oceanogr 36:1555-1577

Bruland KW, Orians KJ, Cowen JP (1994) Reactive trace metals in the stratified central North Pacific. Geochim Cosmochim Acta 58:3171-3182

Campbell PGC (1995) Interactions between trace metals and aquatic organisms: a critique of the free-ion activity model. In: Tessier A Turner DR (eds) Metal speciation and bioavailability in aquatic systems. John Wiley, New York, pp 45-102

Campbell WH (1999) Nitrate reductase structure, function and regulation: bridging the gap between biochemistry and physiology. Annu Rev Plant Physiol Plant Mol Biol 50:277-303

Capone DG, Zehr JP, Paerl HW, Bergman B, Carpenter EJ (1997) *Trichodesmium*, a globally significant marine cyanobacterium. Science 276:1221-1229

Carpenter EJ, Capone DG (1983) Nitrogen in the marine environment. Academic Press, NY

Carpenter EJ, Price NM (1976) Marine *Oscillatoria* (*Trichodesmium*): explanation for aerobic nitrogen fixation without heterocysts. Science 191:1278-1280

Cioni P, Piras L, Strambini GB (1989) Tryptophan phosphorescence as a monitor of the structural role of metal ions in alkaline phosphatase. Eur J Biochem 185:573-579
Coale KH, Flegal AR (1989) Copper, zinc, cadmium and lead in surface waters of Lakes Erie and Ontario. Sci Total Environ 87/88:297-304
Coale KH, Johnson KS, Fitzwater SE, Gordon RM, Tanner S, Chavez FP, Ferioli L, Sakamoto C, Rogers P, Millero F, Steinberg P, Nightingale P, Cooper D, Cochlan WP, Landry MR, Constantinou J, Rollwagen G, Trasvina A, Kudela R (1996) A massive phytoplankton bloom induced by an ecosystem scale iron fertilization experiment in the equatorial Pacific Ocean. Nature 383:495-501
Cole JJ, Lane JM, Marino R, Howarth RW (1993) Molybdenum assimilation by cyanobacteria and phytoplankton in freshwater and salt water. Limnol Oceanogr 38:25-35
Coleman JE, Nakamura K, Chlebowski JF (1983) 65Zn(II), 115Cd(II), 60Co(II), and Mg(II) binding to alkaline phosphatase of *Escherichia coli*. Structural and functional effects. J Biol Chem 258:386-395
Cousins RJ (1996) Zinc. In: Ziegler EE, Filer LJ (eds) Present knowledge in nutrition.. International Life Sciences Institute, Washington, DC, pp 293-306
Cullen JT, Lane TW, Morel FMM, Sherrell RM (1999) Modulation of cadmium uptake in phytoplankton by seawater CO_2 concentration. Nature 402:165-167
de Baar H JW (1994) von Liebig's Law of the Minimum and plankton ecology (1899-1991). Prog Oceanogr 33:347-386
Duce RA, Tindale NW (1991) Atmospheric transport of iron and its deposition in the ocean. Limnol Oceanogr 36:1715-1726
Dugdale RC, Goering JJ (1967) Uptake of new and regenerated forms of nitrogen in primary productivity. Limnol Oceanogr 12:196-206
Entsch B, Sim RG, Hatcher BG (1983) Indications from photosynthetic components that iron is a limiting nutrient in primary producers on coral reefs. Mar Biol 73:17-30
Eppley RW, Garside C, Renger EH, Orellana E (1990) Variability of nitrate concentration in nitrogen-depleted sub-tropical surface water. Mar Biol 107:53-56
Evans JC, Prepas EE (1997) Relative importance of iron and molybdenum in restricting phytoplankton biomass in high phosphorus saline lakes. Limnol Oceanogr 42:461-472
Eyster C (1964) Micronutrient requirements for green plants especially algae. In: Jackson DF (ed) Algae and man. Plenum Press, New York, pp 86-119
Falkowski PG (1997) Evolution of the nitrogen cycle and its influence on the biological sequestration of CO_2 in the ocean. Nature 387:272-275
Falkowski PG, Greene RM, Geider RJ (1992) Physiological limitations on phytoplankton productivity in the ocean. Oceanography 5:84-91
Falkowski PG, Barber RT, Smetacek V (1998) Biogeochemical controls and feedbacks on ocean primary production. Science 281:200-206
Flett RJ, Schindler DW, Hamilton RD, Campbell NER (1980) Nitrogen fixation in Canadian Precambrian Shield lakes. Can J Fish Aquat Sci 37:494-505
Froelich PN (1988) Kinetic control of dissolved phosphate in natural rivers and estuaries: a primer on the phosphate buffer mechanism. Limnol Oceanogr 33:649-668
Fuhrman JA (1999) Marine viruses and their biogeochemical and ecological effects. Nature 399:541-548
Gao Y, Smith GJ, Alberte RS (1993) Nitrate reductase from the marine diatom *Skeletonema costatum*. Biochemical and immunological characterization. Plant Physiol 103:1437-1445
Geider RJ, La Roche J (1994) The role of iron in phytoplankton photosynthesis, and the potential for iron limitation of primary productivity in the sea. Photosynth Res 39:275-301
Gerloff GC, Skoog F (1957) Availability of iron and manganese in southern Wisconsin lakes for the growth of *Microcystis aeruginosa*. Ecology 38:552-556

Gibson RS (1990) Principles of nutritional assessment. Oxford University Press, New York

Glibert PM, Garside C, Fuhrman JA, Roman MR (1991) Time-dependent coupling of inorganic and organic nitrogen uptake and regeneration in the plume of the Chesapeake Bay estuary and its regulation by large heterotrophs. Limnol Oceanogr 36:895-909

Goldman CR (1965) Micronutrient limiting factors and their detection in natural phytoplankton populations. Mem Ist Ital Idrobiol 18 Suppl 77:121-135

Goldman CR (1972) The role of minor nutrients in limiting the productivity of aquatic ecosystems. Special Symposia, Limnology and Oceanography, Nutrients and Eutrophication: The Limiting Nutrient Controversy, W.K. Kellog Biological Station, Michigan State University, 1:21-33

Greene RM, Geider RJ, Kolber Z, Falkowski PG (1992) Iron-induced changes in light-harvesting and photochemical energy conversion processes in eukaryotic marine algae. Plant Physiol 100:565-575

Hallmann A (1999) Enzymes in the extracellular matrix of *Volvox*: an inducible, calcium- - dependent phosphatase with a modular composition. J Biol Chem 274:1691-1697

Hansell DA, Goering JJ (1990) Pelagic nitrogen flux in the northern Bering Sea. Cont Shelf Res 10:501-509

Harrison GI, Morel FMM (1986) Response of the marine diatom *Thalassiosira weissflogii* to iron stress. Limnol Oceanogr 31:989-997

Harrison WG, Head EJD, Conover RJ, Longhurst AR, Sameoto DD (1985) The distribution and metabolism of urea in the eastern Canadian Arctic. Deep-Sea Res 32:23-42

Healey FP (1973) Inorganic nutrient uptake and deficiency in algae. CRC Crit Rev Microbiol 3:69-112

Hecky RE, Kilham P (1988) Nutrient limitation of phytoplankton in freshwater and marine environments: a review of recent evidence on the effects of enrichment. Limnol Oceanogr 33:796-822

Howard WH, Solomonson LP (1982) Quaternary structure of assimilatory NADH:nitrate reductase from *Chlorella*. J Biol Chem 257:10243-10250

Howarth RW, Cole JJ (1985) Molybdenum availability, nitrogen limitation, and phytoplankton growth in natural waters. Science 229:653-655

Howarth RW, Marino R, Lane J, Cole JJ (1988a) Nitrogen fixation in freshwater, estuarine, and marine ecosystems. 1. Rates and importance. Limnol Oceanogr 33:669-687

Howarth RW, Marino R, Cole JJ (1988b) Nitrogen fixation in freshwater, estuarine, and marine ecosystems. 2. Biogeochemical controls. Limnol Oceanogr 33:688-701

Howarth RW, Chan F, Marino R (1999) Do top-down and bottom-up controls interact to exclude nitrogen-fixing cyanobacteria from the plankton of estuaries? An exploration with a simulation model. Biogeochemistry 46:203-231

Huber AL (1986) Nitrogen fixation by *Nodularia spumigena* Mertens (Cyanobacteriaceae). 1: Field studies and the contribution of blooms to the nitrogen budget of the Peel-Harvey Estuary, Western Australia. Hydrobiologia 131:193-203

Hudson RJ, Morel FMM (1990) Iron transport in marine phytoplankton: kinetics of cellular and medium coordination reactions. Limnol Oceanogr 35:1002-1020

Hutchins DA (1995) Iron and the marine phytoplankton community. In: Chapman DJ, Round FE (eds) Progress in phycological research, vol 11. Biopress, Bristol, UK, pp 1-49

Hutchins DA, Bruland KW (1998) Iron-limited diatom growth and Si:N uptake ratios in a coastal upwelling regime. Nature 393:561-564

Hutchins DA, DiTullio GR, Zhang Y, Bruland KW (1998) An iron limitation mosaic in the California upwelling regime. Limnol Oceanogr 43:1037-1054

Ihlenfeldt MJA, Gibson J (1975) Phosphate utilization and alkaline phosphatase activity in *Anacystis nidulans* (*Synechococcus*). Arch Microbiol 102:23-28

Kaplan A, Reinhold L (1999) CO_2-concentrating mechanisms in photosynthetic microorganisms. Annu Rev Plant Physiol Plant Mol Biol 50:539-570

Karl D, Letelier R, Tupas L, Dore J, Christian J, Hebel D (1997) The role of nitrogen fixation in biogeochemical cycling in the subtropical North Pacific Ocean. Nature 388:533-538

Kononen K, Kuparinen J, Makela K, Laanemets J, Pavelson J, Nommann S (1996) Initiation of cyanobacterial blooms in a frontal region at the entrance to the Gulf of Finland, Baltic Sea. Limnol Oceanogr 41:98-112

Kristiansen S (1983) Urea as a nitrogen source for the phytoplankton in the Oslofjord. Mar Biol 74:17-24

Kudo I, Harrison PJ (1997) Effect of iron nutrition on the marine cyanobacterium *Synechococcus* grown on different N sources and irradiances. J Phycol 33:232-240

Lean DRS (1987) Overview of studies on the nutrient status of Lake Ontario. Can J Fish Aquat Sci 44:2042-2046

Lee BG, Fisher NS (1993) Release rates of trace elements and protein from decomposing planktonic debris. 1. Phytoplankton debris. J Mar Res 51:391-421

Lee JG, Roberts SB, Morel FMM (1995) Cadmium: a nutrient for the marine diatom *Thalassiosira weisflogii*. Limnol Oceanogr 40:1056-1063

Lin CK, Schelske CL (1981) Seasonal variation of potential nutrient limitation to chlorophyll production in southern Lake Huron. Can J Fish Aquat Sci 38:1-9

Loveless TM, Saah JR, Bishop PE (1999) Isolation of nitrogen-fixing bacteria containing molybdenum-independent nitrogenases from natural environments. Appl Environ Microbiol 65:4223-4226

Luque I, Flores E, Herrero A (1993) Nitrite reductase gene from *Synechococcus* sp. PCC 7942: homology between cyanobacterial and higher-plant nitrite reductases. Plant Mol Biol 21:1201-1205

Maldonado MT, Price NM (1996) Influence of N substrate on Fe requirements of marine centric diatoms. Mar Ecol Prog Ser 141:161-172

Martin JH, Coale KH, Johnson KS, Fitzwater SE, Gordon RM, Tanner SJ, Hunter CN, Elrod VA, Nowicki JL, Coley TL, Barber RT, Lindley S, Watson AJ, Van Scoy K, Law CS, Liddicoat MI, Ling R, Stanton T, Stockel J, Collins C, Anderson A, Bidigare R, Ondrusek M, Latasa M, Millero FJ, Lee K, Yao W, Zhang JZ, Friederich G, Sakamoto C, Chavez F, Buck K, Kolber Z, Greene R, Falkowski P, Chisholm SW, Hoge F, Swift R, Yungel J, Turner S, Nightingale P, Hatton A, Liss P, Tindale NW (1994) Testing the iron hypothesis in ecosystems of the equatorial Pacific Ocean. Nature 371:123-129

McCarthy JJ (1972) The uptake of urea by natural populations of marine phytoplankton. Limnol Oceanogr 17:738-748

McCarthy MD, Hedges JI, Benner R (1998) Major bacterial contribution to marine dissolved organic nitrogen. Science 281:231-234

McKay RML, Gibbs SP, Espie GS (1993) Effect of dissolved inorganic carbon on the expression of carboxysomes, localization of Rubisco and the mode of inorganic carbon transport in cells of the cyanobacterium *Synechococcus* UTEX 625. Arch Microbiol 159:21-29

McKay RML, La Roche J, Yakunin AF, Durnford DG, Geider RJ (1999) Accumulation of ferredoxin and flavodoxin in a marine diatom in response to Fe. J Phycol 35:510-519

McLachlan J (1973) Growth media — marine. In: Stein JR (ed) Handbook of phycological methods. Culture methods and growth measurements. Cambridge University Press, Cambridge, pp 25-51

Meisch H, Bieling HJ (1975) Effect of vanadium on growth, chlorophyll formation, and iron metabolism in unicellular green algae. Arch Microbiol 105:77-82

Michaels AF, Olson D, Sarmiento JL, Ammerman JW, Fanning K, Jahnke R, Knap AH, Lipschultz F, Prospero JM (1996) Inputs, losses and transformations of nitrogen and phosphorus in the pelagic North Atlantic Ocean. Biogeochemistry 35:181-226

Miller AG, Espie GS, Canvin DT (1990) Physiological aspects of CO_2 and HCO_3^- transport by cyanobacteria: a review. Can J Bot 68:1291-1302

Milligan AJ, Harrison PJ (2000) Effects of non-steady-state iron limitation on nitrogen assimilatory enzymes in the marine diatom *Thalassiosira weissflogii* (Bacillariophyceae). J Phycol 36:78-86

Mobley HLT, Hausinger RP (1989) Microbial ureases: significance, regulation, and molecular characterization. Microbiol Rev 53:85-108

Morel FMM, Westall JC, Rueter JG, Chaplick JP (1975) Description of the algal growth media "Aquil" and "Fraquil". Dept of Civil Engineering, MIT, Cambridge, Technical Note No 16, pp 33

Morel FMM, Hudson RJM, Price NM (1991) Limitation of productivity by trace metals in the sea. Limnol Oceanogr 36:1742-1755

Morel FMM, Reinfelder JR, Roberts SB, Chamberlain CP, Lee JG, Yee D (1994) Zinc and carbon co-limitation of marine phytoplankton. Nature 369:740-742

Murphy TP, Lean DRS, Nalewajko C (1976) Blue-green algae: their excretion of iron-selective chelators enables them to dominate other algae. Science 192:900-902

Nicholas DJ (1963) Inorganic nutrient nutrition of microorganisms. In: Steward FC (ed) Plant physiology III. Academic Press, New York, pp 368-447

Nichols HW (1973) Growth media — freshwater. In: Stein JR (ed) Handbook of phycological methods. Culture methods and growth measurements. Cambridge University Press, Cambridge, pp 7-24

Nimmo M, van den Berg CMG, Brown J (1989) The chemical speciation of dissolved nickel, copper, vanadium, and iron in Liverpool Bay, Irish Sea. Estuarine Coastal Shelf Sci 29:57-74

Nriagu JO, Lawson G, Wong HKT, Cheam V (1996) Dissolved trace metals in Lakes Superior, Erie and Ontario. Environ Sci Technol 30:178-187

O'Kelley JC (1974) Inorganic nutrients. In: Stewart WDP (ed) Algal physiology and biochemistry. Blackwell, Oxford, pp 610-635

Oliveira L, Antia NJ (1984) Evidence of Ni ion requirement for autotrophic growth of a marine diatom with urea serving as nitrogen source. Br Phycol J 19:125-134

Oliveira L, Antia NJ (1986) Nickel ion requirements for autotrophic growth of several marine microalgae with urea serving as nitrogen source. Can J Fish Aquat Sci 43:2427-2433

Oliveira L, Huynh H (1990) Phototrophic growth of microalgae with allantoic acid or hypoxanthine serving as nitrogen source, implications for purine-N utilization. Can J Fish Aquat Sci 47:351-356

Paerl HW, Crocker KM, Prufert LE (1987) Limitation of N_2 fixation in coastal marine waters: relative importance of molybdenum, iron, phosphorus, and organic matter availability. Limnol Oceanogr 32:525-536

Paerl HW, Prufert-Bebout LE, Guo C (1994) Iron-stimulated N_2 fixation and growth in natural and cultured populations of the planktonic marine cyanobacteria *Trichodesmium* spp. Appl Environ Microbiol 60:1044-1047

Palenik B, Henson SE (1997) The use of amides and other organic nitrogen sources by the phytoplankton *Emiliania huxleyi*. Limnol Oceanogr 42:1544-1551

Papen H, Kentemich T, Schmulling T, Bothe H (1986) Hydrogenase activities in cyanobacteria. Biochimie 68:121-132

Paulsen DM, Paerl HW, Bishop PE (1991) Evidence that molybdenum-dependent nitrogen fixation is not limited by high sulfate concentrations in marine environments. Limnol Oceanogr 36:1325-1334

Postgate J (1998) Nitrogen fixation. Cambridge University Press. Cambridge, UK

Price NM, Harrison PJ (1988) Uptake of urea C and urea N by the coastal marine diatom *Thalassiosira pseudonana*. Limnol Oceanogr 33:528-537

Price NM, Morel FMM (1990) Cadmium and cobalt substitution for zinc in a marine diatom. Nature 344:658-660

Price NM, Morel FMM (1991) Colimitation of phytoplankton growth by nickel and nitrogen. Limnol Oceanogr 36:1071-1077

Price NM, Harrison GI, Hering JG, Hudson RJ, Nirel PM, Palenik B, Morel FMM (1989) Preparation and chemistry of the artificial algal culture medium Aquil. Biol Oceanogr 6:443-461

Probyn TA (1988) Nitrogen utilization by phytoplankton in the Namibian upwelling region during an austral spring. Deep-Sea Res 35:1387-1404

Quisel JD, Wykoff DD, Grossman AR (1996) Biochemical characterization of the extracellular phosphatases produced by phosphorus-deprived *Chlamydomonas reinhardtii*. Plant Physiol 111:839-848

Ramalho CB, Hastings JW, Colepicolo P (1995) Circadian oscillation of nitrate reductase activity in *Gonyaulax polyedra*. Plant Physiol 107:225-231

Raven JA (1988) The iron and molybdenum use efficiencies of plant growth with different energy, carbon and nitrogen sources. New Phytol 109:279-287

Raven JA, Evans MCW, Korb RE (1999) The role of trace metals in photosynthetic electron transport in O_2-evolving organisms. Photosynth Res 60:111-149

Ray JM, Bhaya D, Block MA, Grossman AR (1991) Isolation, transcription, and inactivation of the gene for an atypical alkaline phosphatase of *Synechococcus* sp. strain PCC 7942. J Bacteriol 173:4297-4309

Redfield AC (1958) The biological control of chemical factors in the environment. Am Sci 46:205-221

Rees TAV, Bekheet IA (1982) The role of nickel in urea assimilation by algae. Planta 156:385-387

Robson LR, Eady RR, Richardson TH, Miller RW, Hawkins M, Postgate JR (1986) The alternative nitrogenase of *Azotobacter chroococcum* is a vanadium enzyme. Nature 322:388-390

Rueter JG (1982) Theoretical Fe limitations of microbial N_2 fixation in the oceans. Eos 63:445

Rueter JG (1988) Iron stimulation of photosynthesis and nitrogen fixation in *Anabaena* 7120 and *Trichodesmium* (Cyanophyceae). J Phycol 24:249-254

Rueter JG, Ohki K, Fujita Y (1990) The effect of iron nutrition on photosynthesis and nitrogen fixation in cultures of *Trichodesmium* (Cyanophyceae). J Phycol 26:30-35

Rueter JG, Hutchins DA, Smith RW, Unsworth NL (1992) Iron nutrition of *Trichodesmium*. In: Carpenter EJ, Capone DG, Rueter JG (eds) Marine pelagic cyanobacteria: *Trichodesmium* and other diazotrophs. Kluwer, Dordrecht, pp 289-306

Ryther JH, Dunstan WM (1971) Nitrogen, phosphorus, and eutrophication in the coastal marine environment. Science 171:1008-1013

Sahlsten E (1987) Nitrogenous nutrition in the euphotic zone of the Central North Pacific Gyre. Mar Biol 96:433-439

Sahlsten E, Sorensson F, Petterson K (1988) Planktonic nitrogen uptake in the south-eastern Kattegat. J Exp Mar Biol Ecol 121:227-246

Schelske CL (1962) Iron, organic matter, and other factors limiting primary productivity in a marl lake. Science 136:45-46

Schelske CL, Hooper FP, Haertl EJ (1962) Responses of a marl lake to chelated iron and fertilizer. Ecology 43:646-653

Schelske CL, Feldt LE, Santiago MA, Stoermer EF (1972) Nutrient enrichment and its effect on phytoplankton production and species composition in Lake Superior. In: Proc 15th Conf Great Lakes Res, Int Assoc Great Lakes Res, Madison, pp 149-165

Schelske CL, Rothman ED, Simmons MS (1978) Comparison of bioassay procedures for growth-limiting nutrients in the Laurentian Great Lakes. Mitt Int Ver Limnol 21:65-80

Schindler DW (1977) Evolution of phosphorus limitation in lakes. Science 195:260-262

Sclater FF, Boyle EA, Edmond JM (1976) On the marine geochemistry of nickel. Earth Planet Sci Lett 31:119-128

Shah VK, Ugalde RA, Imperial J, Brill WJ (1984) Molybdenum in nitrogenase. Annu Rev Biochem 53:231-257

Sharp JH (1983) The distribution of inorganic nitrogen and dissolved and particulate organic nitrogen in the sea. In: Carpenter EJ, Capone DG (eds) Nitrogen in the marine environment. Academic Press, New York, pp 1-36

Sigg L (1994) Regulation of trace elements in lakes: the role of sedimentation. In: Buffle J, De Vitre RR (eds) Chemical and biological regulation of aquatic systems. CRC Press, Boca Raton, Florida, pp 175-195

Silverman DN (1991) The catalytic mechanism of carbonic anhydrase. Can J Bot 69:1070-1078

So AKC, Espie GS (1998) Cloning, characterization and expression of carbonic anhydrase from the cyanobacterium *Synechocystis* PCC6803. Plant Mol Biol 37:205-215

Solomonson LP, Barber MJ (1990) Assimilatory nitrate reductase: functional properties and regulation. Annu Rev Plant Physiol Plant Mol Biol 41:225-253

Stal LJ, Staal M, Villbrandt M (1999) Nutrient control of cyanobacterial blooms in the Baltic Sea. Aquat Microb Ecol 18:165-173

Stoermer EF, Ladewski BG, Schelske CL (1978) Population responses of Lake Michigan phytoplankton to nitrogen and phosphorus. Hydrobiologia 57:249-265

Subramaniam A, Carpenter EJ, Karentz D, Falkowski PG (1999) Bio-optical properties of the marine diazotrophic cyanobacteria *Trichodesmium* spp. I. Absorption and photosynthetic action spectra. Limnol Oceanogr 44:608-617

Sültemeyer D (1997) Carbonic anhydrase in eukaryotic algae: characterization, regulation, and possible function during photosynthesis. Can J Bot 76:962-972

Sunda WG (1994) Trace metal/phytoplankton interactions in the sea. In: Bidoglio G, Stumm W (eds) Chemistry of aquatic systems: local and global perspectives. ECSC, EEC, EAEC, Brussels, Luxembourg, pp 213-247

Sunda WG (2000) Trace metal-phytoplankton interactions in aquatic systems. In: Lovley DR (ed) Environmental microbe-metal interactions. ASM Press, Washington, pp 79-107

Sunda WG, Huntsman SA (1995a) Cobalt and zinc interreplacement in marine phytoplankton: biological and geochemical implications. Limnol Oceanogr 40:1404-1417

Sunda WG, Huntsman SA (1995b) Iron uptake and growth limitation in oceanic and coastal phytoplankton. Mar Chem 50:189-206

Suzuki Y, Kuma K, Matsunaga K (1995) Bioavailable iron species in seawater measured by macroalga (*Laminaria japonica*) uptake. Mar Biol 123:173-178

Szajn H, Csopak H (1977) Metal ion-induced conformational changes in *Escherichia coli* alkaline phosphatase. Biochim Biophys Acta 480:143-153

Tarapchak SJ, Nalewajko C (1986) Introduction: phosphorus. Plankton Dynamics Symposium. Can J Fish Aquat Sci 43:293-301

Timmermans KR, Stolte W, de Baar HJW (1994) Iron-mediated effects on nitrate reductase in marine phytoplankton. Mar Biol 121:389-396

Tortell PD (2000) Evolutionary and ecological perspectives on carbon acquisition in phytoplankton. Limnol Oceanogr 45:744-750

Twiss MR (2001) Current knowledge of trace metal biogeochemistry in the water column of Lake Erie. In: Ciborowski JJH, Charlton MN, Kreis RG Jr, Reutter JM (eds) Lake Erie at the millenium — changes, trends, and trajectories. Canadian Scholars' Press, Toronto

Twiss MR, Campbell PGC (1998a) Scavenging of ^{137}Cs, ^{109}Cd, ^{65}Zn, and ^{153}Gd by plankton of the microbial food web in pelagic Lake Erie surface waters. J Great Lakes Res 24:776-790

Twiss MR, Campbell PGC (1998b) Modeling the fate of trace metals in the surface waters of Lake Erie: linking ecological and geochemical fates. J Great Lakes Res 24:791-807

Twiss MR, Campbell PGC, Auclair J-C (1996) Regeneration, recycling and trophic transfer of trace metals by microbial food web organisms in the pelagic surface waters of Lake Erie. Limnol Oceanogr 41:1425-1437

Twiss MR, Auclair J-C, Charlton MN (2000) Iron-stimulated phytoplankton productivity in epipelagic eastern and central Lake Erie during thermal stratification: a three year field investigation using trace metal clean techniques. Can J Fish Aquat Sci 53:86-95

Vallee BL, Auld DS (1990) Zinc coordination, function, and structure of zinc enzymes and other proteins. Biochemistry 29:5647-5659

Van den Berg CMG, Nimmo M (1987) Determinations of interactions of nickel with dissolved organic material in seawater using cathodic stripping voltammetry. Sci Total Environ 60:185-195

Varela DE, Harrison PJ (1999) Seasonal variability in nitrogenous nutrition of phytoplankton assemblages in the northeastern subarctic Pacific Ocean. Deep-Sea Res II 46:2505-2538

Vassiliev IR, Kolber Z, Wyman KD, Mauzerall D, Shukla VK, Falkowski PG (1995) Effects of iron limitation on photosystem II composition and light utilization in *Dunaliella tertiolecta*. Plant Physiol 109:963-972

Vergara JJ, Berges JA, Falkowski PG (1998) Diel periodicity of nitrate reductase activity and protein levels in the marine diatom *Thalassiosira weissflogii* (Bacillariophyceae). J Phycol 34:952-961

Vigara AJ, Gomez-Moreno C, Vega JM (1995) The role of flavodoxin in the reaction catalysed by the glutamate synthase from *Monaraphidium braunii*. Bioelectrochem Bioenerg 38:21-24

Vigara AJ, Inda LA, Vega JM, Gomez-Moreno C, Peleato ML (1998) Flavodoxin as an electronic donor in photosynthetic inorganic nitrogen assimilation by iron-deficient *Chlorella fusca* cells. Photochem Photobiol 67:446-449

Vincent JB, Crowder MW, Averill BA (1992) Hydrolysis of phosphate monoesters: a biological problem with multiple chemical solutions. Trends Biochem Sci 17:105-110

Wagner KU, Masepohl B, Pistorius EK (1995) The cyanobacterium *Synechococcus* sp. strain PCC 7942 contains a second alkaline phosphatase encoded by *phoV*. Microbiology 141:3049-3058

Wang W-X, Reinfelder JR, Lee B-G, Fisher NS (1996) Assimilation and regeneration of trace elements by marine copepods. Limnol Oceanogr 41:70-81

Wetzel RG (1966) Productivity and nutrient relationships in marl lakes of northern Indiana. Verh Int Ver Limnol 16:321-332

Whitfield M, Turner DR (1987) The role of particles in regulating the composition of seawater. In: Stumm W (ed) Aquatic surface chemistry. John Wiley, New York, pp 457-493

Wiessner W (1962) Inorganic micronutrients. In: Lewin RA (ed) Physiology and biochemistry of algae. Academic Press, New York, pp 267-286

Wilhelm SW (1995) Ecology of iron-limited cyanobacteria: a review of physiological responses and implications for aquatic systems. Aquat Microb Ecol 9:295-303

Wilhelm SW, Suttle CA (1999) Viruses and nutrient cycles in the sea. BioScience 49:781-788

Wurtsbaugh WA, Horne AJ (1983) Iron in eutrophic Clear Lake, California: its importance for algal nitrogen fixation and growth. Can J Fish Aquat Sci 40:1419-1429

Wyckoff HW (1987) Structure of *Escherichia coli* alkaline phosphatase determined by X-ray diffraction. In: Torriani-Gorini A, Rothman FG, Silver S, Wright A, Yagil E (eds) Phosphate metabolism and cellular regulation in microorganisms. American Society for Microbiology, Washington, DC, pp 118-126

Yee D, Morel FMM (1996) In vivo substitution of zinc by cobalt in carbonic anhydrase of a marine diatom. Limnol Oceanogr 41:573-577

Yoda K, Tachibana K-I, Watanabe S, Yamane K, Yamasaki M, Tamura G (1987) Secretion to periplasm of foreign proteins in *Escherichia coli* by aid of the *phoA*-derived secretion vector Psi. In: Torriani-Gorini A, Rothman FG, Silver S, Wright A, Yagil E (eds) Phosphate metabolism and cellular regulation in microorganisms. American Society for Microbiology, Washington, DC, pp 99-104

Zehr JP, Carpenter EJ, Villareal TA (2000) New perspectives on nitrogen-fixing microorganisms in tropical and subtropical oceans. Trends Microbiol 8:68-73

6 Acclimation of Photosynthetic Light Energy Conversion to the Light Environments

Yoshihiko Fujita[1], Kaori Ohki[2] and Akio Murakami[3]

Contents

[1] Department of Marine Bioscience, Fukui Prefectural University, Obama, Fukui 917, Japan
[2] School of Marine Sciences and Technology, Tokai University, Shimizu, Shizuoka 424, Japan
[3] Kobe University Research Centre for Inland seas, Iwaya, Awajicho, Hyougo 656, Japan

6.1 Introduction

Light variously controls the life of algae. Besides driving photosynthesis, the process supporting algal autotrophic life, light controls algal life as the signals for circadian phenomena, for photomorphogenesis, for tactic behaviour, and so on. However, algal adaptation to light regime seems to occur mostly in photosynthesis; hence, in this chapter, we will explain the light acclimation of algae concentrating on photosynthetic energy conversion processes.

A high density of the environmental medium surrounding aquatic algae, water molecules, provides a characteristic light regime to algal photosynthesis. Because water molecules strongly absorb the light of far red wavelengths, enough red light is not available for algal photosynthesis in most of the algal habitats. Further, water can retain various organic materials such as solutes, which absorb the light of shorter wavelengths. Water also contains suspended materials, which scatter the light penetrating from the surface. Since the scattering effect is greater at the shorter wavelength, the light absorption by organic solutes and scattering action of suspended materials diminish the light of shorter wavelengths, resulting in a blue to green world for such habitats of algae. Water in eutrophic habitats such as ponds, lakes and coastal seas contains great amounts of such materials so that diminution of the shorter-wavelength light becomes greater, thereby making these habitats green to brown. Oligotrophic areas such as most of the tropical open seas are a blue world, due to a lesser amount of solutes and suspended materials. Thus, the light regime for the algal habitats is very different in quality from that for the land plants; the latter is a white world except for special areas. Such an optical feature of aquatic environment also causes a considerable limitation of light quantity. For example, the light penetrating into the sea is attenuated logarithmically and, in the habitats for planktonic algae, generally reduced to 10-30% of the surface intensity. Except for algae living at the very surface of water, the light environment for algae is a dim and coloured world. The light in aquatic environment varies, forming a gradient in quality and quantity, from the surface to the deeper layer. The details of optical features of natural aquatic environments have been nicely explained in a book on aquatic photosynthesis (Kirk 1994). Here, we merely wish to call attention to the features of the light environment in the aquatic world which are different from those of the terrestrial environment.

In natural habitats, therefore, algal photosynthesis is performed with the aid of adaptive mechanisms under a characteristic light gradient. Various adaptive strategies to the biased light regime have been found by many workers. In practice, the phenomena are classified into two types: the first is the adaptation to the light quality, and the second, to the light intensity. The former is the way to improve photosynthetic efficiency under a biased light quality, the latter to reduce light harvesting and to abandon excess light energy. The strategies for the two types are different; the former is an active attitude against the light regime in utilization of light energy, and the latter, a passive attitude in protecting the photosynthetic apparatus from excess light.

In natural habitats and also in laboratory cultures, algal photosynthesis must be regulated by a combined action of the two phenomena. In a majority of studies on algal photosynthesis under laboratory conditions, algae are grown under white light from fluorescent or incandescent lamps, and the intensity is usually chosen so as to keep an optimal and stable growth. Thus, the features of photosynthesis found in laboratory-grown algae must be different from the photosynthetic features under natural environments. To know the photosynthesis under natural environments, we need to learn adaptive responses of algal photosynthesis. In the following we will try to introduce the adaptive mechanisms involved in such responses. Before discussing adaptive phenomena, an outline of photosynthetic mechanism is given in the next section.

6.2 Outline of Photosynthetic Processes

The basic mechanism of oxygenic photosynthesis is common in algae and land plants. The process can be classified into four steps: (1) light-capturing by the photosynthetic pigment system, (2) energy conversion at the two photosystems, (3) electron flow driven by the two photosystems and (4) $NADP^+$ reduction and ATP production at the surface of thylakoids. Assimilatory metabolisms of inorganic carbon, nitrogen and sulfur occur mainly in stroma with consumption of NADPH and ATP (Table 6.1; Fig. 6.1).

Table 6.1. Reaction steps in oxygenic photosynthesis

Step number	Reaction	System involved	Site
1.	Light energy is converted into excitation energy of pigments	Light-harvesting pigment system	In or on thylakoids
2.	Conversion of excitation energy of pigments into redox energy	Photosystem I, II	In thylakoids
3.	Electron transport driven by photosystems I,II from water to ferredoxin	Thylakoid electron carriers; electrons flow through plastoquinons, Cyt b_6-f and plastocyanin in this order	In thylakoids
4.	NADPH and ATP production	Ferredoxin-NADP oxidoreductase ATP synthetase	On thylakoids
5.	Carbon-, nitrogen- and sulfur-assimilatory metabolisms with consumption of NADPH and ATP	Enzymes in stroma	In stroma

Steps 1 to 4 progress on or in thylakoids of chloroplasts or of cyanophyte cells. Step 1 of the photosynthetic system offers phylogenetic connection amongst different groups of algae. Chl a and b are the main components of the pigment system of green algae, a group of the green line, so that they appear green in colour. Pigment systems of the land plants consist of the same Chl species. The pigment system of red algae, the algae of red line, consists of Chl a and phycobiliproteins (PBP), as in cyanophytes (cyanobacteria). Since phycoerythrin (PE), a red-purple pigment, is the dominant species, they are red-purple in colour. The main pigment in some of the red algae and many cyanophytes is phycocyanin (PC), a blue pigment, thus the latter organisms appear blue-green in colour. The algae of the yellow line, diatoms and brown seaweeds, have a pigment system consisting of Chl a and c and an orange xanthophyll, fucoxanthin, so that they are brown in colour. The representatives of dinophytes, another group dominant in the sea, also contain a pigment system comprising Chl a and c and an orange xanthophyll, peridinin. A comparison of spectral characteristics of the algal pigment system with light quality in their habitats indicates that each pigment system of algae in the red and yellow lines can harvest the light energy efficiently in the aquatic environment. The case of green algae appears to be an exception. However, the pigment system of green algae living at the bottom of the euphotic zone in the open sea contains a far larger amount of Chl b than that in algae living in shallow waters, so that they can efficiently absorb the blue light. Cyanophytes and red algae containing PC as a main phycobiliprotein (PBP) live in shallow waters or wetlands.

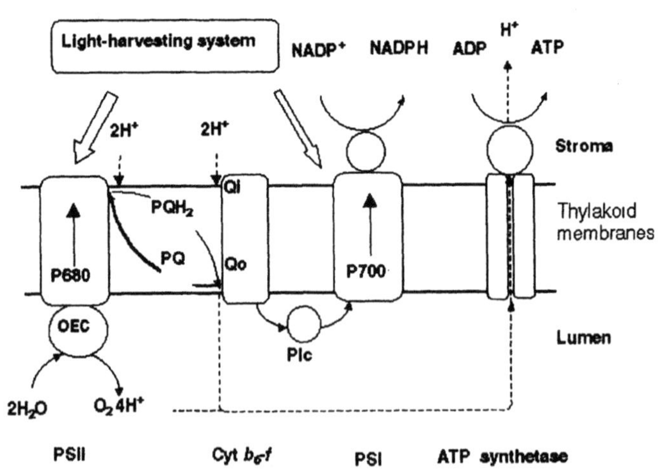

Fig. 6.1. A model of photosynthetic energy conversion processes. *Thick arrows* indicate excitation energy flow; *narrow arrows* electron flow; *broken arrows* proton movement. *OEC* oxygen evolving centre; *P680* reaction centre Chl *a* of PSII; *P700* reaction centre Chl a of PSI; *Plc* plastocyanin; *Qo*, PQH oxidation site of Cyt b_6-*f*; *Qi*, PQ reduction site of Cyt b_6-f. For explanation, see text

In algae of green and yellow lines, chlorophylls and xanthophylls are located in transmembrane proteins penetrating through thylakoid membranes; interestingly, proteins are homologous in both the algal phylogenetic lines. In red algae and cyanobacteria, pigments are bound covalently to proteins to form PBP. PBP forms a supramolecular structure, phycobilisome (PBS), which attaches on the surface of thylakoid membranes. PE and PC are located in the peripheral area of PBS, and allophycocyanin (APC), together with terminal emitters, forms the core of PBS (cf. Fig. 6.2). Pigment proteins act as a large antenna to trap ambient light. Thus, photon energy is converted to the excitation energy of pigment molecules at this step. The excitation energy of antenna pigment molecules converges to the reaction centre Chl a of the two photosystems, photosystem I (PSI) and II (PSII). The energy flow to PSII is generally greater than the flow to PSI. Several adaptive phenomena have been known at this level; complementary chromatic adaptation (CCA; adaptation of pigment species to ambient light quality), state transitions (regulation of energy transfer to the two photosystems) and adjustment of antenna size in response to ambient light quantity are some examples.

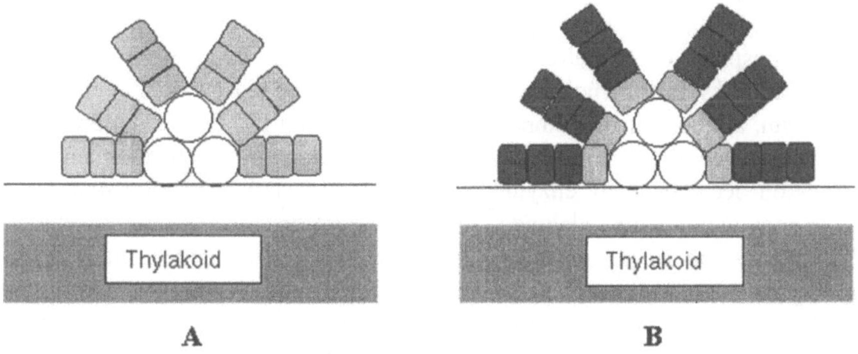

Fig. 6.2. Schematic presentation of hemidiscoidal PBS of cyanophytes. *Open blocks* APC hexamers forming PBS core; *thin shaded blocks* PC hexamers; *dark shaded blocks* PE hexamers. **A** PBS consisting of PC peripheral rods. **B** PBS consisting of PE, PC rods. All APC cores are connected with PSII in thylakoid membranes at state 1

Step 2 is the heart of photosynthesis. The excited reaction centre Chl a of each photosystem reduces the partner pigment, pheophytin a (Pheo a) in PSII and Chl a in PSI, with consumption of excitation energy and resumes ground state, but in an oxidized form. Each photosystem consists of numerous proteins to form a giant protein complex. Within a complex, electron transfers from the reduced partner pigment to the electron acceptor at the outside of the complex and from the outside electron donor to the oxidized reaction centre Chl a occur. As a result, PSII induces water oxidation on one hand and plastoquinone (PQ) reduction, outside the complex, on the other. PSI causes oxidation of plastocyanin, the terminal redox carrier of the electron transport system connecting the two photosystems, and reduction of ferredoxin (Fd), the electron donor of $NADP^+$ reductase. Thus, the excitation energy of pigment molecules is converted to the redox energy within each photosystem.

Step 3 is analogous to the respiratory electron transport system (ETS) in the inner membranes of mitochondria. PQ shuttles electrons from PSII to the Cyt b_6-f complex, the third protein complex in the thylakoid membranes, then another electron shuttle, plastocyanin (in some algae, Cyt c_{553} acts in place of plastocyanin) connects the Cyt complex with PSI. For a smooth electron flow, balanced actions of the two photosystems are required. State transition is a regulatory mechanism for maintaining such a balance. Besides state transition, regulation of photosystem stoichiometry has also been known. The electron flow between the two photosystems causes $NADP^+$ reduction at the outside of thylakoids on one hand; on the other hand, electron flow within thylakoid membranes causes a proton gradient between the outside and the inside of thylakoids. ATP synthetase, the fourth protein complex in thylakoid membranes, synthesizes ATP at the thylakoid surface with consumption of energy of the proton gradient. Thus, redox energy is converted to biochemical energy in the forms of NADPH and ATP, and they are supplied to metabolism occurring in the stroma. Carbon fixation by the reductive pentose phosphate cycle, called the Calvin cycle, is the main event in stroma. Carboxylation by PEP carboxylase also occurs with the aid of the mitochondrial metabolism, although it is of minor importance. Nitrate and SO_4^{2-} reductions also occur with the consumption of NADPH and ATP. Here, light activation and dark inactivation occur in many enzymes so as to keep a dark metabolic state for a smooth cooperation with thylakoid system upon light-on.

The mechanism of photosynthesis has been well documented in many books (e.g. Bryant 1994; Falkowski and Raven 1997; Ort and Yocum 1998; Hall and Rao 1999), and will not be discussed at length in this chapter, which focuses on adaptive strategies for light harvesting and energy conversion in algae and cyanophytes.

6.3 Acclimation to the Light Quality of the Light-Harvesting Process

6.3.1 Complementary Chromatic Adaptation (CCA) — Acclimation of Pigment Species to the Light Quality

The phenomenon called CCA was first demonstrated by Engelmann (1883,1902). He observed change in colour of cyanophyte cells when grown under coloured light; the cells were blue-green in colour under red light and red or reddish brown under green light. After the change in algal colour, photosynthesis became more active under the new coloured light. Since the colour change is almost complementary to the ambient light, the phenomenon has been named complementary chromatic adaptation (Gaidukov 1903). Subsequently, it was shown that the change in algal colour was due to a change in dominance of pigment species of PBP, PE and PC.

A series of studies with *Tolypothrix tenuis* (Fujita and Hattori 1960a,b, 1962) revealed that (1) only PC is synthesized under red light, longer than 600 nm, whereas green light induces PE synthesis and suppresses PC synthesis; (2) a short preillumination with coloured light induces PBP synthesis during the subsequent dark period; red and green pre-illumination, respectively, cause PC and PE synthesis; (3) PBP species is determined by the quality of the light received most recently by the organism when red and green lights are alternately given. Action spectra for CCA (Fujita and Hattori 1962; Diakoff and Scheibe 1973; Haury and Bogorad 1977; Vogelmann and Scheibe 1978; Ohki et al. 1982) have shown that the action maximum for inducing PE and suppressing PC synthesis is located around 540 nm, and that for enhancing PC and suppressing PE synthesis, around 640 nm. The evidence that the light effect is experimentally separable from synthesis of pigment proteins suggests that the process of light-sensing is independent from the synthetic process of pigment protein. However, the nature of such a pigment(s) is still unknown.

Tandeau de Marsac (1977) grew 44 PE-containing cyanobacteria under red or green light, and classified them into three groups based on the features of CCA; strains that do not adapt chromatically (group 1), strains that adapt chromatically by modification of PE synthesis (group 2), and strains that adapt chromatically by modification of both PC and PE syntheses (group 3). Moreover, the ability of CCA was widespread in the cyanophyte genera examined by the above author.

As shown in Fig. 6.2, hexamers of the major PBP, PE and/or PC, form peripheral rods which attach to the core part consisting of allophycocyanin (APC) hexamers and terminal emitters. Rod structure of PBP hexamers and attachment between the core and rods are stabilized by specific colorless proteins called linker proteins (see Table 6.2 for important features of linker proteins). Based on the molecular structure of PBS, CCA is now recognized as the phenomenon that changes PBP species of the peripheral part of PBS in response to the colour of ambient light; under red light, only PC forms the peripheral part, and under green light, PE becomes dominant. During CCA, the synthesis of linker proteins for PC ($L_R^{30.5}$, $L_R^{32.5}$ and $L_R^{9.7}$) and PE ($L_R^{31.8}$, $L_R^{27.9}$ and $L_R^{27.6}$) is also photoregulated (Grossman et al. 1993).

Taking recourse to structural and physiological evidence, the mechanism of CCA has been analysed and developed at the genetic level. The two cyanophytes, *Calothrix* sp.PCC7601 and *Fremyella diplosiphon*, have been mainly used for such research. CCA of these cyanophytes is group 3 type (Tandeau de Marsac 1977). Three sets of *cpcBA* genes encoding α- and β-subunits of PC apoprotein were found in *F. diplosiphon*. Expression of two gene sets are regulated chromatically (Grossman et al. 1993). Expression of a set designated *cpcB1A1* (the gene product, constitutive PC, PCc) is constitutive, whereas genes designated

Table 6.2. Linker proteins stabilizing PBS structure

Linker protein	Molecular mass (kDa)	Gene designation	Function
$L_R^{30.5}$	30.5	*cpcH*	Stabilization of PC hexamers and PC rods
$L_R^{32.5}$	32.5	*cpcI*	
$L_R^{9.7}$	9.7	*cpcD*	
$L_R^{31.8}$	31.8	*cpeC*	Stabilization of PE hexamers and PE rods
$L_R^{27.9}$	27.9	*cpeD*	
$L_R^{27.6}$	27.6	*cpeE*	
L_{RC}		*cpcG*	Interphase between peripheral rods and APC core
$L_C^{7.8}$	7.8	*apcC*	Stabilization of APC cores

cpcB2A2 are transcribed only under red light (the gene product, inducible PC, PCi). The *cpeBA* gene responsible for the synthesis of the α- and β-subunits of PE apoprotein is expressed only under green light. Since the ratio of PC to PE is determined by the abundance of the two mRNAs in chromatically adapting cells, and since the turnover rate of each mRNA remains almost constant irrespective of light colour, the photoregulated expression of the two gene sets, *cpcB2A2* and *cpeBA*, is the main cause of CCA (Oelmueller et al. 1988a,b). Although genes of linker proteins for PE (*cpeCDE*) are clustered and form an operon independent of *cpeBA*, expression of the two operons is well coordinated (Grossman et al. 1993). Genes of linker proteins for PCi (*cpcD2H2I2*) are located downstream of *cpcB2A2*. Again, gene expression for apoprotein is coordinate to gene expression for linker proteins; the ratio *cpcB2A2* mRNA : *cpcB2A2H2I2D2* mRNA is almost equal to the ratio of PCi to their specific linker proteins (Grossman et al. 1993).

Two proteins, which bind specifically to the promoter of the *cpeBA* operon and are designated RcaA and RcaB, were found in the extracts of the cells grown under green light (Sobczyk et al. 1993). The binding activity of RcaA was lost by alkaline phosphatase treatment. This evidence has led to the interpretation that the transcription of *cpeBA* operon is regulated through the phosphorylation/ dephosphorylation of the effector protein(s) similarly to the regulation of transcription in bacteria. However, the base sequence which interacts with RcaA cannot be found in the promoter of the *cpeDHI* operon although *cpeDHI* expression is photoregulated similarly to the expression of *cpeBA* (Mazel et al. 1986).

A protein that binds to the promoter region of *cpcB2A2*, RcaD, was also found in the extracts of red light-grown cells of *Calothrix* sp. PCC7601 (Sobczyk et al. 1994). Further, the binding activity of RcaD is sensitive to alkaline phosphatase treatment. However, analysis of the promoter region of *cpcB2A2* in *F. diplosiphon* using the in vivo expression of a reporter gene connected to the promoter of *cpcB2A2* has indicated that the promoter region effective in regulation of gene expression in *F. diplosiphon* is different from the binding site for *RcaD* in

Calothrix (Casey and Grossman 1994). They also found that one or more than one proteins in the extracts from red light-grown cells interacted with the promoter region of *F. diplosiphon*, but the proteins have not been identified as yet. The promoter region may be different in the two cyanophytes.

Analysis of phenotypic mutants of *F. diplosiphon* has opened up a new phase of research. The FdR class phenotypic mutants, which synthesize an equal amount of PE under either red or green light but cannot synthesize PCi (Bruns et al. 1989), are complemented by *rcaC* gene from the wild type; *rcaC* gene was found to exist as two segments separated by inserted sequence in the mutant (Chiang et al. 1992). The predicted amino acid sequence of RcaC (Chiang et al. 1992) showed a high homology to the sequences of response regulators operating in the two-component system (Parkinsons and Kofoid 1992; Appleby et al. 1996). However, RcaA is unique: (1) it is almost two times larger than the ordinary response regulators, and (2) it contains two, instead of one, conserved aspartate-containing receiver domains: a DNA-binding domain and a histidine-containing transmitter domain located between the two aspartate-containing receiver domains. Aspartate in the C-terminal side of the receiver domain is essential (Kehoe and Grossman 1995). However, the role of aspartate in the N-terminal side receiver domain is unclear.

Complementation of FdBk class phenotypic mutants, which can synthesize both PC and PE irrespective of light quality, has revealed that the mutants lack the expression of the *rcaE* gene (Grossman and Kehoe 1997; Kehoe and Grossman 1997). Again, the *rcaE* gene in the mutant was found to have two segments separated by the insertion sequence. The C-terminal domain of RcaE was similar to that of the bacterial histidine kinase acting in the two-component signal transduction system. Interestingly, the predicted amino acid sequence of RcaE at the N-terminal side (about 140 amino acids) is similar to the sequence around the chromophore-binding site of the phytochrome (Kehoe and Grossman 1996), thereby suggesting that RcaE acts as a photosensor in CCA. However, it is not clear whether RcaE contains a chromophore.

Some of the FdR class mutants cannot be complemented by either *rcaC* or *rcaE*. One of such mutants was complemented with a small gene, *rcaF*, located downstream of *rcaE* (Kehoe and Grossman 1997). The predicted amino acid sequence of RcaF is similar to the amino acid sequence of small response regulators, CheY in *Escherichia coli* or Spo0 in *Bacillus subtilis* (Kehoe and Grossman 1997).

Based on the findings described above, Kehoe and Grossman (1997) proposed a working model that involves a four-step phosphorelay control mechanism (Fig. 6.3). In the model, RcaE is a photosensing protein, and RcaF is a mediator between RcaE and RcaC. Red light induces autophosphorylation of the histidine residue of RcaE. The phosphoryl group of RcaE is then transferred to the aspartate residue in the receiver domain of RcaF; it may be phosphorylated by other phosphate sources also. The phosphoryl group of RcaF is transferred to the

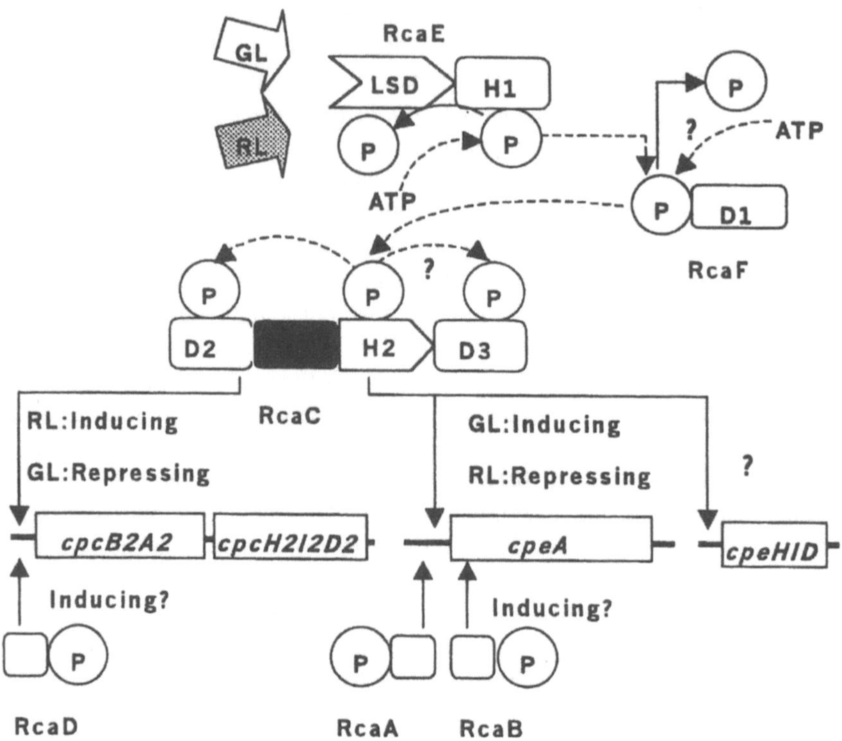

Fig. 6.3. Model for phosphorelay mechanism of signal transduction in CCA. The phosphorelay system consists of three putative components, light-sensor (*RcaE*) and two response regulators (*RcaF* and *RcaC*). Red light activates histidine kinase of RcaE and induces phosphorylation of RcaF and RcaC through the transfer of phosphoryl group as indicated by *dotted arrows*. Phosphorylated RcaC induces *cpcB2A2* and *cpcH2I2D2* expression. Green light suppresses histidine kinase activity and induces phosphtase activity of RcaE and prevents the phosphorelay (*solid arrow*). Non-phosphorylated RcaC induces *cpeBA* and *cpeCDE* expression and represses *cpcB2BA2* and *cpcH2I2D2* expression. *GL* green light; *RL* red light; *LSD* light-sensing domain of *RcaE*; *H1* histidine residue in the transmitter domain of RcaE; *H2* histidine residue in the receiver domain of RcaC; *D1* arginine residue in the transmitter domain of RcaF; *D2*, *D3*, arginine residue in the transmitter domain of RcaC. For further explanation, see text

aspartate residue (probably D51) in the receiver domain of RcaC via histidine in the transmitter domain. Phosphorylated RcaC induces the transcription of *cpcB2A2* and suppresses the transcription of *cpeBA*. Under green light, dephosphorylation of RcaF by the phosphatase activity of RcaE may occur. Unphosphorylated RcaC induces *cpeBA* expression and suppresses *cpcB2A2* expression. Molecular genetic analysis has greatly enhanced our understanding of CCA at the molecular level and helped in predicting a detailed mechanism of signal transduction, as shown in Fig. 6.3. However, most of the findings, especially on regulatory proteins for initiation and repression of the transcription, require biochemical confirmation.

Available evidence suggest that the PBS system of red algae has an ability for CCA (Lopez-Figueroa and Niell 1990; Abe et al. 1994; Sagert and Schubert 1995); PE/PC ratio is greater in thalli grown under green light, and smaller under red light. However, the amplitude of variation is rather small in every case, suggesting that the variation in red algae may be induced by a mechanism different from that for the cyanobacterial CCA. In most of the cases, the applied light conditions, especially the light quantity, were not adequate to examine the effect of the light quality. Therefore, it is possible that the size of the peripheral part of PBS varies in response to the light quantity effective for photosynthesis as explained in the following section (cf. Sect. 6.4); it is smaller under stronger light. PE/PC composition may vary in PBS of different size, since PE is located at the outermost of the peripheral part, and PC, at the innermost.

The blue to green world in deep water must be unfavourable for the Chl a/b system in most of the green algae since the content of blue to blue-green absorbing Chl b is low. However, there are numerous reports showing the occurrence of green seaweeds in waters deeper than 30 m (cf. Luning 1990). Some of these algae contain a high abundance of Chl b (Yokohama and Misonou 1980). A similar Chl a/b system is found in tiny green unicellular algae, often called pico- or ultra-phytoplankton, living in the deep open ocean, in a blue world (Stockner and Antia 1986; Stockner 1988). Light-harvesting Chl a/b proteins isolated from such pico-phytoplankton showed a Chl a/b ratio around 0.6 (Ohki and Honjho 1997). In an extreme case, Chl b content was found to be almost equal to Chl a (Ikeya et al. 1991; K. Ohki and S. Honjho, unpubl.). The composition of pigments in such picophytoplankton seems to be fixed, since their ratio of Chl a to b did not change under white light (K. Ohki and S. Honjho, unpubl.).

Another type of pigment system has been found in siphonaceous seaweeds. Their Chl a/b system contains green-absorbing siphonaxanthin and siphonein, which can transfer their excitation energy to Chl a, so that they can function as light-harvesting pigments for photosynthesis (Kageyama and Yokohama 1978; Hiller et al. 1990). However, such a pigment system does not necessarily occur only in organisms living in deep waters but in all the members of Siphonales (Yokohama et al. 1992).

Two types of special Chl a/b system appear to be genetically fixed. The phenotype may result not from adaptation to the light regime in their habitats but from phylogenetic processes. In this connection, a hypothesis has been proposed that at least the second type is an ancient form of Chl a/b system; such a system was first acquired when algae of the green line were established (Yokohama et al. 1992).

6.3.2 Regulation of Energy Transfer from the Light-Harvesting System to the Two Photosystems — State Transitions

Light regimes exciting the light-harvesting pigment system (light 2) cause a photochemistry bias in favour of PSII. Contrarily, excitation of Chl a by long wavelength (light 1) causes a photochemistry bias in favour of PSI, since the red absorption band of Chl a molecules in PSI is located at longer wavelength than that of PSII Chl a, and Chl a content in PSI is about double of that in PSII (Fujita and Murakami 1987; Cunningham et al. 1990). Under such light conditions, regulation of energy distribution from the light-harvesting system to the two photosystems occurs. The former light regime induces an increased energy transfer to PSI and a decreased transfer to PSII, resulting in energy distribution close to a balanced excitation of the two photosystems; the latter, a reverse regulation of energy distribution. The phenomenon was first found with the PBS system of *Porphyridium cruentum* (Murata 1969) and Chl a/b system of *Chlorella pyrenoidosa* (Bonaventura and Myers 1969). The phenomenon was also reported with the Chl a/c system (Shimura and Fujita 1973). Since the light regime in the former mainly excites PSII, such a situation is called state 2 of the light-harvesting system, and the latter, state 1. The phenomenon is called the state transition of the light-harvesting system (Fig. 6.4).

Although phenomenological appearance is common for different types of light-harvesting systems, the regulatory mechanism may not necessarily be the same. Regulation of Chl a/b system has been most extensively studied, especially in land plants. The Chl a/b light-harvesting system consists of the system for PSII (LHCII) and for PSI (LHCI). State transitions occur in LHCII. The most widely accepted model of this phenomenon involves phosphorylation/dephosphorylation of a 25-kDa peptide component of LHCII that is located in the peripheral area of LHCII complex, and concomitant movement of the component between the thylakoid sites for location of each PS complex. The 25-kDa protein is hydrophobic, but it becomes hydrophilic when phosphorylated. Under light 2, the peptide is phosphorylated and moves from the PSII to the PSI site. Since PSII is located in the area where thylakoids are stacked on each other, the hydrophilic phosphorylated LHC component has to be repulsed and moved to the hydrophilic thylakoid area where PSI is located. Thus, the LHCII component is situated closer to PSI and far from PSII, therefore the energy transfer becomes greater to PSI and smaller to PSII. Under light 1, the component is dephosphorylated by phosphatase and moves back to the hydrophobic area where PSII is located, and the energy transfer to PSII becomes greater (Fig. 6.4).

Phosphorylation/dephosphorylation of the component is mediated by intrinsic enzymes. Light regime causes activation/deactivation of LHCII protein kinase; light 2 activates the enzyme, and light 1 deactivates it. Phosphatase that dephosphorylates the phosphoprotein has been known to be constitutively active. It has long been assumed that the redox state of the PQ pool controls activity of

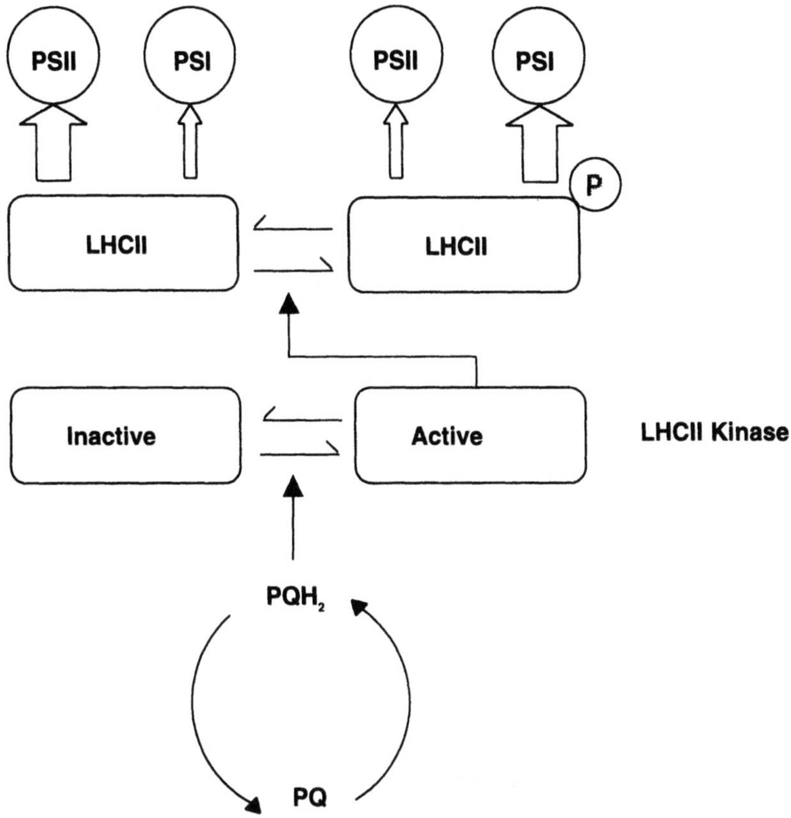

Fig. 6.4. Model for the mechanism of state transitions of Chl a/b system. *PQH_2/PQ* indicates redox state of PQ molecule at Qo site of Cyt b_6/f complex. For explanation, see text

the kinase, since the activation appears to follow the redox characteristic of the PQ pool (Horton et al. 1981). However, the study with mutants lacking Cyt b_6-f showed that the activation depends on this Cyt complex (Gal et al. 1987; Lemaire et al. 1987; Bennett et al. 1988; Coughlan 1988). An elegant experiment by Vener et al. (1997) for the activation mechanism with spinach chloroplasts revealed that the redox state of PQ molecule at the Qo site of the cytochrome is responsible for the activation; even when a majority of PQ in the pool is in an oxidized state, the enzyme is active in so far as the molecule at the Qo site is reduced. The conclusion has been confirmed by Zito et al. (1999) in the green alga *Chlamydomonas reinhardtii*. In the former study, the redox state of PQ molecules in the pool was manipulated so as to discriminate the state of PQ molecule at the Qo site whereas, in the latter, the relationship between mutant strains for Cyt b_6-f and the ability for enzyme activation was analysed. The evidence that the two studies of different type reached the same conclusion is a strong proof in support of the conclusion. It has been speculated that a conformational change of Rieske protein induced by PQH_2/PQ at the Qo site of Cyt b_6-f causes activation/deactivation of the protein kinase (cf. Vener et al. 1998; Zito et al. 1999).

The mechanism for the PBS system remains in controversy (cf. van Thor et al. 1998). There are three working models: the PBS-movement model (model A), the spillover model (model B) and the detachment model (model C) (cf. Fujita et al. 1994; van Thor et al. 1998). Model A predicts lateral movement of PBS between the two photosystems; PBS moves from PSII to PSI under light 2, and the reverse movement occurs under light 1. This is analogous to state transitions of Chl a/b system but, in this case, the entire PBS has to move between the two photosystems. Model B assumes the energy transfer from PSII to PSI; the flow becomes greater under light 2, and smaller under light 1. This idea was proposed by Murata (1969) on the basis of his studies on *Porphyridium cruentum*. In model C, regulation of coupling between PBS and PSII is assumed; PBS is decoupled from PSII under light 2, and coupled with PBS under light 1. This model is based on the evidence that change in the energy transfer to PSII is always greater than the transfer to PSI (Salehian and Bruce 1992).

Either model assumes that the affinity between PBS and the two photosystems determines the flux size of the energy flow to either photosystem like state transitions of Chl a/b system. Since protein phosphorylation can largely change the electronic character of the protein surface, such a regulatory event must involve a phosphorylation/dephosphorylation reaction as for the Chl a/b system. Indeed, Allen and his colleagues (Allen et al. 1985; Sanders and Allen 1987; Sanders et al. 1989) demonstrated phosphorylation/dephosphorylation of two proteins, the first an 18.5-kDa protein identified later as β-subunit of PC and the second a 15-kDa membrane-bound protein in the cyanophyte *Synechococcus* PCC6301 following illumination with light 2 and light 1, respectively. Since a PC-less mutant of *Synechocystis* PCC6803 still possesses the ability for state transitions, at least the 18.5-kDa protein is not involved in the mechanism of state transitions. Contrarily, Biggins et al. (1984a) and Kirschner and Senger (1986) independently reported no protein phosphorylation in *Porphyridium cruentum* during state transitions. They claimed that experimental conditions for *Synechococcus* PCC6301 were not adequate for examining state transitions. Thus, the involvement of protein phosphorylation in the mechanism is not yet clear. The relationship between state transitions and the ETS state is also under controversy. A correlation between the reduced state of the PQ pool and state 2 was found with *Synechochoccus* PCC 6301 (Mullineaux and Allen 1986), and suppression of state shift to state 1 by 2,5-dibromo-3-methyl-6-isopropyl-*p*-benzoquinone (DBMIB), an inhibitor for PQ pool oxidation, was also observed (Mullineaux and Allen 1991). However, Satoh and Fork (1983) and Biggins et al. (1984b) independently reported that shift to state 1 is blocked when cyclic electron flow around PSI is inhibited by carbonylcyanide-*m*-chlorophenyl-hydorazone (CCCP), methyl-viologen (MV) or DBMIB, and argued that the cyclic electron flow changes electrochemical environments around PS complexes causing a difference in affinity of PS complexes to PBS that leads to state 1. As will be explained in the following section (cf. Sect. 6.4), however, the regulation of photosystem stoichiometry involves the process for monitoring the redox state of Cyt b_6-f similarly to state transitions of Chl a/b system so that it is highly probable that either the same or analogous monitoring process works in state transitions of the PBS system also.

Molecular genetic analysis of the PBS system has revealed the detailed features of energy transfer from PBS to the two photosystems and also of state transitions. The PBS core consists of the bulk of APC and terminal emitters, APC-B, $\beta^{18.5}$ and L_{CM}, and mediates the energy transfer from peripheral PE and PC to Chl a of photosystems (Fig. 6.5; also consult van Thor et al. 1998; Ashby and Mullineaux 1999). Genes of three emitters as well as PE, PC and APC are already identified. Products of genes, *apcD, apcE* and *apcF*, are the α-subunit of APC-B, L_{CM}, and $\beta^{18.5}$, respectively. Analysis of mutants lacking genes of respective emitters has been providing new evidence for the mechanism of energy transfer to the two photosystems can be a great help in understanding the mechanism of state transitions. The location of ApcE in the energy-transfer path remains somewhat controversial. However, the model thus elaborated indicates that the energy

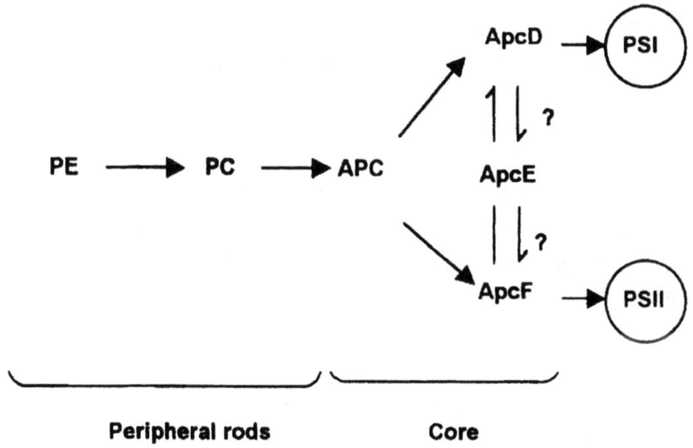

Fig. 6.5. Schematic presentation of the excitation energy flow from PBS to the two photosystems. *Arrows* indicate excitation energy flow. For explanation, see text

collected by PBS is directly transferred to PSI via the terminal emitter ApcD besides transfer to PSII via ApcF. Thus, the detachment model (model C) is impossible, since energy transfer from PBS to PSI is assumed to occur via PSII in this model. Zao et al. (1992) discovered that deletion of *apcD* of *Synechococcus*

PCC7002 abolishes the ability of state transitions leaving the pigment system at state 1 while deletion of *apcF*, the gene of another terminal emitter, seems to fix the pigment system at state 2. Their findings indicate that the regulation of energy distribution to the two photosystems mainly occurs at the coupling between the terminal emitters of PBS and Chl a of photosystems although the molecular mechanism is not clear. The lateral mobility of PBS on thylakoids is reported to be far greater than that of photosystems (Mullineaux et al. 1997). Thus, the state transition of the PBS system is most probably induced by the movement of PBS between the two photosystems (model A). Even when energy spills over from PSII to PSI occurs, the amplitude of regulation may be far smaller.

A recent analysis of mutant of *Synechocystis* PCC6803 has provided the new information that, besides terminal emitters of PBS, protein factor(s) is also involved in regulation of the energy distribution to the two photosystems. A 16-kDa intrinsic protein encoded by *rpaC* is essential for state transitions of this organism (Emlyn-Jones et al. 1999). However, this protein is different from the 15-kDa phosphoprotein found by Allen and his colleagues. The function of this protein is unknown. State transition is a fast phenomenon. Regulation of energy transfer is completed on a time scale of seconds to minutes. Recently, genes of two regulatory proteins designated *rpaA* and *rpaB* were found in *Synechocystis* PCC6803; in the *rpaA*-less mutant the energy distribution was fixed at state 1, while state 2 is fixed in the mutant poor in copy number of *rpcB* (Ashby and Mullineaux 1999). The authors argue that since the two proteins are DNA-binding type, they may act in the regulation of energy distribution to the two photosystems on a much longer time scale. If this is correct, such a system may help in expanding the range of state transitions.

Most of the genetic analyses have been conducted with the PBS system of cyanobacteria which contain hemidiscoidal PBS (cf. MacColl 1998). PBS of many rhodophytes is different in shape. However, the molecular structure of PBS core is most probably common in both PBS systems, so that the mechanism of state transition must also be common.

6.4 Acclimation of the Light-Harvesting Pigment System to the Light Quantity

The size of the light-harvesting system is variable; it becomes smaller under strong light, and larger under weak light. This variation can protect the photosynthetic apparatus from excess light energy under strong light, and collect the light energy under weak light efficiently. This phenomenon is called sun-shade adaptation, and occurs commonly in the light-harvesting system of each type. However, the mechanism for the variation may not necessarily be common. The variation can be followed by determining contents of pigments specific to the light-harvesting system; Chl b in the algae of green line, Chl c, in the algae of yellow line and PE and/or PC, in the algae of red line. An example is shown in Table 6.3. Since Chl a is commonly present in light-harvesting system and photosystems, ratios Chl a/b, Chl a/c and Chl a/PBP, are currently adopted for expressing the size of the light-harvesting pigment system.

Table 6.3. Photosynthetic pigments in cells of the green alga *Dunaliella salina* grown under weak and strong light

Pigment ratio	Weak light (500 μmol m^{-2} s^{-1})	Strong light (1700 μmol m^{-2} s^{-1})
Chl_{total}/cell (mol×10^{-13})	1.8±0.1	0.3±0.05
Chl b/Chl a	0.22	0.06
Carotenoids/Chl a	0.32	0.69
Chl_{total}/PSI	620	1090
Chl_{total}/PSII	440	360

Data recalculated from respective references

6.4.1 Chl a/b System

Extensive studies have been carried out on the light-harvesting system of the green alga *Dunaliella salina* (Escoubas et al. 1995; Durnford and Falkowski 1997) and a working model that includes regulation of gene expression has been proposed (Fig. 6.6). This model is based on experiments examining (1) correlation between the abundance of LHCII and mRNA of *Lhc*, genes of LHCII apoproteins, (2) correlation between ETS state and gene expression and (3) signal transduction mechanism. The model indicates that the state of the electron flow between the two photosystems is again monitored for the regulation. According to the proposed model, the redox state of the PQ pool is monitored. Similarly to state transitions of the Chl a/b system, however, the redox state of PQ at the Qo site of Cyt b_6-f may be monitored by the same or a similar mechanism. Since the rate of electron flow between the two photosystems is limited at the step for PQH_2 oxidation by Cyt b_6-f, the PQ pool, including PQ molecule bound to the Qo site, remains in a reduced state under strong light. Then intrinsic protein kinase is activated and phosphorylates the CPP (chloroplast phosphoprotein) present in stroma. Phosphorylated CPP moves out from chloroplasts to the cytoplasmic area and transfers the phosphoryl group to the DNA-binding protein (CRF). Phosphorylated CRF then moves to the nucleus and binds to the regulatory region of *Lhc* genes, resulting in the repression of their expression. Thus, repression of *Lhc* expression is stronger under a stronger light regime resulting in slower synthesis of apoprotein of LHCII. This model can explain the phenomenon well, but several aspects warrant further research: (1) identification of protein kinase, CPP, CRF and cytoplasmic phosphatase, (2) mechanism of signal transfer to cytoplasmic area across chloroplast envelopes and (3) mechanism of binding of DNA with phosphorylated CRF, and so on. Another major problem is the relationship between Chl biosynthesis and the ambient light quantity. Chl molecules are hierarchically supplied to LHC and two photosystems

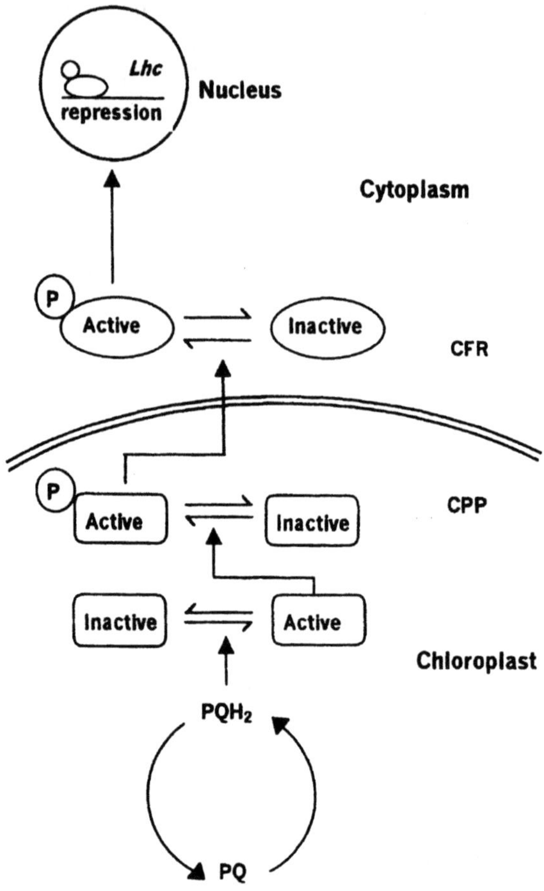

Fig 6.6. A working model for the light-induced repression of *Lhc* gene expression. *PQH$_2$/PQ* indicates redox state of PQ molecule at Qo site of Cyt b_6-f complex. Signals monitored in chloroplasts are transduced and repress *Lhc* expression in nucleus. For explanation, see text

in the order of PSII, PSI and LHC (cf. Fujita et al. 1990). If Chl biosynthesis is regulated in response to changes in light quantity, variation of LHC size is a result of combined regulatory events. Further, the synergetic effect of temperature and nutrient deficiency is well known. Strong light-induced changes in the assimilatory metabolisms of C, N and S may also contribute to the variation in the size of LHC.

No evidence is available for the mechanism of LHC variation of the Chl a/c system. However, evidence that at least in the diatom *Phaeodactylum tricornutum*, apoproteins of LHCII are nuclear encoded (Fawley and Grossman 1986; Friedman and Alberte 1986) and homologous to LHCII of Chl a/b system (Grossman et al. 1990) lead us to the assumption that the regulatory mechanism is the same as or analogous to that of Chl a/b system.

6.4.2 PBS System

PBS, the light-harvesting system of red algae and cyanobacteria, is very different from LHC in the Chl a/b and a/c systems. Thus, the mechanism must also be different from that described above. Variation in PBS size is induced not only by ambient light quantity but also by growth temperature, availability of nutrients such as N and S, etc. (Yamanaka and Glazer 1980; Collier and Grossman 1992). PBS seems to be not only the light-harvesting system but also serves as a nitrogen reserve (cf. Grossman et al. 1993, 1994). Thus, it is probable that the variation is a result of complex metabolic phenomena. However, the evidence available to elaborate the model for the regulatory mechanism has accumulated. Size reduction under N-deficient conditions first occurs at the peripheral part of PBS consisting of PE and/or PC proceeding from the outermost to the basal molecules and then at the core, resulting in reduction of the PBS population (Yamanaka and Glazer 1980; Grossman et al. 1993, 1994). Since the effect of light quantity is enhanced by N deficiency (Grossman et al. 1993, 1994), the above is true when the variation is induced simply by the light quantity. Since a rapid turnover of peripheral PE and PC molecules has been suggested (Ohki and Fujita 1991), the variation may be attributed to acceleration/deceleration of synthesis and/or degradation of peripheral molecules. In this connection, a mutant of *Synechococcus* PCC7942 insensitive to N deficiency has been isolated whose PBS size is not reduced under N-deficient conditions (Collier and Grossman 1994). A complementation experiment has indicated that a small soluble protein designated NblA, without homology to any protease, is essential to induce the reduction of PBS size. *NblA* expression is initiated by nutrient deficiency and strong light, so that it may be an activating factor for the existing protease or for modification of peripheral pigment protein molecules to a form sensitive to protease action if the protein is associated with degradation. Another gene, *NblR*, obtained by the analysis of a different mutant, is a component of the signal transduction system (Schwarz and Grossman 1998). The third gene, *NblB*, constitutive in nature, found in a similar way, is also essential for PBS degradation (Dolganov and Grossman 1999). An assumption for the regulatory mechanism was also made based on these findings (Schwarz and Grossman 1998); the signal from the light-sensing system, as for the system of state transitions, is transduced by a two-component system consisting of NblR, and it regulates *NblA* expression.

Regulation also possibly occurs in the process of formation or assembling of pigment protein molecules. A mutant of *Synechocystis* PCC6714 deficient in PC even under weak light (PC content, about one third of the wild-type strain) was isolated (cf. Nakajima and Ueda 1997). Deletion of the gene of the linker protein for PC seems to have occurred in this mutant (Y. Nakajima, pers. Comm.). If this is correct, such a phenotype suggests that assembling of PBS is also regulated through regulation of formation of linker protein(s). Further, C assimilation generally exceeds N and S assimilation under strong light so that such a biased metabolic feature may also negatively affect the synthesis of pigment proteins. As described above, light-dependent variation in the size of PBS may not occur under a simple regulatory mechanism, but is a result of numerous events occurring, depending on the light quantity.

6.5 Acclimation to the Light Quality at the Process of Energy Conversion — Regulation of Photosystem Stoichiometry

Together with state transitions, the balance between the actions of the two photosystems is adjusted by another regulatory mechanism, regulation of photosystem stoichiometry. Photosystem stoichiometry is not necessarily unity, but varies in response to light regime. Myers et al. (1980) first found that photosystem stoichiometry (PSI/PSII) in *Synechococcus* PCC6301 is greater than 1 in cells grown under fluorescent light absorbed mainly by PBS, while it is almost unity in cells grown under red light absorbed by Chl a. Variations in photosystem stoichiometry induced by light quantity were also found in cyanophytes (Kawamura et al. 1979); the abundance of PSI was variable; under a weak fluorescent light, it was two to three times greater than PSII, and reduced to the same level as PSII under strong white light. Based on this background knowledge, variations in photosystem stoichiometry have been studied in relation to light quality. These studies revealed that PSI/PSII stoichiometry is minimal for cells grown under Chl a-absorbed light that excites mainly PSI (PSI light), and maximal for cells grown under PBS-absorbed light that excites PSII (PSII light) (Fujita et al. 1985; Manodori and Melis 1986). Further, the photosystem stoichiometry under each light regime improves the photosynthetic efficiency under the same light regime (Murakami and Fujita 1988), thereby indicating that variation in photosystem stoichiometry in response to light quality is an adaptation of photosynthesis to light quality. As in state transitions, the balance between the actions of the two photosystems is held under biased light quality but, in this case, through adjustment of photosystem stoichiometry. Changes in photosystem stoichiometry also cause a variation in pigment composition. Since PBS/PSII stoichiometry in cyanobacteria is constant, the proportion of PBS to Chl a is greater in cells of low PSI/PSII while it becomes smaller in cells of high PSI/PSII; no change occurs in pigment species differing from CCA. Chromatic variation in pigment composition and quantum yield was first observed by Yocum and Blinks (1958) in the red alga *Porphyra nereocystis* and by Brody and Emerson (1959) in *Porphyridium cruentum*. Jones and Myers (1964) found the same phenomenon also with *Synechococcus* PCC6301. The reason for the variation remained unknown at that time. However, this phenomenon is now explained by the adaptive variation of photosystem stoichiometry.

Table 6.4. Stoichiometry of thylakoid components in various algae grown under PSI and PSII light

Algal strain	Stoichiometry (mol/mol)		Reference
	Cyt b_6-f/PSII	PSI/PSII	
Cyanophytes			
Synechocystis PCC6714			
PSI light-grown	0.87	1.16	Fujita et al. (1987)
PSII light-grown	0.91	3.02	
Green algae			
Chlamydomonas reinhardtii			
PSI light-grown	0.87	0.45	Melis et al. (1996)
PSII light-grown	0.83	0.88	
Red algae			
Porphyridium cruentum			
PSI light-grown		0.83	Cuninngham et al.
PSII light-grown		3.85	(1990)
Porphyra yezoensis			
PSI light-grown	0.51	0.71	Abe et al. (1994)
PSII light-grown	1.06	5.94	

Data recalculated from respective references.

Variation in photosystem stoichiometry has been extensively studied mainly with *Synechocystis* PCC6714 at the physiological level (cf. Fujita 1997). Quantitative analysis of the two photosystems (Fujita and Murakami 1987; Aizawa et al. 1992) has revealed that the abundance of PSI, but not PSII, is regulated in response to the light quality. Further, the synthesis, but not degradation, of PSI was found to be regulated (Fujita et al. 1988). As shown in Table 6.4, Cyt b_6-f/PSII stoichiometry is always constant not only in cyanophytes but also in the green alga *Chlamydomonas reinhardtii* and probably in land plants. PSI/PSII variation in response to light regime results mainly from the regulation of the synthesis of the terminal of the thylakoid ETS, PSI. Interestingly, the activity of another ETS terminal, cytochrome oxidase, in the cyanophyte *Synechocystis* PCC6714 was found to vary in parallel with variation in PSI abundance (Murakami et al. 1997), thus suggesting that a similar regulation occurs in the ETS terminal for respiration.

As shown in Table 6.4, however, not only PSI but also PSII abundance varies in red algae; the abundance of PSII varies in reverse manner to variation in PSI (Cunningham et al. 1990; Abe et al. 1994). Although such evidence suggests that the abundance of PSII is also regulated in the red algal systems, a recent study on *Porphyridium cruentum* (Fujita 1999) has indicated that the reduction of PSII abundance upon shift from PSI light to PSII light is mainly attributed to photoinactivation of PSII followed by its degradation. Since such a photoinactivation does not occur under PSI light, a reverse light shift can cause increase in PSII abundance up to the maximal level. Since one PBS associates with plural PSII and the number of PSII associating with one PBS is maximal under PSI light, some PSIIs have to be blocked and subsequently destroyed under PSII light, that excites PSII more efficiently than PSI (Fujita 1999). If such a situation is common in the red algal system, the pattern of stoichiometric variation in the above system is a result of at least two events, the active regulation of PSI abundance and the photodestruction of PSII under PSII light. Thus, it is highly probable that variation in photosystem stoichiometry in red algae is achieved by the same mechanism as in green algae and cyanophytes except for an additional event, the photodestruction of PSII. At present, we do not know whether the photosynthetic system of the algae of the yellow line has a similar ability for adaptation to light quality.

Comparison of stoichiometry variation in cells grown under different trophic conditions indicated that variation in PSI abundance correlates with the redox state of electron transport between the two photosystems (Fujita et al. 1987); the level of PSI was elevated upon overreduction of the pool between the two photosystems, and was reduced under the state for overoxidation of the pool. This interpretation was confirmed by a direct analysis of ETS state by flash-induced absorbance change of Cyt f and P700 (Murakami and Fujita 1991a). These features indicate that (1) stoichiometry adjustment results in an open state of both photosystems so that photosynthetic efficiency is improved, and (2) the redox state of the PQ pool and/or Cyt b_6/f is monitored for regulation of PSI abundance.

In the experiments for flash analysis, a rapid Cyt f reduction after the flash was always observed ($T_{1/2}$, 3 ms; ordinary Cyt f reduction, 10 ms) only when cells grown under PSI light were shifted to PSII light. Such a rapid Cyt f reduction after the flash was assumed to correspond with Cyt f reduction through the Q cycle, so that the redox state of Cyt b_6-f, but not the PQ pool, might be monitored for adjustment of photosystem stoichiometry. The evidence that 2-heptyl-4-hydroxlquinoline *N*-oxide (HQNO), an inhibitor for the Q cycle of Cyt b_6-f in *Synechocystis* PCC6714 (Matsuura et al. 1988), slowed down Cyt f reduction to the rate of the PSII-adapted system and suppressed the increase, but not the decrease, in the stoichiometry PSI/PSII (Murakami and Fujita 1993), supports the above assumption and suggests a possibility that the turnover of the Q cycle correlates closely with the signal development (cf. Fujita 1997).

As discussed above, phosphorylation of LHC in state transitions of the land plant system is regulated in response to the redox state of PQ molecule at the Qo site of Cyt b_6-f; the protein kinase for LHC phosphorylation is most probably activated by PQH_2 at the Qo site through a conformational change in Rieske protein. In the two adaptive mechanisms, a putative site for monitoring the electron transport state is common, Cyt b_6-f. However, the mechanism for monitoring the ETS state is different; the redox state of the PQ molecule located at the Qo site is monitored in state transitions while the electron flow through the Q cycle is watched for the regulation of photosystem stoichiometry. Is the monitoring mechanism truly different in the two types of regulation? Since the function of the two regulations is the same, balancing the actions of the two photosystems, it is feasible that the ETS state must be monitored under the same mechanism in both types of regulation. In the study of photosystem stoichiometry, a rapid Cyt f reduction was assumed as a reflection of a high electron flux through the Q cycle (Murakami and Fujita 1991a, 1993) and this was the main evidence in favour of their hypothesis. However, inhibitors suppressing the Q cycle, such as HQNO, cannot affect LHC phosphorylation, which induces the state 2 in a *Lemna* mutant (Gal et al. 1987). Further, Cyt b_6 remained in the oxidized state even when LHC phosphorylation is stimulated (Vener et al. 1997). Although this evidence does not necessarily discount the Q-cycle hypothesis, it is unfavourable to it. The amplitude of the rapid Cyt f reduction after flash oxidation, which correlated with the elevation of PSI/PSII stoichiometry, may correspond to the proportion of the complexes turning over at the highest rate. Since the Qo sites of such molecules must be occupied by PQH_2 before flash, results obtained by Murakami and Fujita (1991a, 1993) can be explained alternatively, for example that elevation of PSI/PSII stoichiometry and formation of PSI complexes are stimulated in proportion to the amount of Cyt b_6-f, the Qo site of which is occupied by PQH_2. If this is correct, the mechanism for monitoring the ETS state is common in the two types of regulation, state transitions and the regulation of photosystem stoichiometry, and protein phosphorylation may be involved in the signal transduction of the latter regulation.

PSI formation may be regulated at the level of either gene expression, processing products of gene expression, translation of mRNAs or assembling of the PSI complex. Decline in the activity for synthesis of PsaA/B proteins, forming the core of PSI, under suppression of gene expression, was found to be the same under either PSI or PSII light, indicating that stability of mRNA is not affected by the light regime. Further, no difference was observed between in vitro translations with RNA isolated from cells grown under PSI and PSII light (Aizawa and Fujita 1997). Only in vivo synthesis of PsaA/B was strongly affected by the light regime; the rate of synthesis under PSII light was twice as fast as that under PSI light (Aizawa and Fujita 1997). Although these findings are not direct evidence, it has been interpreted that PSI formation is regulated at or after the level of translation; the steps from the gene expression to the production of matured mRNAs, including decay of mRNAs, are not responsible for the regulation of PSI formation even when such a step(s) is regulated by the light regime. Aizawa and Fujita (1997) measured the synthesis of stable PsaA/B proteins which were probably bound with Chl a molecules and located in the PSI complex. If the translation product is unstable and quickly degraded prior to its binding with Chl a, their results do not necessarily indicate regulation at the translation level. In a mutant of *Plectonema boryanum* lacking the light-independent protochlorophyllide reductase, a temporary elevation of Chl a supply by either PSI or PSII light after a long dark incubation caused a temporary acceleration of synthesis of stable PsaA/B. The accelerated level was almost the same under either light, although after prolonged illumination the rates of synthesis were slowed down to the levels determined by respective light regimes. This evidence indicates that synthesis of stable PsaA/B proteins depends on Chl a supply (Aizawa and Fujita 1997). Thus, the step for assembling PsaA/B proteins with Chl a molecules or the step(s) in the process of Chl a biosynthesis is the most possible candidate to be controlled, in the regulation of photosystem stoichiometry.

Recently, Pfannschmidt et al. (1999a,b) reported that in seedlings of mustard plants, expression of genes of core peptides of PSI (*psaA/B*) and PSII (*psbA*) are controlled by the redox state of the PQ pool; the light regime inducing a highly reduced state of the PQ pool accelerates transcription rates of *psaA/B* genes, and the transcription rate of *psbA* gene is accelerated under the light regime for the oxidized state of the PQ pool. They proposed a model that photosystem stoichiometry is regulated differently from the model described above at the level of gene expression in response to the redox state of the PQ pool. However, their light shift experiments were conducted with intact cotyledons, then chloroplasts were isolated for determination of rates of gene expression and the abundance of mRNA. It is somewhat difficult to understand how they successfully maintained the state of ETS during the period from preparation of chloroplasts to assay for gene expression. Recent reports have shown that *psbA* gene expression is regulated in response to the redox state of Cyt b_6-f in *Synechocystis* PCC6803 but in a reverse way to the observation by Pfannschmidt et al. (Alfonso et al. 1999, 2000). The relationship between *psbA* expression and the redox state of Cyt b_6-f is controversial. In *Synechocystis* PCC6714, the expression of photosystem genes was not directly determined, so the regulation of gene expression might have been missed. Thus, the investigation for the molecular mechanism is still under controversy. More comprehensive experiments are necessary for evaluating the two models.

6.6 Acclimation to the Light Quantity at Photosystem Level

6.6.1 Regulation of Photosystem Stoichiometry in Response to the Light Quantity

Photosystem stoichiometry is regulated not only by the light quality but also by the light quantity for algal growth. As shown in Table 6.5, a common pattern can be found in photosynthetic systems of cyanophytes, green algae and land plants; the stoichiometry, PSI/PSII, is smaller when algae and plants are grown under strong light. Again, the stoichiometry is adjusted by regulation of the abundance

Table 6.5. Stoichiometry of thylakoid components of various algae and land plants grown under weak and strong light

Organism	Stoichiometry (mol/mol)		Reference
	Cyt b_6-f/PSII	PSI/PSII	
Cyanophytes			
Synechocystis PCC6714			
Weak light-grown	0.9	2.9	Murakami and Fujita
Strong light-grown	0.9	1.3	(1991b)
Green algae			
Dunaliella salina			
Weak light-grown		0.71	Smith et al. (1990)
Strong light-grown		0.33	
Red algae			
Porphyridium cruentum			
Weak light-grown		1.85	Cuninngham et al.
Strong light-grown		1.89	(1990)
Diatoms			
Phaeodactylum tricornutum			
Weak light-grown		1.72	Friedman and Alberte
Strong light-grown		1.56	(1986)
Nitzschia closterium			
Weak light-grown		1.0	Y. Fujita (unpubl.)
Strong light-grown		0.5	
Land plants			
Pisum sativum			
Weak light-grown	1.0	1.0	Leong and Anderson
Strong light-grown	1.0	0.4	(1986)

Data recalculated from respective references are presented.

of PSI. Murakami and Fujita (1991b) assumed that the regulatory mechanism responding to light quality also functions in the synthesis of PSI. When we adopt the Q-cycle mechanism (see Sect. 6.5), the assumption seems to be valid. However, if the Qo-site mechanism (see Sect. 6.3.2) governs the regulation responding to light quality, the formation of PSI should be accelerated in this case. Decrease in PSI is plausibly attributed to some other mechanism, such as accelerated degradation. However, the available data are not sufficient to elaborate a feasible hypothesis.

The pattern common in cyanobacteria and the algae of green line is not necessarily true in the algae of red and yellow line (Table 6.5). However, some of these algae, e.g. *Porphyridium cruentum* and the diatom *Nitzschia closterium* (Y. Fujita, unpubl. data), seem to have the ability to regulate photosystem stoichiometry in response to light quality, as in cyanophytes, (Table 6.4). Algae of the red and yellow lines may have potentially the same ability for regulating photosystem stoichiometry as cyanophytes and green algae, but may be responding to the light quantity with a different strategy, such as reducing photosynthetic unit per chloroplast or per cell and so on.

6.6.2 Photosystem II Repair Cycle Under Strong Light

It is due to the well-known phenomenon called photoinhibition that under strong light, above that saturating to photosynthesis, photosynthetic activity is apt to be reduced. Photoinhibition is observed not only in land plants but also in algae of every phylogenetic line; the phenomenon becomes prominent in algae growing under a weak light (cf. Osmond 1994). Reduction of the activity under strong light has been known to result mainly from inactivation of PSII with photodestruction of the D1 peptide (encoded by *psbA* gene), one of the core peptides of PSII (cf. Ort and Yocum 1998). Under such conditions, the repair of photoinactivated PSII, replacement of the D1 peptide, occurs. The D1 peptide has been known to turn over rapidly (cf. Aro et al. 1993), the turnover becoming faster under strong light. This is an indication of the repair of inactivated PSII. The study of D1 synthesis in the green alga *Chlamydomonas reinhardtii* under strong light revealed regulation at the translation level by protein factors which are activated in the light (Danon and Mayfield 1994). The active factors bind to the *psbA* mRNA at the 5' untranslated region. The binding is essential for formation of the initiation complex of mRNA and ribosome (cf. Bruick and Mayfield 1999). In the light, one of the factors, RB60 protein, a homologue of protein disulfide isomerase (PDSI), is reduced through the ferredoxin-thioredoxin system. The reduced form of this protein then reduces a homologue of poly(A)-binding protein (PABP) in chloroplasts. PDSI and reduced PABP form a complex with at least two other uncharacterized proteins and attach to the untranslated region of *psbA* mRNA. Attachment of *trans* factors causes activation of mRNA for the translation (Fig. 6.7). Thus, the synthesis of D1 becomes faster under stronger light, thereby facilitating the replacement of the damaged D1. The study was performed using the green alga *Chlamydomonas reinhardtii*; evidence for this mechanism has not been available for other algae. However, the basic mechanism may be expected to be the same in algae of other phylogenetic groups. Recently, the expression of *psbA* gene in the cyanophyte *Synechocystis* PCC6803 has been found to be also regulated in response to the redox state of, probably, Cyt b_6-f (Alfonso et al. 1999, 2000). The expression of *psbA* was accelerated markedly when the PQ pool and Cyt b_6-f were reduced in the light. Although degradation of mRNA was simultaneously accelerated, the level of its abundance became higher. Using inhibitors for thylakoid ETS, these authors proposed a model that *psbA* expression is regulated in response to the redox state of the PQ molecule at the Qo site of Cyt b_6-f. Thus, light quantity seems to regulate D1 synthesis at the gene expression level and also at the translation level if the two regulatory mechanisms are commonly equipped, at least in cyanophytes and green algae.

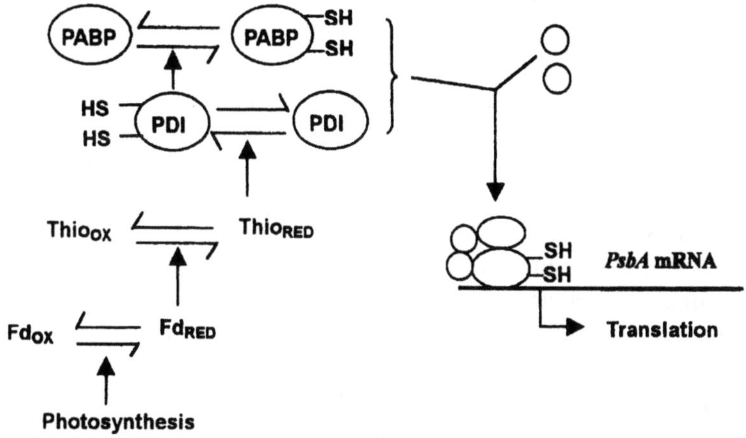

Fig. 6.7 Schematic presentation of the mechanism of light-dependent translation of *psbA* mRNA. Fd_{ox} and $Thio_{ox}$ indicate oxidized forms of ferredoxin and thioredoxin, respectively; Fd_{red} and $Thio_{red}$ reduced forms of ferredoxin and thioredoxin, respectively. For explanation, see text

Such a replacement requires processes other than D1 synthesis, such as movement of inactivated PSII from the stacked area of thylakoids to the unstacked area, removal of damaged D1 from the PSII complex, and so on. Such processes may also be accelerated under strong light (cf. Aro et al. 1993; Andersson and Barber 1996). A recent report (Spetea et al. 1999) has shown that D1 is first fragmented by serine protease into 23-kDa N-terminal fragments. This process depends on GTP being tightly bound to thylakoid membranes, suggesting that the process is a regulatory reaction. Then, fragments are degraded by ATP-and Zn-dependent protease(s).

6.6.3 Xanthophyll Cycle —a Process for Dissipation of Excess Energy

Another mechanism for protecting PSII from strong light is known as the xanthophyll cycle (cf. Yamamoto and Bassi 1996; Faeber and Jahns 1998). This is a mechanism for dissipating the excess energy absorbed by the light-harvesting system. Under strong light, a full turnover of thylakoid ETS, not only for $NADP^+$ reduction, but also those for the Mehler reaction and for cyclic ETS, brings about acidic environment in the inside of thylakoids lumen. Such a pH shift activates violaxanthin de-epoxidase in thylakoids, and conversion from violaxanthin to zeaxanthin and astaxanthin occurs, which can directly or indirectly non-radiatively dissipate the excess energy in Chl a molecules of LHC. The xanthophyll cycle has been known to operate in LHCII for protecting PSII from excess light. Plants grown under strong light have been reported to develop a greater capacity for zeaxanthin formation (cf. Yamamoto and Bassi 1996). Thus, the energy dissipation system of this kind must be regulated in response to the light stress, although the regulatory mechanism is not yet known. The details of molecular mechanism of the cycle, specially at the step(s) for non-radiative dissipation of excess energy in LHC via zeaxanthin and astaxanthin are not very clear so far.

Most of the studies on the xanthophyll cycle have been conducted with land plants. However, zeaxanthin commonly exists in algae belonging to green and red lines including cyanophytes, so that the xanthophyll cycle must be operating in algae of these phylogenetic groups. In algae of the yellow line and in dinophytes, different xanthophyll species such as diadinoxanthin and diatoxanthin may be involved in this cycle (Stransky and Hager 1970; Frank et al. 1996). However, a recent investigation has shown that the violaxanthin cycle also acts under strong light, together with the diadinoxanthin cycle, in the algae of the yellow line and dinophytes (Lohr and Wilhelm 1999).

6.7 ETS State as a Signal for Regulation of the Photosynthetic System

Except for CCA in the light-harvesting system consisting of PBS, the light regime is always sensed through the redox steady state of the thylakoid ETS. As described in the preceding sections, so far as we know, the redox steady state is sensed at two sites; the one at the Cyt b_6-f complex, and the other at Fd, the terminal electron acceptor of the thylakoid ETS. The redox level at the former site reflects the balance between the turnovers of the two photosystems under a weak light limiting the overall process of photosynthesis. Indeed, adaptive responses to the biased light quality, state transitions and adjustment of photosystem stoichiometry, are induced in response to the redox state of Cyt b_6-f. The redox state of the PQ pool also reflects such a balancing state. Since the oxidation of PQH_2 at the Qo site of Cyt b_6-f is the slowest step in the electron flow between the two photosystems, the state of components acting after PQH_2 oxidation, such as plastocyanin, reflects only the balance between PQH_2 oxidation and turnover of PSI. Indirect evidence indicates that the abundance of electron mediator between Cyt b_6-f and PSI, plastocyanin, also varies in parallel with variation in PSI amount (Murakami and Fujita 1991a). The redox state of the PQ pool has long been believed to be sensed for state transitions (cf. Fujita et al. 1994). This assumption is still found in many books. However, the recent studies by Vener et al. (1997) and Zito et al. (1999) seem to resolve such a controversy.

It is highly probable that the ETS site monitored for the regulation of light-harvesting size is also the Qo of *Cyt* b_6-f though the model proposed by Escoubas et al. (1995) involves redox-sensing at the PQ pool. If this is correct, the redox state of Qo site is monitored for the rate of electron flow through PSII in this case. Since the redox reaction at the Qo site is the slowest, the redox state at this site will be affected not only by the biased action of the two photosystems but also by the amplitude of the electron flux through PSII. Thus, the mechanism sensing the redox state of PQ at the Qo site may be common in various regulating processes responding to the light regime (Table 6.6).

Concerning the acceleration of D1 synthesis in the repair cycle, the regulation occurs in response to the state of ETS in two ways. The first is the regulation of *psbA* expression, which occurs probably by a mechanism similar to that for state transition, the regulation of LHC size and the regulation of photosystem stoichiometry. The second is the regulation at the translation level. Since the *trans*-acting protein factor for the translation of *psbA* mRNA becomes active when its disulfide group is reduced by thioredoxin, acceleration of D1 synthesis depends on the electron flux through thioredoxin from the terminal of thylakoid ETS. Thus, the electron flux from the output of thylakoid ETS is another signal for the regulation (Table 6.6). Thioredoxin system functions as an indicator for the amplitude of incident light in this case; the greater light quantity causes the

Table 6.6. ETS sites for monitoring light regime and regulatory events

Monitored ETS site	Monitored light regime	Regulatory event
Q_0 site of Cyt b_6-f (redox state of PQ molecule at Q_0)	Light quality	State transitions of Chl a/b system (activity of LHCII kinase)
		Regulation of photosystem stoichiometry (synthesis of stable Psa A/B)
	Light quantity	Regulation of LHCII size (expression of *Lhc* genes)
		Regulation of photosystem stoichiometry ?
		Regulation of D1 synthesis (expression of *psbA* gene)
Ferredoxin-thioredoxin system (electron flux from the thylakoid ETS output)	Light quantity	Regulation of D1synthesis (initiation of translation of *psbA* mRNA)
	Light on/off	Activation of enzymes (enzymes of Calvin-Benson cycle, ATP synthetase)

greater signal, which induces the faster D1 synthesis. Further, the signal elicited from the thioredoxin system acts as the light switch. In the dark, no electron flux can occur, so that the thioredoxin system is switched off. Light-on drives the electron flux through the photosynthetic ETS and the thioredoxin system is switched on. Because of this switching characteristic, the thioredoxin system regulates the light-induced activation of enzymes of Calvin cycle (Buchanan 1991) and chloroplastic ATP synthetase (Mills et al. 1980). It has been hypothesized that, besides D1 synthesis, translational regulation by this switching mechanism may also act in the synthesis of other functional proteins in chloroplasts (cf. Bruick and Mayfield 1999).

6.8 General Considerations

In this chapter we discussed the flexibility of the photosynthetic energy conversion process. The flexibility appears at every step of energy conversion processes and allows algae to live in light environments of a wide range. Although our understanding of such a flexibility is not complete as yet, recent molecular genetic and biological analyses have greatly accelerated the pace of research progress. Most of the regulatory mechanisms that bring about such a flexibility have a common basic structure, that is three basic processes: (1) the process for monitoring the light regime, (2) the process for transducing the signal elicited by the monitoring system to the one for regulation, and (3) the process for regulating the transcription, the translation of target protein(s) or the activity of the existing enzyme(s). In the first process, the light regime is not directly monitored, but is sensed through the redox state of ETS except for CCA. As explained above, the terminal cytochrome oxidase in cyanophytes appears to be regulated like the abundance of PSI. If it is correct, the regulation of ETS terminal in response to the redox state of ETS might be equipped in an ancient prokaryotic ETS that is much older than the photosynthetic ETS. Further, in most cases, protein phosphorylation/dephosphorylation appears to be involved in the signal transduction, which is the type of two-component system common in bacteria. Some of the proteins involved in signal transduction appear to be homologous to those in bacteria. Such a characteristic may come from the common origin of chloroplasts of algae and land plants. Indeed, most of the regulatory abilities are common in all algal phyla and land plants, so that we can suppose that such regulatory abilities might have been elaborated in ancient cyanophytes which had initiated oxygenic photosynthesis 3000 million years ago.

For convenience in explanation, each regulatory phenomenon has been explained independently of other regulatory events. However, many regulatory events must occur simultaneously, resulting in an adaptive phenomenon fitted for the ambient light regime. For example, biased light quality causes state transitions, on one hand, and induces simultaneous regulation of photosystem stoichiometry on the other. The former is a fast event and is followed by the latter, slower event, indicating that the range of regulation in the former is narrower than the latter. For balancing the actions of the two photosystems, the regulation of photosystem stoichiometry acts as a coarse tuning to the ambient light, and state transitions as a fine tuning. The adaptive feature of the photosynthetic system, in most cases, must be the result of multiple events that maintain a homeostasis of photosynthetic function.

Besides adaptive phenomena which are induced through a specific and active mechanism that actively senses the light regime and shows an adaptive feature like CCA and state transitions, there are phenomena that result passively from a biased state among plural metabolisms due to environmental stress, but show a variation suited for living under such environments. The phenomena induced by strong light, such as variation of PBS size, may be in this category, although an active regulatory mechanism may be involved in the variation. In such cases, the synergistic effect is apt to appear among different environmental factors; the light effect is enhanced by temperature, nutrient stress and so on. These two types can be distinct only under an experimental system. Even under experimental conditions, however, the appearance of adaptive phenomena still involves such a complexity; experimental conditions merely emphasize one of the multiple processes. Research requires such a focus that when we look at some adaptive phenomenon, we should have an insight into the complexity inherent in the phenomenon.

References

Abe S, Murakami A, Ohki K, Aruga Y, Fujita Y (1994) Changes in stoichiometry among PSI, PSII and Cyt b6-f complexes in response to chromatic light for cell growth observed with the red alga *Porphyra yezoensis*. Plant Cell Physiol 35:901-906

Aizawa K, Fujita Y (1997) Regulation of synthesis of PSI in the cyanophytes *Synechocystis* PCC6714 and *Plectonema boryanum* during acclimation of the photosystem stoichiometry to the light quality. Plant Cell Physiol 38:319-326

Aizawa K, Shimizu T, Hiyama T, Satoh K, Nakamura Y, Fujita Y (1992) Changes in composition of membrane proteins accompanying the regulation of PSI/PSII stoichiometry observed with *Synechocystis* PCC6803. Photosynth Res 32:131-138

Alfonso M, Perewoska I, Constant S, Kirilovsky D (1999) Redox control of psbA gene expression in the cyanobacterium *Synechocystis* strain. Photochem Photobiol B:Biol 48:104-113

Alfonso M, Perewoska I, Kirilovsky D (2000) Redox control of psbA gene expression in the cyanobacterium *Synechocystis* PCC6803. Involvement of the cytochrome b6/f complexes. Plant Physiol 122:505-515

Allen JF, Sanders CE, Holmes NG (1985) Correlation of membrane protein phosphorylation with excitation energy distribution in the cyanobacterium *Synechocystis* PCC6301. FEBS Lett 193:271-275

Andersson B, Barber J (1996) Mechanisms of photodamage and protein degradation during photoinhibition of photosystem II. In: Baker NR (ed) Photosynthesis and the environment. Kluwer, Dordrecht, pp 101-121

Appleby JL, Parkinson JS, Burret RB (1996) Signal transduction via the multi-step phosphorelay: not necessarily a road less travelled. Cell 86:845-848

Aro E-M, Virgin I, Andersson B (1993) Photoinhibition of photosystem II. Inactivation, protein damage and turnover. Biochim Biophys Acta 1143:113-134

Ashby MK, Mullineaux CW (1999) Cyanobacterial ycf27 gene products regulate energy transfer from phycobilisomes to photosystems I and II. FEBS Lett 181:253-260

Bennett J, Shaw EK, Michel H (1988) Cytochrome *b6f* complex required for phosphorylation in green plant chloroplasts. Annu Rev Plant Physiol Plant Mol 42:281-311

Biggins J, Campbell CL, Bruce D (1984a) Mechanism of the light state transition in photosynthesis. II Analysis of phosphorylated polypeptides in the red alga, *Porphyridium cruentum*. Biochim Biophys Acta 767:138-144

Biggins J, Campbell CL, Creswell LL, Wood EA (1984b) Mechanism of light state transition in *Porphyridium cruentum*. In: Sybesma C (ed) Advances in photosynthesis research. Martinus Nijhoff, Dordrecht, pp 303-306

Bonaventura C, Myers J (1969) Fluorescence and oxygen evolution from *Chlorella pyrenoidosa*. Biochim Biophys Acta 189:366-383

Brody M, Emerson R (1959) The quantum yield of photosynthesis in *Porphyridium cruentum*, and the role of chlorophyll a in the photosynthesis of red algae. J Gen Physiol 43:251-264

Bruick RK, Mayfield SP (1999) Light-activated translation of chloroplast mRNAs. Trends Plant Sci 4:190-195

Bruns BU, Briggs WR, Grossman AR (1989) Molecular characterization of phycobilisome regulatory mutants of *Fremyella diplosiphon*. J Bacteriol 171:901-908

Bryant DA (ed) (1994) The molecular biology of cyanobacteria. Kluwer, Dordrecht

Buchanan BB (1991) Regulation of CO_2 assimilation in oxygenic photosynthesis: the ferredoxin/thioredoxin system: perspective on its discovery, present status and future development. Arch Biochem Biophys 288:1-9

Casey ES, Grossman AR (1994) In vivo and in vitro characterization of the light-regulated cpcB2A2 promoter of *Fremyella diplosiphon*. J Bacteriol 176:6362-6374

Chiang GG, Schaefer MR, Grossman AR (1992) Complementation of a red-light-in different cyanobacterial mutant. Proc Natl Acad Sci USA 89:9415-9419

Collier JL, Grossman AR (1992) Chlorosis induced by nutrient deprivation in *Synechococcus* sp. Strain PCC 7942: not all bleaching is the same. J Bacteriol 174:4718-4726

Collier JL, Grossman AR (1994) A small polypeptide triggers complete degradation of light--harvesting phycobiliprotein in nutrient-deprived cyanobacteria. EMBO J 13:1039-1047

Coughlan SJ (1988) Chloroplast thylakoid protein phosphorylation is influenced by mutations in the cytochrome bf complex. Biochim Biophys Acta 933:413-422

Cunningham FX, Dennenberg RJ, Jursinic PA, Gantt E (1990) Growth under red light enhances photosystem II relative to photosystem I in the red alga *Porphyridium cruentum*. Plant Physiol 93:888-895

Danon A, Mayfield SP (1994) Light-regulated translation of chloroplast messenger RNAs through redox potential. Science 266:1717-1719

Diakoff S, Scheibe S (1973) Action spectra for chromatic adaptation in *Tolypothrix tenuis*. Plant Physiol 51:382-385

Dolganov N, Grossman AR (1999) A polypeptide with similarity to phycocyanin a-subunit phycocyanobilin lyase involved in degradation of phycobilisomes. J Bacteriol 181:610-617

Durnford DG, Falkowski PG (1997) Chloroplast redox regulation of nuclear gene transcription during photoacclimation. Photosynth Res 53:229-241

Emlyn-Jones D, Ashby K, Millineaux CW (1999) A gene required for the regulation of photosynthetic light harvesting in the cyanobacterium *Synechocystis* 6803. Mol Microbiol 33:1050-1058

Engelmann TW (1883) Farbe und Assimilation. Assimilation findet nur in den farbstoffhaltigen Plasmatheilchen statt. II. Näherer Zusammenhang zwischen Lichtabsorption und Assimilation. Bot Z 41:1-13

Engelmann TW (1902) Über experimentelle Erzeugung zwechmassiger Änderung der Färbung pflanzlicher Chromophylle durch farbiges Licht. Arch Anat Physiol 333-33

Escoubas J-M, Loas M, LaRoche J, Falkowski PG (1995) Light intensity regulation of cab gene transcription is signaled by the redox state of the plastoquinone pool. Proc Natl Acad Sci USA 92:10237-10241

Faeber A, Jahns P (1998) The xanthophyll cycle of higher plants: influence of antenna size and membrane organization. Biochim Biophys Acta 1363:47-58

Falkowski PG, Raven JA (1997) Aquatic photosynthesis. Blackwell Science, Malden

Fawley MW, Grossman AR (1986) Polypeptides of light-harvesting complexes of the diatom *Phaeodactylum tricornutum* are synthesized in the cytoplasm of the cell as precursors. Plant Physiol 81:149-155

Frank HA, Cua A, Chynwat V, Young A, Gosztola D, Wasiewski MR (1996) The lifetimes and energies of the first excited singlet states of diadinoxanthin and diatoxanthin: the role of these molecules in excess energy dissipation in algae. Biochim Biophys Acta 1277:243-252

Friedman AL, Alberte RS (1986) Biogenesis and light regulation of the major light-harvesting chlorophyll-protein of diatoms. Plant Physiol 80:43-51

Fujita Y (1997) A study on the dynamic features of photosystem stoichiometry: accomplishments and problems for future studies. Photosynth Res 53:83-93

Fujita Y (1999) An evidence indicating that a weak orange light absorbed by phycobilisomes causes inactivation of PSII in cells of the red alga *Porphyridium cruentum* grown under a weak red light preferentially exciting Chl a. Plant Cell Physiol 40:942-932

Fujita Y, Hattori A (1960a) Formation of phycoerythrin in pre-illuminated cells of *Tolypothrix tenuis* with special reference to nitrogen metabolism. Plant Cell Physiol 1:281-292

Fujita Y, Hattori A (1960b) Effect of chromatic lights on phycobilin formation in a blue- - green alga, *Tolypothrix tenuis*. Plant Cell Physiol 1:293-303

Fujita Y, Hattori A (1962) Photochemical interconversion between precursors of phycobilin chromoproteins in *Tolypothrix tenuis*. Plant Cell Physiol 3:209-220

Fujita Y, Murakami A (1987) Regulation of electron transport composition in cyanobacterial photosynthetic system: stoichiometry among photosystem I and II complexes and their light-harvesting antennae and Cyt b_6-f complex. Plant Cell Physiol 28:1547-1553

FujitaY, Ohki K, Murakami A (1985) Chromatic regulation of photosystem in the photosynthetic system of red and blue-green algae. Plant Cell Physiol 26:1541-1548

Fujita Y, Murakami A, Ohki K (1987) Regulation of photosystem composition in cyanobacterial photosynthetic system: the regulation occurs in response to the redox state of the electron pool located between the two photosystems. Plant Cell Physiol 28:283-292

Fujita Y, Murakami A, Ohki K, Hagiwara N (1988) Regulation of photosystem composition in cyanobacterial photosynthetic system: evidence indicating that photosystem I formation is controlled in response to the electron transport state. Plant Cell Physiol 29:557-564

Fujita Y, Murakami A, Ohki K (1990) Regulation of the stoichiometry of thylakoid components in the photosynthetic system of cyanophytes: model experiments showing that control of the synthesis or supply of Chl a can change the stoichiometric relationship between the two photosystems. Plant Cell Physiol 31:145-153

Fujita Y, Murakami A, Ohki K (1994) Short-term and long-term adaptation of the photosynthetic apparatus: homeostatic properties of thylakoids. In: Bryant DA (ed) The molecular biology of cyanobacteria. Kluwer, Dordrecht, pp 677-692

Gaidukov N (1903) Die Farbveränderung bei den Prozessen der komplementären chromatischen Adaptation. Ber Dtsch Bot Ges 21:484-492

Gal A, Shahak Y, Schuster G, Ohad I (1987) Specific loss of LHCII phosphorylation in the *Lemna* mutant 1073 lacking the cytochrome b6/f complex. FEBS Lett 221:205-210

Grossman AR, Kehoe DM (1997) Phosphorelay control of phycobilisome biogenesis during complementary chromatic adaptation. Photosynth Res 53:95-108

Grossman AR, Manodori A, Snyder D (1990) Light-harvesting proteins of diatoms: their relationship to the chlorophyll a/b binding proteins of higher plants and their mode of transport into plastids. Mol Gen Genet 224:91-100

Grossman AR, Schaefer MR, Chiang GG, Collier JL (1993) The phycobilisome, a light-harvesting complex responsive to environmental conditions. Microbiol Rev 57:725-749

Grossman AR, Schaefer MR, Chiang GG, Collier JL (1994) The responses of cyanobacteria to environmental conditions. In: Bryant DA (ed) The molecular biology of cyanobacteria. Kluwer, Dordrecht, pp 641-675

Hall DO, Rao KK (1999) Photosynthesis, 6th edn. Cambridge University Press, Cambridge

Hattori A, Fujita Y (1959) Effect of pre-illumination on the formation of phycobilin pigments in a blue-green alga, *Tolypothrix tenuis*. J Biochem 46:159-161

Haury JF, Bogorad L (1977) Action spectra for phycobiliprotein synthesis in a chromatically adapting cyanophyte, *Fremyella diplosiphon*. Plant Physiol 60:835-839

Hiller RG, Anderson JM, Larkum AWD (1990) The chlorophyll-protein complexes of algae. In: Scheer H (ed) Chlorophylls. CRC Press, Boca Raton, Florida, pp 5290-5347

Horton P, Allen JF, Black MT, Bennett J (1981) Regulation of phosphorylation of chloroplast membrane polypeptides by the redox state of plastoquinone. FEBS Lett 125:193-196

Ikeya T, Ohki K, Takahashi M, Fujita Y (1991) Photosynthetic pigment system of picophytoplankton of cyanophytes isolated from subsurface water in the Kuroshio area. J Oceanogr Soc Jpn 47:1-6

Jones LW, Myers J (1964) Enhancement in the blue-green alga, *Anacystis nidulans*. Plant Physiol 39:938-946

Kageyama A, Yokohama Y (1978) The function of siphonein in a siphonous green alga *Dichotomosiphon tuberosus*. Jpn J Phycol 26:151-155

Kawamura M, Mimuro M, Fujita Y (1979) Quantitative relationship between two reaction centres in the photosynthetic system of blue-green algae. Plant Cell Physiol 20:697-705

Kehoe DM, Grossman AR (1995) The use of site directed mutagenesis in the analysis of complementary chromatic adaptation. In: Mathis P (ed) Photosynthesis: from light to biosphere, vol III. Kluwer, Dordrecht, pp 501-504

Kehoe DM, Grossman AR (1996) Similarity of a chromatic adaptation sensor to phytochrome and ethylene receptors. Science 273:1409-1412

Kehoe DM, Grossman AR (1997) New classes of mutants in complementary chromatic adaptation provide evidence for a novel four-step phosphorelay system. J Bacteriol 179:3914-3921

Kirk JTO (1994) Light and photosynthesis in aquatic ecosystems (2nd edn). Cambridge University Press, Cambridge

Kirschner J, Senger H (1986) Thylakoid protein phosphorylation in the red alga *Porphyridium cruentum*. In: Akoyunoglou G, Senger H (eds) Regulation of chloroplast differentiation. Alan R Liss, New York, pp 339-344

Lemaire C, Giard-Bascou J, Wollman FA (1987) Characterization of the b_6/f complex subunits and studies on the LHC-kinase in *Chlamydomonas reinhardtii* using mutant strains altered in the b6/f complex. In: Biggins J (ed) Progress in photosynthesis research, vol IV. Martinus Nijhoff, Dordrecht, pp 655-658

Leong T-Y, Anderson JM (1986) Light-quality and irradiance adaptation of the composition and function of pea-thylakoid membranes. Biochim Biophys Acta 766:533-541

Lohr M, Wilhelm C (1999) Algae displaying the diadinoxanthin cycle also possess the violaxanthin cycle. Proc Natl Acad Sci USA 96:8784-8789

Lopez-Figueroa F, Niell FX (1990) Effects of light quality on chlorophyll and biliprotein accumulation in seaweeds. Mar Biol 104:321-327

Luning K (1990) Seaweeds their environment, biogeography, and ecophysiology. John Wiley, New York

MacColl R (1998) Cyanobacterial phycobilisomes. J Struct Biol 124:311-334

Manodori A, Melis A (1986) Cyanobacterial acclimation to photosystem I or photosystem II light. Plant Physiol 82:185-189

Matsuura K, Murakami A, Fujita Y (1988) Changes in light-induced electron transfer through cytochrome b-563 depending on the oxidation-reduction conditions in intact cells of the cyanobacterium *Synechocystis* PCC6714. Plant Cell Physiol 29:1261-1268

Mazel D, Guglielmi G, Houmard J, Sidler W, Bryant DA, Tandeau de Marsac N (1986) Green light induces transcription of the phycoerythrin operon in the cyanobacterium *Calothrix* 7601. Nucleic Acids Res 14:8279-8290

Melis A, Murakami A, Nemson JA, Aizawa K, Ohki K, Fujita Y (1996) Chromatic regulation in *Chlamydomonas reinhardtii* alters photosystem stoichiometry and improves the quantum efficiency of photosynthesis. Photosynth Res 47:253-265

Mills JD, Michell P, Schuermann P (1980) Modulation of coupling factor ATPase activity in intact chloroplasts. The role of the thioredoxin system. FEBS Lett 112:173-177

Mullineaux CW, Allen JF (1986) The state 2 transition in the cyanobacterium *Synechococcus* 6301 can be driven by respiratory electron flow into the plastoquinone pool. FEBS Lett 205:155-160

Mullineaux CW, Allen JF (1991) State 1-state 2 transitions in the cyanobacterium *Synechococcus* 6301 are controlled by the redox state of electron carriers between photosystems I and II. Photosynth Res 23:297-311

Mullineaux CW, Tobin MJ, Jones GR (1997) Mobility of photosynthetic complexes in thylakoid membranes. Nature 390:421-424

Murakami A, Fujita Y (1988) Steady state of photosynthesis in cyanobacterial photosynthetic systems before and after regulation of electron transport composition: over all rate of photosynthesis and PSI/PSII composition. Plant Cell Physiol 29:305-311

Murakami A, Fujita Y (1991a) Steady state of photosynthetic electron transport in cells of the cyanophyte *Synechocystis* PCC6714 having different stoichiometry between PSI and PSII: analysis of flash-induced oxidation-reduction of cytochrome f and P700 under steady state of photosynthesis. Plant Cell Physiol 32:213-222

Murakami A, Fujita Y (1991b) Regulation of photosystem stoichiometry in the photosynthetic system of the cyanophyte *Synechocystis* PCC6714 in response to light-intensity. Plant Cell Physiol 32:223-230

Murakami A, Fujita Y (1993) Regulation of stoichiometry between PSI and PSII in response to light regime for photosynthesis observed with *Synechocystis* PCC6714: relationship between redox state of Cyt b_6-f complex and regulation of PSI formation. Plant Cell Physiol 34:1175-1180

Murakami A, Kim S-J, Fujita Y (1997) Changaes in photosystem stoichiometry in response to environmental conditions for cell growth observed with the cyanophyte *Synechocystis* PCC6174. Plant Cell Physiol 38:392-397

Murata N (1969) Control of excitation transfer in photosynthesis. I. Light-induced change of chlorophyll a fluorescence in *Porphyridium cruentum*. Biochim Biophys Acta 172:242-251

Myers J, Graham J-R, Wang RT (1980) Light-harvesting in *Anacystis nidulans* studied in pigment mutants. Plant Physiol 66:1144-1149

Nakajima Y, Ueda R (1997) Improvement of photosynthesis in dense microalgal suspension by reduction of light harvesting pigments. J Appl Phycol 9:503-510

Oelmuller R, Conley PB, Federspiel N, Briggs WR, Grossman AR (1988a) Changes in accumulation and synthesis of transcripts encoding phycobilisome components during acclimation of *Fremyella diplosiphon* to different light qualities. Plant Physiol 88:1077-1083

Oelmuller R, Grossman AR, Briggs WR (1988b) Photoreversibility of the effect of red and green light pulses on the accumulation in darkness of mRNAs coding for phycocyanin and phycoerythrin in *Fremyella diplosiphon*. Plant Physiol 88:1084-1091

Ohki K, Fujita Y (1991) Complementary chromatic adaptation in the cyanobacterium *Tolypothrix tenuis*: location of phycoerythrin newly synthesized by green illumination. Plant Cell Physiol 32: 483-488

Ohki K, Honjho S (1997) Oceanic picophytoplankton having a high abundance of chlorophyll b in the major light harvesting chlorophyll protein complex. Photosynth Res 53:121-127
Ohki K, Watanabe M, FujitaY (1982) Action of near UV and blue light on the photocontrol of phycobiliprotein formation; a complementary chromatic adaptation. Plant Cell Physiol 23:651-656
Ort DR, Yocum CF (eds) (1998) Oxygenic photosynthesis: the light reactions. Kluwer, Dordrecht
Osmond CB (1994) What is photoinhibition? Some insights from comparisons of shade and sun plants. In: Baker NR, Bowye JR (eds) Photoinhibition of photosynthesis— from molecular mechanisms to the field. Bios Sci Publ, Oxford, pp 1-24
Parkinson JS, Kofoid EC (1992a) Communication module in bacterial signaling proteins. Annu Rev Genet 26:71-112
Pfannschmidt T, Nilsson A, Allen JF (1999a) Photosynthetic control of chloroplast gene expression. Nature 397:625-628
Pfannschmidt T, Nilsson A, Tullberg A, Allen JF (1999b) Direct transcriptional control of the chloroplast genes psbA and psaAB adjusts photosynthesis to light energy distribution in plants. IUBMB Life 48:271-276
Sagert S, Schubert H (1995) Acclimation of the photosynthetic apparatus of *Palmaria palmata* (Rhodophyta) to light qualities that preferentially excite photosystem I or photosystem II. J Phycol 31:547-554
Salehian O, Bruce D (1992) Distribution of excitation energy in photosynthesis: quantification of fluorescence yield from intact cyanobacteria. J Lumin 51: 91-98
Sanders CE, Allen JF (1987) The 18.5 kD phosphoprotein of the cyanobacterium *Synechoccus* 6301: a component of phycobilisome. In: Biggins J (ed) Progress in photosynthesis research, vol II. Martinus Nijhoff, Dordrecht, pp 761-764
Sanders CE, Melis A, Allen JF (1989) In vivo phosphorylation of ptoteins in the cyanobacterium *Synechochoccus* 6301 after chromatic acclimation to photosystem I or photosystem II light. Biochim Biophys Acta 976:168-172
Satoh K, Fork DC (1983) A new mechanism for adaptation to changes in light intensity and quality in the red alga, *Porphyra perforata*. I. Relation to state 1-state 2 transitions. Biochim Biophys Acta 722:190-196
Schwarz R, Grossman AR (1998) A response regulator of cyanobacteria integrates diverse environmental signals and is critical for survival under extreme conditions. Proc Natl Acad Sci USA 95:11008-11013
Shimura S, Fujita Y (1973) Some properties of the chlorophyll fluorescence of the diatom *Phaeodactylum tricornutum*. Plant Cell Physiol 14:341-352
Smith BM, Morrissey PJ, Guenther JE, Nemson JA, Harrison MA, Melis A (1990) Response of the photosynthetic apparatus in *Dunaliella salina* (green algae) to irradiance stress. Plant Physiol 93:1433-1440
Sobczyk A, Schyns G, Tandeau de marsac N, Houmard J (1993) Transduction of the light signal during complementary chromatic adaptation in the cyanobacterium *Calothrix* sp. PCC7601: DNA-binding proteins and modulation by phosphorylation. EMBO J 12:997-1004
Sobczyk A, Bely A, Tandeau de Marsac N, Houmard J (1994) A phosphorylated DNA-binding protein is specific for the red-light signal during complementary chromatic adaptation in cyanobacteria. Mol Microbiol 13: 875-885
Spetea C, Hundal T, Lohmann F, Andersson B (1999) GTP bound to chloroplast thylakoid membranes is required for light-induced, multienzyme degradation of the photosystem II D1 protein. Proc Natl Acad USA 96: 6547-6552
Stockner JG (1988) Phototrophic picoplankton: an overview form marine and freshwater ecosystems. Limnol Oceanogr 33:765-775
Stockner JG, Antia NJ (1986) Algal picoplankaton from marine and freshwater ecosystems: a multidisciplinary perspective. Can J Fish Aquat Sci 43:2472-2503

Stransky H, Hager A (1970) Das Carotinoidmuster und die Verbreitung des lichtinduzierten Xanthophyllcyclus in verschiedenen Alganklassen. VI. Chemosystematische Betrachtung. Arch Mikrobiol 73:315-323

Tandeau de Marsac N (1977) Occurrence and nature of chromatic adaptation in cyanobacteria. J Bacteriol 130:82-91

van Thor JJ, Mullineaux CW, Matthijs HCP, Hellingwerf KJ (1998) Light harvesting and state transitions in cyanobacteria. Bot Acta 111: 430-443

Vener AV, van Kan PJ, Rich PR, Ohad I, Andersson B (1997) Plastoquinol at the quinol oxidation site of reduced cytochrome bf mediates signal transduction between light and protein phosphorylation: thylakoid protein kinase deactivation by a single-turnover flash. Proc Natl Acad Sci USA 94:1585-1590

Vener AV, Ohad I, Andersson B (1998) Protein phosphorylation and redox sensing in chloroplast thylakoids. Curr Opin Plant Biol 1:217-223

Vogelmann TC, Scheibe J (1978) Action spectrum for chromatic adaptation in the blue-green alga *Fremyella diplosiphon*. Planta 143:233-239

Yamamoto HY, Bassi R (1996) Carotenoids: localization and function. In: Ort DR, Yocum CF (eds) Oxygenic photosynthesis: the light reactions. Kluwer, Dordrecht, pp 539-563

Yamanaka G, Glazer AN (1980) Dynamic aspects of phycobilisome structure: phycobilisome turnover during nitrogen starvation in *Synechococcus* sp. Arch Microbiol 124:39-47

Yocum CS, Blinks LR (1958) Light-induced efficiency and pigment alteration in red algae. J Gen Physiol 41:1113-1117

Yokohama Y, Misonou T (1980) Chlorophyll a:b ratios in marine benthic green algae. Jpn J Phycol 28:219-223

Yokohama Y, Hirata T, Misonou T, Tanaka J, Yokochi H (1992) Distribution of green light-harvesting pigments, siphonaxanthin and siphonein, and their precursors in a marine green alga. Jpn J Phycol 40:25-31

Zao J, Zhou J, Bryant DA (1992) Energy transfer processes in phycobilisomes as deduced from analyses of mutants of *Synechococcus* sp. PCC7002. In: Murata N (ed) Research in photosynthesis, vol 1. Kluwer, Dordrecht, pp 25-32

Zito F, Finazzi G, Delosme R, Nitschke W, Picot D, Wollman F (1999) The Qo site of cytochrome b_6f complexes controls the activation of the LHCII kinase. EMBO J 18:2961-2969

7 Adaptation to UV Stress in Algae

Donat-P. Häder*

Contents

*Friedrich-Alexander-Universität, Erlangen-Nürnberg, Institut für Botanik und Pharmazeutische Biologie, Staudtstr. 5, 91058 Erlangen, Germany

7.1 Introduction: Ozone Layer and UV Radiation

Stratospheric ozone depletion results in an increased transmission of solar ultraviolet (UV) radiation to the Earth's surface. UV radiation has been found to affect many biological and chemical processes and has detrimental effects on the biota. Major concerns deal with increases in the incidence of certain skin cancers and other health effects in humans and animals, impact on terrestrial and aquatic ecosystems, changes in the chemistry of the troposphere (e.g. photochemical smog formation) and effects on biogeochemical cycles (Madronich et al. 1998).

The level of solar UV radiation varies with time of the day and season, cloud pattern and atmospheric pollution as well as latitude and altitude, and strongly depends on the total ozone column. Monitoring of the ozone layer commenced as early as the beginning of this century and has been extended to a number of terrestrial stations since the 1950s. Continuous ozone observations are available from ground-based stations and satellite-based instruments (such as Total Ozone Mapping Spectrometer, TOMS, version 7, aboard the Nimbus 7 satellite or the recent Earth Probe) since 1979.

Several networks have been installed, e.g. by Biospherical Instruments Inc. (San Diego, California) which are responsible for obtaining and distributing irradiance data from the US National Science Foundation UV Spectroradiometer Monitoring Network (http://www.biospherical.com). Data are available from stations in San Diego (California), Ushuaia (Argentina), Barrow (Alaska) and three Antarctic stations (South Pole, Palmer and McMurdo). Considerable efforts are necessary to guarantee quality control of the light measurements (Seckmeyer et al. 1994). One of the most recent monitoring networks is the European Light Dosimeter Network (ELDONET) with over 40 stations spread over Europe and additional stations in India, Japan, Egypt, New Zealand and Argentina. Two dosimeters have been installed at high-altitude stations in the European Alps and the Spanish Sierra Nevada. Most of the stations are terrestrial, but seven instruments are located in the water column to monitor the transparency of major coastal waters where they operate in conjunction with a terrestrial counterpart. The instruments record solar radiation fully automatically in three channels (UV-B, UV-A, PAR), and the data are transmitted to a server in Pisa, where they are available to the public on the internet(http://power.ib.pi.cnr.it:80/eldonet).

Continued decreases in the ozone layer have been observed during the past two decades over Antarctica and later also over Southern and Northern mid-latitudes as well as in the Arctic (Müller et al. 1997; Stolarski 1997) with resulting higher UV-B radiation levels (Kerr and McElroy 1993). Despite earlier doubts, there is no substantial reason to assume other than anthropological causes for the depletion of stratospheric ozone due to the emission of halogen-containing compounds of mostly human activities WMO 1994a). The thickness of the ozone layer is measured in Dobson units (DU) defined as the height in milli-centimetres that the gaseous ozone layer would have if compressed to atmospheric pressure at the Earth's surface at 0 °C (equals 2.69×10^{16} molecules cm^{-2}) (Madronich and Flocke 1997).

There is a strong correlation between decreased total ozone layer and increased UV radiation levels at the Earth's surface.

7.1.1 Biologically Active UV Radiation

Extraterrestrial solar radiation contains a large component of short wavelength radiation <280 nm (UV-C), which is almost quantitatively filtered out by the oxygen in the atmosphere and therefore not ecologically relevant. The majority of the UV-B radiation (280 – 315 nm, CIE definition) is removed by atmospheric ozone, most of which is located in the stratosphere and only a small amount in the troposphere (Fig. 7.1). The longer wavelength UV-A (315 – 400 nm) and the visible or photosynthetic active radiation (PAR, 400 –700 nm) are not affected by the ozone in the atmosphere.

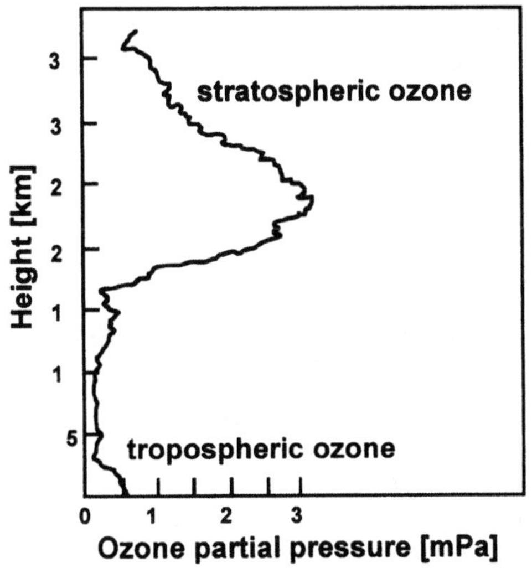

Fig. 7.1. Concentration of ozone (in Dobson units) in the troposphere and stratosphere

Figure 7.2 compares the solar spectral irradiance at the Earth's surface in the UV range with a biological action spectrum for the inhibition of photosynthesis in the cyanobacterium *Nodularia*. The overlap of the two spectra defines the biologically active radiation. Ozone depletion results in an increase in the UV emission of solar radiation and a shift to shorter wavelengths. Both effects result in a high increase in the biologically active radiation.

In order to estimate the biological effects of ozone-dependent UV increases, the spectral sensitivity of the biological effect must be determined. These action spectra have been calculated for a number of physiological functions in several aquatic ecosystems (Häder et al. 1994; Boucher and Prezelin 1996). Multiplication of the physically measured solar UV radiation with the specific action spectrum is defined as the biologically active UV irradiance or exposure UV_{bio},

$$UV_{bio} = \int E(\lambda)\, B(\lambda)\, d\lambda,$$

where $E(\lambda)$ is the spectral irradiance and $B(\lambda)$ is the biological action spectrum (Madronich et al. 1998). The biologically active irradiance depends on the total ozone column; the dependence can be defined by the radiation amplification factor (RAF), expressed as the increase in UV_{bio} that results from a 1% ozone reduction. However, this dependence is non-linear for larger ozone changes. RAFs are good indicators of the sensitivity of a biological process to ozone depletion (large RAF values indicate a strong sensitivity to stratospheric ozone depletion).

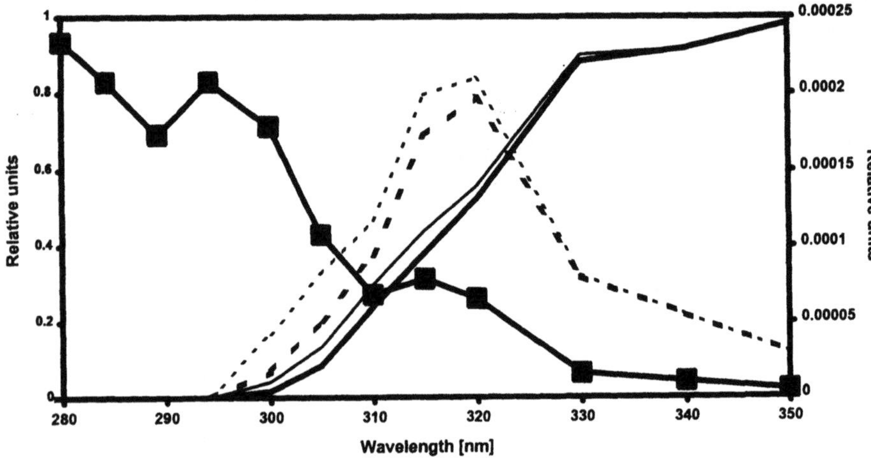

Fig. 7.2. Solar emission spectra $E(\lambda)$ at 250 DU (*fine line*) and 350 DU (*heavy line*) and the biological action spectrum for the inhibition of photosynthesis $P(\lambda)$ in the cyanobacterium *Nodularia* (*heavy line* and *square symbols, right ordinate*) overlap in the UV-B range. The area under the product function $E(\lambda) \times P(\lambda)$ is defined as the biologically active fluence rate (*broken lines*). As the ozone column decreases, the integral under the product function increases correspondingly

7.1.2 Measurement of UV Radiation

The number and quality of UV measurements have been improved considerably during the past few years due to the development of new sensors and instruments as well as calibration procedures. Quality control has been improved by intercalibrations between instruments (Seckmeyer et al. 1994; Kjeldstad et al. 1997; Thompson et al. 1997; Webb 1997; Leszczynski et al. 1998). Today measurement errors of better than 5% in the UV-A range and 10% in the UV-B range can be achieved. While a few years ago the predicted increases in solar UV-B could not be verified over North America and Europe, recent measurements have confirmed the expected levels calculated from ozone depletion values determined from satellites (Seckmeyer et al. 1995). The remaining deviations between satellite and terrestrial measurements may in part be due to tropospheric pollution (WMO 1994a). UV measurements at more sites are necessary to obtain a global reliable picture.

Aerosols can have a substantial effect on the penetration of solar UV to the surface, depending on the concentration, size and chemical and physical properties of the particles. It has been calculated that anthropogenic sulfate aerosols, resulting primarily from fossil fuel burning, have reduced surface UV-B irradiances by 5 – 18% in industrialized regions of the Northern Hemisphere (Varotsos and Kondratyev 1995; Mims 1997). This hypothesis is further confirmed by comparisons between measurements in the Northern polluted and Southern less polluted Hemispheres (Seckmeyer and McKenzie 1992; Seckmeyer et al. 1995) and by measurements along a vertical gradient of mountains (Madronich et al. 1995; Piazena 1996; Blumthaler 1997).

When all other factors are equal, the total ozone concentration is largely controlled by the total ozone column (Bernhard et al. 1997; Kirchoff et al. 1997; Seckmeyer et al. 1997; Taalas et al. 1997). Shorter UV wavelengths are more affected by ozone changes because of the decreasing absorption of ozone toward longer wavelengths. Biologically weighted radiation is also strongly affected by ozone because of the large sensitivity toward shorter wavelengths (Horneck 1995; Rontó et al. 1995; Häder et al. 1997c).

Highest levels of UV-B radiation have been monitored under the Antarctic ozone hole (Seckmeyer et al. 1995), and have sometimes exceeded maximum summer values measured at San Diego, USA (32°N) (Booth et al. 1997). The decreases in stratospheric ozone between 1970 and 1998 are estimated to be about 50% in the Antarctic during the Austral spring, 15% in the Arctic spring, 6% at Northern Hemisphere mid-latitudes in winter and spring, 3% in summer and fall and 5% at Southern Hemisphere mid-latitudes throughout the year (Madronich et al. 1998). The resulting increases in sunburning UV radiation are estimated to be 130, 22, 7, 4 and 6%, respectively. At tropical and subtropical latitudes no significant ozone changes are measured. The size and depth of the Antarctic ozone hole have not changed much since the early 1990s and also little change was found at mid-latitudes during the past 5 years.

Satellite-based measurements provide a global and continuous long-term monitoring. The measurement of UV radiation impinging on the oceans will be necessary to estimate the impact of solar UV on aquatic ecosystems. However, the derivation of surface UV irradiances from satellite monitoring is indirect, since satellite instruments assay the radiation reflected from the atmosphere and the Earth's surface from which the surface incidence is calculated by the use of radiative transfer models which are based on transmission, reflection and atmospheric absorption. In recent years also measurement of cloud cover experienced some progress which has a large impact on UV doses at the surface (Meerkoetter et al. 1997). It may be possible in the future to measure tropospheric aerosol distributions from satellites (Krotkov et al. 1997).

7.1.3 Predicted UV Radiation Levels in the Future

Short-term forecasts based on meteorological analyses are now available in several countries and are made available to the public using the recently developed UV index (WMO 1994b), which is in essence the irradiance weighted by the erythemal action spectrum (McKinlay and Diffey 1987) and multiplied by 40. A UV index of 10 or more indicates very high UV irradiances.

Long-term predictions of future ozone trends are not easy since, e.g. natural perturbations such as volcanic eruptions cannot be foreseen, while they can have a large effect on stratospheric ozone as shown by the 1991 Mt. Pinatubo eruption. Predictions of the production and emission of anthropogenic pollutants are also difficult including greenhouse gases and the emissions from intercontinental and supersonic aircraft. A recent study assumes that the recovery of the ozone layer may be delayed significantly due to interactions with increasing greenhouse gas concentrations (Schindell et al. 1998).

The 1987 Montreal Protocol and its subsequent amendments aim at reducing the production and emission of halocarbons and other ozone-destroying substances. The concentrations of these substances has increased since the 1970s, but have been found to show lower growth or even decreases as a consequence of the Protocol (WMO 1994a). The future scenarios are based on the 1997 Montreal Amendments. Future levels will depend on whether the subscriber countries will comply with the regulations and whether illegal production and trafficking of ozone depleting substances will be substantial. Anyway, the UV radiation is expected to return to pre-1980 levels by the middle of the next century. This scenario is based on the natural time scale for the decay of the halocarbons currently present in the atmosphere. Thus, even under the most optimistic scenarios it will take several decades to return to normal values.

7.1.4 UV Penetration into the Water Column

In order to estimate the effects of solar radiation on marine ecosystems, the penetration of UV and PAR into the water column needs to be determined (Montecino and Pizarro 1995; Björn et al. 1996). Aquatic ecosystems show large temporal and spatial differences in the concentrations of dissolved and particulate absorbing substances. Marine waters have been classified into nine types of coastal and five types of open ocean waters depending on their transmission (Jerlov 1968). The ratio between the 0.1% penetration for UV-B and PAR can be utilized to evaluate the deleterious effects on algae by solar UV-B radiation in the euphotic zone (Piazena and Häder 1997). Recent hardware developments allow an accurate determination of the underwater light field (Morrow and Booth 1997). Gelbstoff (yellow substances), chlorophyll a and other photosynthetic pigments, as well as organic and inorganic particulate material mainly attenuate the UV-B radiation in the water column. The effect of phytoplankton on the underwater light field has been extensively studied, but measurements of the underwater light field in coastal habitats populated by macroalgae are rare, particularly in the UV-B range. Algal canopies modify the light quality by absorption and scattering of the incident light.

Most of the dissolved organic carbon (DOC) in coastal waters is of terrestrial origin. Solar UV radiation degrades DOC photolytically (Naganuma et al. 1996). DOC consists to a large extent of humic substances which are fairly resistant to bacterial consumption, but after photolytic activity the products (e.g. formaldehyde, acetaldehyde, glyoxylate and pyruvate) are degraded by bacterioplankton (Wetzel et al. 1995). Ultraviolet radiation is strongly absorbed by humic substances. Therefore, increased breakdown of DOC and subsequent consumption by bacteria increases the penetration of UV-B into the water. However, near the surface, solar UV radiation affects the bacteria and the activity of the bacterial ectoenzymes to split the low molecular weight DOC. Only when both bacteria and the photolyzed DOC are circulated to deeper layers does the uptake rate increase. Furthermore, DOC photolysis generates photosensitizers, which produce reactive oxygen species or free radicals upon absorption of UV radiation (Herndl 1997).

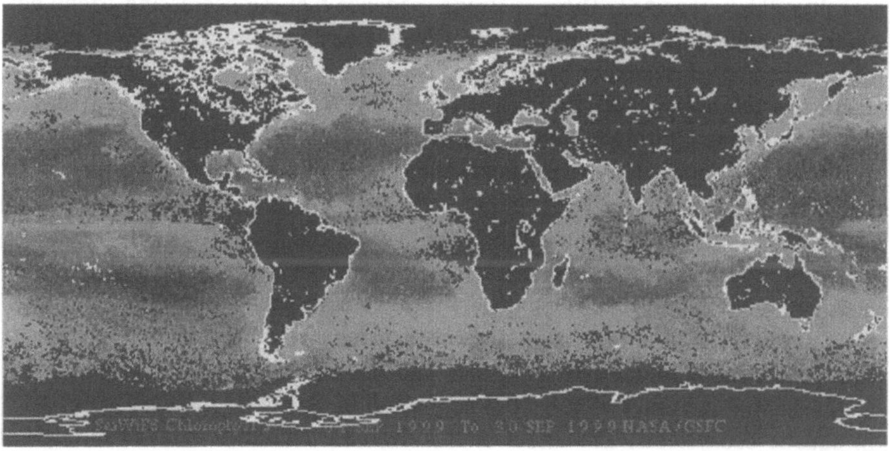

Fig. 7.3. Chlorophyll concentration in the ocean for the month of September 1999. Provided by the SeaWiFS Project, NASA/Goddard Space Flight Centre

Another approach to quantify the underwater light climate is based on satellite data (e.g. CZCS and SeaWiFS; Fig. 7.3). These instruments cover only the visible range, but attempts are being made to extrapolate the data into the UV range.

There have been great efforts to measure algal biomass by using remote sensors. Most studies have focused on quantifying chlorophyll from phytoplankton in surface waters (Brown et al. 1995). One major problem of remote monitoring is the fact that overflying instruments mainly determine the surface signal (Piazena and Häder 1997). Profound knowledge of the vertical distribution of phytoplankton, as well as the composition of the algal communities is necessary to derive a quantitative analysis of biomass productivity.

Behrenfeld and Falkowski (1997a) have developed models to quantify photosynthetic rates derived from satellite-based chlorophyll concentration. In addition, they have evaluated several primary productivity models (Behrenfeld and Falkowski 1997b) and showed that many of these, though different, models show basic agreement. If similar parameterizations are used for satellite-derived chlorophyll and carbon fixation rate, estimates of global annual primary production are found to be mainly based on these parameters.

7.1.5 Effects of Solar UV on Marine Primary Producers

Solar short-wavelength radiation has been found to penetrate to ecologically significant depths in many freshwater and marine ecosystems (USEPA 1987; Smith et al. 1992; Scully and Lean 1994; Häder 1995; Coohill et al. 1996; Booth et al. 1997). Drastic stratospheric ozone depletion over both the Antarctic and Arctic, as well as moderate decreases in total ozone column over high and mid-latitude waters, have been reported. There is strong evidence that the decreasing ozone concentration increases the level of solar UV-B which penetrates into the euphotic zone, where phytoplankton populations are concentrated. There is growing evidence that this effect alters the ratio of UV-B:UV-A:PAR radiation, which may affect the intricate light-dependent responses of aquatic organisms, including photosynthesis, photoorientation, photoinhibition and photoprotection (Smith et al. 1992; Häder et al. 1995; Gerber et al. 1996; Jiménez et al. 1996; Häder 1997a,b). Changes in the spectral composition may cause significant stress for the primary producers (IASC 1995). UV-B and UV-A impair both productivity and growth by affecting a number of mechanisms, and hit several molecular targets. Most organisms possess effective protective and repair mechanisms, but excessive exposure to solar UV radiation may overload their capabilities.

Substantial increases of solar UV may result in decreased biomass productivity in aquatic ecosystems. The effect of this decrease would be relayed through all levels of the intricate food web, resulting in reduced food production (Häder et al. 1995; Häder 1997e; Häder and Worrest 1997), reduced uptake capacity for atmospheric carbon dioxide (Ducklow et al. 1995; Takahashi et al. 1995, 1997). Another consequence might be an altered species composition and ecosystem integrity. The oceanic carbon dioxide sink capacity is of great significance for global warming (Sarmiento and Le Quéré 1996; Thomson 1997). Over the past few decades research has been intensified to assay UV-B-related damage of aquatic ecosystems (Nolan and Amanatidis 1995). Major reviews on UV effects published so far deal with aquatic ecosystems in general (Häder 1997c; Häder and Worrest 1997), the role of mycosporine-like amino acids (MAAs) in marine organisms (Dunlap and Shick 1998), phytoplankton (Cullen and Neale 1997a,b; Häder 1997a), microalgae (Franklin and Foster 1997; Häder and Figueroa 1997), and lake acidification and UV penetration (Williamson 1995, 1996).

7.2 Cyanobacteria (Blue-Green Algae)

Cyanobacteria are prokaryotes that carry out oxygenic photosynthesis. In addition, several of these organisms are capable of fixing atmospheric nitrogen (Sinha and Häder 1997) using the enzyme nitrogenase to reduce atmospheric nitrogen to ammonium ions. Thus, they make nitrogen available for aquatic eukaryotic phytoplankton and higher plants (Kashyap et al. 1991; Sinha et al. 1996; Sinha and Häder 1996; Kumar et al. 1996a). The agricultural potential of cyanobacteria as a biological fertilizer for wet soils such as in rice paddies is substantial (Banerjee and Häder 1996). Cyanobacteria also fix nitrogen in mid-latitude agricultural systems, though not as productive as in paddy fields. These organisms are found in all major habitats of the Earth and possess a high potential of adaptation to extreme environmental factors. A number of physiological and biochemical processes of cyanobacteria such as growth, survival, pigmentation, photosynthetic oxygen production, motility, enzymes of nitrogen metabolism, phycobiliprotein composition and $^{14}CO_2$ uptake have been reported to be susceptible to UV-B (Sinha et al. 1995a,b 1996, 1997; Sinha and Häder 1998; Häder et al. 1998c; Donkor and Häder 1996, 1997). Growth and survival can cease within a few hours of UV-B irradiation.

The photosynthetic pigments are highly sensitive targets of solar UV radiation. Especially the phycobiliproteins are readily bleached and destroyed (Sinha et al. 1995a, 1996; Aráoz and Häder 1997). Destruction of these accessory pigments is far more efficient than that of chlorophyll a or carotenoids (Sinha et al. 1995a). At lower UV doses the transfer of energy to the reaction centre of photosystem II is inhibited (Sinha et al. 1996). However, under mild UV-B stress, an increased synthesis of phycobiliproteins has been observed, which has been interpreted as a protective mechanism since these pigments strongly absorb in the UV-B range and form a peripheral layer around the centroplasm containing the DNA (Aráoz and Häder 1997). The phycobilins are capable of intercepting more than 99% of UV-B radiation before it penetrates to the DNA. The photosynthetic activity has been found to be impaired in a number of marine and freshwater cyanobacteria. In addition to the bleaching of the photosynthetic pigments, RuBisCO (ribulose-1,5-bis-phosphate carboxylase/oxygenase) activity is severely affected by UV-B treatment (Sinha et al. 1996).

Ammonium uptake was reduced by 10% in cultures exposed to solar radiation, and the nitrogen -fixing enzyme nitrogenase is inhibited by UV-B even after a few minutes of exposure. After 35 – 55 min a complete loss of activity was found, depending upon the species (Kumar et al. 1996b), which may be due to the inhibition of ATP synthesis by UV-B. Also the ammonia-assimilating enzyme glutamine synthetase (GS) is inhibited. In contrast, a stimulation of nitrate reductase by UV-B was found in all nitrogen-fixing cyanobacteria investigated so far (Sinha et al. 1995b).

Wetlands constitute important ecosystems both in the tropics and at temperate latitudes. Cyanobacteria are major constituents in microbial mats. These organisms optimize their position in the community by vertical migration in the mat, which is controlled by both visible radiation and UV-B, by which the cyanobacteria counter the impact of excessive radiation. Other adaptive strategies include synthesis of UV-screening pigments and production of chemical scavengers to quench the highly reactive oxidants produced photochemically (Vincent and Roy 1993). Among the screening pigments scytonemin and a number of mycosporine--like amino acids (MAAs, see Sect. 7.3) have been identified, as well as a number of chemically unidentified, water-soluble pigments which have only been characterized spectroscopically (Donkor and Häder 1995; Kumar et al. 1996b). Often, cyanobacterial filaments are embedded in a mucilaginous sheath into which screening pigments such as scytonemins, carotenoids or MAAs can be excreted. These organisms are more tolerant to UV-B irradiation than those that do not produce such protective components. For example, other species of *Scytonema* that do not produce this pigment could not survive 2 h of UV-B (2.5 W m^{-2}). Scytonemin is a yellow-brown, lipid-soluble dimeric pigment located in the extracellular polysaccharide sheath of some cyanobacteria having an absorption maximum at 386 nm (Sinha et al. 1999a,b; Fig. 7.4). In contrast, the screening pigment from *Scytonema hofmannii* is released into the medium during the late stationary phase of growth.

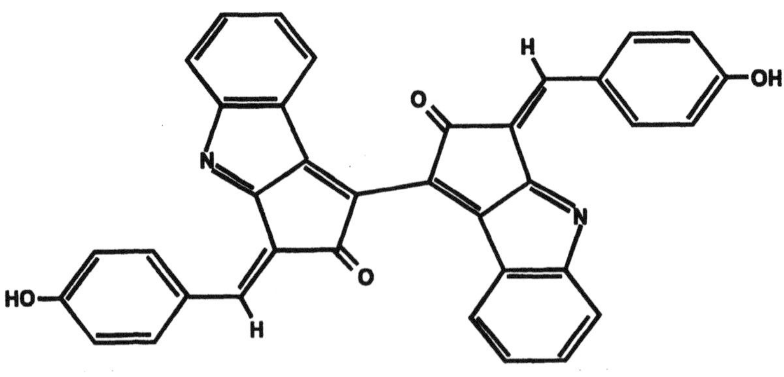

Fig. 7.4. Chemical structure of the UV-absorbing sheath pigment scytonemin

7.3 Phytoplankton

Phytoplankton are the most important biomass producers in marine ecosystems. They occupy the top layers of the water column (euphotic zone); there, they harvest sufficient solar radiation for their photosynthetic processes. The euphotic zone is defined by a net positive carbon incorporation and is typically limited by the penetration of PAR to 0.1% of its surface value. In the euphotic zone phytoplankton are simultaneously exposed to solar UV radiation. Environmental UV-B is an important ecological stress factor that affects growth, survival and distribution of phytoplankton. In order to assess the UV-related stress, information is necessary on the spectral characteristics of solar radiation penetrating into the water column, the biological weighting function (BFW) of the biological effect and an assessment of how vertical mixing controls the irradiance exposures of phytoplankton. We are far from a comprehensive understanding of the complex system, but advances in each of these areas have been made during the past few years (Häder 1997c; Cullen and Neale 1997b; Vernet and Smith 1997).

Biologically weighting functions have recently been determined for several photosynthetic organisms (Cullen and Neale 1996; Boucher and Prezelin 1996; Neale et al. 1998a). All researchers agree that UV-B is the major deleterious component but that the UV-A component also has a significant effect. However, biological weighting functions vary by species, region, mixing characteristics of the water column and a number of other environmental factors. Therefore, more BWFs are necessary for a complete description of a complex ecosystem.

Exposure-response curves (ERC) have recently been measured (Cullen and Neale 1996, 1997a,b). First of all, one must determine whether the measured damage is only a function of the accumulated dose or of the irradiance. This question has an important impact on the design as well as on the interpretation of experiments. The ERC shows a range of experimental variability dependent on the balance between damage and repair (Neale et al. 1998a). Recent investigations have shown both forms of ERC in phytoplankton from different environments making accurate modelling of UV-related impacts more complex. Several models have been developed to estimate the impact of ozone depletion (Arrigo 1994; Cullen and Neale 1994; Boucher and Prezelin 1996; Cullen and Neale 1996a; Neale et al. 1998a). These models are important steps to quantify possible UV impacts and identify the most vulnerable processes as well as the uncertainties of the model. These recent advances demonstrate the difficulty of extrapolating short- -term observations to estimate long-term ecological response (Smith et al. 1992; Vincent and Roy 1993; Bothwell et al. 1994; Cullen and Neale 1994; Holm-Hanson 1997; Neale et al. 1998a).

Solar UV impairs growth and reproduction as well as photosynthesis (Vassiliev et al. 1994; Herrmann et al. 1995a, 1996, 1997; Giacometti et al. 1996; Figueroa et al. 1997; Gieskes and Buma 1997). It affects cellular proteins and bleaches the photosynthetic pigments (Gerber and Häder 1995a,b; Buma et al. 1996a; Peletier et al. 1996; Häder 1997a). Ammonium and nitrate uptake are reduced by solar

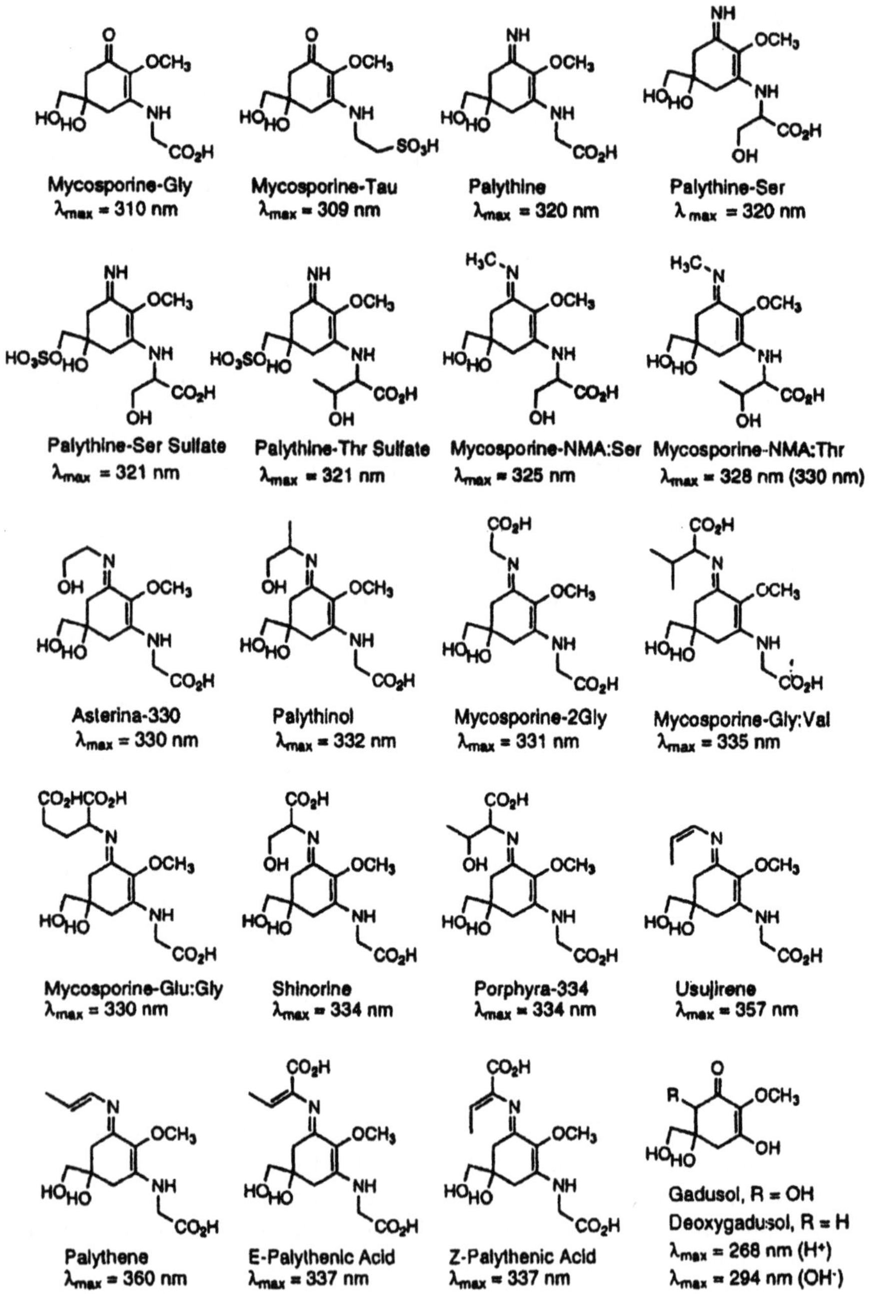

Fig. 7.5. Chemical structures of different MAAs. (After Cockell and Knowland 1999)

radiation in both phytoplankton (Behrenfeld 1995; Döhler 1996, 1997; Döhler and Hagmeier 1997) and macroalgae (Döhler et al. 1995). Phytoplankton organisms have been found to respond with the production of heat-shock proteins, as well as changes in the cellular amino acid pools. The cellular DNA is a major target, which strongly absorbs shorter wavelengths. Solar UV-B induces DNA damage and delay in DNA synthesis in many organisms (Scheuerlein et al. 1995; Buma et al. 1995, 1996b, 1997). UV-B effects have also been studied on the ecosystem level using mesocosms (Santas et al. 1996; Wängberg and Selmer 1997).

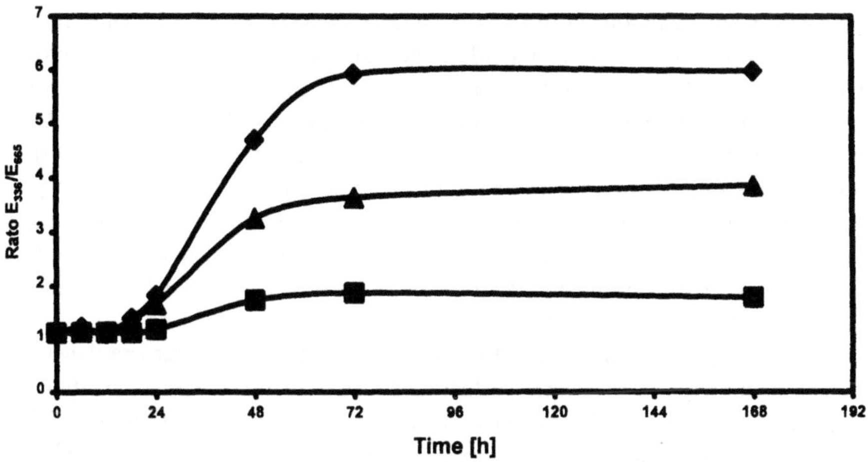

Fig. 7.6. Time course of the increase in the ratio of UV peak absorption (336 nm) to Chl a absorption in methanolic extracts from *G. dorsum* exposed to PAR and UV light with different cutoff filters (395 nm *diamonds*; 320 nm *triangles*; 295 nm *squares*) for up to 168 h. The light source was an array of white light and UV fluorescent tubes with irradiances of 12.31 W m^{-2} PAR (400 – 700 nm), 2.2797 W m^{-2} UV-A (320 – 400 nm) and 0.2122 W m^{-2} UV-B (280 – 320 nm)

The production of mycosporine-like amino acids (MAAs) is interpreted as a major adaptation to mitigate the effects of solar UV radiation in phytoplankton, but also in cyanobacteria and macroalgae. MAAs are water-soluble substances characterized by a cyclohexenone or cyclohexenimine chromophore conjugated with the nitrogen substituent of an amino acid or its imino alcohol, having absorption maxima ranging from 310 to 360 nm (Cockell and Knowland 1999; Fig. 7.5). It is interesting to note that there are important differences between species and even between strains of the same species in different habitats. While most diatoms in the Antarctic ocean possess large amounts of MAAs, these substances are lacking or at lower concentrations in the North Sea. In some organisms, e.g. *Gyrodinium* (Fig. 7.6), the production of MAAs can be induced by short wavelength radiation.

7.4 Macroalgae

Most macroalgae are sessile and therefore restricted to their growth site (Lüning 1990). They show a characteristic pattern of vertical distribution in their habitats. Some macroalgae grow in the supralittoral above the high water mark exposed only to the spray from the surf, whereas others inhabit the intertidal zone (eulittoral), which is characterized by the regular temporal change in the tides (Häder 1997d). Still others are restricted to the sublittoral zone, where they are always submerged. The range in irradiation exposure can be dramatic, from over 1000 W m^{-2} (total solar radiation) at the surface to less than 0.01% of this value reaching the understorey of a kelp habitat (Markager and Sand-Jensen 1994). Different algal species inhabit different depth niches and are adapted to different solar exposure (Häder and Figueroa 1997). They also differ in their ability to tolerate UV radiation (Dring et al. 1996). A broad analysis was carried out to study the different adaptation strategies to solar radiation of ecologically important species of green, red and brown algae from the North Sea, Baltic Sea, Mediterranean, Atlantic, polar and tropical oceans (Markager and Sand-Jensen 1994; Wiencke et al. 1994; Figueroa et al. 1996; Beach and Smith 1996a,b; Kirst and Wiencke 1996; Häder and Figueroa 1997; Porst et al. 1997).

Algae located in the littoral and sublittoral are adapted to lower irradiances than those encountered at the water surface and, consequently, face substantial stress when exposed to higher irradiances (Häder and Figueroa 1997; Franklin and Forster 1997). Solar radiation may cause reversible photoinhibition (Hanelt et al. 1992, 1993) or even irreversible photodamage. Photoinhibition can be interpreted as a reversible decrease of the photosynthetic electron transport chain to protect the plant from overexposure to solar radiation (Krause and Weis 1991; Trebst 1991). The mechanism is based on the damage of the D1 protein in photosystem II which is subsequently degraded by a proteinase. During recovery the damaged protein is replaced by newly synthesized protein on a time scale of minutes to hours. Recovery is faster and more effective in low light conditions than in darkness.

While most of the earlier work on photosynthesis was based on gas-exchange measurements (Häder and Schäfer 1994), the technique of pulse amplitude modulation (PAM) fluorescence was introduced recently into ecophysiology (Schreiber et al. 1986; Krause and Weis 1991). This method provides information on the physiological status of the regulatory processes of photosynthetic energy allocation in vivo based on transient changes of chlorophyll fluorescence. Initial measurements were restricted to the laboratory, but recent miniaturization and computer control allows operation of the equipment in the field (Schreiber and Bilger 1987, 1993; Schreiber et al. 1986). Recently, a diving PAM instrument was developed which allows the user to measure the quantum yield of fluorescence in organisms at their growth site, which promises advances in the knowledge on ecophysiology of macroalgae. This instrument was used, for example, to determine the photosynthetic fluorescence parameters in Atlantic macroalgae in the Canary Islands (Häder et al. 2000a,b).

PAM fluorescence analysis is based on the measurement of the ground fluorescence F_0 induced by low level irradiation of dark-adapted samples, so that all reaction centres are in the open state. A single saturating flash induces maximal fluorescence, F_m; at this time all reaction centres are closed. The difference between F_0 and F_m is called variable fluorescence, F_v. The optimal quantum yield can be calculated from the ratio F_v/F_m (Björkman and Demmig 1987). Adaptation to higher fluence rates leads to a decrease of F_m, then called F_m', and an increase or a decrease in F_0, (F_0'). From these measured fluorescence parameters and the currently encountered fluorescence, F_t, the photochemical quenching q_P can be determined (Schreiber et al. 1994), i.e. the amount of excitation energy guided into the photochemical processes using the following equation:

$$q_P = (F_m' - F_t)/(F_m' - F_0').$$

The non-photochemical quenching q_N is calculated by:

$$q_N = 1 - (F_m' - F_0')/(F_m - F_0).$$

The non-photochemical quenching includes all nonradiative processes dissipating excitation energy, but the underlying molecular mechanism is still controversial. The effective quantum yield of photochemical energy conversion is calculated as $(F_m' - F_t)/F_m'$ (Genty et al. 1989).

The quenching analysis allows one to determine the photochemical and non-photochemical quenching (Büchel and Wilhelm 1993). The increase in non-photochemical quenching is related to the violaxanthin cycle, which is thought to quench excess excitation energy in algae and also in higher plants (Demmig-Adams and Adams 1992; Häder and Figueroa 1997). Even algae harvested from rock pools exposed to extreme irradiances show photoinhibition after extended periods of exposure. Algae growing in deeper water and those adapted to shaded conditions are inhibited even faster when exposed to unfiltered solar radiation (Häder et al. 1996a,b,c, 1997a,b, 1998a,b; Porst et al. 1997; Aguilera et al. 1999). Tremendous differences were also found in the recovery between surface and deep-water species. A sizable proportion of photoinhibition is due to PAR (400 - 700 nm), but exclusion studies showed that also UV-B and UV-A play a considerable role, which is substantially higher than their energy contribution in solar radiation (Herrmann et al. 1995a). Figure 7.7 shows the inhibition in the photosynthetic quantum yield after exposure to solar radiation with and without the UV components in the Atlantic green macroalga *Caulerpa racemosa*. UV-A and UV-B also cause depression of the photosynthetic rate in the brown alga *Laminaria digitata* (Forster and Lüning 1996).

Increasing exposure to solar radiation also resulted in a shift of the compensation point to higher irradiances. The compensation point describes the irradiance at which photosynthetic oxygen production and respiratory oxygen consumption have equal values. Exclusion of UV-B partially decreased the effects, and when

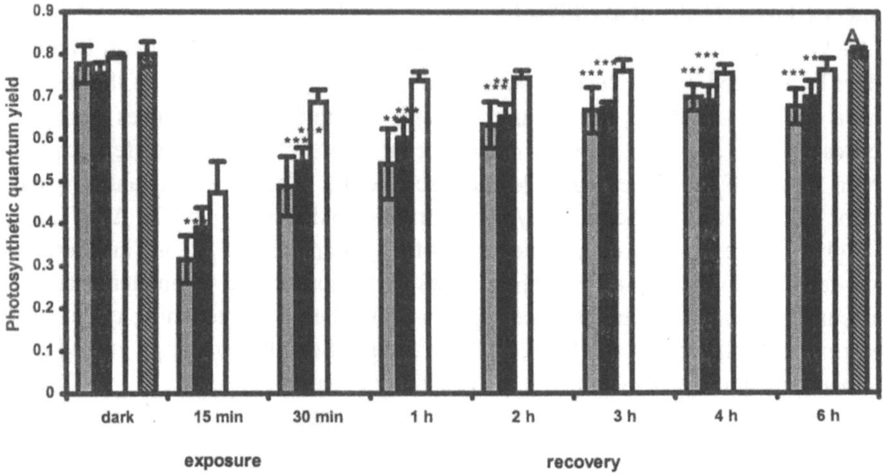

Fig. 7.7. Photosynthetic quantum yield of the green alga *Caulerpa racemosa* measured after 30 min dark adaptation, 15 min exposure and after increasing recovery times in the shade calculated as $(F_m'-F_t)/F_m'$. The specimens were exposed either to unfiltered solar radiation (*grey bars*), radiation filtered through a 320 nm cutoff filter foil (*black bars*) or filtered through a 395 nm cutoff filter foil (*white bars*). Independent controls were subjected to the same treatment except solar exposure and measured after the dark treatment and after the recovery period, respectively (*striped bars*). For *each data point* at least eight measurements were averaged and the standard deviation calculated. The values for unfiltered solar radiation and under the 320 nm cutoff filter treatments are statistically significantly different from the PAR value (395 nm cutoff) in each set with $p < 0.001$ (***) and < 0.01 (**), respectively, as indicated by the Student's *t*-test

about half or all of the UV-A radiation was excluded, the damage was even less (Schott cutoff filters WG 360 and 395). The lowest light compensation point for photosynthesis has been reported in Arctic and Antarctic algae (Gómez et al. 1995; Wiencke 1996; Gómez and Wiencke 1996).

Surface macroalgae, such as several brown (*Cystoseira*, *Padina*, *Fucus*) and green (*Ulva*, *Enteromorpha*) algae, show a maximal oxygen production at or close to the surface (Herrmann et al. 1995b; Häder and Figueroa 1997), while algae adapted to lower irradiances usually thrive best in deeper water (the green algae *Cladophora*, *Caulerpa*, most red algae) (Häder et al. 1996a,b,c, 1997a,b, 1998a,b; Häder and Figueroa 1997; Porst et al. 1997; Aguilera et al. 1999). It is interesting to note that respiration is inhibited to a far lesser degree than photosynthesis.

Chronic photoinhibition occurs when algae are exposed to excessive irradiance for longer periods of time. The inhibition is characterized by photodamage of PS II reaction centres (Critchley and Russell 1994). In contrast, dynamic photoinhibition is readily reversible and follows a diurnal pattern, with the lowest quantum yield around or soon after noon (Hanelt et al. 1994; Häder and Figueroa 1997).

The long-term effects of solar UV on photosynthetic productivity of macroalgae still need to be evaluated, but it is interesting to note that algae are inhibited not only upon sudden exposure to solar radiation at the surface when removed from their habitat. When the photosynthetic quantum yield was measured on site using the diving PAM without removing the algae from their habitat, a significant photoinhibition was found during and shortly after local noon (Fig. 7.8).

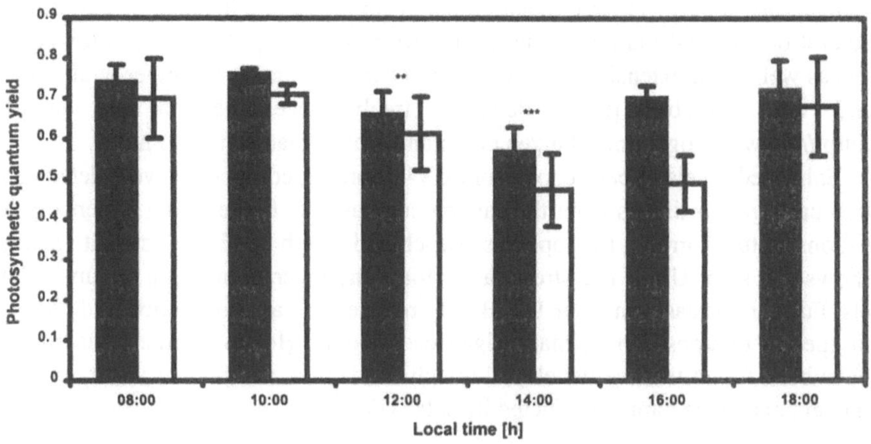

Fig. 7.8. Photosynthetic quantum yield measured at 2-h intervals with the diving PAM on the same thalli of the Atlantic brown alga *Lobophora variegata* at its growth site at 1.5 m depth (*open bars*) and 5 m depth (*closed bars*). For *each data point* at least eight measurements were averaged and the standard deviation calculated. The measured values during the day are statistically significantly different from the first value with $p < 0.001$ (***) and < 0.01 (**), respectively, as indicated by the Student's *t*-test

Different methods for measuring the light absorption in macroalgae have been investigated (Mercado et al. 1996). The absorption determined by using an integrating sphere and by the opal-glass technique in a spectrophotometer in thin macroalgae was compared. García-Pichel (1995) developed a scalar irradiance fibre-optic microprobe for the measurement of UV radiation at high spatial resolution.

Some macroalgae such as the green alga *Ulva lactuca* (Grevby 1996) and the brown alga *Dictyota dichotoma* (Uhrmacher et al. 1995), possess the xanthophyll cycle as a photoprotective mechanism which has also been studied in microalgae (Schubert et al. 1994). Red algae do not seem to have a xanthophyll cycle.

Another adaptive mechanism in macroalgae for protection against UV radiation is the production of screening pigments such as carotenoids or UV-absorbing mycosporine-like amino acids. MAAs have been identified in green, red and brown algae from tropical, temperate and polar regions. These substances are chemically very stable and accumulate in the sediment of lakes. There, they can be used as a permanent record for past ultraviolet radiation levels (Leavitt et al. 1997). A thorough screening of many macroalgae harvested in the field has indicated the presence of various MAAs in a number of species. It is interesting to note that some macroalgae, such as several deep-water red algae, do not possess MAAs and cannot be induced to synthesize these substances even by surface exposure or in the laboratory. Other algae have large amounts of one or several MAAs, but no further synthesis can be induced by exposure to solar visible or UV radiation (Sinha et al. 1998). In a third group of organisms, the MAA content is low when the organisms are kept in UV-free radiation but synthesis can be induced by exposure to natural or artificial radiation. The results from screening of naturally growing algae as well as an extensive literature search can be found in Gröniger et al. (2000) and are available in a database freely accessible on the internet (http://www.biologie.uni-erlangen.de/botanik1/html/maa_database.htm).

Enhanced levels of carotenoids and UV-absorbing compounds were detected in tropical algae in tissues from the canopy compared to tissues from understory locations of turf-forming rhodophytes (Beach and Smith 1996a,b). Present research shows that solar UV-B is a stress factor for many macroalgae even at current levels. Further increases in solar UV-B may reduce biomass production and changes in species composition in macroalgae ecosystems. However, many macroalgae have been found to have developed effective adaptive and protective strategies to prevent extensive damage inflicted by solar UV.

7.5 Ecosystems

The adaptive capabilities of individual specimens or species do not allow conclusions on the behaviour of whole communities or ecosystems to environmental stress factors. The succession of species during the seasons may change as well as the species composition and food web properties. For example, UV-B has been found to be more detrimental for small phytoplankton organisms (Karentz et al. 1994).

7.5.1 Freshwater Ecosystems

The succession of periphytic and limnic algal populations is governed by the complex external stress factors as well as interspecific dependencies (Rai et al. 1996). Freshwater communities are characterized by a high biomass turnover. While the fate of an individual specimen or species may depend on a number of unpredictable factors, the general succession of ecosystem structure follows well-described patterns (Biggs 1996). Increased solar UV-B is an additional stress factor which may affect species composition and biomass productivity even though UV penetration into the water column is lower in freshwater than in the ocean (Williamson 1995, 1996; Häder and Häder 1997; Piazena and Häder 1997).

In a large study dozens of freshwater species of Chlorophyta and Chromophyta were investigated to quantify their UV-B sensitivity (Xiong et al. 1996). The algae were selected from different ecosystems (from high-altitude lakes to thermal springs). In the most sensitive species photosynthetic oxygen production was reduced by 30 – 50% during a 2-h UV-B exposure (2 W m^{-2}). In high mountain locations many resistant species were found. They are often characterized by solid cell walls which may be encrusted with sporopollenin. In another experiment in a high-altitude mountain lake (Halac et al. 1997) no significant differences were found between exposure to full sunlight and the UV-B depleted samples. Also UV-A can have a negative effect on growth and photosynthesis (Kim and Watanabe 1994), while in others it has a beneficial effect counteracting UV-B inhibition (Quesada 1995).

The interaction of UV-B and heavy-metal stress synergistically inhibited nutrient uptake, enzyme activity, carbon fixation, ATP synthesis and oxygen exchange in several phytoplankton species (Rai et al. 1996; Rai and Rai 1997).

In order to understand the effects of stress factors on freshwater ecosystems, the contribution of other components such as the heterotrophic picoplankton has to be considered (Sommaruga and Robarts 1997). In an experiment by Bothwell et al. (1994), better algal growth was observed in an artificial stream under UV-B exposure than in the control. The explanation of this surprising result was that the grazers, larval chironomids, were more sensitive to UV-B radiation than their pray, the algae.

7.5.2 The Polar Aquatic Ecosystems

Productivity in the Antarctic waters shows large-scale spatial and temporal variability (Sullivan et al. 1993; Arrigo 1994). Therefore, it is difficult to identify UV-B-specific effects (Neale et al. 1998a). This may also be the reason for the different and opposing evaluations of the impact of solar UV on single species or whole phytoplankton communities (Karentz and Spero 1995; Davidson et al. 1996). Variability in solar elevation, cloud cover, deep vertical mixing and ice and snow cover significantly interferes with the direct UV effect. The effect of 50% ozone reduction on the biomass productivity in the water column has been estimated to be between <5% (Boucher and Prezelin 1996) and 0.7– 8.5% (Smith et al. 1992; Neale et al. 1998b).

Numerous studies have accumulated convincing evidence of UV-B damage to phytoplankton on a short-term basis, but in order to evaluate long-term effects, adaptation (Villafañe et al. 1995; Lesser 1996; Helbling et al. 1996; Neale et al. 1998a) needs to be determined. Field results clearly show that photosynthesis of Antarctic phytoplankton is inhibited by ambient UV at least in experimental containers (Smith et al. 1992; Helbling et al. 1994; Vernet et al. 1994; Neale et al.

1998b). In the field, vertical mixing is a major problem in the attempt to quantify UV-B effects on phytoplankton (Neale et al. 1998a). Models have been developed (Arrigo 1994; Behrenfeld et al. 1994; Boucher and Prezelin 1996; Neale et al. 1998a) to estimate ecosystem productivity under UV impact. The difficulty stems from the generalization of experimental results to the field where mixing significantly alters the exposure of phytoplankton to UV-B. Near-surface UV strongly inhibits photosynthesis under all modelled conditions, but inhibition of photosynthesis can be modified by vertical mixing. A 50% reduction in stratospheric ozone could, as a worst case, lower photosynthesis in the total water column by as much as 8.5%. Other factors may also have significant effects on photosynthetic biomass productivity such as vertical mixing (±37%), sensitivity of phytoplankton to UV (±46%) and cloudiness (±15%). Thus, ozone depletion is a significant environmental stressor which can inhibit primary productivity in Antarctic open waters, but natural variability, in combination with vertical mixing and cloud cover, plays a major role in determining biomass productivity (Häder et al. 1998c).

Even though increased levels of solar UV have been measured in the Arctic, it differs in many respects from the Antarctic (Weiler and Penhale 1994; Wängberg et al. 1996). The Arctic ocean is a nearly enclosed basin with little water exchange with the adjacent oceans. It receives considerable freshwater inflow which causes pronounced stratification and high concentrations of POC and DOC (Burenkov 1993). Because of the prominent stratification the phytoplankton may experience relatively high levels of solar UV-B. Another characteristic is the larger role of macroalgae in the Arctic. Productivity in the Arctic ocean has also been reported to be higher than in the Antarctic ocean (Springer and McRoy 1993). However, further careful analysis is necessary to quantify UV-B-related inhibition of phytoplankton and macroalgae and possible affects on higher trophic levels.

The Arctic ocean is often nutrient-limited, especially in nitrogen and phosphorus. Nitrogen availability controls the primary productivity of the marine ecosystems as well as oligotrophic lakes and streams. Uptake of nitrogen and phosphorus are sensitive to UV-B (Döhler 1992), which may increase the UV-B sensitivity of Arctic phytoplankton communities. Low doses of UV-B enhance the uptake of phosphate, which is necessary for DNA repair, but it decreases the uptake at higher doses.

7.6 Conclusions and Consequences

Possible consequences of ozone depletion and concomitant higher levels of solar UV-B radiation include loss of biomass, changes in species composition and reduced sink capacity for atmospheric carbon dioxide. The latter may result in the potential augmentation of global warming. There is overwhelming evidence that increased UV-B exposure is harmful to aquatic primary producers such as cyanobacteria, phytoplankton and macroalgae, but effects on ecosystems are still uncertain. Adaptation to changing environmental stress factors such as enhanced levels of UV-B exposure need to be studied not only on the individual level but on whole ecosystems. Shifts in community structure may initially be more common but may result in little detectable differences in ecosystem biomass.

References

Aguilera J, Jimenez C,Figueroa FL, Lebert M, Häder D-P(1999) Effect of ultraviolet radiation on thallus absorption and photosynthetic pigments in the red alga *Porphyra umbilicalis*. J Photochem Photobiol B.Biol 48:75-82

Aráoz R, Häder D-P (1997) Ultraviolet radiation induces both degradation and synthesis of phycobilisomes in *Nostoc* sp.: a spectroscopic and biochemical approach. FEMS Microbiol Ecol 23:301 – 313

Arrigo KR (1994) Impact of ozone depletion on phytoplankton growth in the Southern Ocean: large-scale spatial and temporal variability. Mar Ecol Prog Ser 114:1 – 12

Banerjee M, Häder D-P (1996) Effects of UV radiation on the rice field cyanobacterium, *Aulosira fertilissima*. Environ Exp Bot 36:281 – 291

Beach KS, Smith CM (1996a) Ecophysiology of tropical rhodophytes. I. Microscale acclimation in pigmentation. J Phycol 32:701 – 710

Beach KS, Smith CM (1996b) Ecophysiology of tropical rhodophytes. II. Microscale acclimation in photosynthesis. J Phycol 32:710 – 718

Behrenfeld MJ (1995) Ultraviolet-B radiation effects on inorganic nitrogen uptake by natural assemblages of oceanic plankton. J Phycol 31:25 – 36

Behrenfeld MJ, Falkowski PG (1997a) Photosynthetic rates derived from satellite-based chlorophyll concentration. Limnol Oceanogr 42:1 – 20

Behrenfeld MJ, Falkowski PG (1997b) A consumer's guide to phytoplankton primary productivity models. Limnol Oceanogr 42:1479 – 1491

Behrenfeld MJ, Lee II H, Small LF (1994) Interactions between nutritional status and long-term responses to ultraviolet-B radiation stress in a marine diatom. Mar Biol 118:523 – 530

Berhard G, Mayer B, Seckmeyer G (1997) Measurements of spectral solar UV irradiance in tropical Australia. J Geophys Res 102:8719-8730

Biggs BJF (1996) Patterns in benthic algae of streams. In: Stevenson RJ, Lowe RL, Bothwell ML (ed) Algal ecology. Academic Press, New York, pp 31 – 56

Björkman O, Demmig B (1987) Photon yield of O_2-evolution and chlorophyll fluorescence characteristics at 77 K among vascular plants of diverse origins. Planta 170:489 – 504

Björn LO, Cunningham A, Dubinsky Z, Estrada M, Figueroa FL, Garcia-Pichel F, Häder D-P, Hanelt D, Levavasseur G, Lüning K (1996) Technical discussion. I: Underwater light measurements and light absorption by algae. In: Figueroa FL, Jiménez C, Pérez-Lloréns JL, Niell FX (eds) Underwater light and algal photobiology. Sci Mar 60 (Suppl 1), Barcelona, Spain, pp 59 – 63

Blumthaler M (1997) Braod band detectors for UV-measurments.In: Nato.AST Ser 1 52: Solar ultraviolet radiation modelling, measurements and effects. C S Zerefos, A F Bais(eds) Springer Belin Heidelberg pp 175-185

Booth CR, Morrow JH, Coohill TP, Frederick JE, Häder D-P, Holm-Hansen O, Jeffrey WH, Mitchell DL, Neale PJ, Sobolev I, van der Leun J, Worrest RC (1997) Impacts of solar UVR on aquatic microorganisms. Photochem Photobiol 65:252 – 269

Bothwell ML, Sherbot DMJ, Pollock CM (1994) Ecosystem response to solar ultraviolet-B radiation: influence of trophic level interactions. Science 256:97 – 100

Boucher NP, Prezelin BB (1996) An in situ biological weighting function for UV inhibition of phytoplankton carbon fixation in the Southern Ocean. Mar Ecol Prog Ser 114:223 – 236

Brown CW, Esaias WE, Thompson AM (1995) Predicting phytoplankton composition from space — using the ratio of euphotic depth to mixed-layer depth: an evaluation. Remote Sens Environ 53:172 – 176

Büchel C, Wilhelm C (1993) In vivo analysis of slow chlorophyll fluorescence induction kinetics in algae: progress, problems and perspectives. Photochem Photobiol 58:137 – 148

Buma AGJ, van Hannen EJ, Roza L, Veldhuis MJW, Gieskes WWC (1995) Monitoring ultraviolet-B-induced DNA damage in individual diatom cells by immunofluorescent thymine dimer detection. J Phycol 51:314 – 321

Buma AGJ, Zemmelink HJ, Sjollema K, Gieskes WWC (1996a) UVB radiation modifies protein and photosynthetic pigment content, volume and ultrastructure of marine diatoms. Mar Ecol Prog Ser 147:47 – 54

Buma AGJ, van Hannen EJ, Veldhuis MJW, Gieskes WWC (1996b) UV-B induces DNA damage and DNA synthesis delay in the marine diatom *Cyclotella* sp. Sci Mar 60 (Suppl 1), Barcelona, Spain, pp 101 – 106

Buma AGJ, Engelen AH, Gieskes WWC (1997) Wavelength-dependent induction of thymine dimers and growth rate reduction in the marine diatom *Cyclotella* sp. exposed to ultraviolet radiation. Mar Ecol Prog Ser 153:91 – 97

Burenkov VI (1993) Optical properties of the Laptev Sea near the Lena River delta. Proc Symp on High Latitude Optics. Tromso, Norway, 28 June-2 July 1993

Cockell CS, Knowland J (1999) Ultraviolet radiation screening compounds. Biol Rev 74:311 – 345

Coohill TP, Häder D-P, Mitchell DL (1996) Environmental ultraviolet photobiology: introduction. Photochem Photobiol 64:401 – 402

Critchley C, Russell AW (1994) Photoinhibition of photosynthesis in vivo: the role of protein turnover in photosystem II. Physiol Plant 92:188 – 196

Cullen JJ, Neale PJ (1994) Ultraviolet radiation, ozone depletion, and marine photosynthesis. Photosynth Res 39:303 – 320

Cullen JJ, Neale PJ (1996) Effects of UVon short-term photosynthesis of Antarctic phytoplankton. Abstract s of the 24th annual meeting of the Amer Soc Photobiol. Photochem Photobiol 63: p 12S-13S

Cullen JJ, Neale PJ (1997a) Biological weighting functions for describing the effects of ultraviolet radiation on aquatic systems. In: Häder D-P (ed) The effects of ozone depletion on aquatic ecosystems. Academic Press, RG Landes Company, Austin, pp 97 – 118

Cullen JJ, Neale PJ (1997b) Effects of ultraviolet radiation on short-term photosynthesis of natural phytoplankton. Photochem Photobiol 65:264 – 266

Davidson AT, Marchant HJ, de la Mare,WK (1996) Natural UV exposure changes the species composition of Antarctic phytoplankton in mixed culture. Aquat Microbial Ecol 10:299 – 305

Demmig-Adams B, Adams W III (1992) Photoprotection and other responses of plants to high light stress. Annu Rev Plant Physiol Plant Mol Biol 43:599 – 626

Döhler G (1992) Impact of UV-B radiation (290–320 nm) on uptake of ^{15}N-ammonia and ^{15}N- - nitrate by phytoplankton of the Wadden Sea. Mar Biol 112:485 – 489

Döhler G (1996) Effect of UV irradiance on utilization of inorganic nitrogen by the Antarctic diatom *Odontella weissflogii* (Janisch) Grunow. Bot Acta 109:35 – 42

Döhler G (1997) Impact of UV radiation of different wavebands on pigments and assimilation of ^{15}N-ammonium and ^{15}N-nitrate by natural phytoplankton and ice algae in Antarctica. J Plant Physiol 151:550 – 555

Döhler G, Hagmeier E (1997) UV-effects on pigments and assimilation of ^{15}N-ammonium and ^{15}N-nitrate by natural marine phytoplankton of the North Sea. Bot Acta 110:481 – 488

Döhler G, Hagmeier E, David C (1995) Effects of solar and artificial UV irradiation on pigments and assimilation of ^{15}N ammonium and ^{15}N nitrate by macroalgae. J Photochem Photobiol B: Biol 30:179 – 187

Donkor VA, Häder D-P (1995) Protective strategies of several cyanobacteria against solar radiation. J Plant Physiol 145:750 – 755

Donkor VA, Häder D-P (1996) Effects of ultraviolet irradiation on photosynthetic pigments in some filamentous cyanobacteria. Aquat Microbial Ecol 11:143 – 149

Donkor VA, Häder D-P (1997) Ultraviolet radiation effects on pigmentation in the cyanobacterium *Phormidium uncinatum*. Acta Protozool 36:49 – 55

Dring MJ,Makarov V,Schoshina F, Lorenz M, Luning K (1996) Influenece of ultraviolet radiation on chlorophyll fluorescence and growth in different life history stages of three species of *Laminaria* (Phaeophyta). Mar Biol 126: 183-191

Ducklow HW, Carlson CA, Bates NR, Knap AH, Michaels AF (1995) Dissolved organic carbon as a component of the biological pump in the North Atlantic Ocean. Philos Trans R Soc Lond B: Biol Sci 348:161 – 167

Dunlap WC, Shick JM (1998) Ultraviolet radiation-absorbing mycosporine-like amino acids in coral reef organisms: a biochemical and environmental perspective. J Phycol 34:418 – 430

Figueroa FL, Jiménez C, Pérez-Lloréns JL, Niell FX (1996) Underwater light and algal photobiology. Sci Mar 60 (Suppl 1), Barcelona, Spain, pp 343

Figueroa FL, Jiménez C, Lubián LM, Montero O, Lebert M, Häder D-P (1997) Effects of high irradiance and temperature on photosynthesis and photoinhibition in *Nannochloropsis gaditana* Lubián (Eustigmatophyceae). J Plant Physiol 151:6 – 15

Forster R, Lüning K (1996) Photosynthetic response of *Laminaria digitata* to ultraviolet A and B radiation. In: Figueroa FL, Jiménez C, Pérez-Lloréns JL (eds) Underwater light and algal photobiology. Sci Mar 60 (Suppl 1), Barcelona, Spain, pp 65 – 71

Franklin LA, Foster RM (1997) The changing irradiance environment: consequences for marine macrophyte physiology, productivity and ecology. Eur J Phycol 32:207 – 232

Garcia-Pichel F (1995) A scalar irradiance fiber-optic microprobe for the measurement of ultraviolet radiation at high spatial resolution. Photochem Photobiol 61:248 – 254

Genty BE, Briantais JM, Baker NR (1989) Relative quantum efficiencies of the two photosystems of leaves in photorespiratory and non-photorespiratory conditions. Plant Physiol Biochem 28:1 – 10

Gerber S, Häder D-P (1995a) Effects of enhanced solar irradiance on chlorophyll fluorescence and photosynthetic oxygen production of five species of phytoplankton. FEMS Microbiol Ecol 16:33 – 42

Gerber S, Häder D-P (1995b) Effects of artificial and simulated solar radiation on the flagellate *Euglena gracilis*: physiological, spectroscopical and biochemical investigations. Acta Protozool 34:13 – 20

Gerber S, Biggs A, Häder D-P (1996) A polychromatic action spectrum for the inhibition of motility in the flagellate *Euglena gracilis*. Acta Protozool 35:161 – 165

Giacometti GM, Barbato R, Chiaramonte S, Friso G, Rigoni F (1996) Effects of ultraviolet-B radiation on photosystem II of the cyanobacterium *Synechocystis* sp. PCC 6083. Eur J Biochem 242:799 – 806

Gieskes WWC, Buma AGJ (1997) UV damage to plant life in a photobiologically dynamic environment: the case of marine phytoplankton. Plant Ecol 128:16 – 25

Gómez I, Wiencke C (1996) Photosynthesis, dark respiration and pigment contents of gametophytes and sporophytes of the Antarctic brown alga *Desmarestia menziessi*. Bot Mar 39:149 – 157

Gómez I, Wiencke C, Weykman G (1995) Seasonal photosynthetic characteristics of the brown alga *Ascoseira mirabilis* from King George Island (Antarctica). Mar Biol 123:167 – 172

Grevby C (1996) Organisation of the light-harvesting complex in fucoxanthin-containing algae. PhD Thesis, Göteborg University, Sweden

Gröniger A, Sinha RP, Klisch M, Häder D-P (2000) Photoprotective compounds in cyanobacteria, phytoplankton and macroalgae —a database. J Photochem Photobiol B: Biol (in press)

Häder D-P (1995) Photo-ecology and environmental photobiology. In: Horspool WM, Song P-S (eds) CRC handbook of organic photochemistry and photobiology. CRC Press, Boca Raton, New York, pp 1392 – 1401

Häder D-P (1997a) Effects of UV radiation on phytoplankton. In: Jones JG (ed) Advances in microbial ecology, vol 15. Plenum Press, New York, pp 1 – 26
Häder D-P (1997b) Effects of solar UV-B radiation on aquatic ecosystems. In: Lumsden PJ (ed) Plants and UV-B: responses to environmental change. Cambridge University Press, Cambridge, pp 171 – 193
Häder D-P (1997c) UV-B and aquatic ecosystems. In: Rozema J, Gieskes WWC, van de Geijn SC, Nolan C, de Boois H (eds) UV-B and biosphere. Kluwer, Dordrecht, pp 4–13
Häder D-P (1997d) Penetration and effects of solar UV-B on phytoplankton and macroalgae. Plant Ecol 128:4 – 13
Häder D-P (1997e) Stratospheric ozone depletion and increase in ultraviolet radiation. In: Häder D-P (ed) The effects of ozone depletion on aquatic ecosystems. Academic Press and RG Landes Company, Austin, pp 1 – 4
Häder D-P, Figueroa FL (1997) Photoecophysiology of marine macroalgae. Photochem Photobiol 66:1 – 14
Häder D-P, Schäfer J (1994) In-situ measurement of photosynthetic oxygen production in the water column. Environ Monit Assess 32:259 – 268
Häder D-P, Worrest RC (1997) Consequences of the effects of increased solar ultraviolet radiation on aquatic ecosystems. In: Häder D-P (ed) The effects of ozone depletion on aquatic ecosystems. Academic Press and RG Landes Company, Austin, pp 11 – 30
Häder D-P, Worrest RC, Kumar HD, Smith RC (1995) Effects of increased solar ultraviolet radiation on aquatic ecosystems. Ambio 24:174 – 180
Häder M, Häder D-P (1997) Optical properties and phytoplankton composition in a freshwater ecosystem (Main-Donau Canal). In: Häder D-P (ed) The effects of ozone depletion on aquatic ecosystems. Academic Press and RG Landes Company, Austin, pp 155 – 174
Häder D-P, Worrest RC, Kumar HD, Smith RC (1994) Effects of increased solar ultraviolet radiation on aquatic ecosystems. UNEP Environmental Effects Panel Report, Nairobi, pp 65 – 77
Häder D-P, Herrmann H, Santas R (1996a) Effects of solar radiation and solar radiation deprived of UV-B and total UV on photosynthetic oxygen production and pulse amplitude modulated fluorescence in the brown alga *Padina pavonia*. FEMS Microbiol Ecol 19:53 – 61
Häder D-P, Herrmann H, Schäfer J, Santas R (1996b) Photosynthetic fluorescence induction and oxygen production in corallinacean algae measured on site. Bot Acta 109:285 – 291
Häder D-P, Porst M, Herrmann H, Schäfer J, Santas R (1996c) Photoinhibition in the Mediterranean green alga *Halimeda tuna* Ellis et Sol measured in situ. Photochem Photobiol 64:428 – 434
Häder D-P, Porst M, Herrmann H, Schäfer J, Santas R (1997a) Photosynthesis of the Mediterranean green alga *Caulerpa prolifera* measured in the field under solar irradiation. J Photochem Photobiol B: Biol 37:66 – 73
Häder D-P, Herrmann H, Schäfer J, Santas R (1997b) Photosynthetic fluorescence induction and oxygen production in two Mediterranean *Cladophora* species measured on site. Aquat Bot 56:253 – 264
Häder D-P, Lebert M, Tahedl H, Richter P (1997c) The Erlanger flagellate test (EFT): photosynthetic flagellates in biological dosimeters. J Photochem Photobiol B: Biol 40:23 – 28
Häder D-P, Lebert M, Figueroa FL, Jiménez C, Viñegla B, Perez-Rodriguez E (1998a) Photoinhibition in Mediterranean macroalgae by solar radiation measured on site by PAM fluorescence. Aquat Bot 61:225 – 236
Häder D-P, Porst M, Santas R (1998b) Photoinhibition by solar radiation in the Mediterranean alga *Peyssonnelia squamata* measured on site. Plant Ecol 139:167 – 175
Häder D-P, Worrest RC, Kumar HD, Smith RC (1998c) Effects on aquatic ecosystems. UNEP Environmental Effects Panel Report, pp 86 – 112
Häder D-P, Porst M, Lebert M (2000a) Photosynthetic performance of Atlantic green algae measured in Gran Canaria on site. J Photochem Photobiol B: Biol(in press)

Häder D-P, Porst M, Lebert M (2000b) Photosynthetic performance of the Atlantic brown macroalgae, *Cystoseira abies-marina, Dictyota dichotoma* and *Sargassum vulgare,* measured in Gran Canaria on site. Environ Exp Bot (in press)

Halac S, Felip M, Camarero L, Sommaruga-Wögrath S, Psenner R, Catalan J, Sommaruga R (1997) An in situ enclosure experiment to test the solar UVB impact on plankton in a high-altitude mountain lake. I. Lack of effect on phytoplankton species composition and growth. J Plankton Res 19:1671 – 1686

Hanelt D, Hupperts K, Nultsch W (1992) Photoinhibition of photosynthesis and its recovery in red algae. Bot Acta 105:278 – 284

Hanelt D, Huppertz K, Nultsch W (1993) Daily course of photosynthesis and photoinhibition in marine macroalgae investigated in the laboratory and field. Mar Ecol Prog Ser 97:31 – 37

Hanelt D, Jaramillo MJ, Nultsch W, Senger S, Westermeir R (1994) Photoinhibition as a regulatory mechanism of photosynthesis in marine algae of Antarctica. Ser Cient Inst Antarct Chil 44:67 – 77

Helbling EB, Chalker BE, Dunlap WC, Holm-Hansen O, Villafane VE (1996) Photoacclimation of antarctic marine diatoms to solar ultraviolet radiation. J Exp Mar Biol Ecol 204:85 – 101

Helbling EW, Villafane V, Holm-Hansen O (1994) Effects of ultraviolet radiation on Antarctic marine phytoplankton photosynthesis with particular attention to the influence of mixing. In: Weiler CS, Penhale PA (eds) Ultraviolet radiation in Antarctica: measurements and biological effects, Antarctic Research Series no. 62. American Geophysical Union, Washington, DC, pp 207 – 227

Herndl GJ (1997) Role of ultraviolet radiation on bacterioplankton activity. In: Häder D-P (ed) The effects of ozone depletion on aquatic ecosystems. Academic Press and RG Landes Company, Austin, pp 143 – 154

Herrmann H, Häder D-P, Köfferlein M, Seidlitz HK, Ghetti F (1995a) Study on the effects of UV radiation on phytoplankton photosynthetic efficiency by means of a sunlight simulator. Med Biol Environ 23:36 – 40

Herrmann H, Ghetti F, Scheuerlein R, Häder D-P (1995b) Photosynthetic oxygen and fluorescence measurements in *Ulva laetevirens* affected by solar irradiation. J Plant Physiol 145:221 – 227

Herrmann H, Häder D-P, Köfferlein M, Seidlitz HK, Ghetti F (1996) Effects of UV radiation on photosynthesis of phytoplankton exposed to solar simulator light. J Photochem Photobiol B: Biol 34:21 – 28

Herrmann H, Häder D-P, Ghetti F (1997) Inhibition of photosynthesis by solar radiation in *Dunaliella salina*: relative efficiencies of UV-B, UV-A and PAR. Plant Cell Environ 20:359 – 365

Holm-Hansen O (1997) Short- and long-term effects of UVA and UVB on marine phytoplankton productivity. Photochem Photobiol 65:266 – 268

Horneck G (1995) Quantification of the biological effectiveness of environmental UV radiation. J Photochem Photobiol B: Biol 31:43 – 49

IASC (1995) Effects of increased ultraviolet radiation in the Arctic. IASC Report No 2, IASC Secretariat. J Plant Physiol 148:42 – 48

Jerlov NG (1968) Optical oceanography. Elsevier, Amsterdam

Jiménez C, Figueroa FL, Aguilera J, Lebert M, Häder D-P (1996) Phototaxis and gravitaxis in *Dunaliella bardawil*: influence of UV radiation. Acta Protozool 35:287 – 295

Karentz D, Spero HJ (1995) Response of a natural *Phaeocystis* population to ambient fluctuations of UVB radiation caused by Antarctic ozone depletion. J Plankton Res 17:1771 – 1789

Karentz D, Bothwell ML, Coffin RB, Hanson A, Herndl GJ, Kilham SS, Lesser MP, Lindell M, Moeller RE, Morris DP, Neale PJ, Sanders RW, Weiler CS, Wetzel RG (1994) Report of working group on bacteria and phytoplankton. In: Williamson CE, Zagarese HE (eds) Impact of UV-B radiation on pelagic freshwater ecosystems. Arch Hydrobiol Beih 43 (Spec Issue), pp 31 – 69

Kashyap AK, Pandey KD, Gupta RK (1991) Nitrogenase activity of the Antarctic cyanobacterium *Nostoc* commune: influence of temperature. Folia Microbiol 36:557 – 560

Kerr JB, McElroy CT (1993) Evidence for large upward trends of ultraviolet-B radiation linked to ozone depletion. Science 262:1032 – 1034

Kim DS, Watanabe Y (1994) Inhibition of growth and photosynthesis of freshwater phytoplankton by ultraviolet A (UVA) radiation and subsequent recovery from stress. J Plankton Res 16:1645 – 1654

Kirchoff VWJH, Zamorano F, Casiccia C (1997) UV-B enhancements at Punta Arenas, Chile. J Photochem Photobiol B: Biol 38:174 – 177

Kirst GO, Wiencke C (1996) Ecophysiology of algae. J Phycol 31:181 – 199

Kjeldstad B, Johnsen B, Koskela T (1997) The nordic intercomparison of ultraviolet and total ozone instruments at Iana, October 1996, Final report. Finnish Meteorological Institute Report No 36, Helsinki

Krause GH, Weis E (1991) Chlorophyll fluorescence and photosynthesis: the basics. Annu Rev Plant Physiol Plant Mol Biol 42:313 – 349

Krotkov NA, Barthia PK, Herman J, Celarier E, Eck T (1997) Estimates of spectral UVB irradiance from the TOMS instrument: effects of clouds and aerosols. In: Smith WL, Stamnes K (eds) IRS'96: Current problems in atmospheric radiation, Proc Int Radiation Symp, Fairbanks, Alaska 19-24 August 1996. Deepak Publishing, Hampton Virginia, pp 873-876

Kumar A, Tyagi MB, Srinivas G, Singh N, Kumar HD, Sinha RP, Häder D-P (1996a) UVB shielding role of $FeCl_3$ and certain cyanobacterial pigments. Photochem Photobiol 64:321 – 325

Kumar A, Sinha RP, Häder D-P (1996b) Effect of UV-B on enzymes of nitrogen metabolism in the cyanobacteria *Nostoc calcicola*. J Plant Physiol 148:86 – 91

Leavitt PR, Vinebrooke RD, Donald DB, Smol JP, Schindler DW (1997) Past ultraviolet radiation environments in lakes derived from fossil pigments. Nature 388:457 – 459

Lesser M (1996) Acclimation of phytoplankton to UV-B radiation: oxidative stress and photoinhibition of photosynthesis are not prevented by UV-absorbing compounds in the dinoflagellate *Prorocentrum micans*. Mar Ecol Prog Ser 132:287 – 297 (correction: Mar Ecol Prog Ser 141:312)

Leszczynski K, Jokela K, Ylianttila L, Visuri R, Blumthaler M (1998) Erythemally weighted radiometers in solar UV monitoring: results from WMO/STUK intercomparison. Photochem Photobiol 67:212 – 221

Lüning K (1990) Seaweeds. Their environment, biogeography and ecophysiology. Wiley, New York

Madronich S, Flocke S (1997) Theoretical estimation of biologically effective UV radiation at the Earth's surface. In: Zerefos C (ed) Solar ultraviolet radiation - modeling, measurements and effects, NATO ASI Ser vol 152. Springer, Berlin Heidelberg New York

Madronich S, McKenzie RL, Björn LO, Caldwell MM (1995) Changes in ultraviolet radiation reaching the Earth's surface. Ambio 24:143 – 152

Madronich S, McKenzie RL,Bjorn LO, Caldwell MM (1998) Changes in biologically active ultraviolet radiation reaching the Earth's surface. J Photochem Photobiol B Biol 46: 5-19

Markager S, Sand-Jensen K (1994) The physiology and ecology of light-grown relationship in macroalgae. In: Round FE, Chapman DJ (ed) Progress in phycological research, Biopress, Bristol, pp 209 – 298

McKinlay AF, Diffey BL (1987) A reference action spectrum for ultraviolet induced erythema in human skin. In: Passchier WR, Bosnjakovic BFM (eds) Human exposure to ultraviolet radiation: risks and regulations. Elsevier, Amsterdam

Meerkoetter R, Wissinger B, Seckmeyer G (1997) Surface UV from ERS-2/GOME and NOAA/AVHRR data: a case study. Geophys Res Lett 24:1939 – 1942

Mercado JM, Jiménez C, Niell FCX, Figueroa FL (1996) Comparisons of methods for measuring light absorption by algae and their application to the estimation of the package effect. In: Figueroa FL, Jiménez C, Pérez-Lloréns JL (eds) Underwater light and algal photobiology, Sci Mar 60 (Suppl 1), Barcelona, Spain, pp 39 – 45

Mims FM III (1997) Biological effects of diminished UV and visible sunlight caused by severe air pollution. In: Smith WL, Stamnes K (eds) IRS'96: Current problems in atmospheric radiation, Proc Int Radiation Symp, Fairbanks, Alaska 19-24 August 1996. Deepak Publishing, Hampton Virginia, pp 905 – 908

Montecino V, Pizarro G (1995) Phytoplankton acclimation and spectral penetration of UV irradiance off the central Chilean coast. Mar Ecol Prog Ser 121:261 – 269

Morrow JH, Booth CR (1997) Instrumentation and methodology for ultraviolet readiation measurements in aquatic environments. In: Häder D-P (ed) The effects of ozone depletion on aquatic ecosystems. Academic Press and RG Landes Company, Austin, pp 31 – 44

Müller R, Crutzen PJ, Gross J-U, Brühl C, Russel JM III, Gernandt H, McKenna DS, Tuck AF (1997) Severe chemical ozone loss in the Arctic during the winter of 1995-96. Nature 389:709 – 712

Naganuma T, Konishi S, Inoue T, Nakane T, Sukizaki S (1996) Photodegradation or photoalteration? Microbial assay of the effect of UV-B on dissolved organic matter. Mar Ecol Prog Ser 135:309 – 310

Neale PJ, Cullen JJ, Davis RF (1998a) Inhibition of marine photosynthesis by ultraviolet radiation: variable sensitivity of phytoplankton in the Weddell-Scotia Sea during austral spring. Limnol Oceanogr 43:433 – 488

Neale PJ, Davis RF, Cullen JJ (1998b) Interactive effects of ozone depletion and vertical mixing on photosynthesis of Antarctic phytoplankton. Nature 392:585 – 589

Nolan CV, Amanatidis GT (1995) European Commission research on the fluxes and effects of environmental UVB radiation. J Photochem Photobiol B: Biol 31:3 – 7

Peletier H, Gieskes WWC, Buma AGJ (1996) Ultraviolet-B radiation resistance of benthic diatoms isolated from tidal flats in the Dutch Wadden Sea. Mar Ecol Prog Ser 135:163 – 168

Piazena H (1996) The effect of altitute upon the solar UV-B and UV-A irradiation in the tropical Chilean Andes. Soil Energy 57: 133-140

Piazena H, Häder D-P (1997) Penetration of solar UV and PAR into different waters of the Baltic Sea and remote sensing of phytoplankton. In: Häder D-P (ed) The effects of ozone depletion on aquatic ecosystems. Academic Press and RG Landes Company, Austin, pp 45 – 96

Porst M, Herrmann H, Schäfer J, Santas R, Häder D-P (1997) Photoinhibition in the Mediterranean green alga *Acetabularia mediterranea* measured in the field under solar irradiation. J Plant Physiol 151:25 – 32

Quesada A (1995) Growth of Antarctica cyanobacteria under ultraviolet radiation: UVA counteracts UVB inhibition. J Phycol 31:242 – 248

Rai LC, Tyagi B, Mallick N (1996) Alternation in photosynthetic characteristics of *Anabaena dolium* following exposure to UV-B and Pb. Photochem Photobiol 64:658 – 663

Rai PK, Rai LC (1997) Interactive effects of UV-B and Cu on photosynthesis, uptake and metabolism of nutrients in green alga *Chlorella vulgaris* under simulated ozone column. J General Appl Microbiol 43:281-288

Rontó G, Gróf P, Gáspár S (1995) Biological UV dosimetry — a comprehensive problem. J Photochem Photobiol B: Biol 31:51 – 56

Santas R, Häder D-P, Lianou C (1996) Effects of solar UV radiation on diatom assemblages of the Mediterranean. Photochem Photobiol 64:435 – 439

Sarmiento JL, Le Quéré C (1996) Oceanic carbon dioxide uptake in a model of century-scale global warming. Science 274:1346 – 1350

Scheuerlein R, Treml S, Thar B, Tirlapur UK, Häder D-P (1995) Evidence for UV-B-induced DNA degradation in *Euglena gracilis* mediated by activation of metal-dependent nucleases. J Photochem Photobiol B: Biol 31:113 – 123

Schindell RT, Rind D, Lonergan P (1998) Increased polar stratospheric ozone losses and delayed eventual recovery owing to increasing greenhouse-gas concentrations. Nature 392:589 – 592

Schreiber U, Bilger W (1987) Rapid assessment of stress effects on plant leaves by chlorophyll flourescence measurements. In: Tenhunen JD, Catarino FM, Lange OL, Oechel WD (eds) Plant response to stress. Springer, Berlin Heidelberg New York, pp 27 – 53

Schreiber U, Bilger W (1993) Progress in chlorophyll fluorescence research: major developments during the past years in retrospect. In: Lüttge U, Ziegler H (eds) Progress in botany vol 54. Springer, Berlin Heidelberg New York, pp 151 – 153

Schreiber U, Schliwa U, Bilger W (1986) Continuous recording of photochemical and non-photochemical chlorophyll fluorescence quenching with a new type of modulation fluorometer. Photosynth Res 10:51 – 62

Schreiber U, Bilger W, Neubauer C (1994) Chlorophyll fluorescence as a nonintrusive indicator for rapid assessment of in vivo photosynthesis. In: Schulze ED, Caldwell MM (eds) Ecophysiology of photosynthesis, Ecological Studies, vol 100. Springer Berlin Heidelberg New York,pp 49 – 70

Schubert H, Kroon BMA, Matthijs HCP (1994) In vivo manipulation of the xanthophyll cycle and the role of the zeaxanthin cycle in the photoprotection against photodamage in the green alga *Chlorella pyrenoidosa*. J Biol Chem 269:7267 – 7272

Scully NM, Lean DRS (1994) The attenuation of ultraviolet radiation in temperate lakes. In: Williamson CE, Zagarese HE (eds) Advances in limnology: impact of UV-B radiation on pelagic ecosystems. Arch Hydrobiol Beih 43: 135 – 144

Seckmeyer G, McKenzie RL (1992) Elevated ultraviolet radiation in New Zealand (45°S) contrasted with Germany (48°N). Nature 359:135 – 137

Seckmeyer G, Thiel S, Blumthaler M, Fabian P, Gerber S, Gugg-Helminger A, Häder D-P, Huber M, Kettner C, Köhler U, Köpke P, Maier H, Schäfer J, Suppan P, Tamm E, Thomalla E (1994) Intercomparison of spectral-UV-radiation measurement systems. Appl Optics 33:7805 – 7812

Seckmeyer G, Mayer B, Bernhard G, McKenzie RL, Johnston PV, Kotkamp M, Booth CR, Lucas T, Mestechikina T (1995) Geographical differences in the UV measured by intercompared spectroradiometers. Geophys Res Lett 22:1889 – 1892

Seckmeyer G, Mayer B, Bernhard G, Erb R, Albold A, Jäger H, Stockwell WR (1997) New maximum UV irradiance levels observed in Central Europe. Atmos Environ 31:2971-2976

Sinha RP, Häder D-P (1996) Photobiology and ecophysiology of rice field cyanobacteria. Photochem Photobiol 64:887 – 896

Sinha RP, Häder D-P (1997) Impacts of UV-B irradiation on rice-field cyanobacteria. In: Häder D-P (ed) The effects of ozone depletion on aquatic ecosystems. Academic Press and RG Landes Company, Austin, pp 189 – 198

Sinha RP, Häder D-P (1998) Effects of ultraviolet-B radiation in three rice-field cyanobacteria. J Plant Physiol 153:763 – 769

Sinha RP, Klisch M, Gröniger A, Häder D-P (1998) Ultraviolet-absorbing/screening substances in cyanobacteria, phytoplankton and macroalgae. J Photochem Photobiol B: Biol 47:83 – 94

Sinha RP, Kumar HD, Kumar A, Häder D-P (1995a) Effects of UV-B irradiation on growth, survival, pigmentation and nitrogen metabolism enzymes in cyanobacteria. Acta Protozool 34:187 – 192

Sinha RP, Lebert M, Kumar A, Kumar HD, Häder D-P (1995b) Disintegration of phycobilisomes in a rice field cyanobacterium *Nostoc* sp. following UV irradiation. Biochem Mol Biol Int 37:697 – 706

Sinha RP, Singh N, Kumar A, Kumar HD, Häder M, Häder D-P (1996) Effects of UV irradiation on certain physiological and biochemical processes in cyanobacteria. J Photochem Photobiol B: Biol 32:107 – 113

Sinha RP, Singh N, Kumar A, Kumar HD, Häder D-P (1997) Impacts of ultraviolet-B irradiation on nitrogen-fixing cyanobacteria of rice paddy fields. J Plant Physiol 150:188 – 193

Sinha RP, Klisch M, Häder D-P (1999a) Induction of a mycosporine-like amino acid (MAA) in the rice-field cyanobacterium *Anabaena* sp. by UV irradiation. J Photochem Photobiol B: Biol 52:59 – 64

Sinha RP, Klisch M, Vaishampayan A, Häder D-P (1999b) Biochemical and spectroscopic characterization of the cyanobacterium *Lyngbya* sp. inhabiting mango (*Mangifera indica*) trees: presence of an ultraviolet-absorbing pigment, scytonemin. Acta Protozool 38:291 – 298

Smith RC, Prézelin BB, Baker KS, Bidigare RR, Boucher NP, Coley T, Karentz D, MacIntyre S, Matlick HA, Menzies D, Ondrusek M, Wan Z, Waters KJ (1992) Ozone depletion: ultraviolet radiation and phytoplankton biology in Antarctic waters. Science 255:952 – 959

Sommaruga R, Robarts RD (1997) The significance of autotrophic and heterotrophic picoplankton in hypertrophic ecosystems. FEMS Microbiol Ecol 24:187 – 200

Springer AM, McRoy CP (1993) The paradox of pelagic food webs in the northern Bering Sea. III. Patterns of primary production. Cont Shelf Res 13:575 – 599

Stolarski R (1997) A bad winter for Arctic ozone. Nature 389:788 – 789

Sullivan CW, Arrigo KR, McClain CR, Comiso JC, Firestone J (1993) Distributions of phytoplankton blooms in the Southern Ocean. Science 262:1832 – 1837

Taalas P, Damski J, Kyro E, Ginzberg M, Talamoni G (1997) Effect of stratospheric ozone variations on UV radiation and on tropospheric ozone at high latitudes. J Geophys Res 102:1533 – 1539

Takahashi T, Takahashi TT, Sutherland S (1995) An assessment of the role of the North Atlantic as a CO_2 sink. Philos Trans R Soc Lond B: Biol Sci 348:143 – 152

Takahashi T, Feely RA, Weiss RF, Wanninkhof RH, Chipman DW, Sutherland SC, Takahashi T (1997) Global air-sea flux of CO_2: an estimate based on measurements of sea-air pCO_2 difference. Proc Natl Acad Sci USA 94:8282 – 8299

Thompson A, Early EA, DeLuisi J, Disterhoff P, Wardle D, Kerr J, Rives J, Sun Y, Lucas T et al. (1997) The 1994 North American interagency intercomparison of ultraviolet monitoring spectroradiometers. J Natl Inst Stand Technol 102:279 – 322

Thomson DJ (1997) Dependence of global temperatures on atmospheric CO_2 and solar irradiance. Proc Natl Acad Sci USA 94:8370 – 8377

Trebst A (1991) A contact site between the two reaction center polypeptides of photosystem II is involved in photoinhibition. Z Naturforsch 46:557 – 562

Uhrmacher S, Hanelt D, Nultsch W (1995) Zeaxanthin content and the degree of photoinhibition are linearly correlated in the brown alga *Dictyota dichotoma*. Mar Biol 123:159 – 165

USEPA (US Environmental Protection Agency) (1987) An assessment of the effects of ultraviolet-B radiation on aquatic organisms. In: Assessing the risks of trace gases that can modify the stratosphere. EPA 400/1-87/001C, pp 1 – 33

Varotsos C, Kondratyev KY (1995) On the relationship between total ozone and solar ultraviolet radiation at St. Petersburg, Russia. Geophys Res Lett 22:3481 – 3484

Vassiliev IR, Prasil O, Wyman KD, Kolber ZK, Hanson AK, Prentice JE, Falkowski PG (1994) Inhibition of PS II photochemistry by PAR and UV radiation in natural phytoplankton communities. Photosynth Res 42:51 – 64

Vernet M, Smith RC (1997) Effects of ultraviolet radiation on the pelagic Antarctic ecosystem. In: Häder D-P (ed) The effects of ozone depletion on aquatic ecosystems. Academic Press and RG Landes Company, Austin, pp 247 – 265

Vernet M, Brody EA, Holm-Hansen O, Mitchell BG (1994) The response of Antarctic phytoplankton to ultraviolet radiation: absorption, photosynthesis, and taxonomic composition. In: Weiler CS, Penhale PA (eds) Ultraviolet radiation in Antarctica: measurements and biological effects, Antarctic Research Series no. 62. American Geophysical Union, Washington, DC, pp 207 – 227

Villafane VE, Helbling WW, Holm-Hansen O, Chalker BE (1995) Acclimatization of Antarctic natural phytoplankton assemblages when exposed to solar ultraviolet radiation. J Plankton Res 17:2295 – 2306

Vincent WF, Roy S (1993) Solar ultraviolet radiation and aquatic primary production: damage, protection and recovery. Environ Rev 1:1 – 12

Wängberg S-Å, Selmer J-S (1997) Studies of effects of UV-B radiation on aquatic model ecosystems. In: Häder D-P (ed) The effects of ozone depletion on aquatic ecosystems. Academic Press and RG Landes Company, Austin, pp 199 – 214

Wängberg S-Å, Selmer J-S, Ekelund NGA, Gustavson K (1996) UV-B effects on nordic marine ecosystems. TemaNord, Nordic Council of Ministers, Denmark 1996:515

Webb AR (1997) Advances in solar ultraviolet spectroradiometry. European Commission Air Pollution Research Report 63, Luxembourg

Weiler CS, Penhale PA (1994) Ultraviolet radiation in Antarctica: measurements and biological effects. American Geophysical Union, Washington, DC

Wetzel RG, Hatcher PG, Bianchi TS (1995) Natural photolysis by ultraviolet irradiance of recalcitrant dissolved organic matter to simple substrates for rapid bacterial metabolism. Limnol Oceanogr 40:1369 – 1380

Wiencke C (1996) Recent advances in the investigation of Antarctic macroalgae. Polar Biol 16:231 – 240

Wiencke C, Bartsch I, Bischoff B, Peters AF, Breeman AM (1994) Temperature requirements and biogeography of Antarctic, Arctic and amphiequatorial seaweeds. Bot Mar 37:247 – 259

Williamson CE (1995) What role does UV-B radiation play in freshwater ecosystems? Limnol Oceanogr 40:386 – 392

Williamson CE (1996) Effects of UV radiation on freshwater ecosystems. Int J Environ Stud 51:245 – 256

WMO (World Meteorological Organization) (1994a) Scientific assessment of ozone depletion: 1994. In: Albritton DL, Watson RT, Aucamp PJ (eds) Global ozone research and monitoring project report No 37. World Meteorological Organization, Geneva

WMO (World Meteorological Organization) (1994b) Report of the WMO meeting of experts on UV-B measurements, data quality and standardization of UV indices. World Meteorological Organization Global Atmosphere Watch Report No 95, Geneva

Xiong F, Lederer F, Lukavsky J, Nedbal L (1996) Screening of freshwater algae (Chlorophyta, Chromophyta) for ultraviolet-B sensitivity of the photosynthetic apparatus. J Plant Physiol 148:42 – 48

8 Temperature Stress and Basic Bioenergetic Strategies for Stress Defence

Guenter A. Peschek and Roland Zoder*

Contents

* Molecular Bioenergetics Group, Institute of Physical Chemistry, University of Vienna, Währingerstrasse 42 and Althanstrasse 14, 1090 Wien, Austria

8.1 Introduction

8.1.1 The Energetic and Nutritional Requirements of a Living Cell

Common sense tells us that "life implies work". Therefore, every living cell must extract energy from the environment and convert it into a biologically utilizable form, viz. ATP. This fundamental process of biological energy conversion universally uses energy either from light (phototrophic processes) or from the oxidation of reduced material in the dark (chemotrophic, dissimilatory processes). Mechanistically speaking, chemotrophic energy conversion relies either on substrate level phosphorylation (ATP generation by intramolecular redox processes in a soluble cytosolic system) or on respiration which depends on electron transport processes in biological membranes and needs external oxidants of sufficiently positive redox potential (Schlegel 1975). Also photosynthesis generally depends on the membrane-bound electron transport; the only known exception, halobacterial photophosphorylation through bacteriorhodopsin-mediated light-driven conformational proton-pumping (Oesterhelt et al. 1977; Lanyi 1978; Renthal 1992), nevertheless belongs to the major category of chemiosmotic energy conversion which, like electron transport-dependent photosynthesis and respiration, inevitably needs membrane-bound charge separation systems (see Nicholls and Ferguson 1992).

Apart from energy, a living cell needs a carbon source from the environment in order to be able to synthesize what it essentially consists of, the cell material. This carbon assimilation ensures the possibility of performing the fundamental task of a living cell, the autoreduplication by growth and proliferation. Two categories of assimilable building blocks are used by living cells, either C_1 compounds in autotrophy, or C_n compounds in the form of complex, organic substances in heterotrophy. Autotrophic C_1 substrates are mostly CO_2-assimilated via the Calvin cycle, or reduced C_1 compounds such as CH_4, CH_3OH, HCHO, etc. which are used through peculiar biochemical pathways in methanogens and methylotrophs (see Sect. 8.6).

Besides energy and carbon there is a third inevitable constituent, electrons, which living cells require in order to assimilate the carbon sources taken up and transform them into cell material. Carbon in cell material, generally symbolized by carbohydrate [see Eq. (1)], is on the average at the oxidation level of zero. However, the carbon to be assimilated has a higher oxidation state, cf. +4 in CO_2. Yet, even if the carbon source (nutrient) is at or close to an oxidation level of zero, the process of energy conversion requires that it first be oxidized. Thereafter, assimilation again is reductive, and we can say that possibly all assimilatory processes performed by living cells are reductive in nature. Thus, automatically they are energy-requiring and they also need an external source of some utilizable reductant or, as we might also say, an electron source. In our biosphere such reductants can be inorganic in nature (as is characteristic of lithotrophy), while the alternative is an organic reductant (as is characteristic of organotrophy). Taken together, the above-named three basic constituents of all life processes: energy source, electron source and carbon source, each in its empiric dichotomy, can be produced in eight different ways, from photo-litho-autotrophy to chemo-organo-heterotrophy (or briefly: heterotrophy as such), and make up the so-called ergotrophic hexagon of life (Fig. 8.1). It should be noted that each and every type of living cell on Earth can be consistently accommodated for by at least one of the eight ergotrophic types.

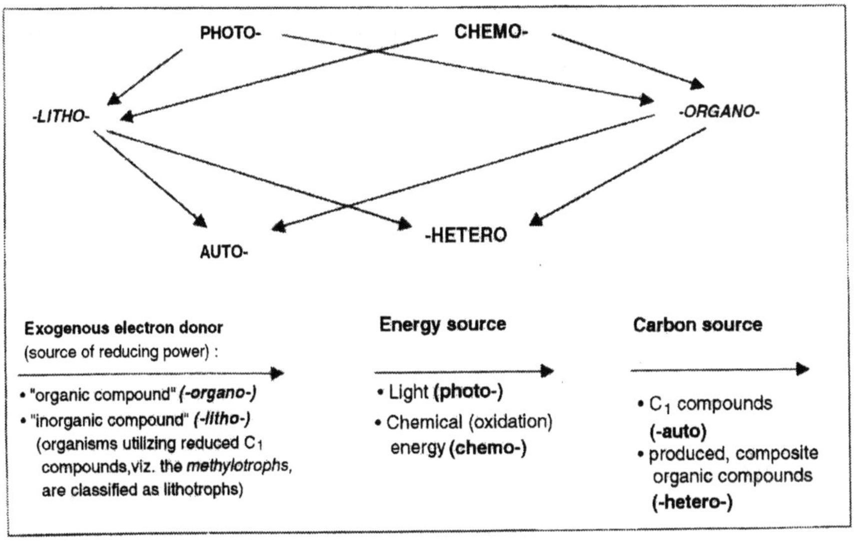

Fig. 8.1. The ergotrophic hexagon. The scheme summarizes all possible combinations between the energy source (either light/photo-, or dark chemical oxidation energy/chemo),the electron source (source of exogenous reducing power for anabolic = assimilatory processes: either inorganic compounds and reduced C_1 compounds/-litho-, or organic C_n compounds/-organo-), and the carbon source (either a C_1 compound/-auto-, or a produced C_n compound/-hetero-). Examples of all eight possible ergotrophic classes of living cells are found in nature. Many living cells are facultative, i.e. they belong to more than one ergotrophic class

8.1.2 Prokaryotic and Eukaryotic Algae

Before we start analyzing the impact of temperature on living processes, let us first briefly consider the basic subject of our discussion, the algae. The unifying concept of the algae is a bioenergetic theme, viz. oxygenic, plant-type photosynthesis (Sect. 8.1.3.), as everybody might agree. We are well aware that there are eukaryotic and prokaryotic algae (cyanobacteria). Despite the significant and absolutely unbridgeable morphological dichotomy between these two groups there is no denying the fact that both "behave like plants", viz. in performing water-splitting, oxygen-liberating photosynthesis. Considering the well-known endosymbiont hypothesis (Mereschkowsky 1905; Stanier 1974; Gray and Doolittle 1982), cyanobacteria (blue-green algae) might even be looked on as a kind of free-living chloroplast in both structural and, even more, functional respect (see Fig. 8.2). Being capable of oxygenic, plant-type photosynthesis and CO_2 assimilation, both prokaryotic and eukaryotic algae belong to the (aerobic) photolithoautotrophic or photophytotrophic (Broda 1975) organisms according to the ergotrophic scheme given earlier (Fig. 8.1). The earlier literature on cyanobacterial cell types was exhaustively reviewed by Geitler (1925), Fritsch (1945) and Desikachary (1959).

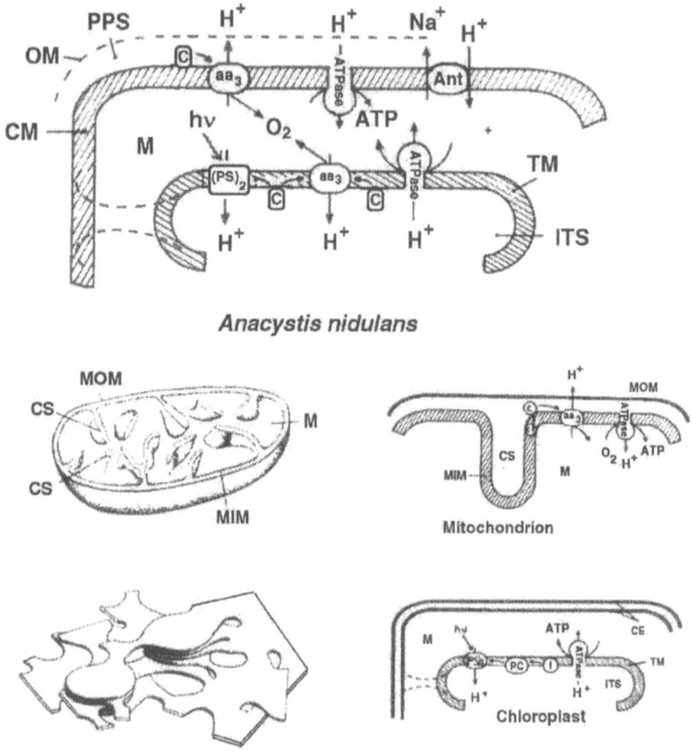

Fig. 8.2. Comparison of bioenergetic membrane functions in a cyanobacterium (*top*), a mitochondrion (*middle*) and a chloroplast (*below*). The *left-hand side* shows hypothetical stereographic views. *CM* Plasma or cytoplasmic membrane; *ICM* (*TM*) intracytoplasmic or thylakoid membrane; *PPS* periplasmic space; *OM* outer membrane; *MOM* mitochondrial outer membrane; *MIM* mitochondrial inner membrane; *ITS* intrathylakoid space; *M* matrix; *Ant* Na^+/H^+-antiporter; *aa_3* cytochrome-c oxidase; *c* soluble c-type cytochrome (= cytochrome c_6, which replaces plastocyanin in certain algae)

In our contemporary eukaryotic biosphere the overwhelming share of (most efficient) biological energy conversion is borne by the mitochondria, the so-called powerhouses of eukaryotic cells. No less important in this respect, of course, are the chloroplasts of eukaryotic algae and plants. Clearly, however, the principle of chemiosmotic energy conversion (see Nicholls and Ferguson 1992) must have first arisen in a prokaryotic cell (Raven and Smith 1976). Figure 8.2 shows the startling relationship of sidedness and vectoriality of electron transfer, transmembrane proton translocation, and ATP synthesis in/on the membranes of a prokaryotic alga, viz. *Anacystis nidulans*, a mitochondrion, and a chloroplast. The scheme is self-explanatory and with a striking similarity and uniformity of all electron transport components. Nature has built up electron transport systems in both prokaryotes and eukaryotes, and in both photosynthesizers and respirers, according to the so-called conversion hypothesis (Broda 1975; Broda and Peschek 1979; Peschek 1981a; also see Baltscheffsky 1974), which lends much support to the monophyletic origin of biological electron transport. The only known exception to the uniform chemical pattern of biological electron transport components primarily comprising iron-sulfur proteins, flavoproteins, quinones and cytochromes, is found in certain members of the archaea (see Danson et al. 1992). Finally, regarding the fact that, apart from the lack of chlorophylls (see Sect. 8.6) the similarities, both analogy and homology (see Dworsky et al. 1995), of bioenergetic primary components (electron transport, ATPase, etc.) between a mitochondrion and a cyanobacterium (see Fig. 8.2) is by no means much less pronounced than that between a chloroplast and a cyanobacterium (Fig. 8.2). Likewise, regarding the extremely low probability for the (first successful and irreversible) original endosymbiont event (see Sect. 8.6), the interesting hypothesis has been put forward that, in the sense of a generalized endosymbiont hypothesis, both chloroplast and mitochondrion could have arisen from an endosymbiotic cyanobacterium which, in the interior of the host cell and in the course of the aeons available, must have evolved, e.g. into a stable endosymbiotic cyanbobacterium (as in *Cyanophora paradoxa,* Sect. 8.6), into a chloroplast, or into a mitochondrion according to prevailing and long-lasting environmental constraints, thus rendering the free-living chloroplast (see above, Fig. 8.2) or even a free-living chloromitochondrion (Peschek 1999).

8.1.3 Plant-Type Photosynthesis

It is generally agreed that oxygenic photosynthesis was invented by the primordial cyanobacteria about 3.2 billion years ago (Barghoorn and Schopf 1965, 1966). For the first time, the process of oxygenic photosynthesis produced substantial amounts of molecular oxygen, O_2, in a hitherto essentially anoxic biosphere. This most decisive event, which might also be circumscribed as the oxygen crisis, has paved the way for the evolution of all higher forms of life, from simple unicellular eukaryotes (algae, protozoa) up to the very *Homo sapiens* (see Fig. 8.5). Thus let us look at this biochemical process more closely [Eq. (1)]:

$$6\,CO_2 + 12\,H_2O = C_6H_{12}O_6 + 6\,H_2O + 6\,O_2\,; \qquad \Delta G_o = 2821.5\ kJ. \qquad (1)$$

This equation, which is identical to the famous Van Niel equation of autotrophic CO_2 fixation with water as the external reductant, represents by far the most important means of biomass (carbohydrate) production on the earth. An estimated 100 billion tons of carbon (in the form of CO_2) per year is converted into biomass by plant-type photosynthesis (photosynthetic primary production of biomass) and the equivalent amount of O_2 is thereby released from water according to Eq. (1). Recent estimates assign between 20 and 30% of this worldwide primary productivity to cyanobacteria, in particular to small unicellular *Synechococcus* species (Waterbury et al. 1979; Glover et al. 1988) and likewise unicellular planktonic prochlorophytes (Chisholm et al. 1988) which, though not especially concentrated anywhere in the euphotic zone of the oceans, are nevertheless extremely widespread in all oceans. Thus, also as terrestrial organisms, they may incidentally serve as excellent examples of how to cope with environmental stress situations such as low levels of nutrients, particularly nitrate and CO_2, and temperature, desiccation, light, etc., which at the same time render them the primary colonists of otherwise uninhabitable (coastal) regions such as intertidal zones, newly born volcanic rocks just in the state of cooling down, arid desert regions, etc. Apart from the evolutionary privilege of having been the first to deal with oxygen gas, ecologically also the blue-green algae assume the privilege of ranking highest amongst photosynthetic primary producers and primary colonists perfectly adapting to a wide variety of different stresses, including temperature stress. Be it recalled that after the cataclysmatic explosion of the volcano Cracatao in 1884 the very first primary colonists of the remaining rocks were identified as (nitrogen-fixing) blue-green algae.

The energy budget of the oxygen cycle as given in Eq. (1) and amounting to almost 3000 kJ is endergonic, energy-requiring (+) in the direction of CO_2 fixation, but exergonic, energy-yielding (-) in the direction of respiration. From gas-exchange rates between, and pool sizes of, oxygen reservoirs in the atmosphere and hydrosphere (Gilbert 1981), it is calculated that the mean residence time (turnover time) of oxygen is 6000 years in O_2, 500 years in CO_2, but 20 million years in H_2O (Broda 1975). Finally, it should be remembered that it is the delicate balance between oxygen consumption and photosynthetic oxygen production worldwide that keeps the wheel of our life revolving.

8.1.4 Some Evolutionary Considerations

Assuming that the advent of the first bulk amounts of free oxygen gas on the Earth dates back to the photosynthetic activity of primordial blue-green algae

(Barghoorn and Schopf 1965, 1966; Broda 1975) the evidence for the gradual change of the oxygen content in the earth's atmosphere as independently described by paleontology, geology and geochemistry is summarized in Fig. 8.3. Considering the highly vulnerable and oxidizable nature of a living cell, and considering that even in an extant, highly aerobic mitochondrion, for example, metabolism in its essence is still anaerobic, everybody might agree that the initial steps in the evolution of living cells (eobionts or protocells) simply must have occurred in the absence of free oxygen. Even more stringently, the preceding abiotic formation of the chemical building blocks of a living cell in a reducing atmosphere according to Miller and Urey (1959) (see Miller and Orgel 1974; Broda 1977b; Zohner and Broda 1979) demands the absence of free oxygen. Thus, irrespective of whether we keep to Oparin's and Haldane's well-established primaeval soup scenario for the early evolution of life on Earth (Oparin 1924; Haldane 1954, 1970), or adopt the more fashionable redox-and-surface concept of Wächtershäuser (1988, 1990), which considers chemotrophic primary production in and around black smokers (see Sect. 8.2) central to the origin of life on Earth (see Drobner et al. 1990), in all events life's early infancy must have taken place in the absence of oxygen and light (including ionizing radiation in those days).

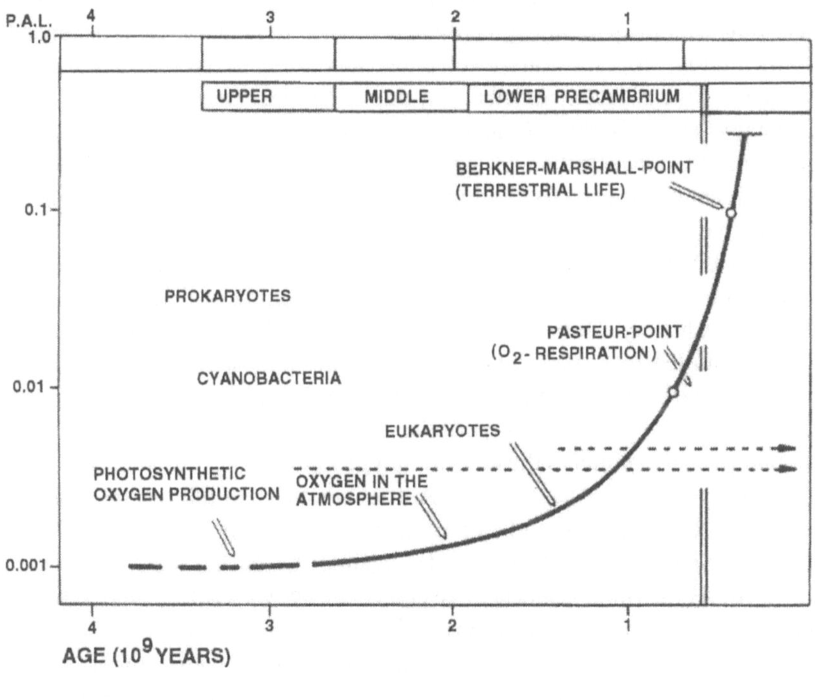

Fig. 8.3. The oxygen content of the Earth's atmosphere during the past four billion years. PAL, present atmospheric level (fractions of 21% v/v). (Diagram compiled from Rutten 1966, Berkner and Marshall 1967, Rhoads and Morse 1970, Schidlowsky 1971, Broda 1975, Towe 1978 and Gilbert 1981)

Figure 8.3 shows the increase in O_2 levels in the Earth's atmosphere during the past four billion years. Before the oxygen gas, initially produced by the primordial cyanobacteria in depths of the oceans where destructive ionizing radiation and short-wave UV light from the space could not reach, was allowed to disembark, i.e. to enter the atmosphere, 1.2 billion years of purely chemical reduction of the dissolved O_2 by the huge amounts of reductants (ferrous iron, sulfide, etc.), dissolved in the waterbodies where the cyanobacteria produced it, had elapsed. However, 0.4 billion years thereafter the first eukaryotic cells appeared, as evidenced by the fossil record (Round 1980; Knoll 1992). Regarding the absence of O_2 from the early Earth, it is interesting to note that there is no primary anaerobe among extant eukaryotes, anaerobic protists or even worms reflecting just the drop of respiration in the course of adaptation to peculiar anaerobic habitats, while to many extant anaerobic bacteria, notably to methanogens (see Balch et al. 1979; Wolfe 1992), free oxygen is still highly and irreversibly toxic (Morris 1975; Cadenas 1989).

With the advent of eukaryotic algae, still confined to critical dark depths of the primeval oceans, a violent increase in the atmosphere's oxygen content ensued. The Pasteur point was reached 0.6 billion years later marking the time when the global energetic efficiency (ATP production) of respiration, viz. oxidative phosphorylation (38 mol $ATPmol^{-1}$ glucose dissimilated to H_2O and CO_2) exceeded that of fermentation, viz. substrate-level phosphorylation (2 mol $ATPmol^{-1}$ glucose dissimilated to, e.g. lactate or ethanol), due to the increasing availability of oxygen. Finally, at the Berkner-Marshall point (see Berkner and Marshall 1967) an atmospheric O_2 concentration of approx. 2% (vol/vol) was reached when photochemical reactions started to build up a substantial ozone layer at a height of 45-55 km above the ground level, thus protecting the surface of the Earth against deleterious UV (200–300 nm) and other ionizing radiation from the outer space, opening up the era of terrestrial life just at the beginning of the Palaeozoic period of the Earth's history. During this time, the initially almost unlimited proliferation of land plants led to a transient overshoot in the atmosphere's O_2 concentration (cf. the massive Palaeozoic vegetation that gave rise to our present-day fossil fuel reserves). During the following periods, a likewise massive development of O_2-consuming (land) animals adjusted the oxygen content of the atmosphere to the contemporary level of 21% (vol/vol). A brief summary of the crucial stages in the evolution of our Universe and our Earth is given in Fig. 8.4, where the whole thing is viewed (on the left-hand side) against the background of the creation myth in the Holy Bible (Genesis) as described by Moses 1,1, who successfully competes with sophisticated astrophysics, including big bang and cosmic strings, for the most comprehensible explanation of how the universe might have come into existence. Incidentally, any other representative of the major creation myths in world religions, such as the mythological Egyptian potters Ptah and Chnum, or the weaver Neith, or the Indian blacksmith Twaschtar, the mythological couple Izanagi and Izanami, who created the Japanese islands, parents of the goddess of the Sun, Amaterasu, a later offspring of which is the Tenno, or even Odin's ravens, Hugin and Munin, etc. might serve the same purpose as potter Jahve in Fig. 8.4.

בְּרֵאשִׁית
בָּרָא אֱלֹהִים
אֵת הַשָּׁמַיִם
וְאֵת הָאָרֶץ׃
וְהָאָרֶץ הָיְתָה
תֹהוּ וָבֹהוּ
וְחֹשֶׁךְ עַל־פְּנֵי
תְהוֹם
וְרוּחַ אֱלֹהִים
מְרַחֶפֶת
עַל־פְּנֵי הַמָּיִם׃

Origin or appearance of	Years ago ($\times 10^9$)
Universe	18
Sun	10
Earth (solid crust)	4.6
Life (prokaryotes)	3.5
CYANOBACTERIA (O_2!)	3.2
Atmospheric O_2	2
Eukaryotes	1.6
Terrestrial biosphere (O_3-shield)	0.4
Homo erectus	0.0007-4
Homo sapiens	0.00005

Fig. 8.4. Highlights in the evolution of the universe and of our Earth. *Left-hand side* Original sentence (Moses 1,1) introducing the creation myth described in the Holy Bible ("At the beginning God made heaven and Earth")

8.2 Temperature and Life

Life is invariably linked to liquid water, which is both external and internal communication medium for/between enzymes, metabolites, individual cells etc. Under normal atmospheric pressure, this corresponds to a temperature interval between 0 and 100 °C. (1 atm/techn. = 1 kp cm^{-2}; 1 atm/phys. = 1.033 kp cm^{-2}, both units roughly corresponding to 760 mm Hg and identified with 1 bar in the old unity system). The modern international pressure unit is Pascal (Pa): 1 bar = 100,000 Pa

= 0.1 MPa with 1Pa = 1 N m^{-2}). That living structures and processes, however, might occur even at temperatures considerably above 100 °C per se is exemplified by the black smokers, i.e. deep sea vents where hot water coming from the Earth's interior, under appropriate hydrostatic pressure, exhibits boiling points higher than 100 °C. Organisms, in particular the primary-producing chemolithoautotrophic bacteria thriving in the immediate vicinity of black smokers, may show optimum growth temperatures of no less than 105-110 °C, as is the case for *Pelodictium occultum* (see Danson et al. 1992). Black smokers, though of no major ecological relevance nowadays, constitute the only known chemoautotrophic system of primary biomass production on Earth.

Originally based on the analysis of a relatively few 16S rRNA sequences, Carl Woese had proposed to separate from eukaryotes and eubacteria the so-called archaebacteria (now systematically known as archaea) as a fundamentally (?) different third kingdom of living organisms (Woese et al. 1978). Consequently, a tripartite living world with a three-domain system was established (Woese et al. 1990; also see Margulis 1996). Two shortcomings are immediately obvious in this system: firstly, both eubacteria and archaea are prokaryotic in cell structure, and the dichotomy between prokaryotes and eukaryotes first recognized by A. Lwoff (see Chatton 1937) is the only irreducible and unreconciled disparity between truly different types of living cells where no missing links have so far been found (Stanier and Van Niel 1962): and secondly, despite certain common traits, such as peculiar membrane lipids (Kates 1992), a cell wall that may consist of nothing but a protein layer (Gram-variable bacteria), and a certain peculiar transfer and 16S rRNAs, it can hardly be denied that the so-called Archaea comprise a bewildering variety of metabolic types including, e.g. strict anaerobes (the methanogens) as well as strict aerobes (the halobacteria). Thus, the name archaea, in spite of possible taxonomic merits, need not necessarily lead one to think of evolutionarily very ancient organisms. Geochemical constraints in the early days, as well as later adaptation to strange habitats, must not be overlooked. Furthermore, and rather embarrassingly, no single member of the extremophiles has ever been detected to display metabolic patterns, gross protein and nucleic acid structures, either in monomer types or in the sequence/primary structure or in secondary and tertiary structures etc., that are significantly different from mesophiles and other organisms living in moderate habitats (Danson et al. 1992). This applies to hyperthermophilic species such as *Pelodictium* (at 110 °C; see above) or hyperhalophilic species such as *Halobacterium* (at >3M NaCl; see Lanyi 1978) or hyper-acido/alkalophilic species such as *Acidianus,* formerly *Desulfurolobus*, at pH <2 (Anemüller et al. 1994; Purschke et al. 1997) or *Bacillus* sp. at pH >11 (Hicks and Krulwich 1995) etc. This would mean that adaptation to growth even at, for example, extraordinarily high temperatures, has never even evolutionarily extended into molecular-mechanistic aspects of the living cell, nor are the basic bioenergetic mechanisms in any way different from what is known from other organisms (Schäfer et al. 1999). Another (negative) characteristic of archaea is that while (anaerobic) chemoautotrophy is common among them, there has been no report so far on any phototrophic archaeon.

The usual microbiological temperature classification distinguishes mesophilic organisms (bacteria) showing optimum growth between 30 and 45 °C, thermophilic species with optimum growth at 55-75 °C (some archaea up to 110 °C, see above), and psychrophilic species growing optimally at 15-18 °C (sometimes still growing, though suboptimally, at 0 °C and dying at 20-25 °C). Among eukaryotic algae, no truly thermophilic species are known, the only thermophilic algae being cyanobacteria with optimum growth temperatures hardly above 73-74 °C (Castenholz 1988). It seems useful to distinguish between thermophilic species which obligately need elevated temperatures (>55 °C) for optimum growth, and thermotolerant species which can just endure (survive) these temperatures, often in a resting state (spores, akinetes, etc.). Clearly, this distinction is also true of the opposite side of the temperature scale, where we should distinguish between truly psychrophilic species (see above) and simply cold-tolerant species (Murata and Wada 1995).

8.2.1 Reaction Rates at Different Temperatures

General experience tells us that any chemical reaction will become faster at elevated temperatures. A well-known biochemical rule of thumb says that the rates of biochemical reactions grossly double when the temperature is raised by 10 °C. An exact physicochemical theory of the temperature dependence of chemical reactions was given by Svante Arrhenius (1859-1927). Basically, the velocity of a chemical reaction encompasses a constant factor k, the rate constant, a term which contains certain exponentials of the concentrations of reactants entering the reaction. These exponentials are determined by the molecularity of the reaction. They will be different for zeromolecular, monomolecular and bimolecular reactions, viz. for reactions of zero order (no concentration dependence at all with respect to the reactant in question), first order and second order, respectively. The order of a reaction, however, is a purely mathematical number (exponential), which formally follows from the kinetic equation describing the reaction. It does not necessarily tell us the true molecularity of the reaction, which atomistically may be defined as the number of reacting molecules forming the active complex or transition state of the reaction. Without going into all the physicochemical details of the Arrhenius theory and its modern adaptations as given by H. Eyring, S. Glasstone and K.J. Laidler (for details see a textbook on chemical kinetics, e.g. Laidler 1965), the following explanations which address biologists rather than physical chemists will try only to mediate a basic biological understanding of the temperature dependence of biochemical reactions in living systems. Such explanations are, of course, also necessary for an understanding of temperature stress and for the mechanisms allowing adaptation to temperature stress, as will be discussed later in this chapter.

8.2.2 Arrhenius Diagrams

The equations originally given by Arrhenius strictly apply to the temperature dependence of the rate constant *k*. However, since *k* is proportional to the reaction rate *v* we may directly apply the equations to reaction rates as such notably when we consider that in biology we are mostly dealing with highly complex chemical reactions (photosynthetic O_2 evolution or CO_2 assimilation, respiratory O_2 uptake or CO_2 release, fermentative acid production, etc.) for which it is usually impossible to elucidate the exact comprehensive kinetic equation. What we would rather like to aim at is a true understanding of the role of temperature in a living cell.

Arrhenius correctly started from the consideration that a reaction between reactants will only occur if there is a collision between them and if the energy temporarily stored in the collision complex (transition state) exceeds a certain critical value, the apparent or Arrhenius' activation energy. Thus (setting *v* for *k*, see above):

$$v = \text{const.A.e,}^{-Ea/RT} \quad (2)$$

where const is a constant, A is the collision coefficient or frequency factor, E_a is the apparent or Arrhenius' activation energy of the reaction, R is the gas constant, and T is the absolute temperature. In a logarithmic scale Eq. (2) can be written as

$$\ln v = \text{const}' - E_a/RT, \quad (3)$$

where const′ includes the frequency factor A. A follows from the collision theory of gas reactions (kinetic gas theory) and it should be noted that A itself is (slightly) temperature-dependent, viz. going with $T^{1/2}$, a factor recurring in the laws of diffusion. Equation (3) is the linearized version of Eq. (2). Graphically therefore, ln *v* (or log *v*) plotted against 1/T, the reciprocal absolute temperature, yields a straight line. Considering ln A = 2.3 log A, from the slope of the straight line, - E_a/R, the apparent or Arrhenius' activation energy E_a can be directly calculated.

Measuring a complex chemical reaction by (often necessarily) following the concentration of only one of the reactants, e.g. measuring the time course of the concentration of O_2 in a respiring (or photosynthesizing) cell suspension at a given temperature, directly concerns only the mechanistic step that reduces (or produces) oxygen. However, only when this step is also the slowest, viz. rate-determining (the bottleneck) enzymatic reaction within the complex overall reaction sequence of, say, respiration or photosynthesis, will it determine the kinetic equation of oxygen uptake/evolution as well as the temperature dependence of the overall reaction. Thus, it is well known that within a wide range of oxygen concentrations down to less than hundredths of the ambient pressure the rate of oxygen uptake by respiring cells and tissues is independent of the O_2 concentration, i.e. a kinetic equation of zero order with respect to the (measured!) oxygen concentration will result, because the oxygen-reducing step proper, i.e. the very step under measurement, is not rate-determining for the overall reaction. Which step indeed might be the rate-determining one, e.g. the bottleneck of respiration as followed by measurement of respiratory O_2 uptake, can never be found out from measurements on more or less integral systems such as whole cells or isolated membranes. Consequently, also the measured temperature dependence of respiratory O_2 uptake might have nothing to do with the oxygen-reducing step proper, and the apparent activation energy derived does not necessarily stem from the terminal oxidase. The fact that temperature coefficients (defined as the rate at (t + 10)°C over the rate at t °C within a given physiological temperature interval) are between 1.3 and 1.5 for photosynthesis but 2.0 and 2.3 for respiration (Baldry et al. 1966) indicates nothing further than that the rate-determining steps are different in photosynthesis and respiration. Further identification of the kinetically relevant rate-determining steps is impossible from whole cell measurements in general.

8.2.3 Temperature and Biomembranes

The invariable constituents of biomembranes are proteins and lipids. The membrane lipids, which comprise phospholipids, glycerolipids, sulfolipids, glycolipids, cholesterol (only rarely in prokaryotes) and C_{20}-C_{20} diacylglycerol diethers (only in archaea; see above; see Danson et al. 1992) etc., form some kind of semifluid matrix in which the membrane proteins (extrinsic, intrinsic and transmembrane proteins) are floating similar to icebergs in the sea. This so-called fluid mosaic model (Singer and Nicolson 1972; Singer 1974) has put the famous lipid bilayer model of Davson, Danielli, Green and others into its final shape. The physical state of the lipid matrix of the biomembrane will greatly depend on the ambient temperature, varying between fluid or semifluid (liquid crystalline) above, and rigid or quasi-crystalline (gel state) below, a certain threshold temperature, viz. the thermotropic membrane lipid phase transition temperature (LPT). Chemical and physical reactions between membrane-bound proteins will, therefore, be influenced by the temperature not only through the intrinsic temperature dependence of the reaction as such but also by the physical state of the membrane lipids which, in itself, is temperature-dependent. Above all, the proper function of a typical membrane-bound electron transport chain, where membrane-embedded electron transfer proteins or complexes must exchange electrons between respective donor and acceptor molecules, and where this process requires a certain minimal distance between the latter, will be severely impaired below the lipid phase transition temperature when the electron transfer proteins are increasingly frozen (= immobilized) in the lipid matrix. The physical principles of biological electron transfer processes are comprehensively described in the Marcus theory (Marcus 1956; Marcus and Sutin 1985) as later refined by Dutton and coworkers, including quantum mechanical tunnelling effects and intraprotein electron transfer (Devault 1980; Moser et al. 1992, 1999). They will not be reiterated here. What is clear from a basic consideration is that the apparent activation energy of a given (though in detail unknown) intramembrane electron transfer process will abruptly increase when the temperature is lowered below the LPT. Since also the LPT as a state transition (cf. liquid → solid) is fairly sharply defined, this will phenomenologically yield a break in the corresponding Arrhenius diagram, i.e. a so-called Arrhenius discontinuity (see Fig. 8.5). The temperature of the Arrhenius discontinuity equals the lipid phase transition temperature of the membrane in which the electron transfer/transport process in question occurs. Since different biomembranes, owing to their characteristic lipid composition, are expected to exhibit distinct LPTs the discontinuity temperatures can be used to localize different electron transport processes (viz. of respiration and photosynthesis) in different intracellular membrane systems (viz. cytoplasmic and thylakoid membranes) in vivo, as has previously been done on intact cells and crude membrane preparations (CM + ICM) of *Anacystis nidulans* (*Synechococcus* PCC6301) long before the nekrochemical separation and purification of cytoplasmic (plasma) and intracytoplasmic (thylakoid) membranes (CM and ICM, respectively) had been successfully performed (see Fig. 8.5; also see Peschek et al. 1981, 1982; cf. Wada et al. 1984 with isolated CM and ICM).

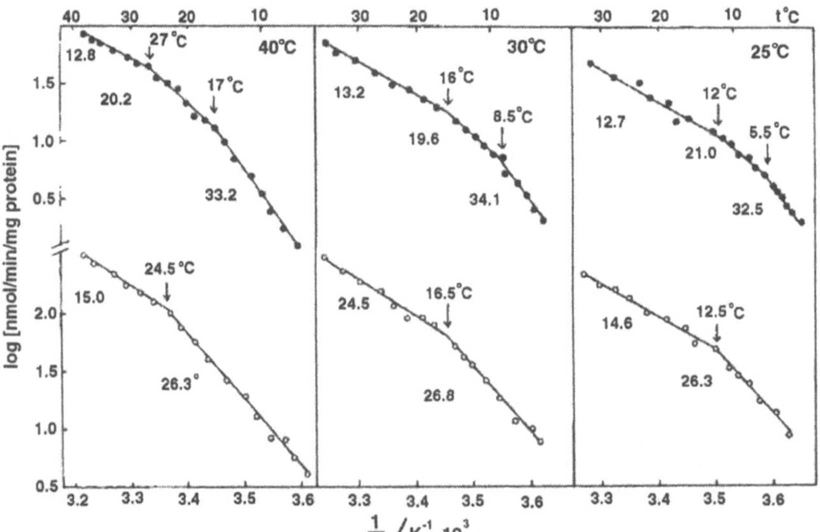

Fig. 8.5. Arrhenius discontinuities (breaks in Arrhenius diagrams) of membrane-bound electron transport reactions. As an example, the temperature dependence of the rates of oxygen uptake supported by ascorbate plus N,N,N′, N′-tetramethyl-*p*-phenylene diamine in the dark (*full circles*) and by diphenylcarbazide in 620 nm light (*open circles*) by membranes prepared from *Anacystis nidulans* after growth at 40, 30 and 25 °C is taken (Peschek et al. 1982)

Physiologically, in the sense of an adaptation to different growth temperatures, microorganisms including prokaryotic and eukaryotic algae can regulate the fluidity of their membranes through changing the lipid composition of the membranes. This has been studied most extensively with bacteria and cyanobacteria. Regarding membrane phenomena the situation in the latter (where we have CM and ICM as the bioenergetically relevant membranes only; see Fig. 8.2) is much clearer and more straightforward than in eukaryotic cells, where the occurrence of a host of different membranes, partly even confined within separate and individually membrane-enclosed intracellular organelles, may severely confuse the interpretation of results obtained on intact cells. Therefore, in the present chapter, which paradigmatically attempts to point out major impacts of temperature on the complex system of an algal cell, cyanobacteria are used as the appropriate examples. Competent reviews on the physiological regulation of membrane fluidity in general and on cyanobacterial membrane lipids, in particular their role in the protection against low temperature stress, and the genetic manipulation of their chemical composition, may be found in Furtado et al. (1979), Murata and Nishida (1987), Quinn (1988) and Gombos et al. (1999).

Focusing on the adaptational regulation of membrane fluidity as might be characterized by distinct LPTs, it has long been known that the fluidity (smoothness) of a biomembrane increases with its content of unsaturated and short-chain esterified fatty acids (Melchior 1982) and that this content physiologically increases with decreasing growth temperature (Satoh et al. 1979). Conversely, growth at higher temperature favours the synthesis and incorporation into membrane lipids of saturated and long-chain fatty acids, thus leading to less fluid (more rigid) membranes and, hence, higher LPTs (see Fig. 8.5 and Tables 8.1 and 8.2). Several independent physical and chemical methods are available to measure thermotropic LPTs of membranes, both artificial and natural, using the following experimental parameters: Arrhenius discontinuities of membrane-bound electron transport reactions (see this chapter; also see Scherer et al. 1981), chlorophyll a absorption (Hoshina 1981) and/or fluorescence intensity (Lee 1975; Murata et al. 1975) in photosynthetic membranes, rotational correlation times of a fatty acid spin probe incorporated into the membrane and measured by EPR spectrometry (Murata et al. 1975; Wada et al. 1984), X-ray diffraction of the membrane lipid bilayer giving a Bragg spacing of 4.2 Å in the frozen (gel) state and a diffuse line corresponding to a spacing of 4.5 Å in the liquid crystalline (fluid) state (Rivas and Luzzati 1969; Tsukamoto et al. 1980), differential scanning calorimetry (Melchior and Steim 1976), photooxidation kinetics (absorption changes) of membrane-bound carotenoids (Ono and Murata 1977; Yamamoto and Bangham 1978; Brand 1979) and, most convincingly because directly observable, freeze-fracture electron microscopy (Verwer et al. 1978; Armond and Staehelin 1979; Furtado et al. 1979). Freeze-fractured surfaces and interfaces of biomembranes incubated at temperatures below the LPT show characteristic (protein) particle-free regions (Fig. 8.6). Doubtless, the best object to study the effects of cell-membrane composition on biochemical activities would be the chemoheterotrophic bacterium *Acholeplasma leidlawii*, a member of the small obligately parasitic and cell wall-less Mycoplasma group which incorporates fatty acids into its plasma membrane just as available from the growth medium. Unfortunately, this organism is an exception applicable only within a rather narrow range of biochemical problems.

Table 8.1. Influence of growth temperature on fatty acid composition of the four major types of lipids in *Anacystis nidulans.* (After Satoh et al. 1979)

Lipid type	Fatty acid	Growth temperature (°C)		
		38	28	22
A. Molar ratio unsaturated/saturated fatty acids				
Monogalactosyldiglyceride	C14	1.40	2.32	5.22
	C16	0.93	1.00	1.24
	C18	1.62	4.60	>10
Digalactosyl- diglyceride	C14	1.82	4.00	3.63
	C16	1.00	1.13	1.13
	C18	2.87	>10	>10
Sulfoquinovosyl- diglyceride	C14	1.00	1.80	4.67
	C16	0.56	0.80	0.81
	C18	>10	7.00	>10
Phosphatidyl glycerol	C14	1.00	2.00	3.12
	C16	0.86	0.94	1.01
	C18	6.69	>10	>10
B. Fatty acid content (mol % of total fatty acids in each lipid class)				
Monogalactosyldiglyceride	C14	2.4	6.3	11.2
	C16	85.6	90.6	88.3
	C18	11.7	2.8	0.4
Digalactosyl- diglyceride	C14	3.1	6.5	8.9
	C16	87.8	91.6	90.1
	C18	8.9	1.6	0.7
Sulfoquinovosyl- diglyceride	C14	0.8	2.9	6.8
	C16	95.2	94.7	92.8
	C18	3.9	2.5	0.2
Phosphatidyl glycerol	C14	0.4	1.5	7.0
	C16	87.0	93.3	91.0
	C18	12.3	4.2	1.3

Table 8.2. Thermotropic membrane lipid phase transition temperatures (LPTs) of isolated plasma membrane (CM) and thylkakoid membrane (ICM) from *Anacystis nidulans* grown at various temperatures in the absence (-) or presence (+) of 0.4 M NaCl. (Molitor et al. 1990a)

Growth temperature (°C)	Type of membrane				Reference
	CM-	CM+	ICM-	ICM+	
25	5-6	-	10-13	-	Peschek et al. (1982)[a]
28	5	-	14	-	Wada et al. (1984)[b]
	-	-	13		Ono and Murata (1979)[c]
30	8.5	22	16	20	Molitor et al. (1990a)[d]
	8-9	-	15-17	-	Peschek et al. (1982)[a]
38	13	-	23	-	Wada et al. (1984)[b]
	-	-	21-24	-	Ono and Murata (1979)[c]
40	19	28	25	28	Molitor et al. (1990a)[d]
	17-18	-	24-27	-	Peschek et al. (1982)[a]

Transition temperatures (°C) were taken equal to corresponding Arrhenius discontinuity temperatures measured for the temperature dependence of respiratory (CM) and photosynthetic (ICM) oxidation of horse heat ferrocyt c (cf. Fig. 4 of Molitor et al. 1990a). For comparison, the table also shows membrane lipid-phase transition temperatures to be found in the literature.

[a] Arrhenius discontinuity temperatures of respiratory (CM) and photosynthetic (ICM) electron transfer reactions determined with crude membranes (mixture of 10-20% CM and 80-90% ICM).

[b] Lipid-phase transition temperatures of isolated CM and ICM as determined with a spin- probe method.

[c] Arrhenius discontinuity temperatures of photosynthetic electron transfer reactions determined with crude membranes.

[d] Cf. Fig. 4 of Molitor et al. (1990a).

Bar (-) indicates determined.

8.3 Temperature and Light

The universal energy source to phototrophic organisms (anoxygenic phototrophic bacteria, algae, and green plants) is light (see Fig. 8.1). C.J. v. Grotthuss and F. Draper in 1818 formulated the first law of photochemistry or the principle of photochemical activation: only the light that is absorbed by a substance is effective in producing a photochemical change. J. Stark in 1908 and again A. Einstein in 1912 coined the second law of photochemistry, the principle of quantum activation: each quantum of radiation (hν) absorbed by a molecule activates one and only one molecule in the primary step of a photochemical process. Clearly, also photosynthesis as a photochemical process must be governed by the laws of photochemistry. Taking into account that sun emits radiation energy amounting to a flux of 3.86×10^{26} J s^{-1} into the space, of which the Earth (above the atmosphere) receives 7.92 J cm^{-2}

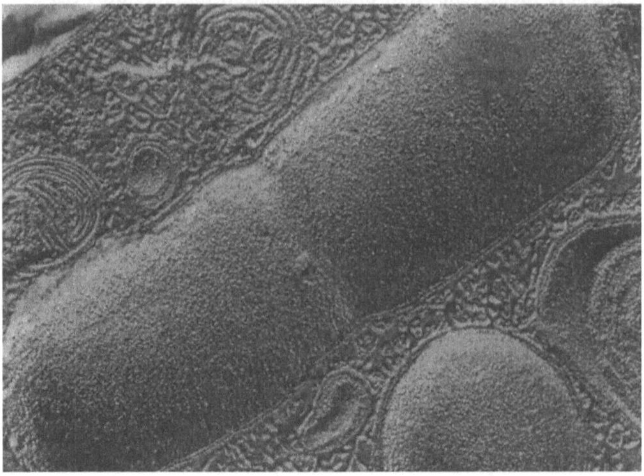

Fig. 8.6. Freeze fracture electron micrograph of glycerinated NaCl-grown *Anacystis nidulans* exposed to 10 °C for 30 min prior to freezing. The particles visible on/in the membrane are intrinsic membrane proteins. The onset of a thermotropic membrane lipid phase transition can be seen on the *lower left-hand side* of the figure (see Molitor et al. 1990b; also see Armond and Staehelin 1979)

min^{-1} (= the solar constant), and taking further into account that 97% of this energy resides in the visible and infrared region, we might understand why photosynthetic organisms in the course of evolution have elaborated an astonishing variety of colourful pigments all of which cooperate in funneling, by way of the Förster resonance transfer mechanism (Förster 1948), the total light energy absorbed by the photosynthetic membrane (plus accessory light-absorbing structures such as phycobilisomes in blue-green and red algae) as efficiently as possible into the photosynthetic reaction centre (RC) where the final and photochemically active recipient invariably is a (protein-bound) chlorophyll a molecule (in plant-type photosynthesis) or bacteriochlorophyll a molecule (in anoxygenic bacterial photosynthesis of purple and green bacteria). Thermodynamically, splitting water (into 2 H^+, 2 e^- and 0.5 O_2) is a much heavier task than splitting H_2S into 2 H^+, 2 e^- and S for the final assimilation of 1 CO_2, viz. approx. 500 vs. 50 kJ/CO_2. Thus, we understand that oxygenic photosynthesis, apart from the crucial oxygen-evolving complex (OEC), has evolved two photosystems, a high-potential PS II and a low-potential PS I working in series to extract four electrons from two molecules of H_2O (releasing as a waste, so to speak, one molecule of O_2) and transfer them via two photochemical excitation events in two photosystems (needing altogether eight light quanta hν) to one molecule of CO_2 to be reduced to 1/6 $C_6H_{12}O_6$ in the Calvin cycle. However, while the simple Hill-Bendall model or Z-scheme of oxygenic photosynthesis (Arnon 1960; Hill and Bendall 1960; Hoch and Kok 1961) would suggest a minimum quantum requirement of eight light quanta finally received by the two RCs of PS I and II for each molecule O_2 evolved (or CO_2-fixed), all measured quantum requirements obtained so far are beyond 12-13. This is commonly explained by additional light quanta which must be absorbed to drive cyclic photophosphorylation in order to supply additional ATP for the Calvin cycle which mechanistically needs 2 NAD(P)H and 3 ATP per molecule CO_2 reduced. Interestingly, and not quite logically, also in the case of anoxygenic (bacterial) photosynthesis, which energetically needs only one photosystem, all measured quantum requirements for 1 CO_2 to be assimilated have always been beyond 12, which might have to do, in a hitherto stoichiometrically unknown fashion, with so-called ATP-driven reverse electron flow which must be used by most chemolithoautotrophic and (anoxygenic) photoautotrophic bacteria in order to synthesize the low-potential reduced pyridine nucleotide (viz. NADH) which, in most anoxyphototrophs, cannot be made by direct photoreduction (viz. of NAD^+), but which is nevertheless a sine qua non for any autotrophic CO_2 fixation (Keister and Yike 1967; Jackson and Crofts 1968; Kelly 1971; Gest 1972; Schlegel 1975; Smith and Hoare 1977).

In the gross light-absorbing step, and considering only the chlorophylls (including accessory, i.e. purely light-harvesting chlorophylls) it had been shown by examining the yield of photosynthesis in flashing light that about 2500 chlorophyll molecules, linked together by resonance energy transfer, must cooperate in the photosynthetic membrane per molecule O_2 released or CO_2 fixed. This is the so-called photosynthetic unit of H. Gaffron and K. Wohl (1936; see Rabinowitch and Govindjee 1969). It explains why the ceiling of light-saturated photosynthesis is unexpectedly low, reaching already long before every potentially absorbing pigment molecule in the cell is hit by a light quantum (see Rabinowitch and Govindjee 1969).

The history of photosynthesis research is perhaps the most fascinating story of all scientific discoveries in modern times. Unfortunately, we cannot follow here in detail the brilliant path of the human spirit that sequentially unravelled the role of the three major chemical ingredients of photosynthesis: O_2, the vital air of Joseph Priestly (1733-1804), also eagerly studied in the context of green plant material by Jan Ingenhousz (1730-1799), the court physician of the Austrian Empress Maria Theresia, and having possibly given rise to the modern chemical theory of combustion first formulated by Antoine Laurent de Lavoisier (1777), who was guillotined during the French Revolution on the eighth of May, 1794; secondly the fixed air (= CO_2) of Jean Senebier (1742-1809); and thirdly the H_2O of Nicolas Theodore de Saussure (1767-1845). For an abridged version of this fascinating story, together with a scientific discussion of the integral topic of photosynthesis, the interested reader is referred to the excellent book Photosynthesis by Rabinowitch and Govindjee (1969). In the author's opinion, all the basic facts of photosynthesis, including historical aspects, have never been presented in a more concise and illustrative way than in this immortal booklet.

Bearing in mind that photosynthetic activity, reflecting a light-dependent biochemical process, naturally will depend on both light and temperature, we leave the history of photosynthesis in 1905, when the British plant physiologist J.J. Blackman suggested that the whole complex of photosynthetic activity consists of two distinct groups of reactions, viz. the light reactions and the dark reactions (Blackman reaction). The former reflects true photochemistry (primarily light absorption and the immediately subsequent steps of light energy distribution and resonance energy transfer through the cooperating pigments according to the Förster mechanism). The light reaction will primarily depend on the availability of light, roughly 2500 light quanta effecting the production of 1 O_2 in accordance with the photosynthetic unit of Gaffron and Wohl (see above). Dealing only with light as such, the light reaction is temperature-independent. Thus, from a certain threshold value onward, increasing light intensity will progressively increase the photosynthetic activity, establishing what is called the proportionality range, a linear dependence of (normalized) photosynthetic activity on the intensity of the light absorbed. The slope of the line should be universally constant for a given photosynthetic mechanism (e.g. the Hill-Bendall model or Z-scheme of plant photosynthesis; see above) and directly related to the quantum yield of photosynthesis, answering the question of how many molecules or mol O_2 will be evolved per quantum of light ($h\nu$) or per Einstein = mol $h\nu$ absorbed by the photochemical machinery, i.e. by the two photosystems (RCs); the reciprocal of the quantum yield is called the quantum requirement. Further increasing the light will result in a saturation of the photosynthetic activity, the ceiling of photosynthesis. Photosynthesis in saturating light, i.e. when the incident light can satisfy every light-absorbing component (pigment) of the light energy-transporting conveyor belt, no longer productively responds to further light. Still higher light intensities will rather damage the photosynthetic structures, first reversibly and then irreversibly, thus leading to an inhibition of photosynthesis by light.

The responses of photosynthetic cells/organisms, viz. of anoxygenic phototrophic bacteria but even more so, of algae and higher plants to environmental stress from strong illumination (photoinhibition), but also to oxygen (photooxidative damage), high salinity (salt stress), desiccation, nutrient starvation, etc. have always been of particular concern to researchers. An incredible wealth (almost 3000 pages) of original scientific results have been compiled and edited, as a final report of research project No. 04273102 on *Molecular mechanisms of responses of the photosynthetic apparatus to the environment* (1992-1995) to the Japanese Ministry of Education, Science and Culture, by Norio Murata. (For an abridged version see Satoh and Murata 1998). Regarding stress responses of algae and other phototrophic organisms much has been written about photoinhibition (see the literature quoted; Chap. 6).

Thus, we will deal only briefly with a plainly basic description of the effect of light on photosynthesis here, concluding our previous discussion concerning the light requirement of photosynthesis. In the ceiling region of photosynthesis, reflecting the dark reactions of photosynthesis, the normalized overall rate of photosynthesis (measured as oxygen evolution or carbon dioxide assimilation) is determined/limited by the intrinsic maximum rate of the enzymes of the Calvin cycle, which is the bottleneck of photosynthesis in saturating light. Since all biochemical reactions are temperature-dependent, it is only light-saturated photosynthesis that will respond to temperature. From the delicate interplay of light and temperature, the well-known biphasic light-saturation curve of photosynthesis results where, at a given temperature, a straight line with universally constant slope, reflecting the quantum yield, passes over into a horizontal line (the ceiling or saturation value) until at still higher light intensities the photosynthetic activity starts to decline due to photodamage.

Speaking about stress from light and temperature, light is usually felt more as a stress at lower temperatures as the various enzymatic repair mechanisms counteracting photodamage necessarily work less efficiently at lower temperatures (Satoh and Murata 1998). An additional stress factor in this context in natural environments is the oxygen the solubility of which in water increases at lower temperature (see Sect. 8.4), thus eventually establishing the photooxidative stress (see Satoh and Murata 1998). In our contemporary highly aerobic biosphere, basically all photosynthetic cells – algae as well as higher plants – are threatened by photooxidation. Major adaptation strategies to photooxidative conditions combine increased pigmentation, shielding excess light, and the various enzymes that remove and detoxify the oxygen and, in particular, its partly reduced and thus even more aggressive reaction products, the ROS (see Sect. 8.4). Particularly harsh and prolonged photooxidative conditions applied to the blue-green alga *Anacystis nidulans* in a previously non-growing state led to the complete, yet reversible disappearance of all pigments from the cells (Peschek and Schmetterer 1978; Schmetterer and Peschek 1981a,b; Schmetterer et al. 1983). Pigment destruction was found to be a special case of the photooxidation of membrane lipids, a major part of which was evolved as ethane and other hydrocarbons (Schmetterer and Peschek 1981a,b; also see Elstner and Pils 1979) and the ongoing process of photooxidation finally leaving white cells (Peschek and Schmetterer 1978; Schmetterer et al. 1983). The bleached (white) *Anacystis* had lost all electron microscopically discernible thylakoid structures, including the phycobilisomes, but remained viable for a certain further time span of photooxidative stress, probably because a part of the respiratory system, including the respiratory chain in the (chlorophyll-free!) CM, had remained intact (Schmetterer et al. 1983) and the generation of maintenance energy (ATP) was partly rescued.

Another factor that seems to lower the stability of photosynthesis towards photoinhibition is the limited turnover of the D1 protein of photosystem II in a membrane environment of lipids with a low degree of unsaturation (Kanervo et al. 1995) which, in turn, are favoured by growth of the organisms at higher temperatures, as we have seen before (see Table 8.1). On the other hand, other cell components such as the low-potential cytochrome c_{550} of cyanobacteria seem to stabilize photosynthetic oxygen evolution against heat inactivation (Koika and Katoh

1979; Hishiyama et al. 1994). Similarly, the membrane-bound and PS II-asssociated cytochrome b_{559} may slow down the rate of photoinhibition in chloroplasts and cyanobacteria more at higher temperatures than at lower temperatures (Nedbal et al. 1992).

Readers interested in further bioenergetic, biochemical or evolutionary and molecular details of photosynthesis in general are referred to the following competent literature: Olson (1970); Broda (1977a); Peschek (1981a); Nitschke and Rutherford (1991); Babcock and Wikström (1992); Blankenship (1992); Fromme et al. (1996); Mulkidjanian et al. (1996); Renger (1997, 1999); and Hoffmann (1998).

8.4 Temperature and Oxygen

As previously pointed out, an unexpected type of generalized stress situation which the whole primordial biosphere must have been facing was the transition from an essentially anaerobic biosphere and atmosphere to an ever increasingly aerobic, viz. oxidizing one that was brought about by the blue-green algae 3.2 billion years ago (see Figs. 8.3 and 8.4). Even though the time span during which this transition took place was fairly long, biochemical mechanisms of adaptation to the worldwide oxygen stress were badly needed for the survival of (still primitive) organisms living in those days, viz. to the originally anaerobic prokaryotes (bacteria). Without any doubt, the blue-green algae, the interior of which was the very site of the first biological oxygen production by water-splitting, oxygenic photosynthesis in the intracellular thylakoid membranes (see Fig. 8.2), at the same time must have been the first to be affected by the (poisonous! see Morris 1975; Cadenas 1989) oxygen gas. Thus, it might be useful to look at the cyanobacteria when trying to identify the most efficient way to adapt to oxygen, at least in an evolutionary sense. The first biochemical strategy for coping with free oxygen and its various, even more toxic partially reduced species (ROS) which inevitably must have resulted from inadvertent and enzymatically not yet streamlined electron transfer onto molecular oxygen, probably was the design of enzymes for the removal (detoxification) of such ROS. The more prominent of the highly reactive dangerous oxygen species are singlet oxygen, superoxide radical anion, the hydroxyl radical, hydrogen peroxide, other reactive (organic) peroxides formed metabolically; their reactivity is normally still multiplied by elevated temperature and by trace amounts of transition metals (heavy metals) invariably present in all natural environments (cf. e.g. Fenton and Haber-Weiss reactions involving iron and/or copper ions, see Hippeli and Elstner 1999). Among the "emergency enzymes" still present in all aerobic cells and serving the removal of ROS are, therefore, superoxide dismutase (SOD), catalase and various other hydroperoxidases. Algae are no exception in this respect, of course. On the contrary, being fairly close to the oxygen evolved they have always been favourite objects for studying SODs and hydroperoxidases (Asada et al. 1975, 1977; Miyake et al. 1991; Asada 1992; Obinger et al. 1998, 1999). Extensive studies have given evidence that algae and cyanobacteria contain various types of soluble and membrane-bound SODs, viz. copper/zinc SOD, iron SOD, manganese SOD and iron/manganese SOD (Bannister et al. 1987; Canini et al. 1992; Scandalios 1993) and that they contain as the primary H_2O_2-scavenging hydroperoxidase the so-called catalase-peroxidase (bifunctional catalase; Obinger et al. 1998, 1999; also see Hochman and Goldberg 1991).

Among >20 cyanobacterial species screened for the occurrence of hydroperoxidases a monofunctional (purely catalytic) catalase was detected quite recently in *Chlorogloeopsis fritschii* only (R. Zoder, B. Khachik, C. Obinger, G. Regelberger and G.A. Peschek unpubl.). Besides the almost ubiquitous catalase-peroxidase (Obinger et al. 1997, 1998, 1999), the only other ecologically significant H_2O_2-removing hydroperoxidases in cyanobacteria may be glutathione peroxidase and thioredoxin peroxidase, the genes for which are present in the totally sequenced genome of *Synechocystis* sp.PCC6803 (http://www.kazusa.or.jp/cyano) though very few biochemical results are available so far. Note that the gene has not been detected in the genome of *Anabaena* sp. PCC 7120, the total sequence of which was also released quite recently and which, though still in a rather disorganized state, is also available now on the internet under the same address as given for *Synechocystis* 6803. At any rate, ascorbate peroxidase, which was repeatedly reported to occur in several cyanobacteria (Tel-Or et al. 1986), may owe itself to experimental artefacts because it seems to play its predominant H_2O_2-scavenging role in the chloroplast of higher plants and eukaryotic algae only, as is also reflected by fairly high concentrations of ascorbate ín chloroplasts in contrast to rather low levels in cyanobacteria (Miyake et al. 1991; Asada 1992).

Apart from O_2- and ROS-scavenging enzymes, which might have been very useful to the living cells adapting to increasing oxygen levels (oxygen stress) in the environment during initial transition period from an anaerobic to an aerobic biosphere (see Figs. 8.4 and 8.5), the most economic and efficient way to detoxify free oxygen must have been the evolutionary transformation of the preexisting anaerobic photosynthetic electron transport chain into an aerobic O_2-reducing respiratory chain. This transition is amply discussed in the so-called conversion hypothesis (Broda and Peschek 1979), the most convincing evidence for which is the striking similarity in both structural and functional respects of all extant electron transport chains in all living cells, prokaryotic and eukaryotic, photosynthetic and respiratory and conforming, of course, to the notion of a monophyletic origin of biological electron transport systems (see Fig. 8.2 and the discussion presented there; also see Broda 1975).

Two further hints may be given for the feasibility of a smooth gradual transition from photosynthesis to respiration according to the conversion hypothesis (also see Broda 1975), always following the Darwinian principle of mutation-and-selection provided there was time enough for evolution which, however, quite a few serious researchers appear to be no more so sure about (see Scherer 1983,

1984): (1) enzymatically, the reductive pentose phosphate pathway (Calvin cycle) of autotrophic CO_2 fixation is very similar to the oxidative pentose phosphate pathway (or cycle), which is an alternative to the Krebs or tricarboxylic or citric acid cycle in many respiring prokaryotes. Cyanobacteria, for example, which, like many chemolithoautotrophic and methylotrophic bacteria (see Fig. 8.1), do not possess a complete Krebs cycle (due to constitutive lack of the gene for 2-ketoglutarate dehydrogenase), use the oxidative pentose phosphate cycle (Warburg-Dickens-Horecker pathway) for aerobic substrate degradation (which, by the way, restricts the range of their respiratory substrates, if any, to carbohydrates; see Pearce et al. 1969; also see Smith and Hoare 1977). (2) Looking at the membrane-bound electron transport chains in photosynthesis and respiration (PET and RET), on the other hand, two basic enzymatic devices representing the missing links between PET and RET are proper dehydrogenases on the low-potential side (the entry site) and a terminal oxidase at the high-potential side (the exit site). With respect to the latter, care had to be taken that the newly available and so excellently suitable electron sink, viz. free oxygen ($E_0' = +820$ mV !) is directly reduced to water in a concerted four-electron transfer step. The enzyme capable of this unusual task must have (at least) four cooperating monovalent redox centres, and partially reduced oxygen species (ROS) must not escape from the enzyme. Anyway the SODs and hydroperoxidases mentioned above would be helpful as emergency enzymes also in this case; and the enzyme to which nature has given final preference as being used as the most widespread and powerful electron sink in the biological combustion of aerobic respiration is the well-known and much-described aa_3-type cytochrome c oxidase. Equation (1) read from right to left illustrates the reaction catalyzed by the terminal respiratory oxidase. The evolutionary scenario discussed before, and the fact that in our laboratory no less than 33 species of cyanobacteria have been unequivocally shown to contain aa_3-type cytochrome c oxidase as the only (or at least predominant) functional terminal respiratory oxidase (see Peschek 1981b,c; 1996 a,b; Peschek et al. 1989a,b,c; 1994a,b for a review) has lent much support to the assumption that primordial blue-greens not only were the first to produce molecular oxygen, but also were the first to elaborate, along the lines of the conversion hypothesis (Broda and Peschek 1979), an aerobic respiratory chain culminating in an aa_3-type cytochrome c oxidase which latter even seems to have been invented by the blue-greens (Peschek et al. 1992). As a major scientific break-through in the field of bioenergetics, high-resolution crystal structures of the cytochrome-c oxidases from the bacterium *Paracoccus denitrificans* (Iwata et al. 1995) and from bovine heart (Tsukihara et al. 1996) have been published.

Finally, it may be mentioned as a curiosity that the formally inverted reaction of the terminal oxidase [Eq. (1) read from left to right] is the water-splitting reaction catalyzed by the oxygen-evolving complex (OEC) in photosystem II of plant-type photosynthesis. Clearly, also here we are dealing with a concerted four-electron transfer, viz. from water to the high-potential reaction centre chlorophyll a in PS II, while in respiration the four electrons are transferred from four molecules of cytochrome c (or two molecules of reduced quinone; see Saraste et al. 1991) to O_2. Again, four cooperating monovalent redox centres will be needed in the OEC, which in this case are the four redox active Mn ions formally corresponding to the S_i-states in Joliot's model of flash photosynthesis (Renger 1999). Electrochemically, (membrane-bound) enzymes such as the terminal respiratory oxidase and the OEC are called capacitor enzymes. They are unique enzymes in living cells playing the decisive role in the vectorial charge separation process across biomembranes which finally enables chemiosmotic (Mitchellian) energy conversion to occur.

A side remark on, and a more trivial explanation for why oxygen, which, as we have seen, has ever been and still is a kind of threat to a living cell, rests in the fact that the solubility of oxygen gas (and other gases) in water increases towards lower temperatures, following the temperature dependence of the Ostwald-Bunsen absorption coefficient. Thus while air-saturated distilled water contains 0.45675 mM O_2 at O °C (and atmospheric pressure) it contains only 0.21656 mM O_2 at 40 °C. Thus, the usual biochemical (or better nekrochemical?) prescription to work, e.g. with isolated proteins or enzymes at 4 °C could sometimes be counterproductive, notably when the protein is oxygen labile due to unprotected sensitive -S-S- bonds, or due to autoxidizable active centres, etc., or when it is eo ipso cold-labile, of course. Speaking of the solubility of gases (viz. O_2) in water, raising the solute concentration has the same effect as raising the temperature. Thus, we might understand why the obligately aerobic halobacteria (Archaea ? see above) thriving in saturated NaCl solutions (up to 33% w/v in concentration), e.g. *Halobacterium maris-mortui* in the Dead Sea, suffering from limited oxygen supply (only one third of the usual!) and, consequently, limited ATP supply from oxidative phosphorylation, have invented their peculiar type of bacteriorhodopsin-based photophosphorylation (Oesterhelt et al. 1977; Lanyi 1978; Renthal 1992). Nevertheless, and possibly strange enough, halobacterial photophosphorylation has turned out to be a dead end in evolution.

8.5 Temperature and Ionic Strength

Growth at elevated ionic strength (high-salt, usually taking NaCl as the salt representative) is quite common among bacteria, microalgae and even higher plants. The halophilic mode of growth, apart from being a matter of the a priori habitat in question —where the organisms naturally must have had a very long time in evolution already to become adapted to the situation, usually has resulted in changes of the cell constituents that have (1) made the barrier between cell interior and environment (viz, the cell or plasma membrane, in plant physiology often also called plasmalemma) less leaky to the salt, and (2) have alleviated the huge discrepancy in the osmotic pressures inside and outside the cell. The synthesis of osmotically compatible solutes in algal cells under salt stress (various soluble carbohydrates, glycine betaine, glucosyl glycerol, etc. has been amply described, in particular for the prokaryotic algae: (Reed 1988; Reed et al. 1986). Yet, since salt stress and salt tolerance prima facie are not directly related to temperature, they will not be considered here at length and the reader is referred to Chapter 11 in this volume.

With respect to the change in the composition of cell membranes in response to salt stress, an interesting fact may be mentioned that seems to relate salt stress somehow to temperature stress. In our studies on stress adaptation of the blue-green alga *Anacystis nidulans* we mostly used salt stress as the model stress. The lipid/protein ratio of membranes isolated from cells adapted to elevated ionic strength (0.4 M NaCl instead of approx. 20 mM NaCl) was roughly twice the normal value (Molitor et al. 1990a) and also other physico-chemical properties (buoyant density, vesicle diameters, surface charge density as reflected by the electrophoretic mobility, and metal content) of membrane vesicles isolated from salt-adapted cells had changed dramatically (Riviere et al. 1990), very similarly to the situation in cells grown at lower temperatures (Sect. 8.2.3). However, a closer analysis of the composition of the esterified fatty acids in membrane lipids in salt-adapted cells revealed that, while the proportion of lipids to proteins went up, the proportion of unsaturated and short-chain fatty acids to saturated and long-chain fatty acids declined. Also LPTs of both CM and ICM were found to increase when determined on membranes isolated from salt-grown cells (Molitor et al. 1990a). As an overall effect of the adaptation to high salt, the increase of lipid-to-protein ratios in the membranes (CM being always much more affected than ICM) with respect to increasing membrane fluidity (smoothness) is nevertheless slightly outweighed by the still steeper increase in the proportion of saturated and long-chain fatty acids (Table 8.1). As a corollary of this it was observed that cold-adapted *A. nidulans* that had been grown for five generations (approx. 40 h) at 20 °C tolerated up to 0.5 M NaCl while cells grown at 40 °C never tolerated more than 0.4 M NaCl (Molitor et al. 1990a and unpubl. Obs.). Thus, the salt stress somehow counteracted low temperature stress, which might be useful under certain conditions of a changing environment in natural habitats (cf. Tables 8.3 -8.4).

At this point, speaking about the influence of stress adaptation on the chemical composition of membrane lipids and particularly emphasizing the (more highly) unsaturated fatty acids which nowadays already make certain cyanobacteria, e.g. *Spirulina* (= *Arthrospira*) have their price as a "health food", recent side results from our group within the framework of a more biotechnological exploitation of blue-green algae —in that case also employing some species which are more "alien" to our laboratory, such as *Spirulina*, *Tolypothrix* and *Agmenellum* (= halophilic *Synechococcus* spp.) —in are shown in Table 8.4. Table 8.4 indicates how the concentration of unsaturated fatty acids in the (total) membrane lipids can be physiologically varied just by adding NaCl (0.4 M final concentration) to the growth medium. Again, also lowering the growth temperature would certainly have the effect of increasing the ratio of unsaturated to saturated esterified fatty acids in the membrane lipids (Sect. 8.2.3) but the likewise lowered growth rate would rather counteract a large-scale biotechnological application. Three general facts are evident from Table 8.4: only a few species are capable of synthesizing highly unsaturated fatty acids (e.g. arachidonic acid in *Tolypothrix* only; see Murata and Nishida 1987); the ratios of esterified total unsaturated to saturated fatty acids in ICM are generally higher than in CM; and in a marine species which is halophilic a priori, e.g. in *"Agmenellum"*, this ratio is considerably higher from the beginning, as expected.

Table 8.3. Contents of esterified fatty acids in total lipids from plasma membrane (CM) and thylakoid membrane (ICM) of *Anacystis nidulans* grown in the absence (-) or presence (+) of 0.4 M NaCl and harvested at a cell concentration of 3-4 μl packed cell ml^{-1} of culture fluid (Peschek et al. 1989c; Molitor et al. 1990a)

Membrane	Fatty acids (nmol mg^{-1} protein)							Total lipids/ protein (w/w)
	14:0	14:1	16:0	16:1	18:0	18:1[a]	18:2[b]	
CM-	31.4	92.2	1413.1	812.0	24.6	211.9	11.6	1.07-1.19
CM+	60.2	148.1	3902.5	1417.3	112.3	1088.3	26.4	2.9-3.1
ICM-	33.0	47.9	725.4	494.5	6.3	146.8	5.4	0.59-0.86
ICM+	42.3	55.7	946.3	517.1	15.0	405.2	9.2	0.83-1.01
	Total				Total			
	Saturated		Unsaturated		C-14	C-16	C-18	
CM-	1469.1		1127.7		123.6	2225.1	248.1	
CM+	4075.0		2680.1		208.3	5319.8	1227.0	
Increase	+178%		+144%		+69%	+139%	+395%	
ICM-	746.7		694.6		80.9	1219.9	158.5	
ICM+	1003.6		987.2		98.0	1463.4	429.4	
Increase	+31%		+46%		+21%	+20%	+171%	

Fatty acids from total lipids were determined by gas chromatography; total lipids as such (including carotenoids, quiones, and in the case of ICM also chlorophyll) were extracted with chloroform/methanol and determined according to Blight and Dyer (1959). All values shown are means from at least three parallel cultures, standard deviations being within ±15% of the corresponding means. Protein content per cell remained constant during slat adaptation (cf. Murata and Nishida 1987)

[a] Presumably a mixture of oleic and vaccenic acids (cf. Murata and Nishida 1987).

[b] No linolenic or other polyunsaturated fatty acids were detected (cf. Murata and Nishida 1987); also linolenic acid (18:2) was undetectable in early logarithmic cells harvested at a cell concentration of <1 μl packed cells ml^{-1} of culture fluid.

Table 8.4. Unsaturated fatty acids in total membrane lipids extracted from control (-) and salt-grown (0.4 M NaCl) cyanobacteria grown photoautotrophically at 20 W m^{-2} warm white fluorescent light and 30-35 °C. (Peschek et al. 1989c)

	18:2	α-18:3	δ-18:3	18:4	Total-unsat./ Total-sat.
Anacystis					
CM-	11.6	0.0	0.0	0.0	0.72
CM+	26.4	0.0	0.0	0.0	0.63
ICM-	5.4	0.0	0.0	0.0	0.91
ICM+	9.2	0.0	0.0	0.0	1.1
Synechocystis					
CM-	12	0.0	22	0.0	0.8
CM+	25	0.0	48	0.0	0.67
ICM-	5	0.0	10	0.0	1.0
ICM+	8	0.0	18	0.0	1.3
Spirulina					
CM-	17	0.0	37	0.0	0.93
CM+	40	0.0	69	0.0	0.71
ICM-	9	0.0	15	0.0	1.2
ICM+	15	0.0	27	0.0	1.7
Tolypothrix					
CM-	13	4	8	9	0.83
CM+	33	10	24	23	0.68
ICM-	5	2	5	6	1.1
ICM+	12	5	8	10	1.9
Agmenellum					
CM	20	0.0	67	0.0	0.98
ICM	13	0.0	44	0.0	1.8

8.6 Brief Interpolation on Stress Definition and Adaptation as an Evolutionary Driving Force

Evolutionarily speaking, and returning to the oxygen crisis caused by the primordial blue-green algae 3.2 billion years ago (Figs. 8.3 and 8.4), a rather trivial way of adapting to oxygen gas was to avoid it. In fact, as is also witnessed by a whole bunch of more or less strict anaerobes still present on our extant Earth, some of the anaerobic bacteria of those days successfully, and forever, so to speak, retreated into oxygen-free ecological niches such as the rotting depths of waterbodies or the stomach of ruminants, the rumen. Good examples of organisms having thus escaped the oxygen threat and avoided the necessity to adapt are the methanogenic bacteria which are killed by low levels of free oxygen (Morris 1975; Balch et al. 1979; Cadenas 1989). In the context of comparative microbial bioenergetics a word might be appropriate here on the second large group of bacteria uniquely capable of metabolizing methane and other reduced C_1-compounds, the methane oxidizers (Higgins et al. 1981). Interestingly, while the methanogens are strictly anaerobic most of the methane oxidizers are strict aerobes and will thus never be found associated with the former in natural habitats. A unique instance of convergent evolution has apparently brought it about that the assimilation of the C_1 compound (see Fig. 8.1), viz. of intermediary CH_2O (= formaldehyde!) two peculiar enzymatic assemblies, the serine pathway and the ribulose monophosphate pathway, respectively, follows exactly the same pattern in both, ecologically most distant, groups of (chemo)autotrophic bacteria. Further, it might be mentioned that even a few (eukaryotic !) yeasts in evolution have learnt to utilize C_1-compounds as food stuff (Wolf et al. 1980).

In this context a fundamental question concerning the term adaptation to stress in general might recur. First, we should distinguish between permissive and nonpermissive stresses. Nonpermissive stresses will by definition lead to cell death, which in this case might also be regarded as some kind of adaptation in a general ontological sense. Whether a stress is permissive or nonpermissive will at least also depend on the intensity and duration of the stress. To speak of an adaptation to stress clearly makes sense only in the case of permissive stresses. Adaptation, with the aim of acclimatization, implies a change in metabolic patterns that enables the cell to continue normal life processes under abnormal conditions. These changes need not necessarily be reversible. Under certain conditions the photolithoautotrophic eukaryotic alga *Euglena gracilis* can be cured of its photosynthesis and may irreversibly change into the chemoorganoheterotroph *Astasia longa*, which is genetically closely related to the former (see Buetow 1968). Also some cyanobacteria might have dropped their photoactive pigmentation and thus abandoned photosynthesis at some stage in evolution, yet might nevertheless have continued on to thrive in the form of so-called apochlorotic blue-greens (Pringsheim and Wiessner 1961; Pringsheim 1963) as obligately chemotrophic representatives of, e.g. the genera *Beggiatoa, Vitreoscilla, Thiothrix,* etc. Both gross-morphological, ultrastructural and metabolic similarities between the latter and their phototrophic counterparts among the blue-green algae, which at that stage might have already acquired a substantial chemoheterotrophic potential, are taken as evidence for such (likewise irreversible) evolutionary transitions. We may view the transition as a kind of large-scale adaptation which must have had something to do with light in general, which was no longer preferred over organic substrates as a source of metabolic energy by those apochlorotic blue-greens. Progressively losing in competition with an increasingly successful eukaryotic algal population in certain (more or less eutrophic, yet still illuminated) habitats might have been another reason for dropping photosynthesis. Other prokaryotic organisms might have sought shelter against certain adverse environmental conditions by retreating to the interior of some host cell which provided a preformed, constant microenvironment for the invading cells which, in the long run, have established a symbiosis with their host which proved to be profitable to both partners and, therefore, was maintained. The very first endosymbiont event most likely gave rise to eukaryotic cells (Mereschkowsky 1905; Margulis 1970, 1996; Stanier 1974), but it necessarily must have occurred between prokaryotes, although no contemporary endosymbiotic system is known that involves two prokaryotes as both host and guest. True, stable endosymbioses obviously have not been established very frequently in evolution. On the other hand, a variety of extant endosymbiotic systems are known, such as *Cyanophora paradoxa* (Wasmann et al. 1987). The same line of adaptation might have led to the formation of tissues (of metaphyta and metazoa) from many independent eukaryotic cells dropping risky independence in favour of increased security and collective efficiency or to the evolution of homeothermic from poikilothermic animals. Finally, even the evolution of human consciousness and intelligence might be conceived of as a result of a general stress adaptation strategy which aims at an improvement of the survival conditions of the population, not necessarily of the individual. What else does Darwinian selection reflect than radical and brutal adaptation? Although such considerations regarding the definition of stress, and the definition of what should be understood by adaptation to stress, as well as the concept of evolution as a series of large-scale, long-term and streamlined adaptation events, might seem futile, experience tells the authors that it is nevertheless very important in a scientific discussion on stress and adaptation first to clearly define the terms and what they mean to each participant of the discussion.

8.7 Bioenergetic Strategy for Stress Defence in Algal Cells: the Role of the Cell Membrane

Remembering the preface to our reflections on algal (temperature) stress physiology, viz: life implies work (see Sect. 8.1.1), this basic axiom will apply even more stringently to living cells exposed to environmental stress, of whatever type the stress might be. All the measures taken by living cells including algae for coping with (permissive! see Sect. 8.4) stress conditions within the biochemical framework of adaptation to the stress are inevitably energy-consuming, be it the de novo synthesis of unsaturated fatty acids in the adaptation to low-temperature stress (Sect. 8.2.3), the increased synthesis of light-absorbing pigments in the adaptation to low-light stress (Sect. 8.3), the increased synthesis of a proton-sodium antiporter, of osmotically compatible solutes, and of membrane lipids in general, in the adaptation to salt stress (Sect. 8.6), increased synthesis of ROS-scavenging enzymes in the adaptation to oxygen stress (Sect. 8.4), etc. All stress adapting processes need ATP and the analysis of the question from where, in algae, this ATP for stress adaptation has to come, leads to the interesting answer that in practically all instances of stress it cannot come from photophosphorylation since the stress primarily inhibits photosynthesis, at least in the first phase of stress impacts of whatever kind. This has been most thoroughly documented for blue-green algae adapting to salt stress (Molitor and Peschek 1986; Molitor et al. 1990a,b), but very similar observations have been made also for eukaryotic algae adapting to (permissive) high and low temperature stresses, to high-light stress and to moderately high salt stress. In the case of nutrient (starvation) stress and heavy-metal stress, more specific interactions between target proteins/enzymes may interfere, and the overall picture might be less clear. Nevertheless, a very comprehensive series of investigations, and a consistent bioenergetic hypothesis, of the adaptation of photoautotrophically growing *Anacystis nidulans* to extremely low levels of phosphate down to 50 nM have been presented by Falkner and coworkers (see Wagner et al. 1995, 1999; Falkner et al. 1996).

Regarding the dichotomy between prokaryotic and eukaryotic algae, it must be noted that in all the pilot experiments on eukaryotic *Chlorella pyrenoidosa*, *Chlamydomonas reinhardtii* and *Euglena gracilis* performed in our laboratory along the lines which had been taken for the investigation of cyanobacterial stress adaptation and stress physiology (see, e.g. this chapter) a very basic fact became clear already from the initial stages of these experiments. In all the cases of environmental stresses experimentally applied to the organisms, viz. from temperature, light, oxygen, salt, etc., it has uniformly turned out that prokaryotic algae were far more stress-tolerant than eukaryotic algae. Without going into details, one example may be mentioned: *Anacystis* and *Chlorella* tolerated prolonged exposure to 47 and 33 °C, 40 and 25 W m^{-2} irradiance, 97 and 55% (vol/vol) O_2 in the growth gas (at 3% CO_2, balance N_2), 0.4-0.5 and 0.13 M NaCl, and <50 nM and 0.9 mM phosphate. The comparison between *Synechocystis* and *Euglena*, though less dramatic, was in the same order of magnitude. Nevertheless, also the eukaryotic algae showed a clearcut syndrome of bioenergetic stress adaptation in particular with respect to increased respiratory oxygen uptake which was typical of all types of stress applied (see Sect. 8.7). On the other hand, it is clear from the foregoing that we are using prokaryotic algae (cyanobacteria) here predominantly for the exemplification of the (bioenergetic) impacts of (temperature and other) stress and the mechanisms of stress adaptation. The initial response to a salt shock was a rapid depletion of the intracellular ATP pool in *A. nidulans* (see Nitschmann et al. 1982). The next step and at the same time the first step of true stress adaptation in the absence of photosynthesis (which usually needs at least one generation time to slowly recover from the primary impact of the stress; see Fig. 8.7) is a significant increase in respiration, a phenomenon which had already been described by H. Lundegardh in 1961 studying salt-stressed plant tissues, who termed it salt respiration. It was interpreted in terms of an increased ATP demand to be satisfied by oxidative phosphorylation under stress conditions when photosynthesis would not work. Similar observations were later made on eukaryotic (Löppert and Broda 1973) and prokaryotic algae (Molitor and Peschek 1986). In accordance with this respiratory stress defence strategy a significant increase in the concentration of membrane-bound RET components (particularly of cytochrome reductase and oxidase) in the course of stress adaptation was measured (Fry and Peschek 1988 ; Peschek et al. 1988a).

Figure 8.7A shows the time course of the activities of (coupled) respiratory oxygen uptake, photosynthetic oxygen evolution and CM-bound Na^+/H^+ antiporter in suspensions of *Anacystis nidulans* adapting to 0.4 M NaCl at 35 °C in the light. Figure 8.7B shows that the adaptation process as such is not completely arrested in the dark though *A. nidulans* is an obligately photoautotrophic species. On the contrary, increased synthesis of respiratory chain components and of the Na^+/H^+ antiporter continues for some time in the resting cells in the dark, probably due to protein rearrangement in the cells until it finally stops due to the absence of de novo energy-generating processes, viz. photosynthesis. Employing chloroamphenicol (CAP), which arrests prokaryotic protein synthesis and which is seen to completely block the adaptation process (Fig. 8.7C), shows that the adaptation process depends on de novo protein (re-)synthesis.

A bioenergetic adaptation pattern strikingly similar to that shown in Fig. 8.7, though slightly less pronounced, was seen when, instead of adding NaCl (salt shock), the temperature of photoautotrophic cultures of *A. nidulans* was rapidly lowered to 19 °C (cold shock) or raised to 47 °C (not shown). That the acclimatization to (permissible) high and low temperatures, apart from affecting photosynthesis and respiration, also affected the Na^+/H^+ antiporter (though to an extent roughly 30-40% less than shown in Fig. 8.7) eo ipso points to increased permeability of the cell membrane to salt which, in turn, gives rise to increased ion, in particular Na^+/H^+-translocating acivity in the CM (see Peschek et al. 1985, 1986, 1988b; Erber et al. 1986; Fry et al. 1986; Nitschmann and Peschek 1986) when the cells are exposed to a temperature stress. Thus, the increased Na^+/H^+-translocating activity (for active sodium extrusion) might be ultimately powered by a CM-bound H^+-translocating respiratory chain and/or a likewise CM-bound H^+-translocating P-type and/or F-type ATPase (See Fig. 8.13 and the explanations given there).

8.7.1 Direct Energization of the Plasma Membrane by Electron Transport

Membrane energization in the chemiosmotic sense means that a transmembrane electrochemical proton potential gradient $\Delta\mu_H{}^+$ or protonmotive force (pmf) comprising the electrical membrane potential $\Delta\psi$ and the difference in proton concentrations ΔpH between both sides of the membrane be built up across the bioenergetically competent or excitable biomembrane (see Nicholls and Ferguson 1982):

$$\Delta\mu_H{}^+ = -F\Delta\psi + 2.3RT\Delta pH \quad \text{(expressed in energy units, e.g. J).} \qquad (4)$$

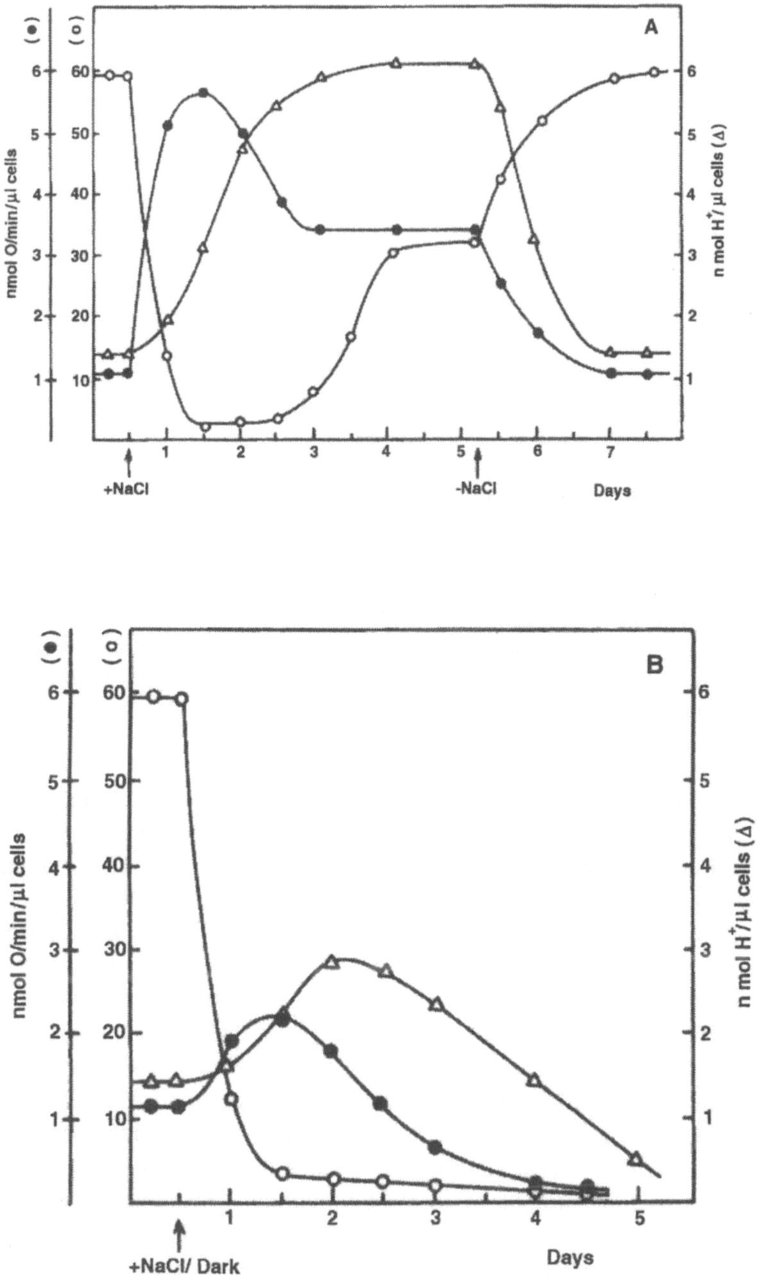
A
B
nmol O/min/µl cells
n mol H+/µl cells (Δ)
+NaCl
-NaCl
+NaCl/ Dark
Days

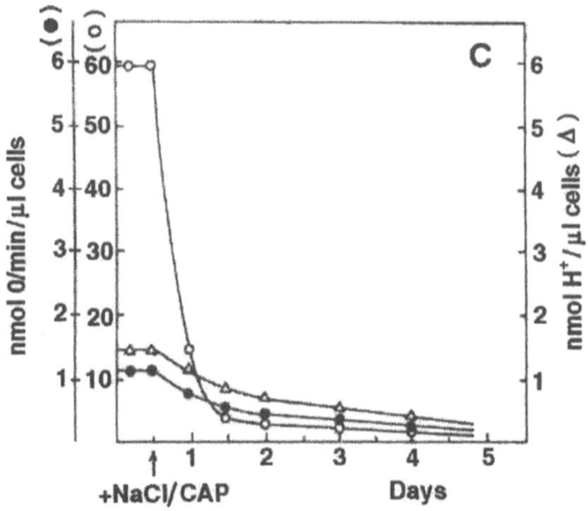

Fig. 8.7. A Changes in activities of dark respiratory O_2 uptake (*full circles*), photosynthetic oxygen evolution (*open circles*), and Na^+/H^+-antiporting activity (*open triangles*) determined on whole cells of *Anacystis nidulans* after addition (+) and withdrawal (-) of 0.4 M NaCl (final concentration; *arrow*) to photoautotrophically growing cultures (35 °C, 20 W m^{-2}). **B** The same experiment performed in darkness (+NaCl/dark), measuring photosynthesis in illuminated samples of harvested cells. **C** The same experiment performed in the light in the presence of 20 µg ml^{-1} chloroamphenicol (+NaCl/CAP). See text for explanation

Herein 1F is the so-called Faraday constant, electrostatically defined as the electrical charge of 1 mol electrons, i.e. 96 500 C (= Coulomb, the physical unit of the electrical charge; the charge of 1 electron being 1.602 x 10^{-19} C), or dynamically (energetically) defined as the electrocaloric equivalent, 1F = 96.5 kJ mol^{-1} V, i.e. the energy = work stored in = applied to 1 mol electrons that are accelerated by a potential difference of 1 V.

The living cell poises the pmf. [Eq. (4)] against the chemical free enthalpy, ΔG, that is needed for the synthesis of 1 mol ATP (from ADP and inorganic phosphate, the so-called phosphorylation reaction; ΔG positive) or, what amounts to the same number, released from 1 mol ATP upon hydrolysis to ADP and P_i (ΔG negative). This energy turnover is 28.5 kJ mol^{-1} ATP under standard conditions (ΔG_o), where physicochemical standard parameters obtain per definitionem, e.g. 25 °C, 1 atm, 1 M concentration of all reaction partners (including H^+, pH 0 thus defining (physicochemical standard conditions!), etc. In realistic (i.e. physiological) conditions where the van't Hoff equation links the adenylate mass action ratio to ΔG according to Eqs. (5) and (6),

$$\Delta G = \Delta G_o + RT.\ln([ATP]/[ADP].[P_i]) \qquad \text{ATP synthesis} \qquad (5)$$

$$\Delta G = \Delta G_o + RT.\ln([ADP].[P_i]/[ATP]) \qquad \text{ATP hydrolysis.} \qquad (6)$$

Calculation of the energy required for the synthesis, or release on hydrolysis, of 1 ATP involves realistic actual concentrations of ATP, ADP and P_i that must be experimentally determined in the bioenergetic system (mitochondrion, chloroplast, living cell) under the environmental conditions to be studied.

Mechanistically, the pmf is usually built up by transmembrane proton translocation effected by the electron transport proteins and other components (e.g. quinones) of the electron flow chain viz. of either photosynthesis or respiration in the membrane (see Fig. 8.2). The eventual driving force for the whole process is either the energy of absorbed light in photosynthesis or the free energy (more strictly enthalpy) difference between respective electron donors and acceptors in the respiratory chain (see Fig. 8.1). The only instance where such chemiosmotic energy conversion (see Nicholls and Ferguson 1992) occurs by light-driven conformational proton-pumping without an alternating red/ox electron transport chain is the halobacterial bacteriorhodopsin-based photophosphorylation (see Oesterhelt et al. 1977; Lanyi 1978; Renthal 1992). In a very few cases in nature, the H^+ as the universal coupling ion seems to be replaced by Na^+ (Dimroth 1987; Skulachev 1992).

8.7.2 Indirect Energization of the Plasma Membrane by H^+-ATPases

The primary task of the cell or plasma membrane, or plasmalemma, of eukaryotic algae is to maintain the crucial semipermeability of the cell acting as an osmotic barrier between outside and inside. As in all living cells, the cell membrane of algae also contains various transport systems (translocases) for solutes including carbon substrates and ions. Bioenergetically speaking, eukaryotic plasma membranes are the site of various transport ATPases (see for review Carafoli and Scarpa 1982). Plasma membrane-bound electron transfer processes, let alone integral electron transport systems, have only rarely been envisaged for eukaryotic cell membranes, the only more prominent example being the plasmalemma redox functions in plants (Möller and Crane 1990; Möller et al. 1992). Energization of the eukaryotic cell membrane in terms of building up a pmf across it (see above), which is necessary in all events to provide a general shield to the cell against various impacts from the environment (viz. against stress) as if surrounded by an electric fence if this anthropomorphic expression might be allowed here, is nevertheless possible by way of various plasma membrane transport ATPases powered by ATP synthesized in one of the bioenergetic organelles, chloroplast or mitochondrion. Clearly, before the ATP from eukaryotic photosynthesis or respiration can reach the plasma membrane it must be exported out of the organelle (see Fig. 8.2) which again costs energy. Therefore, the prokaryotic mechanism for CM energization by H^+-translocating electron transport directly in the CM certainly is the more straightforward one (see Fig. 8.13).

However, be it recalled that also certain anaerobic bacteria, such as Clostridia or lactic acid bacteria of the genus *Streptococcus* which do not posses CM-bound electron flow chains, must energize their CM through ATP which is formed in the cytosol by fermentation (substrate-level phosphorylation), because one or the other type of the universal H^+-ATPase invariably is contained also in the plasma membrane of even strictly anaerobic bacteria. A very similar case of indirect membrane energization occurs in the case of the vacuolar membrane of plants and algae which are known to contain a so-called V-type ATPase (Carafoli and Scarpa 1982).

8.7.3 Plasma Membrane Energization as a Stress Defence Mechanism in Blue-Green Algae

Entirely different from the mechanisms for energization of the plasma membrane in eukaryotic cells or in anaerobic (fermenting) bacteria is the situation for the plasma membrane of cyanobacteria. Investigation of bioenergetically relevant electron transfer and H^+-translocation processes in the cyanobacterial CM has been the main topic of scientific research in our laboratory during the past 30 years. Even one of the very first cases of "nekrochemical" preparative isolation, separation and characterization of fully functional cyanobacterial CM and ICM was achieved in this laboratory (Molitor and Peschek 1986; Molitor et al. 1987). Against rather serious initial objections, after a decade of hard work we could definitely prove the presence of a complete phosphorylating respiratory chain in chlorophyll-free CM prepared from a wide variety of cyanobacteria. At the same time, and at least equally difficult to carry through, we documented the presence of a functionally relevant aa_3-type cytochrome-c oxidase in (both CM and ICM of) all 33 species of blue-green algae analyzed (Peschek 1981b,c; Peschek et al. 1988a, 1989a,b,c, 1994a,b).

In order to avoid misunderstandings from the beginning, the following should be strictly noted: quantitatively, the CM area of a cyanobacterial cell when examined planimetrically on thin section electron micrographs amounts to between 10 and 20% of the total relevant membrane area (CM + ICM) only. Therefrom it follows that, based on the overall contribution to a whole cyanobacterial cell, CM-bound energy-conserving processes are by far outweighed by the ICM-bound processes. In perfect agreement with this, it is the ICMs that are the site of cyanobacterial photosynthesis, and only in *Gloeobacter violaceus* which constitutively does not make thylakoid membranes at all (Rippka et al. 1974) does the CM contain a photosynthetic electron transport chain but the photosynthetic growth rate of *Gloeobacter* is correspondingly low. Thus, it is important to distinguish strictly between overall rates of a metabolic process (e.g. per cell) and specific rates of an electron transport process per mg membrane protein. The specific importance of CM-bound biochemical processes can, therefore, be much higher than judged from the mere per-cell rate. A typical instance of this kind is respiratory electron transport. It has long been known that respiratory rates of blue-green algae are extremely low, and formerly the significance of respiration in blue-greens had been questioned (Holm-Hansen 1968). However, the expressed per-mg membrane protein respiratory rates displayed by CM are frequently much higher than those of ICM (Peschek et al. 1988a, 1989a,b,c; 1994a).

With respect to stress and adaptation, and particularly to the role of respiration in this process, these reflections gain special attention if we consider that it is the cell membrane, and not the intracellular thylakoid membrane that is closest to the environment and, thus, to practically any environmental stress (see Fig. 8.2). It will therefore also be primarily affected. In our discussion of a general bioenergetic stress management we found out that respiratory membrane energization and ATP supply are crucial to a stressed cell which usually shuts down photosynthesis first. It was not surprising therefore that, in *A. nidulans* exposed to a variety of seemingly unrelated stress conditions, we always found a dramatic increase in the activity of the CM-bound cytochrome oxidase (taken as a measure of the rate of electron transport through the whole respiratory chain according to the kinetic considerations given in Sect. 8.2.2) while the ICM-bound respiratory chain was much less affected (Fig. 8.8).

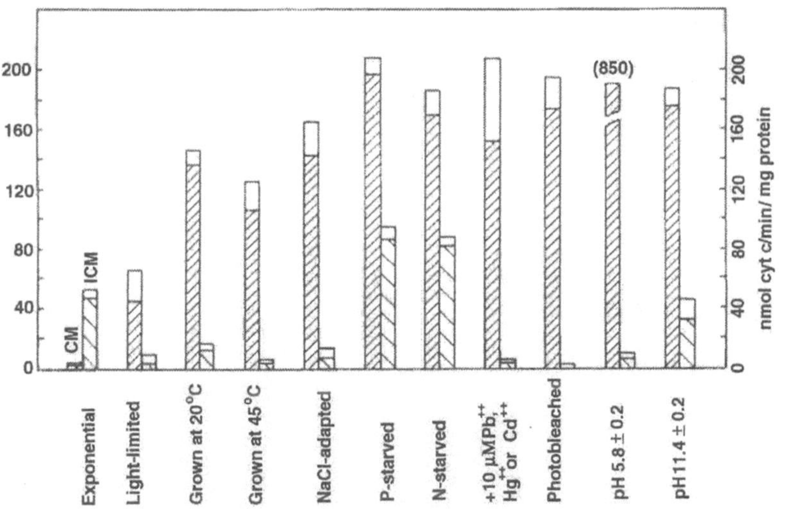

Fig. 8.8. Cytochrome-c oxidase activities measured by dual wavelength spectrophotometry on isolated and purified CM and ICM vesicles prepared from *Anacystis nidulans* exposed to a variety of different environmental stress conditions as indicated *below the figure*. Permissive stresses were allowed to exert their effect for at least seven generations. In the case of non-permissive stresses, i.e. heavy-metal, high-light/photooxidation (150 W m^{-2} warm white fluorescent light) and pH, the non-growing (resting) cells were exposed to the stress for 12-24 h

The effects of stress from high-salt and light limitation on CM- and ICM-bound cytochrome-c oxidase activity of *A. nidulans* are illustrated in Figs. 8.8 and 8.9.

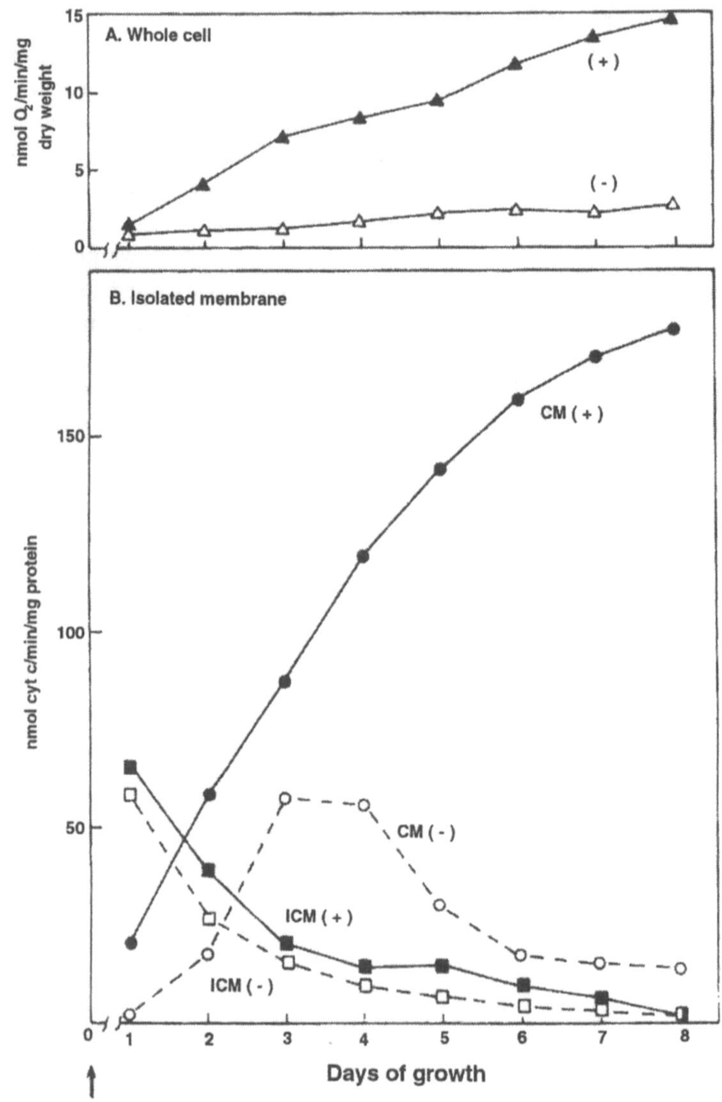

Fig. 8.9. **A** Whole cell respiratory oxygen uptake and **B** cytochrome-c oxidase activities (determined by dual wavelength spectrophotometry on isolated and purified CM and ICM preparations) of *Anacystis nidulans* grown in the absence (-) and presence (+) of 0.4 M NaCl. (experimental details can be found in Peschek et al. 1988a)

Light limitation to an obligately phototrophic alga must be regarded as just another kind of stress, total darkness being a non-permissive stress eventually leading to cell death in any obligate phototroph. Light limitation in an illuminated batch culture is inevitably caused by increasing self-shadowing of the multiplying population (Fig. 8.9). However, that it is not the increasing cell density per se that acts as a stress to the algal cells, even if they might feel more uncomfortable in a dense culture compared to a dilute culture, was shown by adjusting in turbidostat culture various growth rates by varying the incident light intensity, thereby automatically slowing down growth (increasing the doubling time) with decreasing light intensity (Fig. 8.10). Thus it was shown that even in a dilute culture of *A. nidulans* under light limitation, the increased respiratory (viz. cytochrome-c oxidase) activity had to be attributed to the CM and not to the ICM (Fig. 8.10). Interestingly, even prolonged growth at 19 and 47 °C (adaptation to permissive low and high temperature stress; see Fig. 8.8) brought about qualitatively the same effect regarding the allocation of increased COX activity in *A. nidulans* as documented for stress from high-salt or light limitation (see Fig. 8.9) although just in

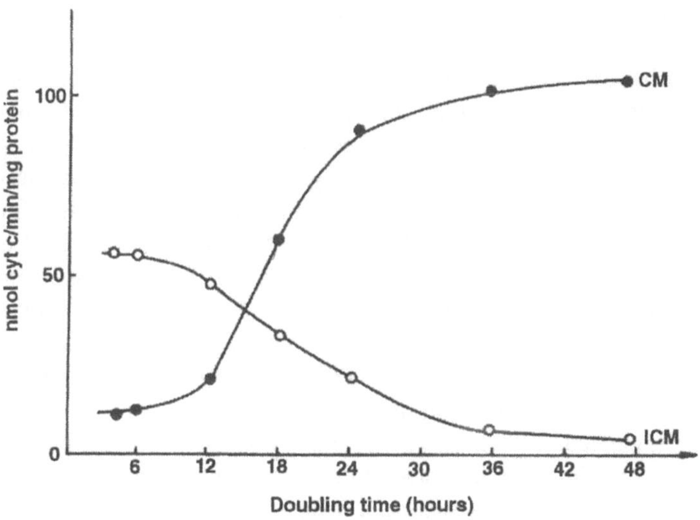

Fig. 8.10. Cytochrome-c oxidase activities of CM and ICM preparations of *Anacystis nidulans* grown in illuminated turbidostat cultures at a constant cell density of 1 μl packed cells ml^{-1}, at 35 °C, and at different incident light intensities (light intensity decreasing, from 25 to 5 W m^{-2} warm white fluorescent light, *from left to right)*

the case of temperature stress the argument (see above) of a closer proximity of the CM to the site of stress action prima facie would not be valid. This argument anyway must be modified in two respects:

1. Although many of the blue-green algae tested, e.g. *Anabaena* spp., *Nostoc* spp., *Plectonema* spp. (see Fig. 8.11A), and in particular *A. nidulans*, which has been studied in great detail, and other species which have been studied in no less detail, viz. *Synechocystis* 6803 and 6714, have been found to display just the opposite topography regarding COX activity in that ICM is the preferred location of ther-espiratory chain (COX) in both normal and stress conditions (Fig. 11B). Figure 8.11A, B illustrates the influence of light-saturated or exponential (E), light-limited (LL) and heterotrophic dark growth (CH) on COX activities in CM and ICM from several unicellular and filamentous cyanobacteria but adaptation to growth at low (19 °C) and high (47 °C) temperatures had effects qualitatively identical to stress from high-salt and light limitation, though in a much less reproducible way, probably because the transition between different temperatures could not be as precisely adjusted as different light intensities or the addition of salt (NaCl).
2. The second caveat with respect to a generalization of a CM-bound respiratory chain as the universal bioenergetic stress defence strategy in blue-green algae is the reasoning that under stress, in the absence of a functional photosynthetic energy conversion, also the enhanced ICM-bound respiratory chain/ATP-production could eventually lead to increased energization of the CM, viz. by way of a proton-translocating CM-bound ATPase (Fig. 8.12B). In both cases increased respiration results in increased CM energization, directly via. the CM-bound proton-translocating respiratory chain without foregoing extra ATP synthesis (*Anacystis, Plectonema,* etc., see above) or through increasingly ATP-powered CM-bound H^+-ATPases whereby the increased ATP-supply comes from an increasingly active (stress-enhanced) ICM-bound respiratory chain (Fig. 12A, panel B). The occurrence of CM-bound both F-type and P-type H^+-ATPases was described in the blue-green algae *A. nidulans*, *Synechocystis* sp. PCC6803, and *Anabaena* sp. PCC7120 (Neisser et al. 1994; Dworsky et al. 1995; Fromwald et al. 1995; also see Peschek et al. 1994a). The scheme in Fig. 8.13 shows the synergistic effect of H^+-translocating electron transport systems (PET and/or RET) and CM-bound H^+-ATPases for the energization of the plasma membrane.

Though the architecture of prokaryotic and eukaryotic cells differs significantly, of course, and the cell membrane of eukaryotic cells never contains an integral electron transport assembly such as a respiratory chain or a photosynthetic electron transport chain, it is nevertheless clear that the energization of the cell membrane as the outermost semi-permeable and metabolically active boundary also of a eukaryotic cell against its environment, from where the stress comes, plays the decisive bioenergetic role for stress defence in both eukaryotic and prokaryotic cells. Unfortunately, the bioenergetic details of the adaptation to environmental stress in eukaryotes have been much less studied than in prokaryotes. One of the reasons might be that eukaryotic cells (and organisms) are usually much less metabolically tolerant to stress than are prokaryotic cells. This conforms to the view that eukaryotes in general are much less versatile than pro

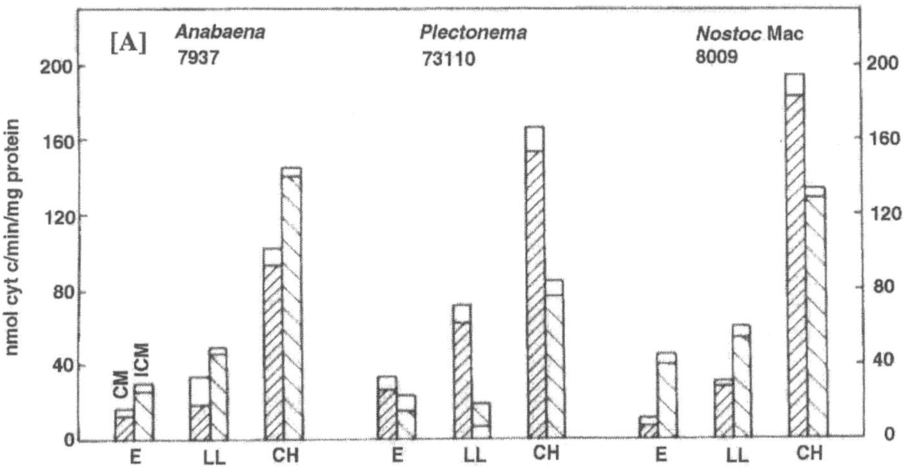

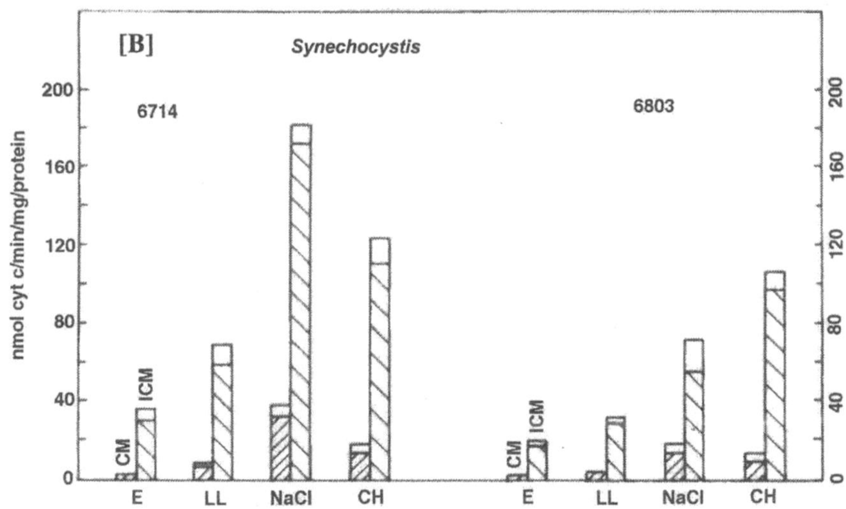

Fig. 8.11 A,B. Cytochrome oxidase activities of CM and ICM preparations of *Anabaena* PCC 7937, *Plectonema* PCC 73110, and *Nostoc* PCC 8009 (**A**) and of *Synechocystis* PCC 6714 and 6803 (**B**) grown photoautotrophically (20 W m^{-2} warm white fluorescent light) in early exponential phase (*E*), in light-limited (late-logarithmic) phase (*LL*), and chemoheterotrophically in complete darkness on 5-10 mM fructose (**A**) and 5 mM glucose (**B**), respectively. Nitrogen source was nitrate and growth temperature was 30 °C throughout. Kratz and Myers medium (**A**) and BG-11 (**B**) were used as the basal mineral media

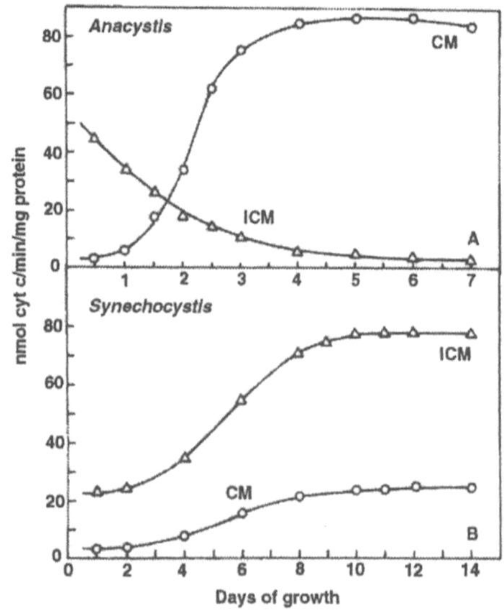

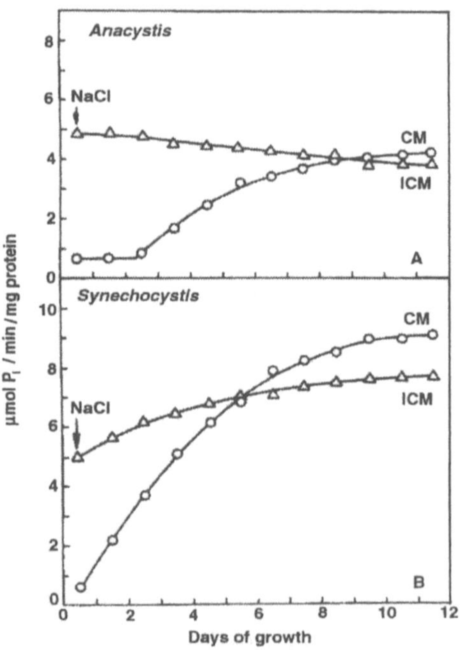

Fig. 8.12. A,B. Activities of cytochrome-c oxidase (**A**) and Mg-dependent total (i.e. P-type and F-type) ATPase (**B**) of CM and ICM vesicles prepared and purified from *Anacystis nidulans* (panels *A*) and *Synechocystis* PCC 6803 (panels *B*) harvested at different times of photoautotrophic growth in the absence or presence of 0.4 M NaCl. (Peschek et al. 1994a; Neisser et al. 1994)

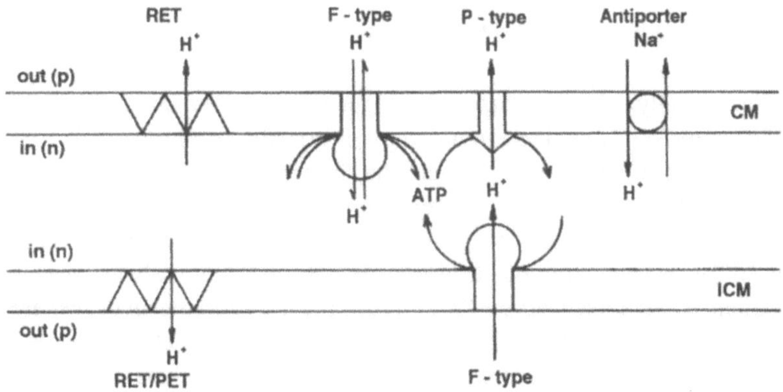

Fig. 8.13. Model of plasma-membrane energization in a blue-green alga either directly by RET in CM through a CM-bound H^+-translocating respiratory chain or by a CM-bound H^+--ATPase at the expense of ATP made by RET (or PET) in the ICM. (Peschek et al. 1994a)

karyotes. In evolution, eukaryotes have put much more effort (if teleology might be allowed for a moment) into getting environmentally adapted in terms of species diversification rather than into simple metabolic adaptation as has been described at length in the previous section for the basic bioenergetic stress adaptation mechanisms in blue-green algae. This does not mean, however, that these mechanisms are not basically valid in eukaryotes also, just like, e.g. all the biochemical mechanisms eukaryotes might be proud of had been elaborated (invented) by prokaryotes already. This is most clearly expressed in the famous unitary principle of biochemistry (and, even more stringent, of bioenergetics) first put forward by Kluyver and Donker (1926). The differences are to be found at the level of a different complexity of interaction of the individual mechanisms at best, rather than the mechanisms themselves. Thus, as we have seen that a prokaryotic alga can energize its plasma membrane directly by (respiratory) electron transport in this membrane while in eukaryotic algae, for likewise necessary plasma membrane energization, the detour via organellar ATP and plasmalemma-bound H^+-ATPases must be taken which, on the other hand, can be found as an alternative also in prokaryotic algae (see Fig. 8.2).

8.7.4 Dinitrogen Fixation as a Physiological Stress

A final trait of bioenergetic stress adaptation, again in cyanobacteria, might be exemplified in the context of dinitrogen fixation. Biological N_2-fixation [Eq. (7); see, e.g. Van Baalen 1987; Gallon and Chaplin 1988) is a process absolutely restricted to prokaryotes, including blue-green algae.

$$N_2 + 8\,[H] + 2\,H^+ + 16\,ATP = 2\,NH_3 + H_2 + 16\,ADP. \qquad (7)$$

The reaction is catalyzed by the enzyme nitrogenase, a molybdenum-containing FeS-enzyme of surprisingly ancient evolutionary origin (Broda and Peschek 1980, 1983). [H] designates metabolic hydrogen, the immediate electron donor to nitrogenase being reduced ferredoxin. Six electrons are transferred to N_2 for the fixation of 2 NH_3 (physiologically NH_4^+, of course), two additional electrons are inevitably used for the reduction of (at least) 2 H^+. H_2 production always accompanies biological N_2 fixation, and an "ATP-requiring H_2-evolving hydrogenase" has been said to be the inborn flaw of nitrogenase diverting substantial amounts of reducing power to the futile evolution of hydrogen which only thereafter will be recycled (oxidatively refixed) by so-called uptake hydrogenases in the membranes of practically all N_2-fixing bacteria (see Adams et al. 1981; Bothe et al. 1999). The requirement for rather huge amounts of ATP energy to power the thermodynamically highly exergonic reduction of dinitrogen to ammonia (cf. the Haber-Bosch process widely practiced in chemical technology for the production of nitrogen fertilizers) seems embarrassing, notably since any explanation on the grounds of high activation energy needed to split the triple-bond of N_2 must remain physicochemically incorrect (Broda and Peschek 1980). The severe energy requirement of biological N_2-fixation, though still unexplained, is a matter of fact and thus also growth at the expense of N_2 as the nitrogen source must be regarded as a stress situation. Correspondingly, it was found that generation times of photoautotrophically growing *Anabaena* spp. were 3.8 h on NH_4Cl, 4.9 h on $NaNO_3$ but 5.2 h on N_2 (Van Baalen 1987). Quite in accordance with what has been said here about the role of respiration as a bioenergetic stress defence mechanism in blue-green algae, and corresponding to the findings in other laboratories (Winkenbach and Wolk 1973), we have found that rates of respiration went up steeply when the species grew on atmospheric N_2 as the sole N source (Wastyn et al. 1988; Peschek et al. 1991). Interestingly, filamentous species where N_2-fixation together with accompanying enhanced respiration is restricted to the heterocysts, prefer ICM as the site of respiratory electron transport (Wastyn et al. 1988; Peschek et al. 1991) while N_2-fixing unicellular species which, in the absence of specialized heterocysts devoid of O_2-evolving PS II, anyway must take special precautions in order to avoid O_2 damage to their nitrogenase as both processes occur in one and the same prokaryotic cell, prefer CM (Peschek et al. 1991). Even the circadian separation of O_2-evolution (day) and N_2-fixation (night), as is commonly envisaged for diazotrophic chroococcal and non-heterocystous filamentous cyanobacteria (Pearson and Howsley 1980; Gallon and Stal 1992), with its disparity of (photosynthetic) O_2 evolution (= CO_2 assimilation) during day time and N_2-fixation in the night in the absence of continuous (photosynthetic!) regeneration of the necessary N-acceptor(s), represents a kind of unnatural physiological uncoupling which, in the long run, must severely depress biomass production by diazotrophic unicellular and non-heterocystous filamentous cyanobacteria such as *Trichodesmium* spp. Very slow growth together with consequently slow photosynthetic O_2-production, which can easily be counteracted and neutralized by rapid respiratory O_2 uptake (respiratory protection; see Yates and Jones 1974), obviously helps to solve the problem (see Peschek et al. 1991; Bergman et al. 1993).

8.8 Genetic Responses: Heat-Shock and Other Stress Proteins

For cyanobacteria genetically triggered primary adaptations, e.g. of the photosynthetic apparatus to light (both light quality, relating to chromatic adaptation, and intensity, involving the extent of pigment synthesis), of the buoyant density (via synthesis of gas vesicles), of assimilatory pathways of inorganic and organic carbon including the glycogen reserve material, of the fixation and assimilation of nitrogen, of sulfur assimilation and starvation, of the resistance to metal ions, nutrient starvation in general, and to heat shock and other stresses have recently been discussed in great detail, though without special reference to bioenergetic mechanisms (Tandeau de Marsac and Houmard 1993). Though in general poorly documented for algae, there exists an increasing literature on the synthesis of stress proteins in blue-green algae following exposure to high temperature (Borbely et al. 1985; Borbely and Suranyi 1988), high salinity (Reed et al. 1986; Reed 1988; Reed and Stewart 1988; Apte and Bhagawat 1989; Apte and Haselkorn 1990), desiccation (Scherer and Potts 1989), ultraviolet light (Nicholson et al. 1987), light deprivation (Borbely et al.1990), nutrient starvation (Green et al. 1989; Duke and Allen 1990), phage infection, etc. The two best-known cyanobacterial heat-shock proteins are the molecular chaperones GroEL and GroES, members of a class of chaperonins found in prokaryotes, mitochondria and chloroplasts (Webb et al. 1990; Lehel et al. 1992; also see Neidhardt et al. 1987). The important role that heat-shock proteins play for the protection of eukaryotic cells against a variety of stress types has been elegantly reviewed by Lindquist and Craig (1988) (*Saccharomyces* spp.) and by Ellis (1990) (plants).

In view of the wealth of the reviewed literature available on the genetic aspects of stress protein synthesis and regulation available, and particularly in view of the focus of the present chapter, which rests on the bioenergetic stress defence mechanisms rather than their more descriptive genetic aspects, it seems unnecessary to reiterate all these details here. Be it also finally noted that, as usual, the countless details of descriptive genetic investigations, as important and necessary as they might be, naturally can at best only trigger future research aiming at the mechanisms of the cellular metabolic network while the bioenergetic concept, as detailed in this chapter, whenever applicable and accessible, leads directly to a comprehensible biological explanation.

Nonstandard abbreviations used:

RET Respiratory electron transport
PET Photosynthetic electron transport
CM Cytoplasmic or plasma membrane (plasmalemma)
ICM Intracytoplasmatic or thylakoid membrane
LPT Thermotropic lipid phase transition temperature
COX Cytochrome-c oxidase
RC (Photosynthetic) reaction centre

References

Adams MWW, Mortenson LE, Chen J-S (1981) Hydrogenase. Biochim Biophys Acta 594:105-176

Anemüller S, Schmidt CL, Pacheco I, Schäfer G, Teixera M (1994) A cytochrome aa_3-type quinol oxidase from *Desulfurolobus ambivalens*, the most acidophilic archaeon. FEMS Microbiol Lett 117:275-280

Apte SK, Bhagawat AA (1989) Salinity stress-induced proteins in two nitrogen-fixing *Anabaena* strains differentially tolerant to salt. J Bacteriol 171:909-915

Apte SK, Haselkorn R (1990) Cloning of salinity stress-induced genes from the salt-tolerant nitrogen-fixing cyanobacterium *Anabaena torulosa*. Plant Mol Biol 15:723-733

Armond PA, Staehelin LA (1979) Lateral and vertical displacement of integral membrane proteins during lipid phase transition in *Anacystis nidulans*. Proc Natl Acad Sci USA 76:1901-1905

Arnon DI (1960) The role of light in photosynthesis. Sci Am 203:104-118

Asada K (1992) Ascorbate peroxidase — a hydrogen-scavenging enzyme in plants. Physiol Plant 85:235-241

Asada K, Takahashi M (1987) Production and scavenging of active oxygen in photosynthesis. In: Kyle DJ, Arntezen CJ, Osmond CB (eds) Photoinhibition. Elsevier, Amsterdam, pp 227-287

Asada K, Kanematsu S, Uchida K (1977) Superoxide dismutase in photosynthetic organisms. Arch Biochem Biophys 179:243-256

Asada K, Miyake C, Sano S, Amako K (1993) Scavenging of hydrogen peroxide in photosynthetic organisms-from catalase to ascorbate peroxidase. In: Welinder KG, Rasmussen SK, Penel C, Greppin H (eds) Plant peroxidases: biochemistry and physiology. University of Geneva Press, Geneva, pp 242-250

Asada K, Yoshikawa K, Takahashi M, Maeda Y, Enmanji K (1975) Superoxide dismutases from a blue-green alga *Plectonema boryanum*. J Biol Chem 250:2801-2807

Babcock GT, Wikström M (1992) Oxygen activation and the conservation of energy in cell respiration. Nature 356:301-308

Balch WE, Fox GE, Magrum LJ, Woese CR, Wolfe RS (1979) Methanogens: reevaluation of a unique biological group. Microbiol Rev 43: 260-296

Baldry CW, Walker DA, Bucke C (1966) The temperature dependence of algal respiration and photosynthesis. Biochem J 101:642-649

Baltscheffsky H (1974) On the evolution of electron transport in photosynthesis and respiration. In: Dose K, Fox SW, Deborin GA, Perovskaya TE (eds) Dynamics of energy-transducing membranes. Elsevier, Amsterdam, pp 21-27

Bannister JV, Bannister WH, Rotilio G (1987) Aspects of the structure, function, and application of superoxide dismutase. CRC Crit Rev Biochem 22:118-180

Barghoorn ES, Schopf JW (1965) Microorganisms from the late Precambrian of Central Australia. Science 150:337-339

Barghoorn ES, Schopf JW (1966) Microorganisms three billion years old from the Precambrian of South Africa. Science 152:758-763

Bergman B, Siddiqui PJA, Carpenter EJ, Peschek GA (1993) Cytochrome oxidase: subcellular distribution and relationship to nitrogenase expression in the nonheterocystous marine cyanobacterium *Trichodesmium thiebautii*. Appl Environ Microbiol 59:3239-3244

Berkner LV, Marshall LC (1967) The rise of oxygen in the earth's atmosphere with notes on the Martian atmosphere. Adv Geophys 12:309-331

Blankenship RE (1992) Origin and early evolution of photosynthesis. Photosynth Res 33:91-111

Bligh EG, Dyer WJ (1959) A rapid method of total lipid extraction and purification. Can J Biochem Physiol 37: 911-917c

Borbely G, Suranyi G (1988) Cyanobacterial heat shock proteins and stress responses. Meth Enzymol 167:622-629

Borbely G, Suranyi G, Korcz A, Palfi Z (1985) Effect of heat shock protein synthesis in the cyanobacterium *Synechococcus* sp. strain PCC 6301. J Bacteriol 161:1125-1130

Borbely G, Suranyi G, Kos P (1990) Stress responses of cyanobacteria and the pleiotropic effects of light deprivation. FEMS Microbiol Ecol 74:141-152

Bothe H, Boison G, Schmitz O (1999) Hydrogenases in cyanobacteria. In: Peschek GA, Loeffelhardt W, Schmetterer G (eds) The phototrophic prokaryotes. Kluwer Academic/Plenum Publishers, New York, pp 589-601

Brand JJ (1979) Spectral changes in membrane fragments and artificial liposomes of *Anacystis nidulans* induced by chilling. Arch Biochem Biophys 193:385-391

Broda E (1975) The evolution of the bioenergetic processes. Pergamon Press, Oxford

Broda E (1977a) The evolution of photosynthesis. Precambrian Res 4:117-132

Broda E (1977b) The length of the transition period from the reducing to the neutral biosphere. Origins of Life 8:87-90

Broda E, Peschek GA (1979) Did respiration or photosynthesis come first? J Theor Biol 81:201-212

Broda E, Peschek GA (1980) Evolutionary considerations on the thermodynamics of nitrogen fixation. BioSystems 13:47-56

Broda E, Peschek GA (1983) Nitrogen fixation as evidence for the reducing nature of the early biosphere. BioSystems 16:1-8

Buetow DE (1968) The biology of *Euglena*. Academic Press, New York

Büttner M, Xie DL, Nelson H, Pinther W, Hauska G, Nelson N (1992) Photosynthetic reaction center genes in green sulfur bacteria and in photosystem I are related. Proc Natl Acad Sci USA 89:8135-8139

Cadenas E (1989) Biochemistry of oxygen toxicity. Annu Rev Biochem 58:79-110

Canini A, Civitareale P, Marini S, Grilli Caiola MG, Rotilio G (1992) Purification of iron superoxide dismutatses from the cyanobacterium *Anabaena cylindrica* Lemm. And the localization of the enzyme in heterocysts by immunogold labelling. Planta 187:438-444

Carafoli E, Scarpa A (eds.) (1982) Transport ATPases. Ann NY Acad Sci, vol. 402, 604pp

Castenholz RW (1988) Thermophilic cyanobacteria: special problems. Methods Enzymol 167: 98-100

Chatton E (1937) Titres et Trouvaux Scientifiques. Sete, France

Chisholm SW, Olson RJ, Zettler ER, Goericke R, Waterbury JB, Welschmeyer NA (1988) A novel free-living prochlorophyte abundant in the oceanic euphotic zone. Nature 334:340-343

Danson MJ, Hough DW, Lunt GG (1992) The Archaebacteria: biochemistry and biotechnology. Biochemical Society Symposium 58, Portland Press, London

Desikachary TV (1959) Cyanophyta. Indian Council of Agricultural Research, New Delhi

Devault D (1980) Quantum mechanical tunneling in biological systems. Q Rev Biophys 13:387-564

Dimroth P (1987) Sodium ion transport decarboxylases and other aspects of sodium ion cycling in bacteria. Microbiol Rev 51:320-340

Drobner E, Huber H, Wächtershäuser G, Rose D, Stetter KO (1990) Pyrite formation linked with hydrogen evolution under anaerobic conditions. Nature 346:742-744

Duke CS, Allen MM (1990) Effect of nitrogen starvation on polypeptide composition, ribulose-1,5-bisphosphate carboxylase/oxygenase, and thylakoid carotenoprotein content of *Synechocystis* sp. strain PCC 6803. Plant Physiol 94:752-759

Dworsky A, Meyer B, Regelsberger G, Fromwald S, Peschek GA (1995) Functional and immunological characterization of both mitochondria-like and chloroplast-like electron/proton transport proteins in isolated and purified cyanobacterial membranes. Bioelectrochem Bioenerg 88:35-43

Ellis RJ (1990) Molecular chaperones: the plant connection. Science 250:954-958

Elstner EF, Pils I (1979) Ethane formation and chlorophyll bleaching in DCMU-treated *E. gracilis* cells and isolated spinach chloroplast lamellae. Z Naturforsch 34c:1040-1043

Erber WWA, Nitschmann WH, Muchl R, Peschek GA (1986) Endogenous energy supply to the plasma membrane of dark aerobic cyanobacterium *Anacystis nidulans*: ATPase-independent efflux of H^+ and Na^+ from respiring cells. Arch Biochem Biophys 247:28-39

Falkner G, Wagner F, Falkner R (1996) The bioenergetic coordination of a complex biological system is revealed by its adaptation to changing environmental conditions. Acta Biotheor 44:283-299

Förster T (1948) Zwischenmolekulare Energiewanderung und Fluoreszenz. Ann Phys 2:55-75

Fritsch FE (1945) The structure and reproduction of the algae, vol II. Cambridge University Press, Cambridge

Fromme P, Witt, HT, Schubert WD, Klukas O, Saenger W, Krauß N (1996) Structure of photosystem I at 4.5 Å resolution: a short review including evolutionary aspects. Biochim Biophys Acta 1275:76-83

Fromwald S, Dworsky A, Peschek GA (1995) F-type and P-type ATPases in cyanobacteria: implications for energy conservation and utilization. In: Mathis P (ed) Photosynthesis: from light to biosphere, vol III. Kluwer, Dordrecht, pp 47-50

Fry IV, Peschek GA (1988) Electron paramagnetic resonance-detectable Cu^{2+} in *Synechococcus* 6301 and 6311: aa_3-type cytochrome-c oidase of cytoplasmic membrane. Methods Enzymol 167:450-458

Fry IV, Huflejt M. Erber WWA, Peschek GA, Packer L (1986) The role of respiration during adaptation of the freshwater cyanobacterium *Synechococcus* 6311 to salinity. Arch Biochem Biophys 244:686-691

Furtado D, Williams WP, Brain APR, Quinn PJ (1979) Phase separations in membranes of *Anacystis nidulans* grown at different temperatures. Biochim Biophys Acta 555:352-357

Gallon JR, Chaplin AE (1988) Nitrogen fixation. In: Rogers LJ, Gallon JR (eds) Biochemistry of the algae and cyanobacteria. Clarendon Press, Oxford, pp 147-173

Gallon JR, Stal LJ (1992) N_2 fixation in the nonheterocystous cyanobacteria: an overview. In: Carpenter EJ, Capone DG, Rueter JG (eds) Marine pelagic cyanobacteria: *Trichodesmium* and other diazotrophs. Kluwer, Dordrecht, pp 115-140

Geitler L (1925) Synoptische Darstellung der Cyanophyceen in morphologischer und systematischer Hinsicht. Beih Bot Zentralbl II 41:163-294

Gest H (1972) Energy conversion and generation of reducing power in bacterial photosynthesis. Adv Microb Physiol 7:243-282

Gilbert DL (1981) Significance of oxygen on Earth. In: Gilbert DL (ed) Oxygen and living processes. An interdisciplinary approach. Springer, Berlin Heidelberg New York, pp 73-101

Glover HE, Prezelin BB, Campbell L, Wyman M, Garside C (1988) A nitrate-dependent *Synechococcus* bloom in surface Sargasso Sea water. Nature 331:161-163

Gombos Z, Tanaka Y, Zsiros O, Varkonyi ZS, Murata N (1999) Lipids of photosyntheticmembranes in cyanobacteria — roles in protection against low-temperature stress. In: Peschek GA, Loeffelhardt W, Schmetterer G (eds) The phototrophic prokaryotes. Kluwer Academic / Plenum Publishers, New York, pp 647-655
Gray MW, Doolittle WF (1982) Has the endosymbiont hypothesis been proven? Microbiol Rev 46:1-42
Green LS, Laudenbach DE, Grossman AR (1989) A region of a cyanobacterial genome required for sulfate transport. Proc Natl Acad Sci USA 86:1949-1953
Grilli Caiola M, Canini A, Galiazzo F, Rotilio G (1991) Superoxide dismutase in vegetative cells, heterocysts and akinetes of *Anabaena cylindrica* Lemm. FEMS Microbiol Lett 80:161-166
Haldane JBS (1954) The origins of life. New Biol 16:12-27
Haldane JBS (1970) The origins of life. In: Cloud P (ed) Adventures in earth history. WH Freeman, San Francisco, pp 377-384
Hicks DB, Krulwich TE (1995) The respiratory chain of alkaliphilic bacteria. Biochim Biophys Acta 1229:303-314
Higgins IJ, Best DJ, Hammond RC, Scott D (1981) Methane-oxidizing microorganisms. Microbiol Rev 45:556-590
Hill R, Bendall F (1960) Function of two cytochrome components in chloroplasts. Nature 186: 136-137
Hippeli S, Elstner EF (1999) Transition metal ion-catalyzed oxygen activation during pathogenic processes. FEBS Lett 443:1-7
Hishiyama H, Hayashi T, Watanabe T, Murata N (1994) Photosynthetic oxygen evolution is stabilized by cytochrome c_{550} against heat inactivation in *Synechococcus* sp. PCC 7002. Plant Physiol 105:1313-1319
Hoch G, Kok B (1961) Photosynthesis. Annu Rev Plant Physiol 12: 155-194
Hochman A, Goldberg I (1991) Purification and Characterization of catalase-peroxidase and a typical catalase from the bacterium *Klebsiella pneumoniae*. Biochim Biophys Acta 1077:299-307
Hoffmann P (1998) Oxygenic photosynthesis — a photon-driven hydrogen generator: the energetic/entropic basis of life. Photosynthetica 35:1-11
Holm-Hansen O (1968) Ecology, physiology and biochemistry of blue-green algae. Annu Rev Microbiol 22:47-70
Hoshina S (1981) Temperature dependence of optical properties of chlorophyll *a* incorporated into phosphatidyl choline liposomes. Biochim Biophys Acta 638:334-340
Iwata S, Ostermeier C, Ludwig B, Michel H (1995) Structure at 2.8 Å resolution of cytochrome-c oxidase from *Paracoccus denitrificans*. Nature 376:660-669
Jackson JB, Crofts AR (1968) Energy-linked reduction of nicotinamide adenine dinucleotides in cells of *Rhodospirillum rubrum*. Biochem Biophys Res Commun 32:909-915
Kanervo E, Aro E-M, Murata N (1995) Low unsaturation level of thylakoid membrane lipids limits turnover of the D1 protein of photosystem II at high irradiance. FEBS Lett 364:239-242
Kates M (1992) Archaebacterial lipids: structure, biosynthesis and function. In: Danson MJ, Hough DW, Lunt GG (eds) The Archaebacteria: biochemistry and biotechnology. Portland Press, London
Keister DL, Yike NJ (1967) Energy-linked reactions in photosynthetic bacteria. Arch Biochem Biophys 121:415-422
Kelly DP (1971) Autotrophy: concepts of lithotrophic bacteria and their organic metabolism. Annu Rev Microbiol 25:177-210
Kluyver AJ, Donker HJL (1926) Das Einheitsprinzip der Biochemie. Chem d Zelle und Gew 13:124-130
Knoll AH (1992) The early evolution of eukaryotes: a geological perspective. Science 256:622-627

Koika H, Katoh S (1979) Heat stabilities of cytochromes and ferredoxin isolated from a thermophilic blue-green alga, *Synechococcus* sp. Plant Cell Physiol 20:1157-1161
Laidler KJ (1965) Chemical kinetics, 2nd edn. McGraw-Hill, New York
Lanyi JK (1978) Light energy conversion in *Halobacterium halobium*. Microbiol Rev 42:682-699
Lee AG (1975) Segregation of chlorophyll a incorporated into lipid bilayers. Biochemistry 14:4397-4402
Lehel C, Wada H, Kovacs E, Török Z, Gombos Z, Horvath I, Murata N, Vigh L (1992) Heat shock protein synthesis of the cyanobacterium *Synechocystis* PCC 6803: purification of the GroEL-related chaperonin. Plant Mol Biol 18:327-336
Lindquist S, Craig EA (1988) Stress tolerance in *Saccharomyces cerevisiae*. Annu Rev Genet 22:631-677
Löppert H-G, Broda E (1973) Salt respiration in *Chlorella fusca*. Z Allg Mikrobiol 13:499-502
Lundegardh H (1961) Salt-stimulated respiration in plant tissues. Nature 192:243-245
Marcus RA (1956) On the theory of oxidation-reduction reactions involving electron transfer. J Chem Phys 24:966-978
Marcus RA, Sutin N (1985) Electron transfer in chemistry and biology. Biochim Biophys Acta 811:265-322
Margulis L (1970) Origin of eukaryotic cells. Yale University Press, New Haven, Connecticut
Margulis L (1996) Archaeal-eubacterial mergers in the origin of eukaryota: phylogenetic classification of life. Proc Natl Acad Sci USA 93:1071-1076
Melchior DL (1982) Lipid phase transition and regulation of membrane fluidity in prokaryotes. Curr Top Membr Transp 17:263-316
Melchior DL, Steim JM (1976) Thermotropic transitions in biomembranes. Annu Rev Biophys Bioenerg 5:205-238
Mereschkowsky C (1905) Über Natur und Ursprung der Chromatophoren im Pflanzenreiche. Biol Zentralbl 25:593-604
Miller SL, Orgel LE (1974) The origins of life on Earth. Prentice Hall, Englewood Cliffs
Miller SL, Urey HC (1959) Organic compound synthesis on the primitive earth. Science 130:245-250
Miyake C, Michihata F, Asada K (1991) Scavenging of hydrogen peroxide in prokaryotic and eukaryotic algae: acquisition of ascorbate peroxidase duriung the evolution of cyanobacteria. Plant Cell Physiol 32:33-43
Molitor V, Peschek GA (1986) Respiratory electron transport in plasma and thylakoid membrane preparations from the cyanobacterium *Anacystis nidulans*. FEBS Lett 195:145-150
Molitor V, Trnka M, Peschek GA (1987) Isolated and purified plasma and thylakoid membranes of the cyanobacterium *Anacystis nidulans* contain immunologically cross-reactive aa_3-type cytochrome oxidase. Curr Microbiol 14:263-268
Molitor V, Trnka M, Erber W, Steffan I, Riviere M-E, Arrio B, Springer-Lederer H, Peschek GA (1990a) Impact of salt adaptation on esterified fatty acids and cytochrome oxidase in plasma and thylakoid membranes from the cyanobacterium *Anacystis nidulans*. Arch Microbiol 154:112-119
Molitor V, Kuntner O, Sleytr UB, Peschek GA (1990b) The ultrastructure of plasma and thylakoid membrane vesicles from the freshwater cyanobacterium *Anacystis nidulans* adapted to salinity. Protoplasma 157:112-119
Möller IM, Crane FL (1990) Redox processes in the plasma membrane. In: Larsson C, Möller IM (eds) The plant plasma membrane. Springer, Berlin Heidelberg New York, pp 93-126
Möller IM, Askerlund P, Widell S (1992) Electron transport constituents in the plant plasma membrane. In: Crane FL, Morre DJ, Loew HE (eds) Oxidoreduction at the plasma membrane: relation to growth and transport, vol II, Plants. CRC Press, Boca Raton, pp 35-56
Morris JG (1975) The physiology of obligate anaerobiosis. Adv Microb Physiol 12:169-246

Moser CC, Keske JM, Warncke K, Farid RS, Dutton PL (1992) Nature of biological electron transfer. Nature 355:796-802

Moser CC, Page CC, Chen X, Dutton PL (1999) Intraprotein electron transfer. In: Peschek GA, Loeffelhardt W, Schmetterer G (eds) The phototrophic prokaryotes. Kluwer Academic/Plenum Publishers, New York, pp 105-111

Mulkidjanian AY, Cherepanov DA, Haumann M, Junge W (1996) Photosystem II of green plants: topology of core pigments and redox cofactors as inferred from electrochromic difference spectra. Biochemistry 35:3093-3107

Murata N, Fork DC (1975) Temperature dependence of chlorophyll a fluorescence in relation to the physical phase of membrane lipids in algae and higher plants. Plant Physiol 56:791-796

Murata N, Nishida I (1987) Lipids of blue-green algae (cyanobacteria). In: Stumpf PK (ed) The biochemistry of plants, vol 9. Academic Press, Orlando, Florida, pp 315-347

Murata N, Wada H (1995) Acyl-lipid desaturases and their importance in the tolerance and acclimatization to cold of cyanobacteria. Biochem J 308:1-8

Murata N, Troughton JH, Fork DC (1975) Relationships between the transition of the physical phase of membrane lipids and photosynthetic parameters in *Anacystis nidulans* and lettuce and spinach chloroplasts. Plant Physiol 56:508-517

Mutsuda M, Ishikawa T, Takeda T, Shigeoka S (1996) The catalase-peroxidase of *Synechococcus* PCC7942: purification, nucleotide sequence analysis and expression in *Escherichia coli*. Biochem J 316:251-257

Nedbal L, Samson G, Whitmarsh J (1992) Redox state of a one-electron component controls the rate of photoinhibition of photosystem II. Proc Natl Acad Sci USA 89:7929-7933

Neidhardt FC, Van Bogelen FC (1987) Heat shock response. In: Neidhardt FC, Van Bogelen (eds) *Escherichia coli* and *Salmonella typhimurium*. Cellular and molecular biology. ASM, Washington, DC, pp 1334-1345

Neisser A, Fromwald S, Schmatzberger A, Peschek GA (1994) Immunological and functional localization of both F-type and P-type ATPases in cyanobacterial plasma membranes. Biochem Biophys Res Commun 200:884-892

Nicholls DG, Ferguson SJ (1992) Bioenergetics 2. Academic Press, London

Nicholson P, Osborn RW, Howe CJ (1987) Induction of protein synthesis in response to ultraviolet light, nalidixic acid and heat shock in the cyanobacterium *Phormidium laminosum*. FEBS Lett 221:110-114

Nitschke W, Rutherford AW (1991) Photosynthetic reaction centers: variations on a common structural theme? Trends Biochem Sci 16:241-245

Nitschmann WH, Peschek GA (1985) Modes of proton translocation across the cell membrane of respiring cyanobacteria. Arch Microbiol 141:330-336

Nitschmann WH, Peschek GA (1986) Oxidative phosphorylation and energy buffering in cyanobacteria. J Bacteriol 168:1205-1211

Nitschmann WH, Schmetterer G, Muchl R, Peschek GA (1982) Active sodium extrusion reduces bet efficiencies of oxidative phosphorylation in the strictly photoautotrophic cyanobacterium *Anacystis nidulans*. Biochim Biophys Acta 682:293-296

Obinger C, Regelsberger G, Strasser G, Burner U, Peschek GA (1997) Purification and characterization of a homodimeric catalase-peroxidase from the cyanobacterium *Anacystis nidulans*. Biochem Biophys Res Commun 235:545-552

Obinger C, Regelsberger G, Pircher A, Sevcik-Klöckler A, Strasser G, Peschek GA (1998) Scavenging of superoxide and hydrogen peroxide in blue-green algae (cyanobacteria). Physiol Plant 104:693-698

Obinger C, Regelsberger G, Pircher A, Sevcik-Klöckler A, Strasser G, Peschek GA (1999) Hydrogen peroxide removal in cyanobacteria. Characterization of a catalase-peroxidase from *Anacystis nidulans*. In: Peschek GA, Loeffelhardt W, Schmetterer G (eds) The phototrophic prokaryotes. Kluwer Academic/Plenum Publishers, New York, pp 719-731

Oesterhelt D, Gottschlich R, Hartmann R, Michel H, Wagner G (1977) Light energy conversion in halobacteria. In: Haddock BA, Hamilton WA (eds) Microbial energetics. Cambridge University Press, London, pp 333-349

Olson JM (1970) The evolution of photosynthesis. Science 168:438-446

Ono T, Murata N (1977) Temperature-dependence of the delayed fluorescence of chlorophyll a in blue-green algae. Biochim Biophys Acta 502:477-485

Ono T, Murata N (1979) Temperature dependence of the photosynthetic activities in the thylakoid membranes from the blue-green alga *Anacystis nidulans*. Biochim Biophys Acta 545:69-76

Oparin AI (1924) The origin of life (Translated in J D Bernal 1967). Plenum Press, London

Pearce J, Leach CK, Carr NG (1969) The incomplete tricarboxylic acid cycle in the blue-green alga *Anabaena variabilis*. J Gen Microbiol 55:371-378

Pearson HW, Howsley R (1980) Concomitant photoautotrophic growth and nitrogenase activity by cyanobacterium *Plectonema boryanum* in continuous culture. Nature 288: 263-265

Peschek GA (1981a) Phylogeny of photosynthesis and the evolution of electron transport: the bioenergetic backbone. Photosynthetica 15:543-555

Peschek GA (1981b) Occurrence of cytochrome aa_3 in *Anacystis nidulans*. Biochim Biophys Acta 635:470-475

Peschek GA (1981c) Spectral properties of a cyanobacterial cytochrome c oxidase: evidence for cytochrome aa_3. Biochem Biophys Res Commun 98:72-79

Peschek GA (1996a) Cytochrome oxidase and the *cta* operon of cyanobacteria. Biochim Biophys Acta 1275:27-32

Peschek GA (1996b) Structure-function relationships in the dual-function photosynthetic-respiratory electron-transport assembly of cyanobacteria (blue-green algae). Biochem Soc Trans 24:729-733

Peschek GA (1999) Photosynthesis and respiration of cyanobacteria. In: Peschek GA, Loeffelhardt W, Schmetterer G (eds) The phototrophic prokaryotes. Kluwer Academic/ Plenum Publishers, New York, pp 201-209

Peschek GA, Schmetterer G (1978) Reversible photooxidative loss of pigments and of intracytoplasmic membranes in the blue-green alga *Anacystis nidulans*. FEMS Microbiol Lett 3:295-297

Peschek GA, Muchl R, Schmetterer G (1981) Temperature dependence of horse heart cytochrome c oxidation by membrane preparations of *Anacystis nidulans*: characterization of two distinct Arrhenius discontinuities. Curr Microbiol 6:233-237

Peschek GA, Muchl R, Kienzl PF, Schmetterer G (1982) Characteristic temperature dependences of respiratory and photosynthetic electron-transport activities in membrane preparations from *Anacystis nidulans* grown at different temperatures. Biochim Biophys Acta 679:35-43

Peschek GA, Czerny T, Schmetterer G, Nitschmann WH (1985) Transmembrane proton electrochemical gradients in dark aerobic and anaerobic cells of the cyanobacterium (blue-green alga) *Anacystis nidulans*. Plant Physiol 79:278-284

Peschek GA, Hinterstoisser B, Riedler M, Muchl R, Nitschmann WH (1986) Exogenous energy supply to the plasma membrane of dark aerobic cyanobacterium *Anacystis nidulans*: thermodynamic and kinetic characterization of the ATP synthesis effected by an artificial proton motive force. Arch Biochem Biophys 247:40-48

Peschek GA, Molitor V, Trnka M, Wastyn M, Erber W (1988a) Characterization of cytochrome-c oxidase in isolated and purified plasma and thylakoid membranes from cyanobacteria. Methods Enzymol 167:437-449

Peschek GA, Nitschmann WH, Czerny T (1988b) Respiratory proton extrusion and plasma membrane energization. Methods Enzymol 167:361-379

Peschek GA, Wastyn M, Trnka M, Molitor V, Fry IV, Packer L (1989a) Characterization of the cytochrome-c oxidase in isolated and purified plasma membranes from the cyanobacterium *Anacystis nidulans*. Biochemistry 28:3057-3063

Peschek GA, Wastyn M, Molitor V, Kraushaar H, Obinger C, Matthijs HCP (1989b) Self-contained or accessory respiration in the phototrophic cyanobacteria (blue-green algae)? In: Kotyk A, Skoda J, Paces V, Kosta V (eds) Highlights of modern biochemistry, vol 1. VSP Publishers, Zeist, The Netherlands, pp 893-902

Peschek GA, Hinterstoisser B, Wastyn M, Kuntner O, Pineau B, Missbichler A, Lang J (1989c) Chlorophyll precursors in the plasma membrane of a cyanobacterium, *Anacystis nidulans*. J Biol Chem 264:11827-11832

Peschek GA, Villgratter K, Wastyn M (1991) Respiratory protection of the nitrogenase in dinitrogen-fixing cyanobacteria. Plant Soil 137:17-24

Peschek GA, Niederhauser H, Obinger C (1992) Cyanobacterial cytochrome oxidase: a candidate for the primordial enzyme? EBEC Short Rep, Elsevier, Amsterdam, vol 7, 48 pp

Peschek GA, Obinger C, Fromwald S, Bergman B (1994a) Correlation between immuno-gold labels and activities of the cytochrome-c oxidase (aa_3-type) in membranes of salt stressed cyanobacteria. FEMS Microbiol Lett 124:431-438

Peschek GA, Obinger C, Sherman DM, Sherman LA (1994b) Immunochemical localization of the cytochrome-c oxidase in a cyanobacterium, *Synechococcus* PCC 7942 (*Anacystis nidulans*) Biochim Biophys Acta 1187:369-372

Pringsheim EG (1963) Farblose Algen. Ein Beitrag zur Evolutionsforschung. VED Gustav Fischer, Jena

Pringsheim EG, Wiessner W (1961) Colourless blue-green algae. Arch Mikrobiol 40:231-241

Purschke WG, Schmidt CL, Petersen A, Schäfer G (1997) The terminal quinol oxidase of the hyperthermophilic archaeon *Acidianus ambivalens* exhibits a novel subunit structure and gene organization. J Bacteriol 179:1344-1353

Quinn PJ (1988) Thermotropic lipid phase segregation in biological membranes. In: Aloia RC, Curtain CC, Gordon LM (eds) Physiological regulation of membrane fluidity, vol 3. Alan R Liss, New York

Rabinowitch E, Govindjee (1969) Photosynthesis. John Wiley, New York

Raven JA, Smith FA (1976) The evolution of chemiosmotic energy coupling. J Theor Biol 57:301-312

Reed RH (1988) Osmotioc adjustment: organic solutes. Methods Enzymol 167:528-534

Reed RH, Stewart WDP (1988) The response of cyanobacteria to salt stress. In: Rogers LJ, Gallon JR (eds) Biochemistry of the algae and cyanobacteria. Oxford University Press, Oxford, pp 217-231

Reed RH, Borowitzka LJ, Mackay MA, Chudek JA, Fister R, Warr SRC, Moore DH, Stewart WDP (1986) Organic solute accumulation in osmotically stressed cyanobacteria. FEMS Microbiol Rev 39:51-56

Regelsberger G, Obinger C, Zoder R, Altmann F, Peschek GA (1999) Purification and characterization of a hydroperoxidase from the cyanobacterium *Synechocystis* PCC6803: identification of its gene by peptide mass mapping using matrix assisted laser desorption ionization time-of-flight mass spectrometry. FEMS Lett 170:1-12

Renger G (1997) Photosynthetic water cleavage. In: Gräber P, Milazzo G (eds) Treatise on bioelectrochemistry, vol. 2: bioenergetics. Birkhäuser, Basel, pp 310-358

Renger G (1999) Studies on structure and mechanism of photosynthetic water oxidation. In: Peschek GA, Loeffelhardt W, Schmetterer G (eds) The phototrophic prokaryotes. Kluwer Academic/Plenum Publishers, New York, pp 35-50

Renthal R (1992) Bacteriorhodopsin. In: Ernster L (ed) Molecular mechanisms in bioenergetics. Elsevier, Amsterdam, pp 75-101

Rhoads DC, Morse JW (1970) Evolutionary and ecological significance of oxygen-deficient marine basins. Lethaia 4:413-428 (Oslo)

Rippka R, Waterbury J, Cohen-Bazire G (1974) A cyanobacterium which lacks thylakoids. Arch Mikrobiol 100:419-436
Rivas E, Luzzati V (1969) Polymorphisme des lipides polaires et des galacto-lipides de chloroplastes de mais, en présence d'eau. J Mol Biol 41:261-275
Riviere M-E, Arrio B, Steffan I, Molitor V, Kuntner O, Peschek GA (1990) Changes of some physical properties of isolated and purified plasma and thylakoid membrane vesicles from the freshwater cyanobacterium *Synechococcus* 6301 (*Anacystis nidulans*) during adaptation to salinity. Arch Biochem Biophys 280:159-166
Round FE (1980) The evolution of pigmented and unpigmented unicells — a reconsideration of the protista. BioSystems 12:61-69
Rutten GM (1966) Geologic data on atmospheric history. Paleogeogr Paleoclimatol Paleoecol 2:47-57
Saraste M, Holm L, Lemieux L, Lübben M, van der Oost J (1991) The happy family of cytochrome oxidases. Biochem Soc Trans 19:608-612
Satoh K, Murata N (1998) Stress responses of photosynthetic organisms. Molecular mechanisms and molecular regulations. Elsevier, Amsterdam
Satoh N, Murata N, Miura Y, Ueta N (1979) Effect of growth temperature on lipid and fatty acid composition in the blue-green algae, *Anabaena variabilis* and *Anacystis nidulans*. Biochim Biophys Acta 572:19-28
Scandalios JG (1993) Oxygen stress and superoxide dismutases. Plant Physiol 101:7-12
Schäfer G, Engelhard M, Müller V (1999) Bioenergetics of the archaea. Microbiol Mol Biol Rev 63:570-620
Scherer S (1983) Basic functional states in the evolution of light-driven cyclic electron transport. J Theor Biol 104:289-299
Scherer S (1984) Transmembrane electron transport and the neutral theory of evolution. Origins Life 14:725-731
Scherer S, Potts M (1989) Novel water stress protein from a desiccation-tolerant cyanobacterium. J Biol Chem 264:12546-12553
Scherer S, Stürzl E, Böger P (1981) Arrhenius plots indicate localization of photosynthetic and respiratory electron transport in different membrane regions of *Anabaena*. Z Naturforsch 36c:1036-1040
Schidlowsky M (1971) Probleme der atmosphärischen Evolution im Präkambrium. Geol Rundsch 60:1351-1384
Schlegel HG (1975) Mechanisms of chemo-autotrophy. In: Kinne O (ed) Marine ecology, vol II, Part I. John Wiley, London, pp 9-60
Schmetterer G (1982) Formation of hydrocarbons by photobleaching cyanobacterium *Anacystis nidulans*. Z Naturforsch 37c: 205-209
Schmetterer G, Peschek GA (1981a) Reversible photooxidative bleaching of *Anacystis nidulans*. In: Akoyunoglou G (ed) Photosynthesis V. Chloroplast development. Balaban Int Science Services, Pennsylvania, pp 721-726
Schmetterer G, Peschek GA (1981b) Treatments effecting reversible photobleaching and thylakoid degradation in the blue-green alga *Anacystis nidulans*. Biochem Physiol Pflanz 176:90-100
Schmetterer G, Peschek GA, Sleytr UB (1983) Thylakoid degradation during photooxidative bleaching of the cyanobacterium *Anacystis nidulans*. Protoplasma 115:202-207
Singer SJ (1974) The molecular organization of membranes. Annu Rev Biochem 43:805-835
Singer SJ, Nicolson GL (1972) The fluid mosaic model of the structure of cell membranes. Science 175:720-731
Skulachev VP (1992) Chemiosmotic systems and the basic principles of cell energetics. In: Ernster L (ed) Molecular mechanisms in bioenergetics. Elsevier, Amsterdam, pp 37-73
Smith AJ, Hoare DS (1977) Specialist phototrophs, lithotrophs, and methylotrophs: a unity among a diversity of prokaryotes? Bacteriol Rev 41:419-448

Stanier RY (1974) The origins of photosynthesis in eukaryotes. Symp Soc Gen Microbiol 24:219-242

Stanier RY, van Niel CB (1962) The concept of a bacterium. Arch Mikrobiol 42:17-35

Tandeau de Marsac N, Houmard J (1993) Adaptation of cyanobacteria to environmental stimuli: new steps towards molecular mechanism. FEMS Microbiol Rev 104:119-190

Tel-Or E, Huflejt ME, Packer L (1986) Hydroperoxide metabolism in cyanobacteria. Arch Biochem Biophys 246:396-402

Towe KM (1978) Early precambrian oxygen: a case against photosynthesis. Nature 274:657-661

Tsukamoto Y, Ueki T, Mitsui T, Ono T, Murata N (1980) Relationship between growth temperature of *Anacystis nidulans* and phase transition temperature of its thylakoid membranes. Biochim Biophys Acta 602:673-675

Tsukihara T, Aoyama H, Yamashita E, Tomizaki T, Yamaguchi H, Shinzawa-Itoh K, Nakashima R, Yaono R, Yoshikawa S (1995) Structures of metal sites of oxidized bovine heart cytochrome-c oxidase at 2.8 Å. Science 269:1069-1074

Tsukihara T, Aoyama H, Yamashita E, Tomizaki T, Yamaguchi H, Shinzawa-Itoh K, Nakashima R, Yaono R, Yoshikawa S (1996) The whole structure of the 13-subunit oxidezed cytochrome c oxidase at 2.8 Å Science 272:1136-1144

Van Baalen C (1987) Nitrogen fixation. In: Fay P, Van Baalen C (eds) The cyanobacteria. Elsevier, Amsterdam, pp 187-198

Verwer W, Ververgaert PHJT, Leunissen-Bijvelt J, Verkleij AJ (1978) Particle aggregation in photosynthetic membranes of the blue-green alga *Anacystis nidulans*. Biochim Biophys Acta 504:231-234

Wächtershäuser G (1988) Before enzymes and templates: theory of surface metabolism. Microbiol Rev 52:452-484

Wächtershäuser G (1990) Evolution of the first metabolic cycles. Proc Natl Acad Sci USA 87:200-204

Wada H, Hirasawa R, Omata T, Murata N (1984) The lipid phase of thylakoid and cytoplasmic membranes from the blue-green algae (cyanobacteria), *Anacystis nidulans* and *Anabaena variabilis*. Plant Cell Physiol 25:907-911

Wagner F, Falkner R, Falkner G (1995) Information about previous phosphate fluctuations is stored via an adaptive response of the high-affinity phosphate uptake system of the cyanobacterium *Anacystis nidulans*. Planta 197:147-155

Wagner F, Sahan E, Falkner G (1999) The complex relation between phosphate uptake and photosynthetic CO_2 fixation in the cyanobacterium *Anacystis nidulans*. In: Peschek GA, Loeffelhardt W, Schmetterer G (eds) The phototrophic prokaryotes. Kluwer Academic/Plenum Publishers, New York, pp 739-744

Wasmann CC, Löffelhardt W, Bohnert HJ (1987) Cyanelles: organization and molecular biology. In: Fay P, Van Baalen C (eds) The cyanobacteria. Elsevier, Amsterdam, pp 303-324

Wastyn M, Achatz A, Molitor V, Peschek GA (1988) Respiratory activities and aa_3-type cytochrome oxidase in plasma and thylakoid membranes from vegetative cells and heterocysts of the cyanobacterium *Anabaena* ATCC 29413. Biochim Biophys Acta 935:217-224

Waterbury JB, Watson SW, Guilard RRL, Brand LE (1979) Widespread occurrence of a unicellular, marine, planktonic cyanobacterium. Nature 277:293-294

Webb R, Reddy KJ, Sherman LA (1990) Regulation and sequence of the *Synechococcus* sp. strain PCC 7942 *groESL* operon, encoding a cyanobacterial chaperonin. J Bacteriol 172:5079-5088

Winkenbach F, Wolk CP (1973) Activities of enzymes of the oxidative and the reductive pentose phosphate pathways in heterocysts of a blue-green alga. Plant Physiol 52:480-484

Woese CR, Fox GE (1977) The archaebacteria —a third kingdom of life? Proc Natl Acad Sci USA 74:5088-5090

Woese CR, Magrum LJ, Fox GE (1978) Archaebacteria. J Mol Evol 11:245-252
Woese CR, Kandler O, Wheelis ML (1990) The tripartite system of living cells. Proc Natl Acad Sci USA 87:4576-4579
Wolf HJ, Christiansen M, Hanson RS (1980) Ultrastructure of methanotrophic yeasts. J Bacteriol 141:1340-1349
Wolfe RS (1992) Biochemistry of methanogenesis. In: Danson MJ, Hough DW, Lunt GG (eds) The Archaebacteria: biochemistry and biotechnology. Portland Press, London
Yamamoto HY, Bangham AD (1978) Carotenoid organization in membranes. Thermal transition and spectral properties of carotenoid containing liposomes. Biochim Biophys Acta 507:119-127
Yates MG, Jones CW (1974) Respiration and nitrogen fixation in *Azotobacter* spp. Adv Microb Physiol 11:97-135
Zohner A, Broda E (1979) Urey-Miller experiments on the origin and stability of nitrate in a reducing atmosphere. Origins Life 9:291-298

9 Acidophilic and Acidotolerant Algae

H. Gimmler*

Contents

* Julius-von-Sachs Institute of Bioscience, University of Würzburg, Julius-von-Sachs Platz 2, 97082 Würzburg, Germany

9.1 The Little Difference: Acidophilic versus Acid-Tolerant

Many microbes are able to grow at pH values as low as 4, but the majority can also grow at neutral and slightly alkaline pH. Such organisms are usually entitled to be acid-tolerant. This chapter, however, concentrates on acidophilic organisms, which by definition are those microbes which are able to grow at pH values smaller than 3, but are unable to thrive at neutral pH (Fig. 9.1). Some of them grow well even at pH values close to 0. This is not a privilege of prokaryotes, as occasionally claimed (Fig. 9.1A,B). Compared to the total number of acidotolerant organisms, very few prokaryotic and eukaryotic species are obligate acidophiles. The most prominent genera among bacteria are *Thiobacillus,*

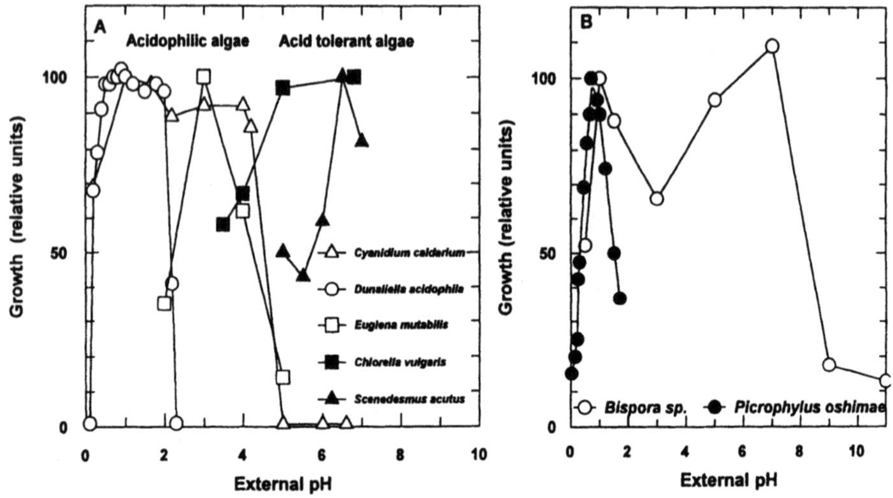

Fig. 9.1. **A.B**. Acidophily and acid tolerance of some microalgae (**A**) and an archaebacterium and a filamentous fungus (**B**). Data of **A** were normalised and redrawn after Brock (1978), Nakatsu and Hutchinson (1988), Fuggi et al. (1988a), Nalewajko et al. (1997) and Rai et al. (1996); those of **B** after Schleper et al. (1995b) and H. Gimmler (unpubl.)

Bacillus, Sulfolobus and *Thermoplasma* (Schleper et al. 1995a). The pH profile of the recently discovered archaebacterium *Picrophilus* is depicted in Fig. 9.1B (Schleper et al. 1995b). Physiology and biochemistry of bacterial acidophiles have been reviewed relatively frequently (see for example Brock 1978; Langworthy 1978; Cobley and Cox 1983; Krulwich and Guffanti 1983; Bakker 1990; Matin 1990; Norris and Ingledew 1992; Norris and Johnson 1998; Johnson 1998). Among the eukaryotes a large number of acid-resistant yeasts and fungi have been described, particularly from acid mine drainage (Cooke 1966; Lopez-Arcilla et al. 1995). They partially withstand even extreme pH values (Starkey and

Waksman 1984; Carandang and Gimmler 1998; Gimmler 2000a). Since these acid-tolerant yeasts and fungi often grow very well at neutral pH values and often have their pH optima of growth close to neutrality, by definition they do not belong to the acidophiles. A typical example for such a pH profile of growth for a recently discovered strain of the filamentous fungus *Bispora* sp. is given in Fig. 9.1B. Comparably fewer reviews have been dedicated to acidophilic algae (Brock 1978; Langworthy 1978; Gimmler and Weis 1992; Albertano 1995; Pick 1998) and some of them covered only one or the other physiological aspect or focused on selected algal species. The difference in interest in bacterial and algal acidophiles is very likely due to the fact that most bacterial acidophiles are also resistant to high temperature, whereas most algal acidophiles are mesophilic or even slightly psychrophilic. Therefore, only enzymes of the former group are of interest for biotechnology. Furthermore, bacterial acidophiles, but not algae, are of importance for biomining (see Brierley 1990; Mann et al. 1989; Johnson 1998 and references therein).

This chapter tries to bridge this gap, emphasising recent biochemical and bioenergetic aspects of acid-resistant algae. It concentrates on properties of acidophilic algae living at pH values between 0 and 3, because under these extreme conditions the essential mechanisms of acidophily can be better elaborated. Some care must be taken in generalising such data for algae thriving well in the pH range between 3 and 5. Some mechanisms are indeed identical, but some properties must be definitely different, at least gradually, due to difference in dissociation of dissolved anionic solvents or anionic constituents of the plasma membrane in these two pH ranges.

9.2 Small Ecological Niches — But All Around the World

Extreme acid environments are of either natural geothermal or anthropogenic origin. The acidity in volcanic waters and fumaroles is largely of biotic origin. Such habitats are usually rich in elemental sulfur or sulfides, which are oxidised to sulfate by bacteria like *Thiobacillus thiooxidans.* Also during the oxidation of iron by *Thiobacillus ferrooxidans* acid is generated. The number of natural acid sites in the world is more or less constant, whereas the increase in industrial acidic waters is a matter of considerable concern. Typical industrial sites are coal and metal mines and associated waste effluents. Furthermore, lakes and rivers are increasingly polluted by acid precipitation and effluents of chemical industry. The increasing interest in acid-resistant algae is certainly related to this man-made pollution of the environment.

Most acid-resistant algae are distributed worldwide, but always restricted to relatively small habitats characterised by pH values lower than 3 (Table 9.1). The typical habitats of acidophilic algae are aquatic, but occasionally acidophilic algae

Table 9.1. Biogeography of selected acidophilic and acidotolerant cyanobacteria and algae

Algal division	Algal species	Site and reference	pH
Cyanobacteria	*Spirulina*?, *Oscillatoria*?, *Limnothriox?*	Lichtenauer See (Lausitz, Germany) (3)[a]	2.9
	Synchechococcus,	(2)	4
	Mastigocladus	*Sphagnum* bogs (20)	4
	Chroococcus turgidus		
Chlorophyta	*Chlamydomonas acidophila*	Fumigated tundra pools (Canada, 1); acid streams (England, 14)	1.8-3.6 < 3
	Chlamydomonas sp.	Sulfuric waters (Italy, 5); Highly acid soils (Italy, 5)	
	Chlorella sp.	Highly acid soils (Italy, 5)	< 3
	Chorycystis sp	Highly acid soils (Italy, 5)	< 3
	Dunaliella acidophila	Sulfuric acid lakes (Italy, 5); Czechoslovakia (12,15); highly acid soils (Italy, 5)	<1.5
	(= *Spermatozopsis acidophila*)		< 3
	Eremosphaera viridis	Bogs (19)	3.5-4.5
	Hormidium rivulare	Acid streams (England, 14)	2.6-3.1
	Hormidium fluitans	Acid mine drainage water (Northern Australia, Papua-New Guinea, 17)	
	Stichococcus sp.	Highly acid soils (Italy, 5)	< 3
	Stichococcus bacillaris	Acid streams (England, 14)	2.6-3.1
	Zygogonium spp.	Acid thermal springs, acid soils (2)	2.0- 3.1
Rhodophyta	*Cyanidium caldarium*	Acidic hot springs (world wide. 2,11)	< 1.5
	C. caldarium, forma A and forma B	Highly acid soils (Italy, 5), sulfur thermal sites (Italy, 6), fumaroles (Mount Lawu, Java, 11), Acid hot springs, North and Central America, 7, 8; Japan,16)	1.5-5
	Cyanidioschyzon sp.	Highly acid soils (Italy, 5)	< 3
	C. merolae	Fumaroles (Yellowstone National Park, USA,10; Mount Lawu, Java, Indonesia,11)	
	Pleurococcus sulphurarius (=*Galdieria sulphuraria*)	Fumaroles (Mount Lawu, Java,7) Acid hot springs (Italy, 8; North and Central America, 11)	<1.5, 1.5-5
Heterokontophyta	*Eunotia arcus*	Fumigated tundra pools (Canada, 1)	1.8-3.6
	Eunotia glacialis	Fumigated tundra pools (Canada, 1)	1.8-3.6
	Eunotia lunaris, Surirella sp.	Volcanic lakes (Japan, 18)	3.4-3.8
	Nitzschia communis	Fumigated tundra pools (Canada., 1)	1.8-3.6
	Pinnularia sp.	Sulfuric waters (Italy, 5), Highly acid soils (Italy, 5)	< 3
Euglenophyta	*Euglena mutabilis*	Fumigated tundra pools (Canada,1,13), acid streams (England, 14)	1.8-3.6
	Euglena sp.	Sulfuric waters (Italy, 5)	<3

Chrysophyta	*Ochromonas vulcanus*	Volcanic spring, Kunashir Island (Russia, 9)

[a]References: 1 Havas et al. (1982); 2 Brock (1978); 3 Steinberg et al. (1998); 4 Pinto (1993); 5 Pinto and Taddei (1978); 6 Pinto and Taddei (1975); 7 De Luca et al. (1981); 8 De Luca et al. (1979); 9 Gromov et al. (1988); 10 De Luca et al. (1978); 11 Merola et al. (1981); 12 Kalina (1965); 13 Nakatsu and Hutchinson (1988); 14 Hargreaves and Whitton (1976); 15 Albertano et al. (1981); 16 Nagashima and Fukada (1981); 17 Madgewick and Ralph (1977); 18 Satake et al. (1995); 19 Deveau et al. (1998); 20 Fogg et al. (1973)

are also found in acid soils (Table 9.1) or on the surface of rocks. Some grow even cryptoendolithic (Gross et al. 1998). Worldwide-distributed natural acid habitats also include bogs, which may exhibit pH values as low as 3.5 – 4.5. Bogs are additionally characterised by extremely low concentrations of minerals and a low penetration of light. In spite of these unfavourable conditions, bogs are inhabited by a considerable number of microalgae, including cyanobacteria, desmids and chlorophytes. A typical alga for this habitat is *Eremosphaera viridis,* which became a favourite subject of algal electrophysiology because of its huge cell size (Deveau et al. 1998 and references therein).

9.3 Who Is Who in Acid Habitats?

Some time ago, Pinto and Taddei (1975, 1978) estimated the number of acid-tolerant algae to be close to 20. Since then, this figure has increased only slightly (Table 9.1). Acid-resistant algae belong to six different algal divisions, including cyanobacteria. This demonstrates that acid resistance is a phenomenon of polyphylogenetic origin.

Text books usually do not report cyanobacteria in acid or very acid environments (< pH 4.8) (cf. also Brock 1973, 1978; Knoll and Bauld 1989; Albertano 1995). Brock checked a huge number of different acid habitats in the world for the presence of cyanobacteria. Results, confirmed by culture experiments, were negative. A pH limit of 4.8 was postulated for proliferation of cyanobacteria. The reason for the supposed exclusion of cyanobacteria from acidic habitats cannot be autotrophy, since chemolithotrophic bacteria are quite common in acid environments. The suggestion of Knoll and Bauld (1989) that the exclusion is due to the direct communication between the exterior of the blue-green algae and the intrathylakoid space is not convincing, since it is not consistent with the light-induced acidification of the intrathylakoid space at alkaline pH (Raven 1990). On the other hand, there are also some reports showing that vital cyanobacteria are indeed present in acid lakes (pH 4.1–5) and even become dominating with decreasing pH (Almer et al. 1974; Kwiatkowski and Rolff 1976). This was confirmed by a recent study which demonstrated the existence of filamentous cyanobacteria in acid lakes (pH 2.9) in Germany (Steinberg et al. 1998). This puts a question mark behind the postulated lower pH limit for cyanobacteria. The postulate of a lower pH limit might be valid for unicellular blue-green algae, but not for filamentous cyanobacteria.

Taxonomy of acidophilic algae was the subject of several revisions in the past, causing a lot of confusion among algal physiologists. In respect to the acid-tolerant Rhodophyta, it is clear, since the studies of Merola et al. (1981), that three unicellular red algal species, often living in mixed populations in thermal acidic environments and generally confused with *Cyanidium caldarium*, must be distinguished: (1) *Galdieria sulphuraria* (formerly *Pleurococcus sulphuraria* or *Cyanidium caldarium*, forma B), (2) *Cyanidium caldarium* and (3) *Cyanidioschyzon merolae*. The first genus belongs to the family Galdieriaceae, the latter two to the family Cyanidiaceae. Based on ribosomal RNA sequences both families can be regarded as early evolved algal eukaryotic groups, representing a transitional phylogenetic position of these acidophiles between Cyanobacteria and Rhodophyta (= prerhodophytes) (Seckbach 1992, 1995; Seckbach et al. 1992, 1993). It is proposed that the simplest eukaryotic genus, *Cyanidioschyzon,* may have evolved from an association between a *Thermoplasma*-like archaebacterium and a thermophilic cyanobacterium, whereas subsequent non-symbiotic compartmental steps may have taken place from *Cyanidioschyzon* to evolve first *Cyanidium* and finally *Galdieria.* The latter genus represents the origin of the recent unicellular red algae and already possesses a vacuole. Recently, the plastid ATPase operons of *Galdieria* were cloned, sequenced and compared to corresponding sequences of other red algae, higher plants and autotrophic prokaryotes (Kostrzewa and Zetsche 1993). Results imply a relatively large phylogenetic distance between red algae and higher plants and support the theory that higher plants have evolved from either red algae with plastids containing chlorophyll a and c or common prokaroytic endosymbionts (Enami et al. 1995).

Also, the acidophilic unicellular green flagellate *Spermatozopsis acidophila* (Kalina 1965; Albertano et al. 1981) had to be renamed. The thorough cytological studies of Melkonian and Preisig (1984) demonstrated that this alga was, in fact, a member of the genus *Dunaliella.* Since then, the synonym *Dunaliella acidophila* has been used by modern systematists and algal physiologists as well. However, its affiliation to the Volvocales (Chlorophyta) has remained unaffected.

9.4 Fighting the Protons

High H^+ concentrations in acidic environments cause both physicochemical and biological effects. Physicochemical effects are, for example, the high osmotic potentials accompanying low pH values, high sulfate concentrations (in sulfuric waters) and high electrical conductivity, solubilisation of elements that are normally not available at neutral pH, protonation of organic acids leading to the formation of phosphoric acid instead of phosphate and CO_2 instead of bicarbonate and to the neutralisation of the anionic carboxyl and phosphate groups at the external side of the plasma membrane. All these effects are unavoidable and cells can only respond to them with the strategy of stress tolerance.

Regarding the biological effects of high H^+ concentrations acidophilic algae have theoretically the choice between the strategies of stress avoidance or stress tolerance. In the former case they are expected to exhibit (1) a low conductance of the plasma membrane for H^+, (2) a high H^+ export capacity, and (3) a high buffer capacity. In cooperation these parameters permit a pH homeostasis with a cytoplasmic pH close to neutrality. Special acid resistance of intracellular enzymes would not be required. In the case of stress tolerance parameters 1-3 are not required or are needed only in a restricted manner. However, algae must pay a high energetic price for meeting the requirement of extreme acid-resistant enzymes. This chapter summarises experimental evidence that acidophilic and acidotolerant algae unanimously follow the strategy of stress avoidance.

9.4.1 pH Homeostasis

A breakthrough for the non-invasive determination of cytosolic pH in acidophilic algae was caused by the introduction of the ^{31}P NMR technique. This method allows the determination of the internal pH from the location of the inorganic phosphate signal within ^{31}P-NMR spectra. Non-vacuolated, unicellular, acidophilic algae are able to keep their internal pH constant at values close to neutrality over a wide range of external proton concentrations (Fig. 9.2) (Lane and Burris 1981; Enami et al. 1986; Gimmler et al. 1988). Thus, they do not behave differently in this respect from neutrophilic algae (Lane and Burris 1981; Gimmler et al. 1988) or from bacterial acidophiles (Fig. 9.2; Michels and Bakker 1985; Booth 1985; Matin 1990). Figure 9.2 also demonstrates that with former methods for the determination of internal pH (for example determination of the distribution of ^{14}C-labelled lipophilic weak acids between cells and the medium) underestimations of the cytosolic pH were obtained at external pH values lower than 5 (Bakker 1990). Most of the extreme acidophiles do not possess large vacuoles, which also could take part in pH homeostasis. An exception is the red alga *Galdieria*, but pH values of its vacuole have not been measured so far. However, the effect of the external pH on the highly vacuolated moderate acidophilic green alga *Eremosphaera viridis* has been investigated (Fig. 9.2B). Adaptation of this alga to pH values down to 3 did not affect the cytoplasmic or vacuolar pH. This implies that under steady-state conditions in this moderate acidotolerant alga H^+ transfer across the plasma membrane is of greater importance for overall acid resistance than H^+ accumulation in the vacuole. This does not exclude that, for example, under sudden acid stress proton translocation into the vacuole takes place and is a part of fine regulation during pH homeostasis. Some marine brown algae, for example some species of the genus *Desmarestia*, accumulate sulfuric acid in vacuoles until pH values slightly lower than 1 are reached (McClintock et al. 1982). Surprisingly, this brown algal genus is neither acidophilic nor acid-tolerant. Hence, vacuolar acid accumulation is not part of a reaction required for acid resistance. Rather, it reflects an exotic type of osmoregulation and sulfate storage in the vacuole of this marine algal genus. Nevertheless, it demonstrates that the tonoplast membrane can tolerate extremely low pH values.

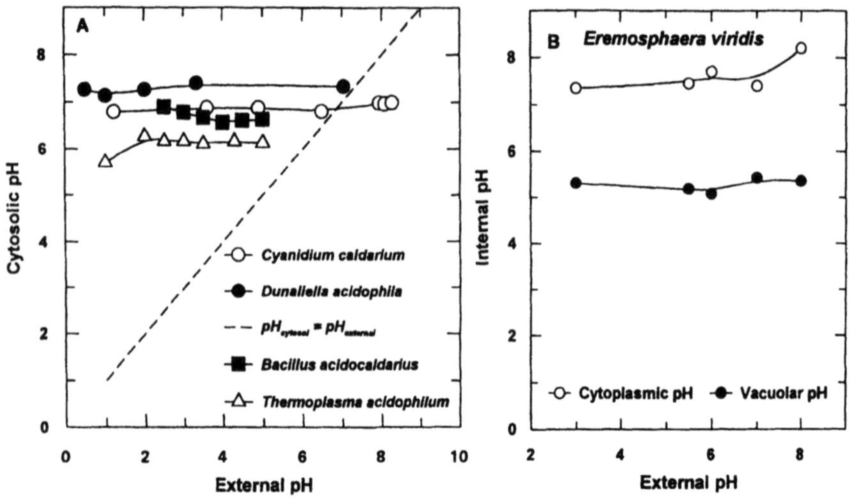

Fig. 9.2. A. Cytoplasmic pH of two non-vacuolated, acidophilic algae (obtained by ^{31}P-NMR techniques) and two acidophilic bacteria (calculated from the distribution of pH probes) as a function of the external pH. Data redrawn from Gimmler and Weis (1992) (*Dunaliella acidophila,* 25 °C), Enami et al. (1986) (*Cyanidium caldarium,* 40 °C) and Michels and Bakker (1985) (*Bacillus acidocaldarius*, 40 °C; *Thermoplasma acidophilum*, 59 °C). Aerobic conditions, dark. *Dotted line* Theoretical equilibrium between internal and external pH. **B** Cytoplasmic and vacuolar pH of the moderate acid-tolerant green alga *Eremosphaera viridis* (^{31}P-NMR technique). (H. Gimmler, unpubl. data)

In summary, acidophiles are able to maintain an internal pH of 7 at an external pH of 1. All data available in this respect refer to steady-state conditions, whereas a considerable lack of information exists as to how fast cells respond to sudden changes in the pH of the medium or internal acid loads. However, the velocity of the equilibration of the internal pH following external or internal pH disturbances is a measure for the strength of pH homeostasis. The lack of data is caused by experimental limitation. In general, the time resolution of the various methods applied for the determination of the internal pH is too low. Nevertheless, internal pH values measured under steady-state conditions indicate an H^+ gradient of 10^6 across the plasma membrane. This raises the question as to how such a gradient is maintained.

9.4.2 Proton Conductance of the Plasma Membrane

For the calculation of the permeability coefficient of the plasma membranes for H^+ (P_{H+} values, dimension $m\ s^{-1}$), H^+ influxes (in terms of mol $H^+\ m^{-2}\ s^{-1}$) and the driving forces for the H^+ influx are required, which are the H^+ gradient across the

Table 9.2. Permeability coefficients of the plasma membrane of *Dunaliella acidophila* and reference systems for protons. Values were calculated from measured H^+ fluxes using the indicated methods and membrane potentials

Species	Method	pH+ value (ms^{-1})
Dunaliella acidophila	Dark relaxation of light-induced H^+ export into the medium at pH 1	2.9×10^{-9}
	Minimal H^+ export from light-induced internal pH changes measured by ^{31}P-NMR techniques (pH 1)	3.7×10^{-9}
Charophytes	Electrical (pH 7) (1)[a]	100×10^{-9}
Elodea nuttallii	Acridine orange (optical) (2)	650×10^{-9}
Lipid bilayers	Electrical (pH 7) (1)	$40 - 170 \times 10^{-9}$

[a]References: 1) Raven (1976); 2) Miedema et al. (1996)

plasma membrane and the membrane potential, ψ. H^+ influxes are difficult to determine under acid conditions. However, for the acidophilic algae *Cyanidium caldarium* and *Dunaliella acidophila,* a light-induced proton efflux into the medium, followed by a recovery (alkalisation) in the subsequent darkness was detected (Kura-Hotta and Enami 1981; Enami and Kura-Hotta 1984; Remis et al. 1994). The latter can be treated as the minimal number for H^+ influx. Alternatively, also from the kinetics of light-induced changes of internal pH minimal proton influxes can be estimated, taking into account the buffer capacity of the cells. In *Dunaliella acidophila* for example an average influx of 70×10^{-9} mol $H^+\ m^{-2}\ s^{-1}$ was calculated at an external pH of 1. Using the membrane potential ψ (see below) and the H^+ gradient across the plasma membrane under these conditions (see above) an apparent permeability coefficient of the plasma membrane for H^+ in the order of $3 \times 10^{-9}\ m\ s^{-1}$ was obtained (Table 9.2). This value is considerably lower than corresponding values obtained for the plasma membranes of neutrophilic algae, higher plants or lipid bilayers (Table 9.2). Also for bacterial acidophiles, low permeability coefficients of the plasma membrane for protons have been postulated by Booth (1985), but attempts to verify this hypothesis are rare (Rius et al. 1994).

9.4.3 Costs of Maintenance of the Proton Gradient Across the Plasma Membrane

The energetic costs of a pH homeostasis of acidophilic algae can be estimated using the concept of maximal or minimal mechanistic power of algal cells, which has been put forward by Raven (1982). In this concept the minimum energy input (ATP consumption) of a reaction or transport process is quantified and compared with the maximal energy output of algal cells (ATP production) during photosynthetic, respiratory or glycolytic conditions (using -46 kJ mol^{-1} as Δg_{ATP}). The mechanistic power is expressed in terms of W (J s^{-1} $cell^{-1}$) or in the dimension of fluxes (mol ATP m^{-2} s^{-1}). The latter allows the direct comparison of ATP production of algal cells with H^+ influxes into the cell. In order to keep the internal pH constantly close to neutrality, protons have to be reexported into the environment by the plasma membrane H^+-ATPase at the expense of one molecule of ATP per proton.

With appropriate data for the conductance of the plasma membrane of *Dunaliella acidophila* and the H^+ gradient across the plasma membrane, an estimation can be made of the costs of the maintenance of such a gradient, by plotting theoretical H^+ influxes (assuming different P_{H^+} values) as a function of the external pH (Fig. 9.3). The resulting set of curves can be compared with the maximal available mechanistic power under different metabolic conditions (horizontal lines in Fig. 9.3). At an external pH of 1 (inset in Fig. 9.3), the curve for proton fluxes according to a P_{H^+} value of 10^{-8} m s^{-1} of *D. acidophila* (solid line) is located below the maximal mechanistic power of photosynthesis and respiration, but above the range of glycolysis. In other words, photosynthesis or respiration produce sufficient energy to reexport all protons which diffuse into the cell according to the chemical gradient of protons. However, under glycolytic conditions the system would require more energy for proton export than it can produce by metabolic reactions. This theoretical consideration matches the experimental observations that laboratory cultures die rapidly under anaerobic conditions as visualised by fast phaeophytinisation of chlorophyll. Figure 9.3 also demonstrates that an even more positive membrane potential (+ 0.1 V instead of 0.05 V; dotted line) would be of benefit for the costs of pH homeostasis, whereas a negative membrane potential (for example –0.05 V, hatched line) would cause a complete consumption of all available photosynthetic or respiratory energy for reexport of protons. Since a lot of energy is also required for other metabolic processes of algae, a negative membrane potential ψ can be definitely excluded for all acidophilic algae. Since the experimentally determined P_{H^+} value is, in fact, 3 x 10^{-9} m s^{-1} (Table 9.2), even lower than the theoretical value of 10^{-8} m s^{-1} discussed above, it can be estimated that the true energy demand for reexport of protons from *D. acidophila* into the medium requires less than 3% of the total energy budget.

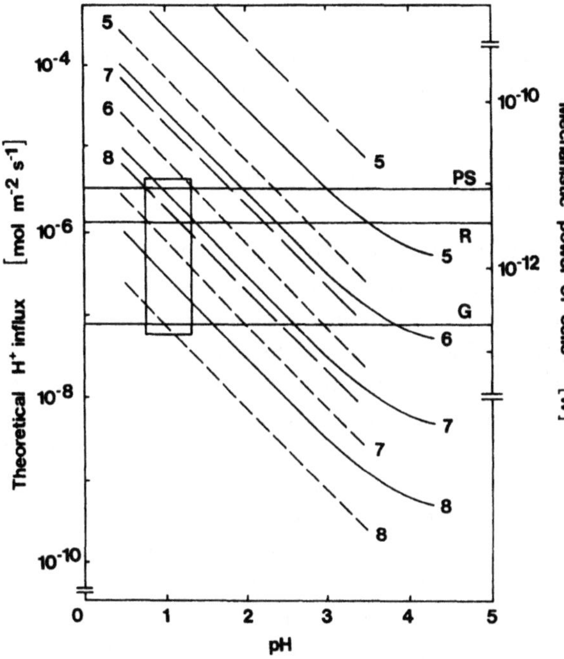

Fig. 9.3. Theoretical H^+ influx into *Dunaliella acidophila* (*left y-axis*) as a function of the external pH (*x-axis*) at different assumed P_{H^+} of the plasma membrane (*numbers attached to lines* represent the negative decadic logarithm of P_{H^+} values). *Solid lines* present P_{H^+} values for aerobic conditions in darkness (Ψ = + 0.05 V, experimentally determined value). *Broken lines* stand for cells under anaerobiosis (dark) and/or with an inhibited H^+-ATPase (Ψ = + 0.1 V), *hatched lines* for P_{H^+} values for cells with a negative Ψ (-0.05 V), typical for neutrophilic algae or *D. acidophila* cells suspended in a medium of pH 7. *Horizontal lines* mark the maximal available energy of cell (mechanistic power, *right y-axis*) under photosynthetic (*PS*), respiratory (*R*) and glycolytic (*G*) conditions. H^+ influx rates *above horizontal lines* are energetically excluded, because they cannot be compensated by H^+ export for prolonged periods via the plasma membrane H^+-ATPase during photosynthesis (light, air), respiration (dark, air) or glycolysis (dark, anaerobic). The *intersections* of the set of curves within the vertical lines within the small box indicate the supposed range of P_{H^+} values for which at pH 1 the maximal available energy of *D. acidophila* cells matches the energy demand for H^+ export at a given conductance of the plasma membrane of this alga for H^+. However, true values must be significantly lower, since energy is also required for growth and other maintenance processes

9.4.4 Internal Buffer Capacity

A high internal buffer capacity of acidophilic algae may cause a transient protection against external H^+-stress. In fact, *Dunaliella acidophila* cells exhibit approximately the double buffer capacity (100 mmol H^+ l^{-1} pH $unit^{-1}$) than neutrophilic, salt-resistant *Dunaliella* species (Gimmler and Weis 1992). Likewise, higher buffer capacities were measured in acidophilic bacteria than in neutrophilic species (Krulwich et al. 1985; Rius et al. 1994).

9.4.5 Proton Export Capacity of Intact Cells

There is no doubt that the reexport of protons from the cytoplasm of acidophilic alga is carried out by a plasma membrane-bound H^+-ATPase. In contrast to neutrophilic algae, where light-induced alkalisation of the medium due to photosynthetic uptake of bicarbonate obscures the picture, in acidophilic algae the activity of the H^+-ATPase can be monitored in vivo by measuring the light-induced acidification of the medium by *Cyanidium caldarium* (Kura-Hotta and Enami 1981; Enami and Kura-Hotta 1984) and *Dunaliella acidophila* (Remis et al. 1994). The light-induced proton export in both systems, operating against a strong H^+ gradient, is sensitive to uncouplers and ionophores (such as FCCP, nigericin and gramicidin) and energy-transfer inhibitors (such as DES, DCCD and omeprazole). It was stimulated by the ATP-elicitor fusicoccin (Remis et al. 1994). All the results indicate that the process is catalysed by a plasma membrane-bound H^+-ATPase. A discrepancy was observed in respect to the source of ATP for the light-induced proton export. Results with *Cyanidium caldarium* suggested that cyclic photophosphorylation supplies the majority of the required ATP. Light-induced proton export was insensitive to DCMU and the action spectrum suggested the involvement of photosystem I (chlorophyll a), whereas the action spectrum of photosynthesis resembled the absorption spectrum of phycocyanin. In *Dunaliella acidophila*, light-induced proton export was completely inhibited by DCMU, thus suggesting the involvement of noncyclic photophosphorylation. Measured rates of H^+ export in both the cases were on an average about 50% of comparable photosynthetic rates. When the rates were converted in terms of fluxes, values between 3 and 15 x 10^{-8} mol H^+ m^{-2} s^{-1} were obtained for *D. acidophila* which match the experimentally determined P_{H^+} value (Table 9.2) and the theoretical considerations of Fig. 9.3.

9.4.6 Properties of the Plasma Membrane H^+-ATPase (PMHA)

Thorough studies on the properties of the plasma membrane ATPase (PMHA) of an acidophilic alga were carried out by the group of Pick (Sekler et al. 1991, 1993, 1994; Pick 1992). Properties of the isolated PMHA of *Dunaliella acidophila* resembled in many aspects those of the corresponding ATPase of higher plants. The PMHA of *D. acidophila* was sensitive to ATPase inhibitors, such as vanadate, DES, DCCD and erythrosine, but was stimulated by the fungal ATPase elicitor fusicoccin and by K^+, and not affected by inhibitors of anion transport like SITS and DIDS (Sekler et al. 1991; Pick 1992). It showed maximal activity at pH 6 and exhibited a ten times higher specific activity and higher affinity to ATP ($K_{m\ ATP}$ = 60 μM) than the higher plant PMHA. Its stability in nonionic detergents such as Triton X (Pick 1992) is of particular interest. Trypsin activated the *D. acidophila* PMHA by elimination of an autoinhibitory domain at the C-terminal end of the PMHA which carries a phosphorylation site (Sekler et al. 1994). The PMHA of *D. acidophila* is strongly inhibited by omeprazole (Pick 1992; Sekler et al. 1993). Omeprazole is an acid-activated sulfhydryl reagent and an inhibitor of the gastric H^+/K^+- ATPase of mammals. Unlike vanadate, it can be tested at external acid pH values in vivo, because its acid activation permits the blocking of SH groups of the PMHA which face the outer phase of the plasma membrane. Therefore, it also inhibits the light-induced proton efflux of intact cells catalysed by the PMHA (Sekler et al. 1993; Remis et al. 1994).

It has also been demonstrated that the expression of PMHA in *Dunaliella acidophila* is far higher than in the neutrophilic *Dunaliella salina* (Sekler et al. 1991; Pick 1992; Weiss and Pick 1996). A higher PMHA activity in acid-tolerant strains than in acid-intolerant strains was also demonstrated for *Chlorella vulgaris* (Rai et al. 1993). This is in agreement with the observation that acid-tolerant bacteria express higher H^+-ATPase activities than acid-intolerant bacteria (Miwa et al. 1997).

9.4.7 Molecular Biology of PMHA

The plasma membrane H^+-ATPase of two different acidophilic algae, *Dunaliella acidophila* (Weiss and Pick 1996) and *Cyanidium caldarium* (Ohta et al. 1997) have been cloned and sequenced by means of PCR and Southern hybridisation techniques. Alignment of the deduced amino acid sequence of the PMHA of *Dunaliella acidophila* with that of various P-type H^+-ATPases of higher plants and fungi revealed a high homology (Weiss and Pick 1996). The putative amino acid sequences of *D. acidophila* and the halotolerant *D. salina* are 75% identical. A special feature of all *Dunaliella* ATPase is an extended C-terminal domain, playing presumably a regulatory role. The latter causes a higher molecular weight of the *Dunaliella* PMHA (105 kDa) compared with the corresponding PMHA of higher plants and fungi. Of interest is the observation that the overall calculated positive net charge contributed by all extrinsic loops of the *D. acidophila* PHMA is + 11 (Weiss and Pick 1996). This is consistent with a positive zeta potential in *D. acidophila* (Gimmler et al. 1991a).

The *Cyanidium* PMHA was found to consist of 955 amino acids (predicted molecular weight 105 kDa) and the deduced amino acid sequence was found to be more homologous to the corresponding P-type H^+-ATPases from higher plants than to that of the halotolerant green alga *Dunaliella bioculata* (Ohta et al. 1997).

Ziegler et al. (1995) cloned and sequenced a V-ATPase from *Cyanidium caldarium,* the function of which was suggested to be H^+ translocation into the vacuole. This is rather unlikely, since there is no electron microscopic evidence for a large vacuole or several tiny vacuoles in *Cyanidium*; only the related genus *Galdieria* possesses a vacuole. More likely, this H^+-ATPase is localised in Golgi vesicles or ER-derived vesicles of the cells.

9.4.8 Reversed Electrical Potentials — an Unusual Response to an Unusual Environment

Electrical potentials of cells are important driving or repelling forces during the uptake of ions. In general, neutrophilic algae exhibit negative electrical potentials in the range between –40 and –180 mV (Raven 1976). Negative electrical potentials facilitate the uptake of cations such as K^+, NH_4^+ or H^+, but hamper the entry of anions such as HCO_3^-, $H_2PO_4^{2-}$, NO_3^- or Cl^-.

The zeta potential of *Dunaliella acidophila*, a measure for the charge distribution at the external side of the plasma membrane, can be calculated from the electrophoretic mobility of cells determined by means of free flow electrophoresis (Gimmler et al. 1991a). Results demonstrate a positive zeta potential (+ 5 to + 20 mV) (Fig. 9.4), caused by complete protonation of anionic groups of membrane constituents exposed to the acid medium while cationic groups remained unaffected. Since the cytosolic pH of acidophilic algae is close to 7 (Fig. 9.2), this implies an asymmetrically charged plasma membrane of cells suspended in an acid medium, excess of positive charges outside and, excess of negative charges on the cytoplasmic side. Such determinations have been carried out so far only with *D. acidophila* and are supported by the fact that this alga does not possess a rigid cell wall. However, it is reasonable to assume that all acidophilic algae have positive zeta potentials, because these potentials are caused by physicochemical laws, not due to biological adaptations. The latter is confirmed by the observation that upon transfer of cells to media with neutral pH the zeta potential changes immediately to negative values (- 30 mV) (Gimmler et al. 1991a).

Likewise, the membrane potential ψ of *D. acidophila* is positive at acid pH (+30 to +65 mV) (Fig. 9.4), but switches to negative values at pH values higher than 5 (Gimmler et al. 1989; Remis et al. 1992). Light causes a slight but significant decrease in the positive membrane potential (hyperpolarisation), whereas in the presence of omeprazole ψ depolarises. Apparently, the major component of the positive membrane potential is the diffusion potential of H^+ (about +350 mV). Another contribution comes from the K^+, H^+-cotransporter of *D. acidophila* (see below). PMHA activity is expected to hyperpolarise ψ. Since light stimulates and omeprazole inhibits PMHA-mediated H^+ efflux, experimental data on the effect of light and omeprazole on ψ confirm this explanation.

It is reasonable to assume that other acidophilic algae also have reversed electrical potentials (internally positive), even though it has not been demonstrated experimentally, yet. However, as far as bacterial acidophiles are concerned, the distribution of reversed potentials at acid pH is common (Michels and Bakker 1985; Bakker 1990; Matin 1990; cf. Fig. 9.4).

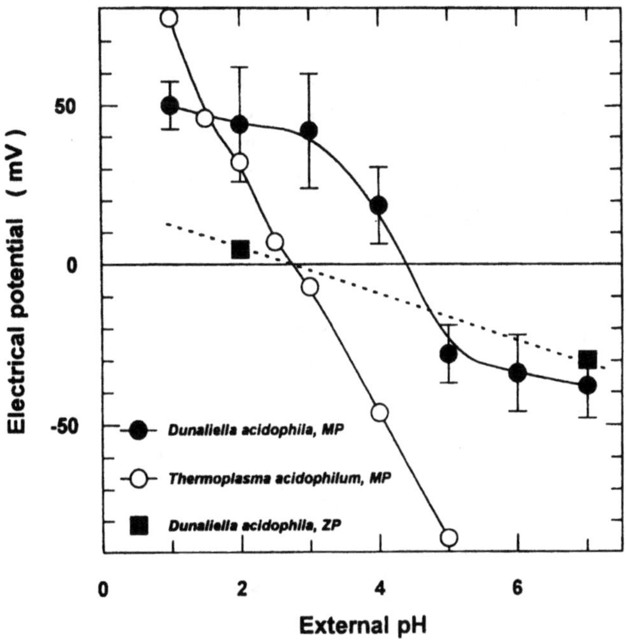

Fig. 9.4. Membrane potential ψ (*solid lines*) and zeta potential (*ZP, dotted line*) as a function of the external pH in the acidophilic alga *Dunaliella acidophila* (data redrawn after Remis et al. 1992; Gimmler et al. 1991a) and the acidophilic bacterium *Thermoplasma acidophilum* (data redrawn after Michels and Bakker 1985)

9.4.9 Adaptation of Membrane Architecture to High Proton Concentration?

Theoretical considerations on the molecular nature of thermophily suggest that the plasma membrane of thermophilic microbes, and therefore also of other microbes, which are both thermophilic and acidophilic at the same time, is a critical cellular component (Brock 1978). This applies certainly to the acidophilic and moderate thermophilic alga *Cyanidium caldarium* (growth characteristics: $T_{min} = 20\ ^{o}C$; $T_{opt} = 48\ ^{o}C$, $T_{max} = 58\ ^{o}C$, Brock 1978), but not to the acidophilic, slightly psychrophilic *Dunaliella acidophila* ($T_{opt} = 24\ ^{o}C$, $T_{max} = 30\ ^{o}C$, Albertano et al. 1981; Fig. 9.5). In order to find out whether a special composition of membrane constituents causes a low proton conductance of acid-resistant algae, the lipid composition of both *Cyanidium caldarium* and *Dunaliella acidophila* have been determined (Boenzi et al. 1977; Enami and Fukada 1977; Pollio et al. 1988; Della Greca et al.1989; see also Brock 1978 for earlier literature). Since no method is presently available for the isolation of a plasma membrane fraction from *Cyanidium,* in all experiments with this alga, total lipid extracts have been analysed, which limits the interpretation of analytical results in respect to the lipid composition of the plasma membrane. *Cyanidium* cells contain smaller amounts of unsaturated fatty acids than higher plants or mesophilic algae, but this is most likely the consequence of higher temperature and has nothing to do with acidophily. Physical treatment of *Cyanidium* by sonification reduces the acid resistance of this cyanobacterium. This supports the view that the mechanism

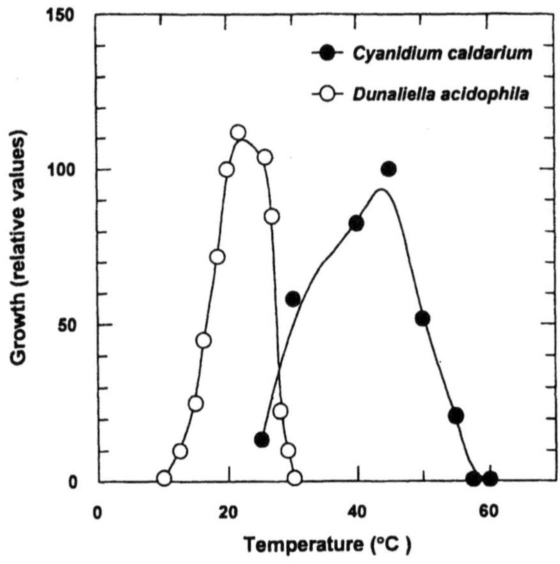

Fig. 9.5. Growth (normalised values) of the moderate thermophilic, acidophilic alga *Cyanidium caldarium* (after Brock 1978) and the slightly psychrophilic acidophilic alga *Dunaliella acidophila* (H. Gimmler, unpubl.) as a function of temperature

underlying acidophily requires the normal, intact structure of the organism, in particular of the plasma membrane and the cell wall (Enami and Fukada 1975; Enami et al.1975). Acidophily of *Cyanidium* was especially sensitive to digitonin treatment, indicating the involvement of the plasma membrane (Enami and Fukada 1977).

Methods for the isolation of pure plasma membrane fractions of *D. acidophila* are available, but these techniques were not utilised for the analysis of the lipid composition. Thus, it remains an open question whether the lipid analysis carried out by Della Greca et al. (1989) and Pollio et al. (1988) is of importance for the interpretation of acidophily of *Dunaliella acidophila.* Definitely of importance for the resistance of the plasma membrane, however, is its stability in nonionic detergents such as Triton X-100 (Pick 1992).

9.5 The Energy Metabolism — Well Protected Against the Acid Exterior

Mitochondria and chloroplasts of acid-resistant algae are embedded in a strongly buffered cytoplasm of neutral pH (Fig. 9.2) and organelles are very well protected against the acid exterior of cells by a plasma membrane with a low proton conductance. Therefore, there is no basic difference between respiration, oxidative phosphorylation, photophosphorylation and photosynthetic electron transport of acid-resistant and neutrophilic algae. A similar situation exists for the soluble enzymes also (Enami 1978); they are not particularly resistant to elevated H^+ concentrations. Consequently, most studies dealing, for example, with the photosynthetic apparatus of *Cyanidium* do emphasise phylogenetic rather than functional aspects (Kostrzewa and Zetsche 1993; Enami et al. 1995). Therefore, there is no need to discuss energy metabolism of acidophilic algae in a separate chapter. Instead, we refer to the review articles by Gimmler and Weis (1992) covering *Dunaliella acidophila* and by Cobley and Cox (1983) and Bakker (1990) dealing with related aspects of bacterial acidophiles.

9.6 Highly Resistant to Protons, but Nutritional Problems

The positive electrical potentials of acidophilic algae and the low conductance of the plasma membrane reflect excellent mechanisms against damage caused by internal acidification of the cell and subsequent inhibition of growth. However, the same mechanisms which protect cells against H^+ also prevent the uptake of those cations which are required for growth such as K^+ or NH_4^+. On the other hand, the mechanisms referred to above are expected to facilitate the uptake of anions. This may be beneficial as far as the uptake of nitrate or sulfate is concerned, but might be a disadvantage in the case of anions such as chloride.

9.6.1 Potassium and Sodium Uptake

Experimental evidence supports the view that the plasma membrane of *Dunaliella acidophila* contains, similarly to acidophilic bacteria (Michels and Bakker 1985; Bakker 1990) and acidophilic yeasts (Gläser et al. 1990), a K^+, H^+ symport (Gläser et al. 1990; Carandang et al. 1992). Such a symporter enables *D. acidophila* to take up potassium against an electrochemical concentration gradient by utilising the huge proton gradient existing across the plasma membrane (Fig. 9.8). The coupled influx of H^+ and K^+ supports a positive membrane potential which is crucial for acidophiles. Another advantage of such a cotransport compared to a theoretically feasible ATP driven, primary K^+/H^+ countertransport is the decoupling between proton extrusion and potassium accumulation. Only that way cells are able to survive under low K^+ concentrations in their environment.

The cytoplasmic Na^+ levels of *D. acidophila* are low (5 – 15 mM) in comparison to the K^+ levels (120 –160 mM). The Nernst potentials are –107 mV (K^+) and –54 mV (Na^+), respectively (Gimmler and Weis 1992). Transfer from dark to light causes a rapid decrease of the endogenous Na^+ pool, which is reversible upon darkening (Remis et al. 1994). The velocity of the light-induced changes in the Na^+ pool and the light-induced H^+ efflux are inhibited by amiloride, an inhibitor of the Na^+/H^+ antiporter of the plasma membrane. Data demonstrate that the well-known Na^+/H^+ antiporter is also present in the plasma membrane of *D. acidophila* (Fig. 9.8). However, it must be strongly regulated in order to avoid internal acidification by the influx of H^+. When the activity of the antiporter is inhibited by amiloride, less H^+ has to be exported by the PMHA.

9.6.2 Ammonium and Nitrate Uptake

A similar problem applies to the uptake of NH_4^+, although this has not been investigated experimentally so far. It is known that the dominating inorganic N source in acid habitats is ammonium (Brock 1978), which is present at pH 1 exclusively in the form of NH_4^+. This excludes any uptake of NH_3, as it can be measured with neutrophilic algae. *Cyanidium caldarium* and *Dunaliella acidophila* exhibit comparable growth on NH_4^+ and nitrate as nitrogen source (Albertano et al. 1981; Fuggi et al. 1981; Di Martino Rigano et al. 1987; Fuggi 1989). An example for *D. acidophila* is shown in Fig. 9.6. The K_m values of growth of *D. acidophila* for nitrogen were 280 μM (nitrate) and 400 μM (ammonium). Assimilation rates for both substrates were higher in the light than in the dark and occurred only under aerobic conditions. The uptake mechanisms of these ions must be entirely different. In *D. acidophila* the internal NH_4^+ concentration is close to 5 mM at an external concentration of 20 mM. So the chemical gradient across the plasma membrane is downhill and only counteracted by a positive ψ of + 50 mV. Also, the corresponding gradient for nitrate is downhill, but the membrane potential is an additional driving force, if nitrate is the species taken up. For nitrate uptake in *C. caldarium* an electroneutral H^+/NO_3^- cotransport has been suggested (Fuggi et al. 1981) using the H^+ gradient across the plasma membrane as the driving force for the uphill translocation of nitrate. The internal nitrate level is regulated by a sophisticated pump and leak mechanism, involving low- and high-affinity uptake systems for nitrate (Fuggi 1989). *C. caldarium* is able to grow at low concentrations of nitrite also, which is usually very toxic, especially in its protonated form (Fuggi 1993). The latter is the species taken up at acid pH. In order to prevent internal acidification, chloroplastic nitrite reduction must be much faster than nitrite entry. Growth of neutrophilic algae (and also that of higher plants) is often different with both inorganic N sources, because ammonium and nitrate uptake cause a differential acid load and higher pH values. Obviously, acidophilic algae do not have this problem, either due to different uptake mechanisms for these ions and/or to a stronger H^+ export capacity.

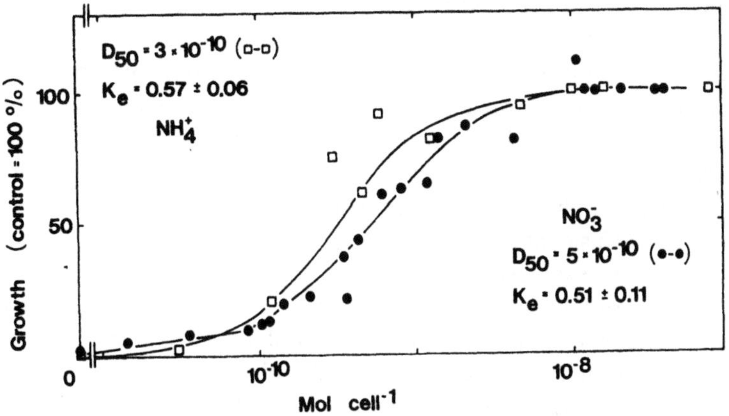

Fig. 9.6. Growth of *Dunaliella acidophila* as a function of nitrate or ammonium concentration (plotted as ratio mole nitrate or ammonium cell^{-1}). K_e Growth constant; D_{50} concentration at which half maximal growth occurs. (H. Gimmler, unpubl.)

9.6.3 Sulfate and Chloride Uptake

The uptake of the essential S source sulfate by acidophilic algae is difficult to determine, because the natural environment is extremely rich in this anion. However, the uptake of the chloride anion by *D. acidophila* has been measured (Hirsch et al. 1992). In contrast to sulfate, the chloride anion is of potential toxicity for enzymes. However, there is no uncontrolled uptake of chloride due to the positive electrical potential of *D. acidophila*. The available data suggest that chloride uptake by this alga is a catalysed diffusion, which is primarily independent of the supply of ATP. In contrast, chloride efflux is coupled to an active pump located in the plasma membrane. Results suggest a tight cooperativity between chloride uptake and efflux, with the cytoplasmic pH and the electrical potentials being important regulators. *D. acidophila* cells show a specific Cl^- requirement for growth. They tolerate up to 300 mM chloride, but keep the internal Cl^- level in excess of external Cl^- close to 30 mM (Hirsch et al. 1992).

9.6.4 Uptake of Phosphate and DIC

Phosphate and DIC (dissolved inorganic carbon) have in common that the uptake of both is essential for normal growth and that their dissociation is very different at neutral and acid pH. At neutral pH the dominant phosphate species are HPO_4^{2-} and $H_2PO_4^-$, at pH 1 phosphorus is present exclusively in the form of H_3PO_4. Bicarbonate is the abundant species at pH 7, at pH 1 only CO_2 exists. This implies different mechanisms of uptake for the uncharged species at pH 1 and the anion at 7. One of the reasons that acidophilic algae do not grow at neutral pH values may, in fact, be due to the inability of the algae to switch over efficiently between the uptake systems required at pH 1 and those required for pH 7.

At low pH and low phosphate concentration the high proton gradient across the plasma membrane is utilised for uptake of H_3PO_4 in *Dunaliella acidophila* (mechanistically not distinguishable from $H^+/H_2PO_4^-$ cotransport) (Hirsch et al. 1993). At higher phosphate concentrations (> 10 mM), experimental results permit a pure diffusional uptake of H_3PO_4. At neutral pH values a Na^+, phosphate symport cannot be excluded.

The analysis of DIC uptake in *D. acidophila* revealed that CO_2 uptake by cells adapted to air is not compatible with the view of a pure lipid solubility-dependent CO_2 uptake, but indicates a catalysed uptake of the CO_2 species (Gimmler and Slovik 1995). At elevated CO_2 concentrations, not typical for natural conditions, a non-catalysed diffusion across the plasma membrane would be sufficient for photosynthesis; but even under such conditions a catalysed uptake of CO_2 cannot be excluded. The K_m values of photosynthesis for CO_2 were 20 µM (air) and 60 µM (air enriched with CO_2), respectively. Also CO_2 uptake of the acid-tolerant alga *Eremosphaera viridis* is electrically silent and not the result of a symport or antiport coupled to electrochemical gradients of ions across the plasma membrane. Therefore it must be facilitated by a specific CO_2 transporter (Deveau et al. 1998). The environmental regulation of CO_2-concentrating mechanisms in microalgae has been reviewed recently by Beardall et al. (1998).

Because of the abundance of CO_2 in the environment of acidophilic algae, it is not surprising that they contain far lower amounts of carbonic anhydrase (CA) than neutrophilic algae (Raven 1995; Geib et al. 1996). This is valid for both the CA of the plasma membrane (external CA) and the internal CA. Photosynthesis of *D. acidophila* is not affected by the non-permeant CA inhibitor acetazolamide, indicating a low importance of this enzyme for photosynthesis. Nevertheless, since a certain amount of CA is present in the plasma membrane of *D. acidophila*, the question has to be raised how such an enzyme can still be active in the presence of 100 mM H^+.

Ribulose-1,5-bisphoshate carboxylase (Rubisco) isolated from the acidophilic and thermophilic algae *Galdieria partita* and *Cyanidium caldarium* exhibits an exceptionally high affinity for CO_2 (K_m = 6-7 µM) (Uemura et al. 1997). It is not known at present whether this high affinity is related to acidophily or to thermophily. In general, K_m values of Rubisco for CO_2 are somewhat higher in algae (20-230 μM) than in higher plants (10-20 μM) (Gimmler and Slovik 1995).

9.7 Upper Limit for H^+ Concentration or Limit for Osmotic Stress?

Culture media with pH values close to 0 and which are tolerated by acidophilic algae (Fig. 9.1), unavoidably have a high osmotic pressure between 0.5 and 1 osmol kg^{-1}. This raises the question of whether the growth limit at low pH reflects an effect of high H^+ concentrations or is caused by an upper limit of the osmotic potential of the medium. *Dunaliella acidophila* is extremely resistant to increasing osmotic concentrations of the medium, its growth is not affected by glucose, Na_2SO_4 and NaCl equivalent to 1 osmol kg^{-1} (Fuggi et al. 1988a). At higher osmolarities salt effects were observed. Growth was reduced to 50% by Na_2SO_4 and NaCl equivalent 2 osmol kg^{-1}, but not at all affected by 2 osmol glucose kg^{-1}. The pairs H_2SO_4 – Na_2SO_4 and NaCl – HCl yielded similar inhibition of growth, suggesting that growth was limited at low pH by the osmotic potential of the medium and corresponding osmoregulation rather than by the high H^+ concentration (Fuggi et al. 1988b; Gimmler and Weis 1992). Glycerol was found to be the internal osmoticum of *Dunaliella acidophila* and its concentration increased proportionally to the osmotic potential of the medium (Fuggi et al. 1988b). Corresponding investigations with other acidophilic algae have not been carried out.

Also, the upper limit of the pH profile for the growth of *D. acidophila* seems to be due to hypoosmotic effects, causing rapid swelling and eventually bursting of cells, rather than due to a decrease in the H^+ concentration (Gimmler and Weis 1992). Under constant osmotic pressure the tolerated pH range was much broader than that observed when the change of the osmotic potential with increasing pH was negligible. However, under no circumstances did cells survive at neutral pH values. The question arises whether the unusually sharp pH profile of the extreme acidophilic bacterium *Picrophylum* (Schleper et al. 1995b) is caused by a similar overlapping of osmotic effects and effects of protons in spite of the presence of a rigid cell wall.

9.8 High Resistance to Protons, but not to Weak Acids

Acidophiles are well equipped against deteriorating effects of high H^+ concentrations in their environment. In spite of this, they are more sensitive to inorganic or organic acids, which are completely protonated at acid pH, than comparable neutrophilic organisms (Fig. 9.7). Most of such low molecular weight acids are lipophilic in their non-dissociated form and accumulate according to

Henderson-Hasselbalch in alkaline compartments. Since a huge pH gradient exists across the plasma membrane of acidophiles, a large accumulation of organic acids in the cytoplasm will occur. This reflects a large acid load by indirect proton transfer. If the acid load cannot be counteracted by the activity of the PMHA, internal acidification with subsequent inhibition of metabolism and growth will take place. In addition, the accumulation of accompanying anions may cause inhibitory effects on enzymes also.

The sensitivity of acidophiles against protonated lipophilic acids is not only of academic interest, but may have ecophysiological importance (Gimmler 2001). Some acidophilic chemolithotrophic bacteria release organic acids into their environment which might accumulate in the environment up to toxic levels. Heterotrophic acidophiles utilise these organic acids as carbon sources and thereby keep their external concentrations at non-toxic levels.

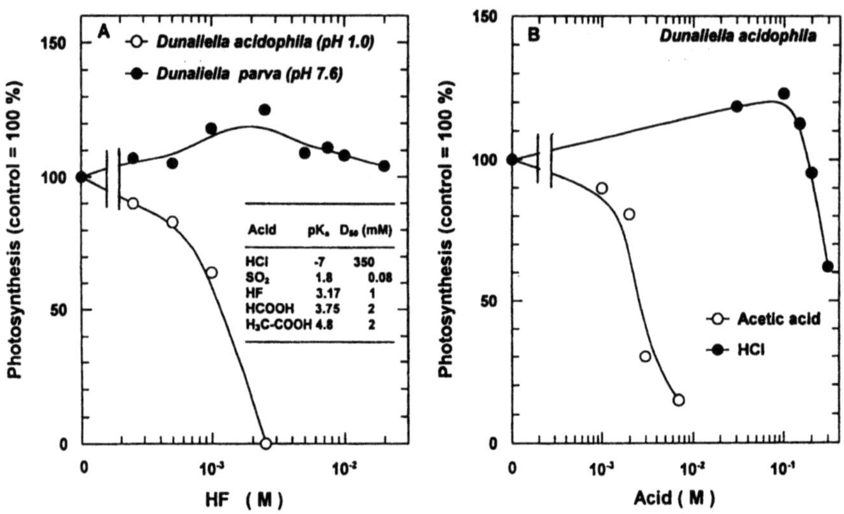

Fig. 9.7 A.B. Sensitivity of photosynthesis of the extreme acidophilic alga *Dunaliella acidophila* against low concentrations of permeant organic acids. **A** Comparison of sensitivity of photosynthesis to HF of *D. acidophila* (pH 1) and *D. parva* (pH 7.6). The *table inset* summarises the effect of some other weak acids on photosynthesis of *D. acidophila*. **B** Differential inhibition of photosynthesis of *D. acidophila* (pH 1) by acetic acid (completely protonated at pH 1) and HCl (totally dissociated at pH 1)

9.9 Fit Against Protons and Heavy Metals

Many papers on acidophilic and acidotolerant microbes deal with the coupling between acid resistance and resistance against heavy metals. Natural or man-made habitats are often rich in heavy metals, because most of them, excepting lead, show a better solubility in acid environments than at neutral or even alkaline pH values. The enhanced concentrations of heavy metals will reduce microbial colonisation of acidic habitats, if these organisms are not endowed with appropriate strategies to encounter metal toxicity.

Several mechanisms, that do not exclude each other, may contribute to the resistance of acid-tolerant algae to heavy metals (HM): (1) Binding of HM to cell wall, (2) Reduced transport of HM across the plasma membrane, (3) Active efflux of HM, (4) Compartmentation inside algal vacuoles, (5) Internal sequestering (complexation) by special proteins (metallothioneins), organic acids and sulphides (cf. Fig. 9.8 and see Chap. 12).

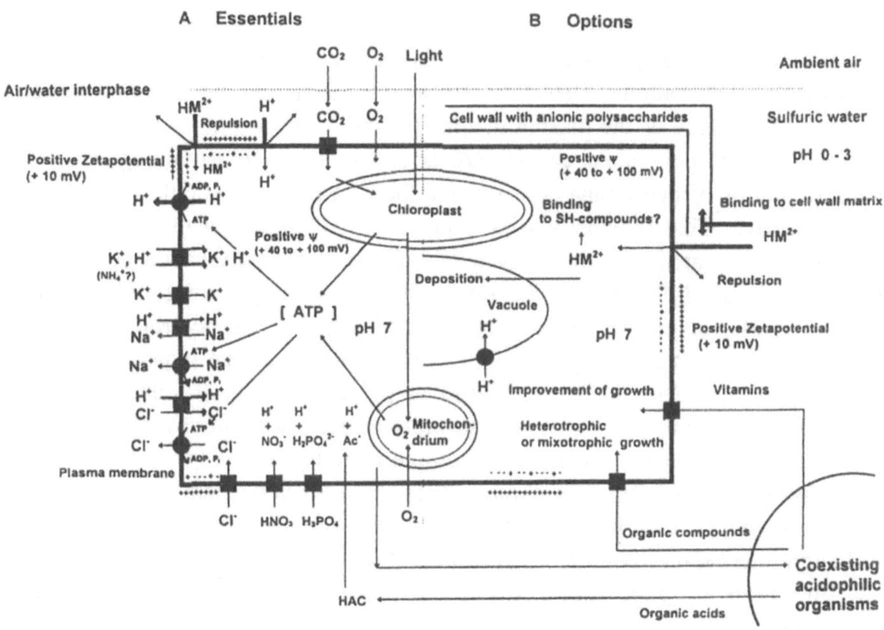

Fig. 9.8 A.B. Summarising scheme of essential (**A**, *left side*) and optional properties (**B**, *right side*) of acidophilic algae as outlined in this chapter. *Bold squares* Carrier or channel; *bold circles* primary pump; *arrows* indicate direction of transport. The scheme also emphasises that acidophilic algae live in association with other acidophiles. Inhabitants of acid ecological niches are interconnected by a variety of commensalistic or mutualistic relations, which improve their viability in a harsh environment

Dissolved cationic heavy metals get into contact with acidophiles first at the level of cell wall and then at the level of the plasma membrane, before they can enter the cell and finally interfere with metabolism. The algal cell wall behaves as a polyfunctional cation exchanger with negatively charged ligands. The cation exchange capacity depends on the pH and the metal species. Adsorption onto the cell wall decreases with decreasing pH and will approach virtually zero at pH values lower than 3. Therefore, cationic heavy metals in extreme acidic environment have best chances to approach the surface of the plasma membrane before intercepted in the cell wall. In the pH range between 3 and 5 considerable amounts of heavy metals will be bound to the cell wall material. The higher the pH, the higher is the binding of heavy metals to the cell wall material. For *Klebshormidium fluitans* (= *Hormidium fluitans*) a particular alkali-soluble polysaccharide fraction was isolated from the cell wall with a high copper-binding capacity (Strong et al. 1982). The properties of this polymer resemble those of alginate.

As shown above, zeta potentials of extreme acidophiles are positive, meaning an excess of positive charges above negative charges at the external side of the plasma membrane. A positive zeta potential is a repelling force for the uptake of di- or polyvalent cations. Therefore less heavy metals can be expected to come close to the surface of the plasma membrane and to be bound to it in acidophilic algae. If di- or polyvalent cations are bound to the few remaining accessible anionic groups of membrane constituents, the zeta potential should become even more positive. That is exactly what takes place if the cell wall-less *Dunaliella acidophila* is treated with di- or trivalent cations or the poly-cation poly-l-lysine (Gimmler et al. 1991a,b). At the same time, these poly-cations did not affect photosynthesis or growth of this alga, whereas that of the neutrophilic relative *Dunaliella parva* was strongly inhibited. The anion tungstate hyperpolarised the zeta potential of *Dunaliella acidophila* at pH 2 to negative values (–7 mV), inhibited photosynthesis of this alga strongly at pH 1 without any effect at pH 7. Electrically silent heavy metals, for example mercury (in the form of $HgCl_2$), do not affect the zeta potential or photosynthesis in a pH-dependent manner.

Results imply that heavy metals are not taken up into acidophilic algae, although internal inactivation or deposition of heavy metals could not be excluded by this type of experiment. Results explained mechanistically the earlier observation of Capasso and Pinto (1982) on the resistance of the acidophilic alga *Spermatozopsis acidophila* (= *Dunaliella acidophila*) to a large number of cationic heavy metals. Later studies showed that metal resistance is a special case neither for *Dunaliella acidophila* nor for an alga lacking a rigid cell wall. From the present view it must be generalised that all prokaryotic and eukaryotic acidophilic or acid-tolerant microbes exhibit an elevated resistance to heavy metals. Further examples of a coupling between acid resistance and heavy metal resistance in algae have been presented for the acid-resistant strains of *Chlorella vulgaris* (Rai et al. 1993, 1994a,b, 1996) and *Scenedesmus acutus* (Nalewajko et al. 1997), *Hormidium fluitans* (Madgwick and Ralph 1977), and the extreme acidophilic algae *Cyanidium caldarium* (Wood and Wang 1983) and *Euglena mutabilis* (Nakatsu and Hutchinson 1988). The results can be summarised as follows:

1. Acid-tolerant strains of *Chlorella vulgaris* grew better and showed higher

metabolic activities at lower pH values than the acid-sensitive wild types.
2. Cu^{2+} and Ni^{2+} uptake of acid-resistant strains are lower than in the acid-sensitive wild types.
3. Acid-tolerant strains overexpress PMHA with decreasing pH (Rai et al. 1993, 1994a b, 1996).
4. Cation (NH_4^+, K^+, Na^+) uptake decreases and anion uptake (HPO_4^{2-}, Cl^-) increases with increasing H^+ concentration.
5. Cu^{2+}-and Ni^{2+}-induced cation efflux (Na^+, K^+) is more pronounced in the acid-sensitive wild type than in the acid-resistant strains.

Besides reduced uptake of heavy metals by acidophilic algae, additional mechanisms of detoxification may operate whenever some heavy metals make their way into the cells. Similar to bacteria, fungi, and neutrophilic algae, the acid-resistant algae may display HM induced activation of HM efflux (mechanism 3), compartmentation inside vacuoles (mechanism 4) or internal sequestering (complexation) by special proteins (metallothioneins), organic acids and sulphides (mechanism 5). However, the occurrence of mechanisms 3 to 4 with reference to acidophily does not seem to have been tested so far.

9.10 Acidophiles Are Designed to Survive in Nature — and Not in Aseptic Glass Tubes

Acidophilic algae cannot simply considered to be mini plastic bags, impermeable for H^+, inside equipped with a sophisticated bioenergetic machinery for the reexport of H^+ and fit for the survival at acid pH under the aseptic conditions of test tubes. They are organisms evolutionary designed to survive under the multifactorial conditions of nature. This means, for example, that they do not occur alone in acidic habitats, but in association with other organisms. It means also, that they are exposed to a large number of abiotic factors, which differ from the conditions of a test tube. Typically, conditions in the field are subject to considerable variations. Acidophiles may fall prey to eukaryotic predators. They have to survive in spite of the presence of competing acidophiles or may benefit from excretion products of other acidophiles (commensalism). Finally, mutualistic relations may exist between various acidophiles (Nakatsu and Hutchinson 1988; Gimmler and Carandang 1998;Gimmler 2001) and improve their survival in a harsh environment.

All acidophilic and acid-tolerant algae studied so far are obligate aerobes, and so are acidophilic bacteria (Johnson 1998). The best probable explanation for this phenomenon is that the maximal mechanistic power supplied by glycolysis or fermentation is not sufficient for the maintenance of a strong proton gradient across the plasma membrane (cf. Fig. 9.3). This does not mean that acidophilic algae are necessarily autotrophs. For example, *Galdieria sulphuraria* grows mixotrophically or heteretrophically (Gross and Schnarrenberger 1995) and so can

Euglena mutabilis. *Dunaliella acidophila* grows auxotrophically (Gimmler and Carandang 1998). Also among bacterial acidophiles we find representatives of both nutrition styles, autotrophs and heterotrophs. Remarkable for some acidophilic algae is a tendency for a sessile or "benthic" life style (*Cyanidium, Galderia, Euglena mutabilis*). The range of tolerated temperature divides acidophilic algae into mesophiles (T_{opt} 20-40 °C) and moderate thermophiles (T_{opt} 40-60 °C) (Fig. 9.5), whereas extreme thermophiles (T_{opt} >60 °C) are found only among bacteria and archaebacteria. However, temperature in the field is not constant compared to laboratory conditions. In the effluents of hot springs, pronounced longitudinal and lateral temperature gradients develop (Brock 1978), which affect the distribution of species within the microbial community as well as the development of different temperature strains. The author believes that a mere study of membrane transport, membrane stability and bioenergetics of acidophilic algae may not be sufficient to unravel their fascinating secrets. One should also thoroughly scrutinise the complex ecological interactions of these microbial communities.

9.11 Essentials and Options — What Makes an Ideal Acidophilic Alga?

Essential and optional properties of acid-resistant algae and acidophilic microbial communities are compiled in Fig. 9.8. The scheme comprises most of the information presented in this chapter, but in order to avoid confusion all details are not included. Like every scheme, it presents an oversimplification, neglecting many complex intracellular interactions of a given species or interactions between different algal species in an acid habitat. The scheme matches properties of acidophilic algae probably better than those of acid-resistant algae, which are gradually distinct from their acidophilic relatives.

In the following, some aspects are summarised which should be given more attention in the future. (1) The complex ecological interactions within acidophilic communities require extended analysis. (2) In order to find out whether there is a special "acidophilic" architecture of the plasma membrane of acidophilic algae, the chemical composition of the plasma membrane should be analysed. This requires techniques to isolate the pure plasma membrane fraction. In order to avoid any interference with thermophily, mesophilic algal species should be preferred. (3) Studies with artificial lipid vesicles can elucidate the mechanisms which determine the intrinsic permeability of phosphobilayers for protons (water wires? long fatty acid chains as H^+ carriers?: Gutknecht 1988; Miedema et al. 1996 and references therein). (4) A problem requiring much more attention is how plasma membrane-bound enzymes (partially exposed to the high proton concentration of the medium), such as the H^+-ATPase or the carbonic anhydrase and even "extracellular enzymes" (acid phosphatase?) can function at extremely low pH values. (5) Extremely useful in this respect would be the development of methods for the determination of the surface pH and the surface potential of acidophilic algae. From the thermodynamic point of view, it is quite clear that the pH at the surface of the plasma membrane of acidophiles is definitely higher than that of the bulk medium, but the extent is not known. (6) For vacuolated, extreme acidophilic algae the contribution of the vacuoles to pH homeostasis has to be evaluated. (7) No data are available so far about the effect of those heavy metals which succeed to enter acidophilic algae in spite of a plasma membrane quite impermeable to these elements. Do cells respond to the uptake of heavy metals by increased formation of SH compounds, as it can be observed with acidophilic fungi? (8) Since copper, cobalt, zinc and nickel are essential heavy metals, it should be investigated as to how acidophilic algae manage to take up sufficient essential heavy metals to satisfy the demand when only low concentrations of these elements are present in the environment. (9) In comparison to acid-tolerant and acidophilic bacteria (Hall et al. 1995), there is a general lack of information about molecular responses of algae to H^+ stress, that is how pH stress regulates gene expression. Our present knowledge about mechanisms of acidophily in bacteria has advanced by the use of bacterial mutants. The meagre availability of mutants of acid-resistant algae and cyanobacteria is definitely hampering progress in the elucidation of cardinal mechanisms of acidophily in these organisms.

References

Albertano P (1995) Microalgae from sulphuric acid environments. In: Wiessner W, Schnepf E, Starr RC (eds) Algae, environment and human affairs. Biopress, Bristol, pp19-39

Albertano P, Pinto G, Santisi S, Taddei R (1981) *Spermatozopsis acidophila* Kalina (Chlorophyta, Volvocales), a little-known alga from highly acidic environments. G Bot Ital 115:65-76

Almer B, Dickson W, Ekström C, Hörnström E, Miller U (1974) Effects of acidification on Swedish Lakes. Ambio 3:30-36

Bakker EP (1990) The role of alkali-cation transport in energy coupling of neutrophilic and acidophilic bacteria: an assessment of methods and concepts. FEMS Microbiol Rev 75:319-334

Beardall J, Johnston A, Raven J (1998) Environmental regulation of CO_2-concentrating mechanisms in microalgae. Can J Bot 76:1010-1017

Boenzi D, Luca PD, Taddei R (1977) Fatty acids in *Cyanidium.* G Bot Ital 111:129-134

Booth IR (1985) Regulation of cytoplasmic pH in bacteria. Microbiol Rev 49:359-378

Brierley JA (1990) Acidophilic thermophilic archaebacteria: potential application for metal recovery. FEMS Microbiol Rev 75:287-292

Brock TD (1973) Lower pH limit for the existence of blue-green algae: evolutionary and ecological implications. Science 179:480-483

Brock TD (1978) Thermophilic microorganisms and life at high temperatures. Springer, Berlin Heidelberg New York

Capasso L, Pinto G (1982) Resistance of the alga *Spermatozopsis acidophila* Kalina (Chlorophyta, Volvocales) to heavy metals. G Bot Ital 116:275-282

Carandang JS, Pick U, Sekler I, Gimmler H (1992) K^+-fluxes and K^+-content in *Dunaliella acidophila*, an alga with positive electrical potentials. I. Low affinity uptake systems. J Plant Physiol 139:413-421
Carandang JSR VI, Gimmler H (1998) Vitamin uptake by the green alga *Dunaliella acidophila*. Manila J Sci 1:5- 8
Cobley JG, Cox JC (1983) Energy conservation in acidophilic bacteria. Microbiol Rev 47:579-595
Cooke WB (1966) The occurence of fungi in acid mine drainage. Proc Ind Waste Conf 21:58-274
Della Greca M, Monaco P, Pinto G, Pollio A, Previtera L (1989) Lipid composition of the acidophilic alga *Dunaliella acidophila*. II. Molecular species of galactolipids. Biochim Biophys Acta 1004: 271-273
De Luca P, Moretii A, Taddei R (1978) Presence of *Cyanidioschyzon merolae* in extra-European acidic environments: USA and Indonesia. Delpinoa 18-19:69-76
De Luca P, Gambardella R, Merola AA (1979) Thermoacidophilic algae of North America and Central America. Bot Gaz 140:418-427
De Luca P, Musacchio A, Taddei R (1981) Acidophilic algae from the fumaroles of Mount Lawu, Java, locus classicus of *Cyanidium caldarium*. G Bot Ital 115:1-10
Deveau JST, Khosravani H, Lew R, Colman B (1998) CO_2 uptake mechanism in *Eremosphaera viridis*. Can J Bot 76:1161-1164
Di Martino Rigano V, Vona V, Di Martino C, Rigano C (1987) Regulatory aspects of NH_4^+ utilization in the acidophilic thermophilic unicellular red alga *Cyanidium caldarium*. New Phytol 105:247-254
Enami I (1978) Mechanisms of the acidophily and thermophily of *Cyanidium caldarium*. Part 5. Acid and heat stabilities of the soluble proteins. Plant Cell Physiol 19:869-876
Enami I, Fukada I (1975) Mechanism of the acido- and thermophily of *Cyanidium caldarium* Geitler. I. Effect of temperature, pH and light intensity on the photosynthetic oxygen evolution of intact and treated cells. Plant Cell Physiol 15:211-220
Enami I, Fukada I (1977) Mechanism of the acido- and thermophily of *Cyanidium caldarium* Geitler. Part 3. Loss of these characteristics due to detergent treatment. Plant Cell Physiol 18:671-680
Enami I, Kura-Hotta M (1984) Effect of intracellular ATP levels on the light-induced proton efflux from intact cells of *Cyanidium caldarium*. Plant Cell Physiol 25:1107-1114
Enami I, Nagashima H, Fukada I (1975) Mechanism of the acido- and thermophily of *Cyanidium caldarium* Geitler. II. Physiological role of the cell wall. Plant Cell Physiol 15:221-231
Enami I, Akutsu H, Kyogoku Y (1986) Intracellular pH regulation in an acidophilic unicellular alga, *Cyanidium caldarium*: ^{31}P-NMR determination of intracellular pH. Plant Cell Physiol 27:1351-1359
Enami I, Murayama H, Ohta H, Kamo M, Nakazato K, Shen JR (1995) Isolation and characterization of a photosystem II complex from the red alga *Cyanidium caldarium*: association of cytochrome c-550 and a 12-kDa protein with the complex. Biochim Biophys Acta 1232:208-216
Fogg GE, Stewart WDP, Fay P, Walsby AE (1973) The blue-green algae. Academic Press, London, 255 pp
Fuggi A (1989) Short-term regulation of nitrate uptake by a pump and leak mechanism in the acidophilic nonvacuolated alga *Cyanidium caldarium*. Biochim Biophys Acta 974:141-148
Fuggi A (1993) Uptake and assimilation of nitrite in the acidophilic red alga *Cyanidium caldarium* Geitler. New Phytol 125:351-360
Fuggi A, Di Martino Rigano V, Vona V, Rigano C (1981) Nitrate and ammonium assimilation in algal cell suspensions and related pH variations in the external medium, monitored by electrodes. Plant Sci Lett 23:129-138

Fuggi A, Pinto G, Pollio A, Taddei R (1988a) Effects of NaCl, Na_2SO_4, H_2SO_4 and glucose on growth, photosynthesis, and respiration in the acidophilic alga *Dunaliella acidophila* (Volvocales, Chlorophyta). Phycologia 27:334-339

Fuggi A, Pinto G, Pollio A, Taddei R (1988b) The role of glycerol in osmoregulation of the acidophilic alga *Dunaliella acidophila* (Volvocales, Chlorophyta). Effect of solute stress on photosynthesis, respiration and glycerol synthesis. Phycologia 27:439-446

Geib K, Golldack D, Gimmler H (1996) Is there a requirement for an external carbonic anhydrase in the extreme acid-resistant green alga *Dunaliella acidophila*? Eur J Phycol 31:1-11

Gimmler H (2001) Mutualistic relations between algae and fungi (excluding lichens). Prog Bot 62: 194-214

Gimmler H, Carandang JS (1998) Do vitamins play a role in commensalistic relationships between microorganisms coexisting in extremely acid habitats? Acta Manilana 46:1-7

Gimmler H, Slovik S (1995) The mode of uptake of dissolved inorganic carbon in the extremely acid-resistant green alga *Dunaliella acidophila*. Theoretical considerations and experimental observations. Plant Physiol Biochem 33:655-664

Gimmler H, Weis U (1992) *Dunaliella acidophila* - life at pH 1.0. In: Avron M, Ben Amotz A (eds) *Dunaliella*. physiology, biochemistry and biotechnology. CRC Press, Boca Raton, Florida, pp 99-133

Gimmler H, Kugel H, Leibfritz D, Mayer A (1988) Cytoplasmic pH of *Dunaliella parva* and *Dunaliella acidophila* as monitored by in vivo ^{31}P-NMR spectroscopy and the DMO method. Physiol Plant 74:521-530

Gimmler H, Weis U, Weiss C, Kugel H, Treffny B (1989) *Dunaliella acidophila* (Kalina) Masyuk —an alga with a positive membrane potential. New Phytol 113:175-178

Gimmler H, Schieder M, Kowalski M, Zimmermann U, Pick U (1991a) *Dunaliella acidophila* – an alga with a positive zeta potential at its optimal pH for growth. Plant Cell Environ 14:261-269

Gimmler H, Treffny B, Kowalski M, Zimmermann U (1991b) The resistance of *Dunaliella acidophila* against heavy metals: the importance of the zeta potential. J Plant Physiol 138:708-716

Gläser HU, Sekler I, Pick U (1990) Indications for a K^+/H^+ cotransport system in plasma membranes from two acidophilic microorganisms. Biochim Biophys Acta 1019:293-299

Gromov BV, Mamkaeva KA, Bobina VD (1988) Acidophilic *Ochromonas* (Chrysophyceae) from a sulfurous spring on Kunashir island (Russian SFSR USSR). IZV. Akad Nauk SSSR Ser Biol 0(2): 293-296

Gross W, Schnarrenberger C (1995) Heterotrophic growth of two strains of the alga *Galdieria sulphuraria* (Rhodophyta). Plant Cell Physiol 36:633-638

Gross W, Kuever J, Tischendorf G, Bouchaala N, Büsch W (1998) Cryptoendolithic growth of the red alga *Galdieria sulphuraria* in volcanic areas. Eur J Phycol 33:25-31

Gutknecht J (1988) Proton conductance caused by long-chain fatty acids in phospholipid bilayers. J Membr Biol 106:83-93

Hall HK, Karem KL, Foster JW (1995) Molecular response of microbes to environmental pH stress. Adv Microb Physiol 37:229-272

Hargreaves JW, Whitton BA (1976) Effect of pH on growth of acid stream algae. Br Phycol J 11:215-223

Havas M, Hellebust JA, Hutchinson TC, Sheath RG (1982) Effects of long term natural acidification on the algal communities of tundra ponds at the Smoking Hills, Northwest Territories Canada. Can J Bot 60:58-72

Hirsch R, Carandang JS, Gimmler H (1992) Cl^- fluxes and Cl^- content of *Dunaliella acidophila* — an alga with a positive membrane potential. J Exp Bot 43:887-896

Hirsch R, De Guis M, Falkner G, Gimmler H (1993) Flexible coupling of phosphate uptake in *Dunaliella acidophila* at extremely low pH values. J Exp Bot 44:1321-1330

Johnson DB (1998) Biodiversity and ecology of acidophilic microorganisms. FEMS Microbiol Ecol 27:307-317

Kalina T (1965) Zur Morphologie und Taxonomie der Gattung *Spermatozopsis Korschikow* (Volvocales) - *Spermatozopsis acidophila* sp. n. Preslia 37:9-12
Knoll AH, Bauld J (1989) The evolution of ecological tolerance in prokaryotes. Trans Roy Soc Edinb, Earth Sci 80:209-223
Kostrzewa M, Zetsche K (1993) Organization of plastic-encoded ATPase genes and flanking regions including homologues of infB and tsf in the thermophilic red alga *Galdieria sulphuraria*. Plant Mol Biol 23:67-76
Krulwich TA, Guffanti AA (1983) Physiology of acidophilic and alkalophilic bacteria. Adv Microbial Physiol 24:173-214
Krulwich TA, Agus R, Schneier M, Guffanti AA (1985) Buffering capacity of bacilli that grow at different pH ranges. J Bacteriol 162:768-772
Kura-Hotta M, Enami I (1981) Light-induced efflux from intact cells of *Cyanidium caldarium*. Plant Cell Physiol 22:1175-1183
Kwiatkowski RE, Rolff JC (1976) Effects of acidity on the phytoplankton and primary production of selected northern Ontario lakes. Can J Bot 54:2546-2561
Lane AE, Burris JE (1981) Effects of environmental pH on internal pH of *Chlorella pyrenoidosa, Scenedesmus quadricauda* and *Euglena mutabilis*. Plant Physiol 58:439-442
Langworthy TA (1978) Microbial life in extreme pH values. In: Kushner DJ (ed) Microbial life in extreme environments. Academic Press, New York, pp 279-315
Lopez-Arcilla AI, Marin I, Amils R (1995) Microbial ecology of an acid river: biotechnological applications. In: Vargas T, Jerez CA, Wiertz JV, Toledo H (eds) Biohydrometallurgical processing vol II. University of Chile, Santiago, pp 63-74
Madgwick JC, Ralph BJ (1977) The metal tolerant alga *Hormidium fluitans* (Gay) Heering from acid mine drainage waters in Northern Australia and Papua-New Guinea. In: Schwartz W (ed) Conference, bacterial leaching. Verlag Chemie, Weinheim, pp 85-91
Mann H, Fyfe WS, Kerric R, Wiseman M (1989) Retardation of toxic heavy metals dispersion from nickel-copper mine tailings, Sudbury District, Ontario: role of acidophilic microorganisms I. Biological pathway of metal retardation. Biorecovery 1:155-172
Matin A (1990) Keeping a neutral cytoplasm; the bioenergetics of obligate acidophiles. FEMS Microbiol Rev 75:307-318
McClintock M, Higinbotham N, Uribe E, Cleland RE (1982) Active irreversible accumulation of extreme levels of H_2SO_4 in the brown alga *Desmarestia*. Plant Physiol 70:771-774
Melkonian M, Preisig HR (1984) An ultrastructural comparison between *Spermatozopsis* and *Dunaliella*. (Chlorophyceae). Plant Syst Evol 146:31-46
Merola AA, Castaldo R, De Luca P, Gambardella R, Musacchio A, Taddei R (1981) Revision of *Cyanidium caldarium* three species of acidophilic algae. G Bot Ital 115:189-196
Michels M, Bakker EP (1985) Generation of a large, protonophore-sensitive proton motive force and pH differences in the acidophilic bacteria *Thermoplasma acidophilum* and *Bacillus acidocaldarius*. J Bacteriol 161:231-237
Miedema H, Staal M, Prins HBA (1996) pH induced proton permeability of plasma membrane vesicles. J Membr Biol 152:159-167
Miwa T, Esaki H, Umemori J, Hino T (1997) Activity in ruminal bacteria with special reference to acid tolerance. Appl Environ Microbiol 63:2155-2158
Nagashima H, Fukada I (1981) Morphological properties of *Cyanidium caldarium* and related algae in Japan. Jpn J Phycol 29:237-242
Nakatsu C, Hutchinson TC (1988) Extreme metal and acid tolerance of *Euglena mutabilis* and an associated yeast from Smoking Hills, Northwest Territories, and their apparent mutualism. Microb Ecol 16:213-231
Nalewajko C, Colman B, Olaveson M (1997) Effects of pH on growth, photosynthesis, respiration, and copper tolerance of three *Scenedesmus* strains. Environ Exp Bot 37:153-160
Norris PR, Ingledew WJ (1992) Acidophilic bacteria: adaptations and applications. In: Herber SA, Sharp RJ (eds) Molecular biology and biotechnology of extremophiles. Blackie, Glasgow, pp 115-142

Norris PR, Johnson DB (1998) Acidophilic microorganisms. In: Horiskoshi K, Grant WD (eds) Extremophiles. Wiley-Liss, New York, pp 133-153

Ohta H, Shirakawa H, Uchida K, Yoshida M, Matuo Y, Enami I (1997) Cloning and sequencing of the gene encoding the plasma membrane H^+-ATPase from an acidophilic red alga *Cyanidium caldarium*. Biochim Biophys Acta 1319:9-13

Pick U (1992) ATPases and ion transport in *Dunaliella*. In: Avron M, Ben Amotz A (eds) *Dunaliella* physiology, biochemistry and biotechnology. CRC Press, Boca Raton, Florida, pp 99-133

Pick U (1998) *Dunaliella* : a model extremophilic alga. Isr J Plant Sci 46:131-139

Pinto G (1993) Acid-tolerant and acidophilic algae from Italian environments. G Bot Ital 127:400-406

Pinto G, Taddei R (1975) New stations for *Cyanidium caldarium*. New record in Italy. Delpinoa 14-15:125-140

Pinto G, Taddei R (1978) Algae of Italian acidic waters and soils. Delpinoa 18-19:77-106

Pollio A, Della Greca M, Monaco P, Pinto G, Previtera L (1988) Lipid composition of the acidophilic algae *Dunaliella acidophila* (Volvocales, Chlorophyta). I. Non-polar lipids. Biochim Biophys Acta 963:53-60

Rai PK, Mallick N, Rai LC (1993) Physiological and biochemical studies on an acid-tolerant *Chlorella vulgaris* under copper stress. J Gen Appl Microbiol 39:529-540

Rai PK, Mallick N, Rai LC (1994a) Effect of nickel on certain physiological and biochemical behaviour of an acid-tolerant *Chlorella vulgaris*. Biometals 7:193-200

Rai PK, Mallick N, Rai LC (1994b) Effect of Cu on growth, mineral uptake, photosynthesis and enzyme activities of *Chlorella vulgaris* at different pH values. Biomed Environ Sci 7:56-67

Rai LC, Rai PK, Mallick N (1996) Regulation of heavy metal toxicity in acid-tolerant *Chlorella*: physiological and biochemical approaches. Environ Exp Bot 36: 99-109

Raven JA (1976) Transport in algal cells. In: Pirson AD, Zimmermann H (eds) Encyclopedia of plant physiology vol 2 A. New Series, Springer, Berlin Heidelberg New York, pp129-188

Raven JA (1982) The energetics of freshwater algae; energy requirements for biosynthesis and volume regulation. New Phytol 92:1-20

Raven JA (1990) Sensing pH? Plant Cell Environ 13:721-729

Raven JA (1995) Photosynthetic and non-photosynthetic roles of carbonic anhydrase in algae and cyanobacteria. Phycologia 34:93-101

Remis D, Simonis W, Gimmler H (1992) Measurements of the transmembrane electrical potential of *Dunaliella acidophila* by microelectrodes. Arch Microbiol 158:350-365

Remis D, Treffny B, Gimmler H (1994) Light-induced H^+ transport across the plasma membrane of the acid-resistant green alga *Dunaliella acidophila*. Plant Physiol Biochem 32:1-10

Rius N, Sole M, Francia A, Loren JG (1994) Buffering capacity and membrane H^+ conductance of lactic acid-bacteria. FEMS Microbiol Lett 120:291-295

Satake K, Oyagi A, Iwao Y (1995) Natural acidification of lakes and rivers in Japan: The ecosystem of Lake Usoriko (pH 3.4-3.8). Water Air Soil Pollut 85:511-516

Schleper C, Puehler G, Kühlmorgen B, Zillig W (1995a) Life at extremely low pH. Nature 375:741-742

Schleper C, Puehler G, Holz I, Gambacorta A, Janekovic D, Santarius U, Klenk HP, Zillig W (1995b) *Picrophilus* gen.nov., fam nov.: a novel aerobic, heterotrophic, thermoacidophilic genus and family comprising archaea capable of growth around pH 0. J Bacteriol 177:7050-7059

Seckbach J (1992) The Cyanidiophyceae and the "anomalous symbisosis" of *Cyanidium caldarium*. In: Ressier W (ed) Algae and symbiosis. Biopress, Bristol, pp 399-426

Seckbach J (1995) The first eukaryotic cells: acid hot-spring algae. Evolutionary paths from prokaryotes to unicellular red algae via *Cyanidium caldarium* (pre-Rhodophyta) succession. Biol Phys 20:335-345

Seckbach J, Gonzales E, Wainwright IM, Gross W (1992) Peroxisomal function in the Cyanidiophyceae (Rhodophyta): a discussion of phylogenetic relationship and the evolution of microbodies. Nova Hedwigia 55:99-109

Seckbach J, Ikan R, Ringelberg D, White D (1993) Sterols and phylogeny of the acidophilic hot springs algae *Cyanidium caldarium* and *Galdieria sulphuraria*. Phytochemistry 34:1345-1349

Sekler I, Gläser HU, Pick U (1991) Characterization of a plasma membrane H^+-ATPase from the extremely acidophilic alga *Dunaliella acidophila*. J Membr Biol 121:51-57

Sekler I, Remis D, Gimmler H, Pick U (1993) Inhibition of the plasma membrane H^+-ATPase from *Dunaliella acidophila* by omeprazole. Biochim Biophys Acta 1142:88-92

Sekler I, Weiss M, Pick U (1994) Activation of the *Dunaliella acidophila* plasma membrane H^+-ATPase by trypsin cleavage of a fragment that contains a phosphorylation site. Plant Physiol 105: 1125-1132

Starkey RL, Waksman SA (1984) Fungi tolerant to external acidity and high concentrations of copper sulfate. J Bacteriol 45:509-519

Steinberg CEW, Schäfer H, Beisker W (1998) Do acid-tolerant cyanobacteria exist? Acta Hydrochim Hydrobiol 26:13-19

Strong JRP, Madgwick JC, Ralph BJ (1982) Metal-binding polysaccharides from the alga *Klebshormidium fluitans*. Biotechnol Lett 4:239-242

Uemura K, Anwaruzzaman Miyachi S, Yokota A (1997) Ribulose-1,5-bisphosphate carboxylase/oxygenase from thermophilic red algae with strong specificity for CO_2. Biochem Biophys Res Commun 233:568-571

Weiss M, Pick U (1996) Primary structure and effect of pH on the expression of the plasma membrane H^+-ATPase from *Dunaliella acidophila* and *Dunaliella salina*. Plant Physiol 112:1693-1702

Wood JM, Wang HK (1983) Microbial resistance to heavy metals. Environ Sci Technol 17:582-590

Ziegler K, Hauska G, Nelson N (1995) *Cyanidium caldarium* genes encoding subunits A and B of V-ATPase. Biochim Biophys Acta 1230:202-206

10 Alkaliphilic and Alkali-Tolerant Algae

H. Gimmler and B. Degenhardt*

Contents

*Julius-von-Sachs Institute of Bioscience, University of Würzburg, Julius-von-Sachs Platz 2, 97082 Würzburg, Germany

10.1 Why Study Alkaliphilic and Alkali-Tolerant Algae?

Microorganisms which exhibit good growth at pH values between 10 and 11, but grow poorly at neutral pH values are defined as alkaliphiles (occasionally also as alkalophiles) (Kroll 1990). Figure 10.1 gives an example for the pH profile of growth of the alkaliphilic and salt-tolerant cyanobacterium *Spirulina platensis* (Fig. 10.1A) and the alkaliphilic, but far less salt-tolerant bacterium *Bacillus firmus* (Fig. 10.1B). The alkali-tolerant microbes grow and survive at pH values

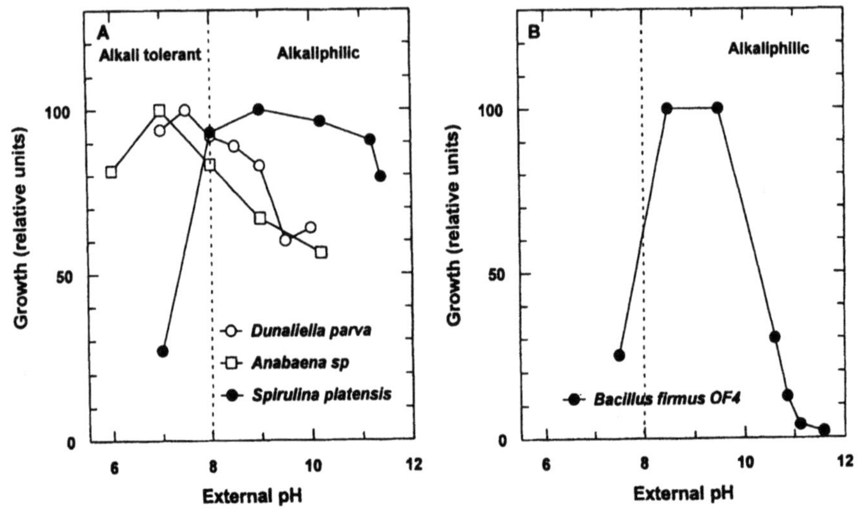

Fig. 10.1A. B. Growth of an alkali-tolerant alga and alkaliphilic cyanobacteria (**A**) and an alkaliphilic bacterium (**B**) as a function of the external pH. Data after Belkin and Boussiba (1991) (*Anabaena* sp., *Spirulina platensis*), Krulwich et al. (1998) (*Bacillus firmus* OF4) and H. Gimmler (*Dunaliella parva*, unpubl.)

higher than 9, but exhibit optimal growth at neutrality or below. In Fig. 10.1 two examples of this type are depicted, one for a prokaryote (*Anabaena* sp., Cyanobacteria) and one for a eukaryote (*Dunaliella parva*, Chlorophyta) (Fig. 10.1A). Bacteria thriving in alkaline environments are diverse and often adapted to multiple environmental stresses, for example high salinity or high temperatures, in addition to the low proton concentration (Krulwich et al. 1988; Grant et al. 1990). Similar adaptations may occur in cyanobacteria and algae also. Thus, it becomes difficult to distinguish properties reflecting adaptations to high pH from those caused by other environmental stress factors. With bacteria, the basic experimental strategy in this respect was to analyse species or strains which are not obligatory extremophiles, in respect to either alkaline pH or high temperature or high salinity. In fact, most of the available information about the mechanisms of alkali tolerance in bacteria was derived from studies of non-marine bacilli species. This is a very important point to remember when in this chapter alkali-tolerant algae and blue-green algae are occasionally compared with alkaliphilic bacteria. The latter group is dominated by organisms tolerating high salt and high pH at the same time. As described below, this requires different bioenergetic responses for the maintenance of appropriate H^+ and Na^+ levels in the cytoplasm than in bacteria.

Alkaliphilic bacteria are of interest for both biotechnological and scientific reasons. The biotechnological interest in alkaliphilic bacteria originates from the fact that these organisms contain alkaline hydrolases, for example proteinases, which can serve as biological detergents. The current proportion of total world enzyme production for the laundry detergent market is for example about 25% (Grant et al. 1990). Additionally, there is an increasing number of applications of enzymes isolated from alkaliphilic bacteria in food, leather tanning and waste-management industries. Corresponding applications of alkali-tolerant algae and blue-green algae are rare. However, some alkaliphilic cyanobacteria such as *Spirulina* play an important role in the food chain of ecosystems such as soda lakes. Blue-green algae of these habitats exhibit a high photosynthetic productivity and also serve as human food (single-cell protein) of high nutritional value in various parts of the world (Ciferri 1983; Ciferri and Tiboni 1985; Richmond 1986, 1988).

Last but not least, there is also an increasing academic interest in alkaliphiles, even though in relation to the early scientific interest in their counterparts at the acidic side of the pH spectrum, the acidophiles, intensive studies of alkaliphiles started relatively late. The breakthrough came with the introduction of the chemiosmotic theory of energy generation (Grant and Horikoshi 1989). In prokaryotes, including cyanobacteria (formerly blue-green algae), the coupling membrane, which couples proton translocation through an ATP synthetase (coupling factor) to ATP synthesis, is the plasma membrane (cf. Fig.10.8A). Respiration causes the generation of an electrochemical potential ($\Delta\mu_{H^+}$) by active extrusion of H^+ across the plasma membrane. In neutrophilic prokaryotes both the H^+ gradient (Δ pH, externally more acid than inside) and the electrical membrane potential gradient ($\Delta\psi$) are the driving forces for ATP synthesis catalysed by the ATP synthetase and coupled ion and solute transport (in the form of co- or countertransport). There is reason to assume that the cytoplasmic pH of alkaliphilic bacteria is regulated to values close to neutrality. This means the existence of a reversed Δ pH (externally more alkaline than inside). The reversed Δ pH causes a drastic decrease in the driving force $\Delta\mu_{H}+$. The basic question of all alkali-tolerant prokaryotes is how to overcome this problem in terms of the chemiosmotic theory.

In alkaliphilic eukaryotes the situation is somewhat different, because mitochondrial and thylakoid membranes, but not the plasma membrane represent the coupling membranes (cf. Fig. 10.8B). However, also at the algal plasma membrane an important H^+ export takes place, catalysed by the plasma membrane-bound H^+ ATPase (PMHA). This enzyme exclusively consumes ATP and is not able to synthesise it. The $\Delta\mu_{H^+}$ generated by the PMHA is utilised for ion exchange. Also here, the reversed Δ pH is of disadvantage and again the question must be raised as to how alkali-tolerant algae counteract the decrease in $\Delta\mu_{H^+}$. In this chapter we often use the term alkaline pH without further reflection. In order to judge what an external pH of 11 really means for alkaliphiles in our conventional thinking, consider a shell around a spherical prokaryotic cell (diameter 4 μm), which has the same volume as the cell. A rough estimation, using Avogadro's number, demonstrates that at pH 11 in the vicinity of such a prokaryotic cell as well as inside the cell we can expect at the most two single protons! This thermodynamic consideration teaches us that one has to be careful in treating high pH values in volumes of microscopic dimensions in the same way as we are used to do with neutral or acid pH values in larger volumes, being larger cells or the bulk medium.

Besides energetic problems there are also some nutritional problems which have to be solved by alkali-resistant algae and cyanobacteria. These additional problems are partially the same as for alkaliphilic bacteria and partially different. At alkaline pH for example, phosphate might become limiting because it tends to precipitate in the presence of divalent cations such as Mg^{2+} and Ca^{2+}. DIC (dissolved inorganic carbon), required for photosynthesis, exists exclusively in the form of bicarbonate and carbonate at alkaline pH. There is essentially no CO_2 available at such pH values. This requires special adaptations during uptake. On the other hand, in alkaline habitats there is a never-ending excess of DIC. How do algae respond to this excess?

In the following the major physiological problems of alkali-tolerant and alkaliphilic cyanobacteria and algae will be reviewed with emphasis on the bioenergetic problems of algae originating from marine or hypersaline habitats. In respect to bacterial alkaliphiles we refer to reviews of Horikoshi and Akiba (1982), Krulwich and Guffanti (1983), Krulwich et al. (1988, 1998), Grant and Horikoshi (1989), Grant et al. (1990), Ivey et al. (1998) and Horikoshi (1998).

10.2 Alkaline Habitats Contain High Na^+ Concentrations

Properties and genesis of terrestrial and aquatic alkaline habitats and their distribution in the world have been reviewed by Grant and Tindall (1986), Wood and Talling (1988), Javor (1989) and Grant et al. (1990). Basically, habitats of stable and transient alkaline conditions have to be distinguished (Table 10.1). The best-known habitats with stable alkaline conditions are soda lakes, distributed worldwide. They contain Na_2CO_3 and NaCl in molar concentrations, but have low Ca^{2+} and sulfate. The pH values of these lakes vary between 10 and 11.5, and salinity and alkalinity are subject to variations. The genesis of soda lakes has been summarised by Grant and Tindall (1986) and Grant et al. (1990). The formation of soda lakes requires the combination of three factors: (1) Geological conditions have to favour the formation of alkaline drainage water. (2) The topography of the area has to limit outflow of the basin. (3) Climatic conditions must permit high rates of evaporation and thereby sustain continuous concentrations of the solutes in the waterbody. Low Ca^{2+} soda lakes are dominated by phototrophs which occur permanently or in blooms. In these habitats the majority of all alkaliphiles are found.

Table 10.1. Types, properties and distribution of alkaline habitats based on information given by Brock (1978), Imhoff et al. (1979), Grant and Tindall (1986), Javor (1989), Grant et al. (1990), Schiller et al. (1997) and Arp et al. (1998)

Type of habitat	Properties	Distribution	
Stable alkaline conditions			
Soda lakes	Na_2CO_3, NaCl, low Ca^{2+} and sulfate; major ions present in molar concentrations pH 10-11.5	Africa:	East African Rift Valley Wadi Natrun (Egypt) Fezzan (Lybia) Lake Chad (Central Africa)
		America:	Owens Lake, Mono Lake, Searles Lake, Borax Lake (USA, California) Ragtown Soda lake (USA, Nevada) Union Pacific Lake (USA, Wyoming) Albert Lake (USA, Oregon) Lake Texcoco (Mexico) Lagunilla Valley (Venezuela) Antofagasta (Chile)
		Asia:	Chita and Baikal region (Russia) Araxes Plain Lakes (Armenia) Lake Looner (India) Various "nors" (China) Soda lakes of Jehol and Liao-Ning (China) Soda lakes (Inner Mongolia)
		Europe:	Caspian Sea region (Russia) Szegedin district (Hungary)
Alkaline groundwater springs	High Ca^{2+}, low CO_3^{2-}, major ions present at 10 mM. Reducing conditions due to the presence of Fe^{2+}; pH 11.2 - 11.4	America: Asia: Europe:	California Oman, Jordan Cyprus, Yugoslavia,
Thermal alkaline springs	Carbonate springs, *Chloroflexus* mats, pH 9-10.5	America: Europe:	Yellowstone (USA) Iceland

Transient alkaline conditions		
Oceans	$32^{o}/_{oo}$ salinity, transient alkaline pH values up to 9 due to oxygenic photosynthesis	Worldwide, large varieties of neutrophilic and alkali-tolerant cyanobacteria and Algae
Small lakes and waterbodies, rock pools	Chemistry depending on the type of waterbodies, transient alkaline pH values up to 10 due to oxygenic photosynthesis and /or other biological activities such as ammonification and reduction of sulfate	Worldwide, large varieties of neutrophilic and alkali-tolerant cyanobacteria and Algae

The second type of alkaline habitats, the alkaline groundwater springs, have much higher concentrations of Ca^{2+}, but lower carbonate concentrations and lower saltiness. Because of the presence of Fe^{2+}, the redox potential of such waters is high. Such habitats, exhibiting pH values between 11 and 11.5, have been described for Oman, Jordan, Cyprus, California and Yugoslavia. For obvious reasons, the size of such habitats is far smaller than that of soda lakes. The microbiology of such springs has not been studied very well so far, but they are dominated by bacteria, and not by oxygenic phototrophs, which are of minor interest for this chapter.

For California and Iceland, the thermal alkaline springs, which represent a special case, have been described (Brock 1978). The pH values of the thermal springs are significantly lower (9-10.5) than those of soda lakes or alkaline groundwater springs. The *Chloroflexus* mats of these thermal springs are inhabited by cyanobacteria (Doemel and Brock 1977).

Habitats with transient alkaline conditions are also distributed worldwide. They are inhabited by a large variety of alkali-tolerant algae and cyanobacteria which can survive for some time at pH values up to 10, but thrive well at neutral pH values also. In contrast to soda lakes, pH values of these habitats are not caused by geology, climate or topography, but are due to biological activities. The largest habitats of this type, also rich in sodium, are represented by the oceans. Due to intensive oxygenic photosynthetic activities, the surface pH of the oceans transiently reaches up to 9. Even though the pH values in the oceans are by far not as high as in soda lakes, they have to be considered in this chapter because of the large dimensions, huge biological productivity and importance for the food chain. Furthermore, microbes in oceans also encounter the basic physiological problem of alkaliphilic algae, that is, how to maintain a low internal sodium level in a saline and alkaline environment.

Besides the oceans, a large variety of smaller aquatic biotopes of different chemistry have been described, ranging from small lakes to rock pools, in which pH values go transiently (often diurnal) up to 10 as a consequence of oxygenic photosynthesis and/or other biological activities such as ammonification and reduction of sulfate. Since in these habitats elevated sodium concentrations often do not occur (except marine rock pools), we shall not consider algae living in such habitats in this article. Not included in Table 10.1 and not discussed in the following are terrestrial alkaline habitats such as alkaline deserts and soils, even though they occur worldwide over large areas.

Since most alkaline habitats are rich in salts, microorganisms growing in such environments are confronted not only with alkali stress, but also with salt and osmotic stress. It goes without saying, therefore, that all alkaliphic phototrophs contain high concentrations of organic osmotica, such as sucrose, trehalose, glucosylglycerol, glycine-/glutamatebetaine (Joset et al. 1996) or glycerol (Pick 1992). However, problems of osmoregulation will not be discussed here.

10.3 Typical Alkaliphilic Phototrophs Belong to Cyanobacteria

In Table 10.2 some typical alkaliphilic cyanobacteria and algae are enumerated. The former include unicellular and filamentous genera and species with and without heterocysts. The list contains genera from Chlorophyta, Heterokontophyta, Pyrrophyta and Chrysophyta. However, Table 10.2 does not give an idea of the abundance of the listed organisms in terms of biomass. The algal flora of soda lakes is relatively poor in comparison to freshwater lakes, containing in most cases only very few species, with cyanobacteria being most abundant. Very often, soda lakes are monospecific, inhabited by *Spirulina platensis* only (Ciferri and Tiboni 1985). Figure 10.2 gives an example of how the abundance of alkaliphilic microbes in terms of biomass changes with the chemistry of a soda lake. *Spirulina platensis* competes most of its alkalophilic, phototrophic competitors out at elevated salinity, whereas *Synechocystis salina* disappears at 6 g salt l^{-1}. The higher competitive abilities of *Spirulina* and other cyanobacteria at high pH and high salinity in comparison to unicellular green algae has been also demonstrated under laboratory conditions (Richmond et al. 1982; Caraco and Miller 1998), and in biotechnological outdoor cultures (Vonshak et al. 1983).

10.4 Excess of DIC Causes High Photosynthetic Productivity in Alkaline Lakes

Although the extreme pH values of soda lakes impose severe bioenergetic problems on the microbial communities, the primary production of these type of lakes is generally extraordinarily high (Talling et al. 1973; Wood and Talling 1988; Grant et al. 1990). Productivity rates of up to 10 g carbon m^{-2} of surface day^{-1} are not unusual in some such lakes, which are 20-fold higher than in

Table 10.2. Examples of alkaliphilic cyanobacteria and algae and their geographical distribution

	Algae (References)	Geographical distribution	pH of habitat
Cyanobacteria	*Anabaenopsis arnoldii* (*Cyanospira*?) (1, 2, 4)*	Wadi Natrun (Egypt), Chad	9.2-11.5
	Anabaena oryzae (8)	Tropical rice fields (India)	8.0-9.0
	Aphanocapsa spp. (6)	Soda lakes (Inner	8.1-10.0
	Chrooccocidiopsis spp. (1, 2)	Mongolia)	9.7-10.2
	Chroococcus spp. (1, 2)	Chad	9.6
	Cyclotella spp. (4)	Salt lakes (Ethiopia)	9.8-10.3
		East African Rift Valley	9.2-11.5
	Oscillatoria limnetica (3)	Wadi Natrun (Egypt)	9.9-11
	Oscillatoria spp. (5,6)	East African Rift Valley	8.1-10.0
		Soda Lakes (Inner	
		Mongolia)	9.2-11.5
	Phormidium sp. (3,6)	Wadi Natrun (Egypt)	8.1-10.0
		Soda lakes (Inner	
		Mongolia)	10.2-10.4
	Spirulina platensis, S. maxima	Lake Rombou, Lake Bodu	
	(1, 2, 3, 4, 5,)	(Chad)	<11.0
		Lake Texcoco (Mexico)	9.8-10.3
		East African Rift Valley	9.2-11.5
		Wadi Natrun (Egypt)	8.1-10.0
	Spirulina spp. (6)	Soda lakes (Inner	
		Mongolia)	9.2-11.5
	Synechocystis spp. (3)	Wadi Natrun (Egypt)	9.8-10.3
	Synechococcus spp. (4)	East African Rift Valley	
Algae	*Amphora coeffeaeformis* (3)	Mono Lake (USA)	
	Chlamydomonas spp. (3)	Mono Lake (USA) (3)	
	Monoraphidium minutum (1, 2)	Salt Lakes, Ethiopia	10.5-11
	Cryptomonas spp. (1)	East African Rift Valley	10.0-10.8
	Diatoms (1,9)	East African Rift Valley	9.8-10.3
	Dunaliella salina (3)	Wadi Natrun (Egypt)	9.2-11.5
	Nannochloris (Coccomyxa?) (3)	Mono Lake (USA)	
	Navicula halophila (5)	East African Rift Valley (5)	9.8-10.3
	Nitzschia spp. (4)	East African Rift Valley (4)	9.8-10.3
	Nitzschia communis (3)	Mono Lake (USA) (3)	
	Oocystis (4)	East African Rift Valley (4)	9.8-10.3
	Chrysophyceae (Stomatocysts) (7)	Alkaline lakes, British Columbia, Canada	7.6-8.4

*References: 1 Ciferri (1983); 2 Richmond (1986); 3 Javor (1989); 4 Grant and Tindall (1986); 5 Richmond (1988); 6 Arp et al. (1998); 7 Duff and Smol (1994); 8 Sethi and Kaushik (1993); 9 Hecky and Kilham (1973)

normal rivers and freshwater lakes of the world. From Fig. 10.2 it follows that this extremely high production is mainly due to the photosynthesis of *Spirulina platensis*. The reasons for high productivity are: (1) unlimited availability of DIC in the carbonate-laden water, (2) elevated temperatures and (3) monospecific occupation by *Spirulina*. Hence, Grant et al. (1990) have very rightly considered soda lakes as the most productive natural aquatic habitat of the world.

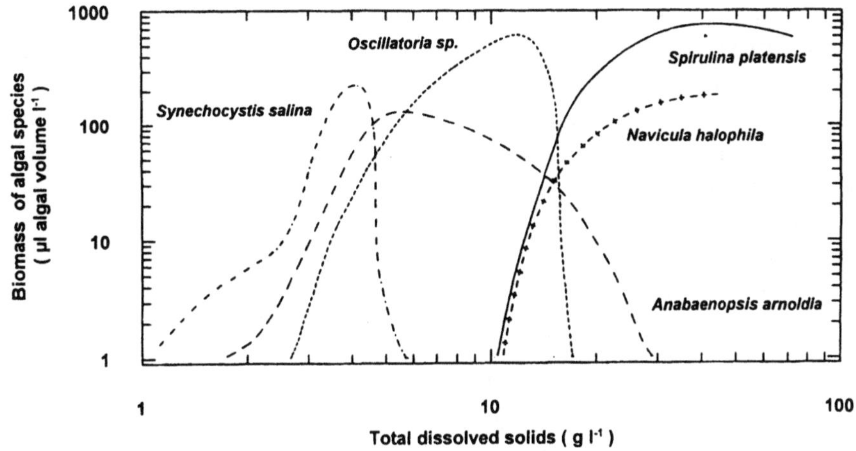

Fig. 10.2. Abundance of alkaliphilic phototrophs (expressed as total biovolume l^{-1} of lake water) in Lake Kanem (Chad) (pH 9.7-10.2) as effected by salinity. (After Iltis 1974)

10.4.1 DIC Uptake at Alkaline pH Is Uptake of Bicarbonate

In nature, the atmospheric CO_2 is in equilibrium with carbonic acid of aquatic environments according to the Eq.:

$$CO_2 + H_2O \leftrightarrow H_2CO_3. \quad (1)$$

Carbonic acid equilibrates to bicarbonate according to Eq. (2):

$$H_2CO_3 \leftrightarrow HCO_3^- + H^+. \quad (2)$$

The pK_a of the reaction is 6.4 under standard conditions. Thus, bicarbonate is the predominant DIC species at neutral pH, but still significant amounts of CO_2 are present. This accounts both for the external medium as well as for the cytoplasm whose pH is usually close to 7.

At alkaline pH the second dissociation step of the DIC system is observed (pK_a 10.3):

$$HCO_3^- + H^+ \leftrightarrow CO_3^{2-} + 2\,H^+. \quad (3)$$

Thus, at alkaline pH values in the environment bicarbonate is still the predominant DIC species and only at extreme alkaline pH values does carbonate become the prevalent species. However, under the conditions of Eq. (3), virtually no CO_2 is present in the system. Significant amounts of carbonate are very unlikely to occur within the cells due to their neutral cytoplasmic pH.

The substrate of Rubisco in the carboxysome or the chloroplast is CO_2:

$$6\,CO_2 + 6\,H_2O \rightarrow C_6H_{12}O_6 + 6\,O_2. \quad (4)$$

The question arises as to how DIC enters the cell prior to photosynthetic reduction of CO_2 [Eq. (4)]. It has been demonstrated that cyanobacteria and algae possess a carbon-concentrating mechanism (CCM) which is localised in the plasma membrane (Lucas and Berry 1984 and references therein; Price and Badger 1989; Thielmann et al. 1990; Badger and Price 1992; Shiraiwa et al. 1993; Beardall et al. 1998). The DIC species translocated through the plasma membrane can be either CO_2 or bicarbonate. Thermodynamically uptake of DIC into cyanobacteria or algae adapted to air is active (against a free energy gradient) (Raven 1984). It is likely to be coupled to the operation of an ATPase. It was calculated that on a molar basis the uptake of bicarbonate requires more energy than the uptake of CO_2. CCM is inducible at low CO_2 and depressed at high CO_2 concentration. A CO_2 pump is usually expressed at pH values between 5 and 8, whereas the bicarbonate pump operates at pH values between 7 and 11 (Shiraiwa et al. 1993). There is some evidence that a bicarbonate pump does exist in algae also at the level of the chloroplast envelope (Shiraiwa et al. 1993). *Synechococcus* possesses the ability to switch between the two modes of DIC uptake under the effect of sodium: at low Na^+ concentration (0.1 mM) CO_2 is the preferred species of uptake, whereas at 20 mM Na^+ DIC is taken up as bicarbonate (Miller and Canvin 1987). As described in Section 10.2, alkaline habitats always contain high Na^+ concentrations, thus favouring the operation of the bicarbonate pump. The presence of an active transport system for bicarbonate ensures that internal DIC concentrations in cyanobacteria and algae are high under conditions when CO_2 in the environment is scarce and bicarbonate is the only DIC species available. This is the case in all alkaline habitats, particularly in soda lakes. The fact that blue-green algae are better competitors at high pH and high salinity than green algae (Richmond et al. 1982; Vonshak et al. 1983; Caraco and Miller 1998) is very likely due to their better efficiency for bicarbonate uptake.

10.4.2 Calcification by Alkaliphilic Cyanobacteria and the Global Carbon Dioxide Budget

Cyanobacteria thriving in saline and alkaline environments precipitate calcium carbonate and are nucleation sites for mineral formation (Thompson and Ferris 1990; Merz 1992; Ferris 1993; Ferris et al. 1994; Yates and Robbins 1995; Merz et al. 1995). A calcification model, based on the model of Price and Badger (1989) for DIC accumulation was proposed by Yates and Robbins (1995). The model suggests that under alkaline and saline conditions DIC is taken up into cyanobacteria (see above) and transported into the carboxysome preferentially as bicarbonate. There it dissociates to CO_2 and H_2O. Rates of bicarbonate uptake and transport are assumed to be faster than the capacity of Rubisco for CO_2 fixation providing excess supply of CO_2. Consequently, only a part of the carboxysomal CO_2 is used for photosynthesis. The rest of CO_2 is reexported into the medium where it dissociates at alkaline pH to carbonate and protons [cf. Eqs. (1-3)]. In the presence of Ca^{2+} carbonate precipitates as $CaCO_3$:

$$CO_3^{2-} + Ca^{2+} \rightarrow CaCO_3 \downarrow. \qquad (5)$$

The precipitation of calcium carbonate depends not only on reactants of Eq. (5), but also on the pH, the concentration of divalent cations and the presence of nucleation sites. Merz et al. (1995) measured the pH in the sheath of the calcifying filamentous cyanobacteria. They found that at an external pH of 8.2 the cytoplasmic pH was close to 7, whereas the inner, noncalcified layer of the sheath exhibited a pH of 7.2 to 7.9 and the outer, calcified layer a pH similar to that of the bulk medium. Data confirmed that $CaCO_3$ precipitates only at sites of alkaline pH. The relatively lower pH in the immediate vicinity of the cells is of benefit to algae under alkaline conditions by providing more bicarbonate ions as substrate for the bicarbonate pump.

There is evidence that calcification by cyanobacteria under alkaline and saline conditions is of geological importance (Thompson and Ferris 1990; Ferris 1993; Ferris et al. 1994; Arp et al. 1998). It leads, for instance, to the formation of spring mounds, pinnacles and stromatolites, and takes part in the formation of banks and lake sediments. Particularly the weathering of silicate minerals on bedrock is biogeochemically coupled to the deposition of carbonate minerals by cyanobacteria (Ferris et al. 1994). The type of carbonate minerals formed varies between calcite and magnesite, depending on the chemistry of lake water and groundwater. The overall process involving oxygenic photosynthesis of cyanobacteria can be summarised as:

$$(Ca, Mg)SiO_3 + 2\,H_2CO_3 + H_2O \rightarrow (Ca, Mg)CO_3 + CH_2O + H_4SiO_4 + O_2. \qquad (6)$$

Organic matter is symbolised by CH_2O, carbonic acid is ultimately derived from the atmospheric CO_2, which equilibrates in the aquatic environment. It is assumed that equal amounts of C are partitioned into calcites or magnesites on one side and organic material synthesised by oxygenic photosynthesis on the other. Since there is a high primary production in saline and alkaline habitats (about 1.4 10^9 t C a^{-1} worldwide, Schlesinger 1991; see also Sect. 10.4), alkaliphilic cyanobacteria have the potential to precipitate up to one billion tonnes of carbonate carbon a^{-1} (Ferris et al. 1994). This is roughly in the order of magnitude of the present imbalance in the global atmospheric CO_2 budget. Cyanobacterial calcification may reflect an important global sink for atmospheric CO_2. For an update of oceanic sinks for atmospheric CO_2, see Raven and Falkowski (1999).

10.5 Alkaliphilic and Alkali-Tolerant Cyanobacteria and Algae Require Sodium

It has been well known for a long time that there is a specific Na^+ requirement for growth of marine and hypersaline cyanobacteria, but the exact reason for this is a subject of speculation (Markarova et al. 1997). Even the minimal required levels vary from species to species. Also, alkaliphilic cyanobacteria require sodium. If *Spirulina platensis*, for example, is suspended in media without Na^+, photosynthetic O_2 evolution stops immediately and the cell lyses within 1 h, a process which is accelerated by light (Schlesinger et al. 1996) (Fig. 10.3). A 150-250 mM Na^+ is required for optimal growth. Among prokaryotes the specific Na^+ requirement is not restricted to cyanobacteria. The alkaliphilic bacilli also require Na^+ (Horikoshi and Akiba 1982; Krulwich et al. 1988, 1990), but the maximal tolerated salinity is far lower in bacilli (Fig. 10.3). The marine alga *Dunaliella tertiolecta* does not grow below 20-50 mM NaCl and exhibits optimal growth at

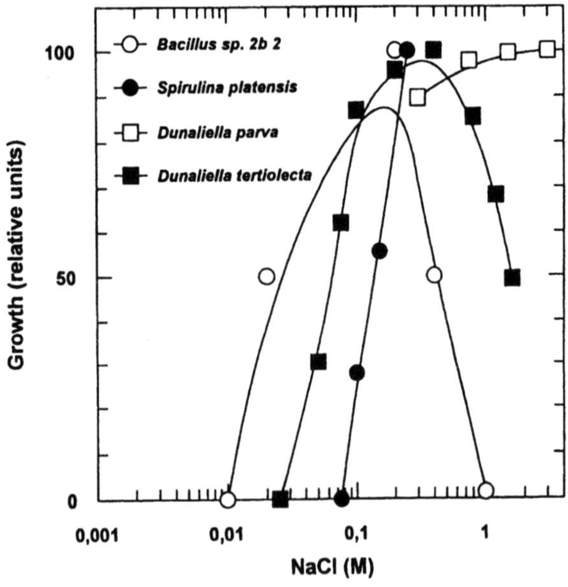

Fig. 10.3. Sodium requirement for the growth of selected alkali-tolerant microbes. Data adapted from Horiskoshi and Akiba (1982) (*Bacillus* sp. 2b 2, pH 9.6), Schlesinger et al. (1996) (*Spirulina platensis*, pH 11), H. Gimmler (*Dunaliella parva*, pH 8, unpubl.) and Hellebust (1988) (*Dunaliella tertiolecta*, pH 8)

400 mM NaCl (Hellebust 1988), but is more sensitive to higher salinity than its hypersaline relative, *Dunaliella parva* (Fig.10.3). Also, the alkali-tolerant and extreme salt-tolerant alga *Dunaliella salina* (growth up to pH 11) cannot be cultured in the absence of Na^+ (Pick 1992). Data suggest a perfect adaptation of alkaliphilic and alkali-tolerant microorganisms to elevated sodium concentration in their environment and the utilisation of Na^+ gradients across the plasma membrane for solute and ion transport and pH homeostasis.

A low requirement for Mg^{2+} and Ca^{2+} is also typical for alkaliphiles. Elevated concentrations of these ions even inhibit their growth. The extreme salt-resistant alga *Dunaliella salina*, for example, survives the molar Mg^{2+} concentrations of the Dead Sea, but it grows much faster if the Mg^{2+} concentration is reduced to 20 mM.

Both heterotrophic prokaryotes and mixotrophic eukaryotes are able to take up organic substrates such as glucose from an environment with a neutral pH by means of an H^+, glucose cotransporter. Since at alkaline pH the H^+ gradient across the plasma membrane cannot serve as the driving force for the uptake of organic substrates, in many alkaliphilic bacteria the cotransport with H^+ is simply replaced by cotransport of organic solutes with Na^+. It is of interest for future research, whether such cotransporters can also be demonstrated for those alkali-tolerant cyanobacteria or algae which can utilise organic substrates.

10.6 Bioenergetic Problems at High pH

In alkaline environments the ratio between sodium ions and protons is high. Depending on the habitats, values up to 3×10^{10} are reached. However, the corresponding ratio in the cytosol of both neutrophilic and acidophilic species is to the tune of 2×10^5 only. How is this ratio in alkaliphilic and alkali-tolerant cyanobacteria and algae maintained?

10.6.1 Alkaliphiles Regulate Their Cytoplasmic pH Close to Neutrality

There is no difference in respect to the maintenance of a neutral cytoplasmic pH between acidophilic, neutrophilic and alkaliphilic algae and blue-green algae. As shown in Fig. 10.4, alkali-tolerant algae such as *Tetraselmis viridis* or *Dunaliella parva* adjust their internal pH close to 7 at external pH values between 7 and 9. Alkaliphilic bacteria and cyanobacteria do the same. Only at pH values above 9 may a slight increase in cytoplasmic pH be observed and at external pH values exceeding 11 pH homeostasis may finally collapse. Thus, in typical environments of alkali-tolerant and alkaliphilic phototrophs the H^+ gradient across the plasma membrane varies between 100 and 1000 (inside more acid than outside), the direction of the pH gradient across the plasma membrane is reversed compared to

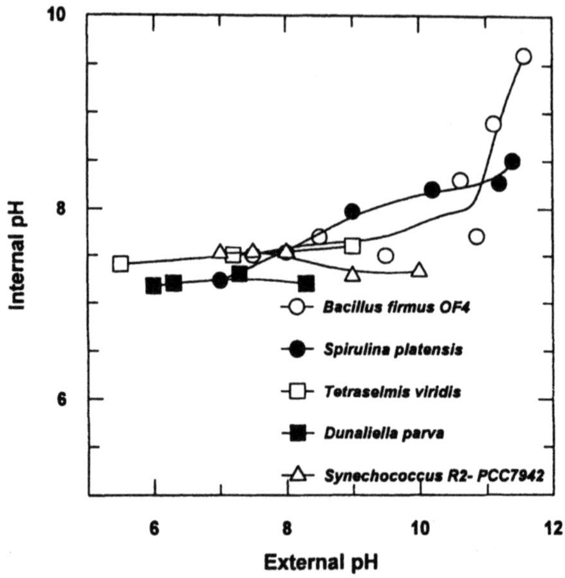

Fig. 10.4. pH homeostasis in alkali-tolerant algae (*Dunaliella salina* and *Tetraselmis viridis*), the alkali-tolerant cyanobacterium *Synechococcus*, an alkaliphilic cyanobacterium *Spirulina platensis* and a bacterium *Bacillus firmus*. Data after Gimmler et al. (1988), Balnokin et al. (1997), Ritchie (1991), Belkin and Boussiba (1991) and Krulwich et al. (1998)

algae growing at external pH values lower than 7. In alkaliphilic bacilli pH homeostasis collapses in the complete absence of Na^+, which indicates an essentiality of Na^+ translocation for the above process (Krulwich et al. 1988).

10.6.2 An Apparent Paradox: Alkaliphiles Require and Exclude Sodium

As described in Section 10.5, alkaliphilic and alkali-tolerant algae and cyanobacteria require sodium for growth. On the other hand, they are sodium excluders. This means that these organisms maintain a constant cytoplasmic Na^+ pool on a level which is significantly lower than that of the environment. In Fig. 10.5A examples are given for the alkali- and extreme salt-tolerant alga *Dunaliella salina* (constant internal Na^+ levels up to molar salt concentrations outside), the alkaliphilic and the salt-tolerant cyanobacterium *Spirulina platensis* (Na^+ homeostasis up to molar NaCl concentrations) and the alkali-tolerant, but only moderate salt-tolerant cyanobacterium *Synechococcus* (Na^+ homeostasis up to 100 mM NaCl outside). Na^+ concentrations across the plasma membrane vary between 10 and 40, depending on the species and the pH and Na^+ concentration of the medium (Fig. 10.5B). Also, alkaliphilic bacteria exclude sodium from the cytoplasm. The ratio of Na^+ ions to protons in the cytoplasm is roughly the same as in neutrophilic or even acidophilic algae. The bioenergetic question is how this is achieved at high external OH^- and sodium concentrations.

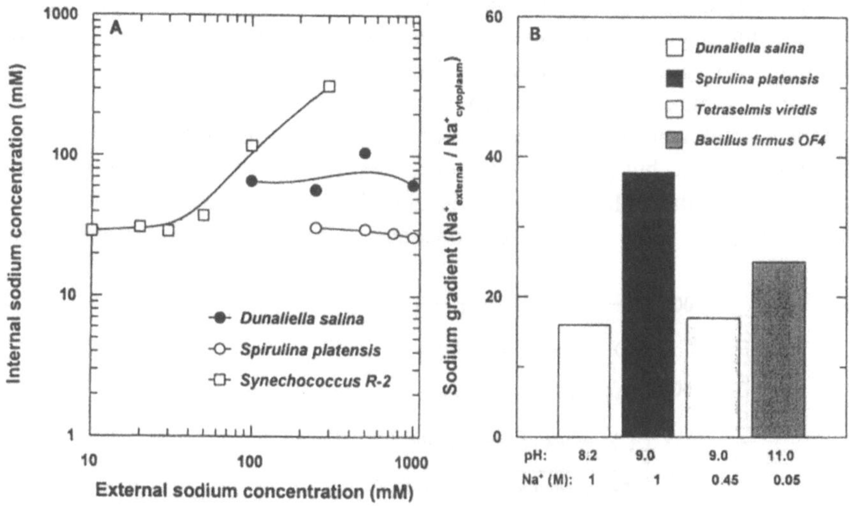

Fig.10.5 A.B. Sodium exclusion in alkali-tolerant algae and cyanobacteria (**A**) and sodium gradient across the plasma membrane (**B**). Data after Bental et al. (1988), Vonshak et al. (1988), Krulwich et al. (1990), Ritchie (1991) and Balnokin et al. (1997). *Dunaliella salina* and *Spirulina platensis* are salt-tolerant, but *Synechococcus* is sensitive to elevated salt concentrations

10.6.3 Membrane Potentials of Salt-Resistant Alkaliphiles are Not High Enough to Compensate Completely for Inversed pH Gradients

As outlined already in Section 10.1, the basic problem of microorganisms living in an alkaline environment is the low protonmotive force (pmf) $\Delta\ \mu_{H^+}$. The latter has two components, the ΔpH and the transmembrane potential $\Delta\psi$. The interrelation is described by the Eq.

$$\Delta\mu_{H^+} = -\Delta\psi - Z\,\Delta pH\,, \qquad (7)$$

where Z is a constant to convert pmf into electrical units. Proton export proceeds via a H^+-ATPase (algae) and/or a redox chain (blue-green algae). From Fig. 10.4 it follows that ΔpH is reversed and thereby not a driving force for the retranslocation of protons. The latter is required for chemical work (ATP synthesis, prokaryotes), osmotic work (solute accumulation, alkaliphilic bacteria) or mechanical work. Thus, $\Delta\ \mu_{H^+}$ will be reduced if the decrease is not compensated for by a corresponding increase of $\Delta\psi$. Indeed, at pH values between 9 and 10 alkaliphilic, non-marine bacilli exhibit higher membrane potentials (up to – 200mV) than at pH 7 (–50 to –130 mV) (Krulwich et al. 1988), but the increase is energetically less than the reversed ΔpH, expressed in electrical units. Consequently, there is a considerable decrease of $\Delta\mu_{H+}$. In principle, there is no doubt that a similar phenomenon takes place in alkali-tolerant cyanobacteria and algae also, although experimental data on Δψ and pmf are limited for such organisms. Figure 10.6 gives an example of the performance of the cyanobacterium *Synechococcus* as a function of external pH (Ritchie 1991). Δψ is relatively constant, but pmf is continuously decreasing with increasing pH. This also applies to the obligate alkaliphile *Bacillus firmus* RAB (Krulwich et al. 1988). Figure 10.6 also demonstrates that the moderately salt-

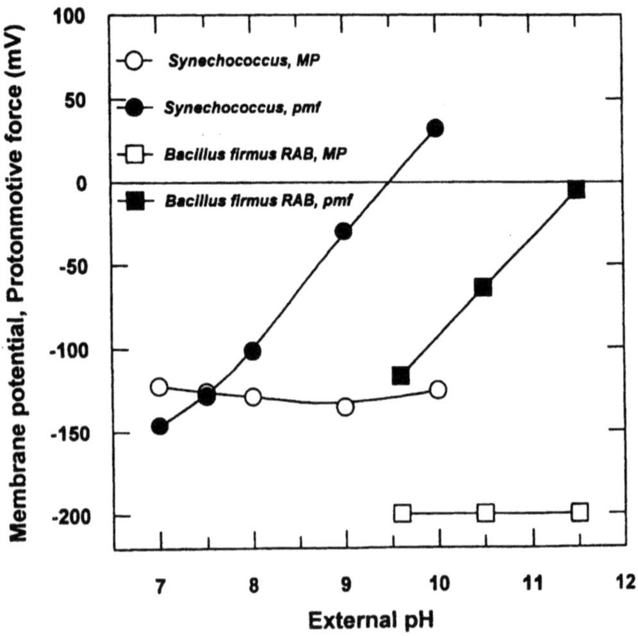

Fig. 10.6. Membrane potential (*MP*) and protonmotive force (*pmf*, $\Delta\mu_{H}+$) in the cyanobacterium *Synechococcus* (under photosynthetic condition) and the bacterium *Bacillus firmus* RAB as a function of external pH. Data after Ritchie (1991) and Krulwich et al. (1988)

resistant blue-green alga *Synechococcus* has a lower membrane potential (–125 mV) than the bacterium *Bacillus* (–200 mV). In extreme salt-tolerant algal species such as *Dunaliella parva* or *Dunaliella bardawil* membrane potentials drop even to values smaller than –50 mV (pH 8) (Remis et al. 1992). Values of pmf approach zero in salt-tolerant microbes at lower pH values (for *Synechococcus* at pH 10) than in non-marine species (for *Bacillus firmus* RAB, pH 11.5) . It is obvious that in saline and hypersaline alkaliphilic or alkali-tolerant organisms $\Delta\mu_{H^+}$ can be disregarded as the driving force for the secondary transport.

One possible strategy of alkaliphiles to overcome a shortage in pmf could be to utilise the Na^+ gradient across the plasma membrane ($\Delta\mu_{Na^+}$) instead of $\Delta\mu_{H^+}$ for energy-demanding processes such as ion uptake, solute translocation or cell movement (Skulachev 1985, 1988, 1994). This would be energetically of advantage, but would lead to an increase of Na^+ influx into the cells. Therefore, a system for the reexport of Na^+ is required (the concept of active Na^+-dependent cycle). No evidence for primary Na^+ extrusion systems coupled to respiration is available for the non-marine alkali-tolerant bacilli. However, there is plenty of information about the existence of primary extrusion systems in marine bacteria (Tokuda and Unemoto 1981, 1984; Skulachev 1985, 1988, 1994; Tokuda et al. 1985; Dibrov et al. 1986a,b; Ken-Dror et al. 1986; Hayashi and Unemoto 1987; Unemoto and Hayashi 1993; Park et al. 1996). Even some of the genes responsible for the expression of such extrusion systems have been cloned and sequenced recently (Beattie et al. 1994). Na^+ extrusion can be driven by a respiration- coupled Na^+ pump, a Na^+-linked ATPase or an Na^+-motive NADH-quinone reduction decarboxylase. The latter could be reconstituted in vitro (Pfenninger-Li et al. 1996). All systems have in common that in vivo Na^+-export is stimulated by conditions which inhibit $\Delta\Psi$. The presence of an active Na^+-dependent cycle is supported by the fact that alkaliphilic bacteria are generally very motile (Na^+ is required for motility as a coupling ion) (Krulwich et al. 1988) and that for a large number of different solutes these organisms bypass the problem of a low $\Delta\mu_{H^+}$ by utilising Na^+ as the coupling ion for solute transport (Horikoshi and Akiba 1982; Krulwich et al. 1988, 1990; Krulwich and Guffanti 1989).

The question is, whether marine alkali-tolerant bacteria and salt-tolerant alkaliphilic or alkali-tolerant cyanobacteria and algae follow similar strategies in this respect. Not all organisms of the latter group are mobile. All of them are obligate phototrophs which do not require special systems for the uptake of organic solutes. Nevertheless, the coupling of uptake of inorganic solutes to the $\Delta\mu_{Na^+}$ would reflect a feasible alternative. Indeed uptake of phosphate, carbonate and nitrate is specifically stimulated by Na^+ at alkaline pH in cyanobacteria and algae (Ullrich-Eberius and Yingchol 1974; Ullrich and Glaser 1982; Lara et al. 1993), which implies uptake via Na^+ cotransporters.

10.6.4 Na^+ Influx into Alkaliphiles is Downhill, Efflux is Uphill

Since the internal Na^+ level of alkali-resistant cyanobacteria and algae is far lower than the external concentration and the membrane potential is negative, the uptake of Na^+ does not require energy. The apparent permeability coefficient P_{Na^+} of the plasma membrane of cyanobacteria is low (3×10^{-9} m s^{-1}) and not affected by the external Na^+ concentration (Ritchie 1992). This is in the order of magnitude measured for other marine algae. The entry of Na^+ could proceed via simple diffusion or by any type of catalysed diffusion, such as Na^+ counter-transport (for example a Na^+/ H^+ antiporter) or Na^+ cotransport (for example an Na^+, phosphate symporter).

In contrast, Na^+ export out of the cells is uphill and requires energy. In fact, active Na^+ extrusion was postulated in cyanobacteria long ago (Batterton and van Baalen 1971; Paschinger 1977). It could proceed via a primary pump (for example a Na^+-ATPase) or by a secondary energised system (for example an antiporter). In the latter case, a primary pump or a redox chain would be expected to create a H^+ gradient (outside acid), which subsequently is utilised for exchange against Na^+. If the antiporter is electrically silent (exchange ratio $H^+/Na^+ = 1$), Na^+ export could proceed uphill provided the H^+ gradient is larger than the Na^+ gradient. If the antiporter is electrogenic (for example $2H^+/Na^+$) it will depend on both $\Delta\psi$ and the ion gradients across the plasma membrane.

10.6.5 Involvement of an Antiporter in the Na^+ Cycle of Alkaliphilic Prokaryotes?

It is generally assumed that a secondary energised antiporter is responsible for Na^+ extrusion out of those extreme alkaliphilic prokaryotes which are not of marine origin and therefore of limited salt tolerance (Krulwich et al. 1988, 1990, 1997). The situation is more complicated in cyanobacteria. Most authors agree that respiration is directly involved in the maintenance of low intracellular Na^+ levels. In the presence of high external Na^+ concentrations an H^+ gradient is believed to be generated by the activity of a plasma membrane cytochrome oxidase, which can function as an H^+ pump. This gradient is assumed to be utilised by an Na^+/H^+ antiporter for the extrusion of Na^+ out of the cells (Nitschmann and Peschek 1985; Peschek 1985, 1987; Fry et al. 1986; Packer et al. 1987). This idea, initially developed for freshwater blue-green algae like *Anacystis,* was later also used to explain the phenomena of salt adaptation in hypersaline cyano-bacteria like *Spirulina* (Gabbay-Azaria et al. 1992). However, the demonstration of the presence of a Na^+/H^+ antiporter in the plasma membrane of cyanobacteria is not identical with its function during the extrusion of Na^+ upon salt shock. Active Na^+ extrusion has already been postulated in cyanobacteria (Batterton and van Baalen 1971; Paschinger 1977). It could proceed via a secondary energised system (for example an antiporter) as discussed above, but also via a primary pump (for example a Na^+-ATPase). In the latter case, a primary pump or a redox chain would be expected to create a H^+ gradient (outside acid), which subsequently is utilised for exchange against Na^+. If the antiporter is electrically silent (exchange ratio $H^+/Na^+ = 1$), Na^+ export could proceed uphill provided the H^+ gradient is larger than the Na^+ gradient. If the antiporter is electrogenic (for example $2H^+/Na^+$), it will depend on both $\Delta\psi$ and the ion gradients across the plasma membrane. A careful bioenergetic analysis of ion transport in *Synechococcus* R2 (*Anacystis nidulans*) led Ritchie (1991, 1992) to the conclusion that $\Delta\mu_{H^+}$ could not drive the active extrusion of Na^+ from this cyanobacterium under physiological ranges of the external pH and Na^+ concentration and at the experimentally determined membrane potentials. This conclusion was drawn under the assumption that the Na^+/H^+ antiporter of the plasma membrane is electrically silent. Since there is scarcely any evidence for an electrogenic Na^+/H^+ antiporter, he postulated that internal Na^+ is actively extruded by a DCCD-sensitive Na^+ pump, in this case a Fo/F1 ATPase. In other words, in this *Synechococcus* strain the sodium circuit rather than the proton circuit appears to be the primary active transport event at the level of the plasma membrane (Ritchie 1992).

10.6.6 Evidence for Primary Sodium Pumps in Salt- and Alkali-Tolerant Algae

For a long time primary sodium pumps were believed to be responsible for the reexport of sodium under saline conditions in algae and higher plants, in spite of the fact that experimental evidence for this hypothesis was not enough. Scientists were misled by the assumption that an algal or plant primary Na^+ pump should have properties very similar to the well investigated mammalian Na^+/K^+-ATPase. One of these properties was the ouabain sensitivity of the Na^+/K^+ -ATPase. Most attempts to demonstrate in vivo or in vitro ouabain sensitivity of physiological reactions in salt-tolerant algae failed; an example is given in Fig. 10.7. Only later was it understood that ouabain sensitivity is caused by binding to a specific, regulatory amino acid domain of the Na^+/K^+-ATPase rather than to the catalytic site. If this sequence is missing in homologous enzymes, ouabain sensitivity will disappear without any effect on transport function.

Na^+/H^+ antiporters, as discussed above for prokaryotes, are present in plasma membranes of hypersaline algae such as *Dunaliella salina*, and its properties have been thoroughly studied (Katz et al. 1986, 1989; Popova and Balnokin 1992; Gimmler 2000). Of particular interest in respect to the topic of this chapter is the failure to demonstrate electrogenicity of the Na^+/H^+ antiporter. Soon, it was realised that an Na^+/H^+ antiporter cannot export sodium out of *Dunaliella salina* against the large chemical sodium gradient existing across the plasma membrane if the corresponding proton gradient is unity or below (Weiss and Pick 1990; Katz et al. 1991, 1992; Pick 1992). Also, the results on the Na^+ transport in intact cells of *Dunaliella maritima* (Balnokin and Medvedev 1984) did not match the operation of a secondary energised antiporter at alkaline pH. Therefore, it was suggested that the Na^+/H^+ antiporter plays a role in pH homeostasis of *Dunaliella* rather than in Na^+ homeostasis. The latter was postulated to be carried out by the primary sodium pump.

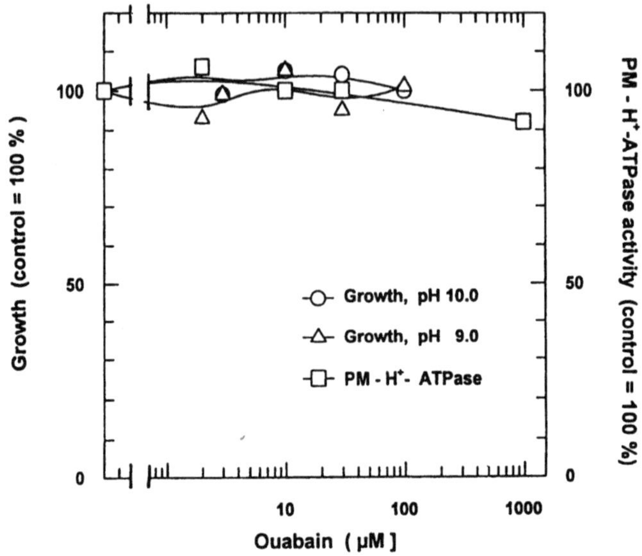

Fig. 10.7. Effect of ouabain on growth (pH 9 and 10) of *Dunaliella parva* at 1.5 M NaCl and on ATP hydrolysis of isolated plasma membranes (pH 7) (H. Gimmler, unpubl.)

Balnokin et al. (1997) and Gimmler (2000) reviewed sodium and proton gradients, as well as membrane potentials of the marine alga *Tetraselmis viridis* and the hypersaline species *Dunaliella parva* in respect to the need for a primary sodium pump. Thermodynamic considerations led the authors to the conclusion that above pH 7 both algae require a primary sodium pump, provided the Na^+/H^+ antiporter, present in these algae, is electrically silent. However, if the antiporter were electrogenic (stoichiometry H^+/Na^+ 2 or 3) it could catalyse sodium export up to external pH values of 7.2. Since both algae grow well at pH values up to 10, they require for thermodynamic reasons a primary sodium pump if thriving above pH 8. Table 10.3 comprises recent information supporting the presence of a primary sodium pump in at least two marine algae, *Heterosigma akashiwo* and *Tetraselmis viridis.* All the data originated from a Japanese group working with the fish-killing alga *Heterosigma* (also known *as Olisthodiscus luteus*) (*Cryptophyta*) and a Russian/German group analysing *Tetraselmis* (= *Platymonas*) *viridis* (*Prasinophyceae*) (Table 10.3). First, it was demonstrated that the isolated plasma membrane vesicles of these two algae catalyse an Na^+-dependent, vanadate- and DCCD-sensitive, but ouabain-resistant ATP hydrolysis (Wada et al. 1989; Shono et al. 1995; Balnokin et al. 1997). ATP hydrolysis required Mg^{2+} and exhibited maximal rates at alkaline pH (8 to 9), which is in contrast to the well-known pH optimum of H^+-ATPase between 6.5 and 7. In the case of Na-

Table 10.3. Data supporting the presence of a primary sodium pump in the plasma membrane of the unicellular marine algae *Tetraselmis (Platymonas) viridis* and *Heterosigma akashiwo*. Both species are salt-tolerant and tolerate pH values up to 9.

Properties of sodium pump	*Tetraselmis viridis (Platymonas viridis)*	*Heterosigma akashiwo*
Na^+-dependent ATP hydrolysis	(6)[a]	(1, 5)
Stimulation	Mg^{2+} (6)	Mg^{2+} (1, 5)
	Na^+ (6)	Na+, K_m 12 mM (5)
	ATP (6)	ATP, K_m 0.88 mM (5)
pH profile	Optimum at pH 8.0 (6)	Maximum at pH 8-9 (5)
Inhibition		Vanadate (5), DCCD (1)
No effect		Ouabain (1)
Molecular properties	(7,8)	(3)
Molecular mass	100 kDa (7, 8)	140 kDa (3)
Homology of amino acid sequence		Identical epitope to Na^+/K^+-ATPase at the amino-terminal region (3)
ATP-dependent Na^+ transport	(4, 6)	(2)
pH optimum	8.0 (6)	
K_m ATP		0.4 mM (liposomes) (2)
K_m Na^+		7 mM (liposomes) (2)
Inhibition by protonophores	ClCCP: No (4, 6) Monensin: No (4)	FCCP: No (2)
Inhibition by amiloride	No (4)	
Effect of permeable ions	Stimulation by permeable ions such as nitrate or chloride (6)	Inhibition by permeable cations (TBA) (2)
Inhibition by vanadate	Yes (6)	Yes (2)
Existence of a Na^+-dependent phosphorylated intermediate	Yes, 100 kDa (7, 8)	Yes, 140 kDa (1, 5)
Inhibition of phosphorylation		
By vanadate	Yes (7, 8)	
By hydroxylamine	Yes (7, 8)	
By ADP	Yes (7, 8)	
By KCl		Yes (5)
No effect on phosphorylation by	DCCD (9), Stauroporin (9)	
Rapid turnover of phosphorylation	Yes (8)	
Stimulation of phosphorylation by Na^+	K_m 12 mM (8)	5 mM required (5)
Effect of Mg^{2+} on phosphorylation	µmolar concentration required, mmolar concentrations inhibit (7, 8)	

[a]References (numbers in parentheses): 1 Wada et al. (1989); 2 Shono et al. (1996); 3 Wada et al. (1992); 4 Balnokin and Popova (1994); 5 Shono et al. (1995); 6 Balnokin et al. (1997); 7 Popova et al. (1998); 8 Popova et al. (1999); 9 H. Gimmler and L. Popova (unpubl.).

dependent ATP hydrolysis, the K_m values for ATP was 0.88 mM and that for Na^+ 12 mM (Shono et al. 1995). The molecular mass of the supposed Na^+-ATPase was 100 kDa (*Tetraselmis*, Popova et al. 1998, 1999; Gimmler 2000) and 140 kDa, respectively (*Heterosigma*, Wada et al. 1992). In *Heterosigma* the enzyme exhibited an epitope at the amino-terminal region identical to that known for the mammalian Na^+/K^+-ATPase (Wada et al. 1992). In the second step, ATP-dependent sodium transport into isolated plasma membrane vesicles or artificial liposomes was demonstrated by using $^{22}Na^+$ as tracer. It was shown that protonophores like ClCCP, FCCP and monensin, which collapse H^+ gradients, did not inhibit ATP-dependent Na^+ uptake. Under some circumstances, the uptake of Na^+ is even stimulated (Balnokin and Popova 1994; Shono et al. 1996; Balnokin et al. 1997). Na^+ uptake was resistant to the inhibitor of the Na^+/H^+ antiporter, amiloride, but was inhibited by membrane permeant anions (such as nitrate or chloride) or cations (such as TBA, tetra-*n*-butylammonium). Both working groups could demonstrate, by the use of γ-^{32}P-ATP, that, comparably to the animal Na^+/K^+-ATPase, an Na^+-dependent phosphorylation takes place during the uptake of Na^+ (Wada et al. 1989; Shono et al. 1995; Popova et al. 1998, 1999). The phosphorylation of the 100-kDa (140-kDa) intermediate is inhibited by vanadate, ADP and KCl, but is not affected by the inhibitor of protein kinases, stauroporin (Table 10.3). Hydroxylamine caused cleavage of ^{32}P-labelling, indicating the involvement of acyl-phosphate linkages. The K_m of the phosphorylation for Na^+ varied between 5 and 12 mM. Results indicated a rapid turnover of phosphorylation, which is accelerated by high concentrations of ATP and Mg^{2+}.

All data agree with the view that in the plasma membrane of the marine algae *Heterosigma* and *Tetraselmis* an electrogenic Na^+-ATPase exists which has the properties of a P, (E_1E_2) type of ATPase (Popova et al. 1998, 1999; Gimmler 2000) according to the Albers-Post scheme (Stein 1986). However, there is no evidence so far that the enzyme exchanges sodium against potassium as the animal Na^+/K^+-ATPase. Data suggest a flexible coupling of the sodium cycle in salt-tolerant algae, as has been also observed for salt-tolerant prokaryotes. At neutral pH the secondarily energised Na^+/H^+-antiporter exports sodium, whereas at alkaline pH the Na^+-ATPase takes over. However, the factors regulating the functioning of both Na^+-exporting systems remain unknown. It should also be proved whether all algae thriving in saline environments at alkaline pH possess a primary sodium pump. Theoretically, hypersaline algal species such as *Dunaliella* require such a pump even more urgently than marine algae.

Since halophytic plants also often encounter high salinity and alkaline pH at the same time, it is our strong belief that roots of higher plants do contain a primary Na^+ pump. Since salinity is a serious worldwide problem of agricultural importance, the search for a primary sodium pump in plants should be speeded up. The obvious experimental strategy should be first to consolidate the present knowledge about biochemistry and molecular biology of algal primary Na^+ pumps and the cloning and sequencing of this transporter. Then, using immunological techniques, a screening for this pump in the plasma membrane of roots of halophytic plants should be carried out. Such studies have to be complemented with transport studies using purified plasma membrane vesicles isolated from rhizodermal protoplasts.

10.7 Phosphate Availability and Uptake at Alkaline pH

Surface layers of natural waters, for example seawater, contain generally little inorganic phosphorus (10 – 2000 nM) and often a great part of total dissolved P is bound to organic compounds (dissolved organic phosphate, DOP). In order to exploit this phosphorus source and to avoid phosphate limitation of growth, algae often express excess of phosphatases. The environmental conditions determine which of the genetically available various inducible isomerases of these monoesterase are expressed. In general, the activity of phosphatases is inversely related to the P content of the cell and that of the environment. In acid to neutral waters acid phosphatases dominate, whereas in alkaline habitats more alkaline phosphatases (EC 3.1.3.1, molecular mass 190 kDa) are expressed. Alkaline phosphatase is induced by increasing salinity (Hernandez et al. 1992, 1996) as has been also found for higher plants (Dubey and Sharma 1989). It requires divalent cations, preferentially Ca^{2+}. Recently, it has been observed that properties of the expressed alkaline phosphatase are affected also by the ambient temperature (Shukla et al. 1997). Alkaline phosphatase activity (APA) increases with P-deficiency and therefore is generally treated as an indicator of P deficiency (Quisel et al. 1996 and references therein; Garcia-Ruiz et al. 1997; Nausch 1998). High APA also correlates with high phosphate uptake. About 70% of phosphate originating from cleavage of DOP by APA is immediately taken up by algae (Hernandez et al. 1996). This suggests a close structural relationship between alkaline phosphatase and the phosphate uptake system. In fact, APA was shown to be localised predominantly in the plasma membrane (Rivkin and Swift 1980) or the cell wall of algae (Quisel et al. 1996 and references therein) and in the periplasm of bacteria. However, free extracellular APA has also been demonstrated (Walther and Fries 1976). In vitro APA increases with decreasing proton concentration, optimal activities occur between pH 8.8 and 10 (Hernandez et al. 1992, 1996; Quisel et al. 1996). Thus, the pH profile is adapted to the external pH rather than to the cytoplasmic pH. The surface of the ocean easily approaches pH 9 under full sunlight and at high photosynthetic activities. The K_m values of the enzyme for various substrates occurring in the natural environment varied between 100 and 300 µM (Quisel et al. 1996).

High APA is of potential benefit not only for marine algae, but also for alkaliphilic organisms thriving in soda lakes, since in the latter phosphate concentrations are low (Wood and Talling 1988). Moreover in alkaline waters phosphate tends to precipitate in the presence of divalent cations and thereby phosphate availability decreases. No systematic study exists about the combined effects of increasing pH and salinity on APA in soda lakes. However, as mentioned in Section 10.6.3, in the presence of sufficient Na^+ at alkaline pH, the sodium-phosphate cotransporter in algae, including blue-greens, guarantees an efficient uptake of phosphate even at low concentrations (Ullrich-Eberius and Yingchol 1974; Ullrich and Glaser 1982) as soon as phosphate has been liberated from organic phosphate esters through the activity of the alkaline phosphatase.

10.8 Membrane Composition and Alkaline pH

In general, plasma membrane plays an important role in strategies of microbes to encounter stress. The question arises whether there are specific responses of the plasma membrane to high pH. However, very little information is available about the effect of alkaline pH on the composition of the plasma membrane of cyanobacteria and algae. This contrasts to numerous corresponding studies in heterotrophic bacteria (Grant and Tindall 1986; Krulwich et al. 1988 and references therein). In respect to polar lipids, alkalophilic bacteria contain exclusively phospholipids, whereas glycolipids are absent (Grant and Tindall 1986). Furthermore, a special role for the retention of endogenous compounds in alkalophilic bacilli is assigned to plasma membrane isoprenoids (Krulwich et al. 1988). In blue-green algae, Funteu et al. (1997) observed a strong increase of desaturase activity with increasing pH.

Most of the studies on lipids of algae and cyanobacteria concentrate on the effect of salinity on membrane lipid composition. As described in this chapter, high salinity often runs parallel with increasing pH. Thus, the described effects of high salt on membrane composition may also occur in alkali-tolerant or alkali-resistant cyanobacteria or algae, and may not be a specific effect of alkaline pH. In general, high salinity increases the fluidity of the plasma membrane, the amount of phosphatidylglycerides and the portion of unsaturated fatty acids, whereas it decreases the permeability to Na^+ and H^+ and the protein/lipid ratio (Joset et al. 1996). The increase of the polar lipids is assumed to enhance the surface charge of the plasma membrane which plays an important role in the control of cation permeability.

10.9 Conclusion and Further Research

Considering the information described in this chapter we have summarised the prominent properties of alkali-tolerant cyanobacteria and algae in a scheme (Fig.10.8). Future studies on the regulation of ion transport, i.e. the Na^+,H^+ antiporter in halotolerant algae and on the lipid composition of alkaliphilic or alkali-tolerant phototrophs will complement our present knowledge on how

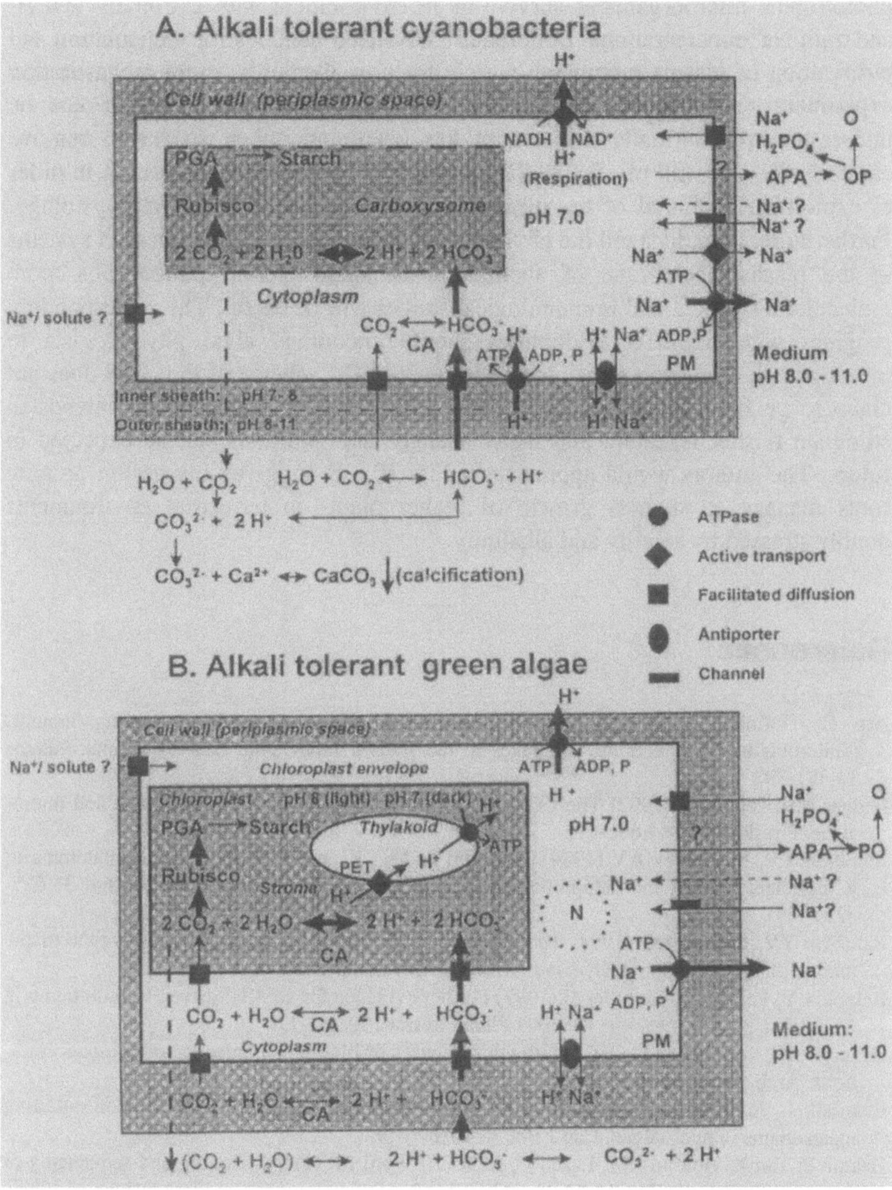

Fig. 10.8 AB. A schematic presentation of the properties of alkali-resistant or alkali-tolerant cyanobacteria (**A** internal membrane system neglected) and algae (**B**) as described in this chapter (cf. in particular Price and Badger 1989; Shiraiwa et al. 1993; Yates and Robbins 1995; Krulwich et al. 1998). *APA* Alkaline phosphatase; *CA* carbonic anhydrase; *N* nucleus; *P* phosphate; *PET* photosynthetic electron transport; *PM* plasma membrane; *O* organic solutes arising from organic phosphate esters; *PO* organic phosphate esters; *dashed arrows* diffusion; *thickness of arrows* reflects activity of the transport system under alkaline conditions

phototrophic microorganisms survive in an environment with extremely low H^+ and high Na^+ concentrations. Both require advanced methods for the isolation and purification of plasma membrane vesicle fraction. Certainly, more reconstitution experiments are needed, but also studies with intact cells should not be underestimated. Virtually no attempt has been undertaken so far to remove carefully the cell wall of alkaliphilic phototrophs by enzymatic digestion in order to exploit the potential of the patch clamp technology in transport physiology. Furthermore, to understand the phylogenetic development of the transport systems in the plasma membrane of alkaliphilic or alkali-tolerant phototrophs more molecular biological and immunological studies will be useful. The corresponding progress with bacterial acidophiles should encourage algal physiologists to investigate the transport physiology of mutants. The scheme of Fig. 10.8 does not claim to be final; rather it should serve as a basis for discussion. It intends to stimulate further research, and many changes and additions can be expected in future. The authors would appreciate if Fig. 10.8 B triggered discussion on how roots manage to support growth of higher plants in terrestrial environments doubly stressed by salinity and alkalinity.

References

Arp G, Hofmann J, Reitner J (1998) Microbial fabric formation in spring mounds ("microbialites") of alkaline salt lakes in the Badain Jaran Sand Sea, PR China. Palaios 13:581-592

Badger MR, Price GD (1992) The CO_2 concentrating mechanism in cyanobacteria and microalgae. Physiol Plant 84:606-615

Balnokin YV, Medvedev AV (1984) Transport of Na^+, K^+ and H^+ across the plasmalemma of K^+-deficient cells of the halophilic alga *Dunaliella maritima*. Sov J Plant Physiol 31:625-629

Balnokin YV, Popova LG (1994) The ATP-driven Na^+-pump in the plasma membrane of the marine unicellular alga, *Platymonas viridis*. FEBS Lett 343:61-64

Balnokin Y, Popova L, Gimmler H (1997) Further evidence for an ATP-driven sodium pump in the marine alga *Platymonas viridis*. J Plant Physiol 150:264-270

Batterton JC, van Baalen C (1971) Growth responses of blue-green algae to sodium concentrations. Arch Microbiol 76:151-165

Beardall J, Johnston A, Raven J (1998) Environmental regulation of CO_2-concentrating mechanisms in microalgae. Can J Bot 76:1010-1017

Beattie P, Tan K, Bourne RM, Leach D, Rich PR, Ward FB (1994) Cloning and sequencing of four structural genes for the Na^+-translocating NADH-ubiquinone oxidoreductase of *Vibrio alginolyticus* FEBS Lett 356:333-338

Belkin S, Boussiba S (1991) Resistance of *Spirulina platensis* to ammonia at high pH values. Plant Cell Physiol 32:953-958

Bental M, Degani H, Avron M (1988) ^{23}Na NMR studies of the intracellular sodium ion concentration in the halotolerant alga *Dunaliella salina*. Plant Physiol 87:813-817

Brock TD (1978) Thermophilic microorganisms and life at high temperatures. Springer, Berlin Heidelberg New York

Caraco NF, Miller R (1998) Effects of CO_2 on competition between a cyanobacterium and eukaryotic phytoplankton. Can J Fish Aquat Sci 55:54-62

Ciferri O (1983) *Spirulina*, the edible microorganism. Microbiol Rev 47:551-578

Ciferri O, Tiboni O (1985) The biochemistry and industrial potential of *Spirulina*. Annu Rev Microbiol 39:503-526

Dibrov PA, Kostyrko VA, Lazarova RL, Skulachev VP, Smirnova IA (1986a) The sodium cycle. I. Na^+ dependent motility and modes of membrane energisation in the marine alkalotolerant *Vibrio alginolyticus*. Biochim Biophys Acta 850:449-457

Dibrov PA, Lazarova RL, Skulachev VP, Verkhovskaya ML (1986b) The sodium cycle. II. Na^+ - coupled oxidative phosphorylation in *Vibrio alginolyticus*. Biochim Biophys Acta 850:458-465

Doemel WN, Brock TD (1977) Structure, growth and decomposition of laminated algal bacteria mats in alkaline hot springs. Appl Environ Microbiol 34:433-452

Dubey RC, Sharma KN (1989) Acid and alkaline phosphatases in rice seedlings grown under salinity stress. Indian J Plant Physiol 32:217-223

Duff KE, Smol JP (1994) Chrysophycean cyst flora from British Columbia (Canada) lakes. Nova Hedwigia 58:353-389

Ferris FG (1993) Microbial biomineralization in natural environments. Earth Sci 47:233-250

Ferris FG, Wiese RG, Fyfe WS (1994) Precipitation of carbonate minerals by microorganisms: implications for silicate weathering and the global carbon dioxide budget. Geomicrobiol J 12:1-13

Fry IV, Huflejt M, Erber WWA, Peschek GA, Packer L (1986) The role of respiration during adaptation of the freshwater cyanobacterium *Synechococcus* 6311 to salinity. Arch Biochem Biophys 244:686-691

Funteu F, Guet C, Wu B, Tremolieres A (1997) Effects of environmental factors on the lipid metabolism in *Spirulina platensis*. Plant Physiol Biochem 35:63-71

Gabbay-Azaria R, Schonfeld M, Tel-Or S, Messinger R, Tel-Or E (1992) Respiratory activity in the marine cynobacterium *Spirulina subsalsa* and its role in salt tolerance. Arch Microbiol 157:183-190

Garcia-Ruiz R, Hernandez I, Lucena J, Niell FX (1997) Preliminary studies on the significance of alkaline phosphatase activity in the diatom *Phaeodactylum tricornutum* Bohlin. Sci Mar 61:517-525

Gimmler H (2000) Primary sodium plasma membrane ATPase in salt-tolerant algae—facts and fiction. J Exp Bot 51:1171-1178

Gimmler H, Kugel H, Leibfritz, Mayer A (1988) Cytoplasmic pH of *Dunaliella parva* and *Dunaliella acidophila* as monitored by in vivo ^{31}P-NMR spectroscopy and the DMO method. Physiol Plant 74:521-530

Grant WD, Horikoshi K (1989) Alkaliphiles. In: Da Costa MS, Duarte FC, Williams RAD (eds) The microbiology of extreme environments and its biotechnological potential. Elsevier, Amsterdam, pp 346-366

Grant WD, Tindall WJ (1986) The alkaline saline environment. In: Herbert RA, Codd GA (eds) Microbes in extreme environments. Academic Press, London, pp 25-54

Grant WD, Mwatha WE, Jones BE (1990) Alkaliphiles: ecology, diversity and applications. FEMS Microbiol Rev 75:255-270

Hayashi M, Unemoto T (1987) Subunit components and their roles in the sodium-transport NADH:quinone reductase of a marine bacterium, *Vibrio alginolyticus*. Biochim Biophys Acta 890:47-54

Hecky RE, Kilham P (1973) Diatoms in alkaline saline lakes: ecology and geochemical implications. Limnol Oceanogr 18:53-71

Hellebust JA (1988) A comparative study of sodium and osmotic requirements for growth and nutrient uptake of two related green flagellates, *Dunaliella tertiolecta* and *Chlamydomonas pulsatilla*. Arch Microbiol 143:11-14

Hernandez I, Niell FX, Fernandez JA (1992) Alkaline phosphatase in *Porphyra umbilicalis* L. Kützing. J Exp Mar Biol Ecol 159:1-13

Hernandez I, Niell FX, Fernandez JA (1996) Alkaline phosphatase activity of the red alga *Corallina elongata* Ellis et Solander. Sci Mar 60:297-306

Horikoshi K (1998) Alkaliphiles. In: Horikoshi K, Grant WD (eds) Extremophiles. Wiley-Liss, New York, pp 155-179

Horikoshi K, Akiba T (1982) Alkalophilic microorganisms. Japan Scientific Societies Press, Tokyo

Iltis A (1974) Phytoplankton of natroned waters at Kanem Chad. Part 7. Structure of population. Cah ORSTOM Ser Hydrobiol 8:51-76

Imhoff JF, Sahl HG, Soliman GSH, Truper HG (1979) The Wadi Natrun (Egypt). Chemical and microbial mass developments in alkaline brines of eutrophic desert lakes. Geomicrobiol J 1:219-234

Ivey DM, Zemsky J, Guffant AA, Sturr MG, Hicks DB, Krulwich TA, Gilmour R, Ito M (1998) Alkaliphile bioenergetics. In: Horikoshi K, Grant WD (eds) Extremophiles. Wiley-Liss, New York, pp 181-210

Javor B (1989) Hypersaline environments. Springer, Berlin Heidelberg New York

Joset F, Jeanjean R, Hagemann M (1996) Dynamics of the response of cyanobacteria to salt stress: deciphering the molecular events. Plant Physiol 96:738-744

Katz A, Kaback HR, Avron M (1986) Antiport in isolated plasma membrane vesicles from the halotolerant alga *Dunaliella salina*. FEBS Lett 202:141-144

Katz A, Pick U, Avron M (1989) Characterization and reconstitution of the antiporter from the plasma membrane of the halotolerant alga *Dunaliella*. Biochim Biophys Acta 983:9-14

Katz, A, Bental M, Degani H, Avron M (1991) In vivo pH regulation by an antiporter in the halotolerant alga *Dunaliella salina*. Plant Physiol 96:110-115

Katz A, Pick U, Avron M (1992) Modulation of Na^+/H^+ antiporter activity by extreme pH and salt in the halotolerant alga *Dunaliella salina*. Plant Physiol 100:1225-1229

Ken-Dror S, Preger R, Avi-Dor Y (1986) Functional characterization of the uncoupler-insensitive Na^+-pump of the halotolerant bacterium BA_1. Arch Biochem Biophys 244:122-127

Kroll RG (1990) Alkalophiles. In: Edwards C (ed) Microbiology of extreme environments. McGraw-Hill, New York, pp 55-92

Krulwich TA, Guffanti AA (1983) Physiology of acidophilic and alkalophilic bacteria. Adv Microb Physiol 24:173-214

Krulwich TA, Guffanti AA (1989) Alkalophilic bacteria. Annu Rev Microbiol 43: 435-463

Krulwich AK, Hicks DB, Seto-Young D, Guffanti AA (1988) The bioenergetics of alkalophilic bacilli. Crit Rev Microbiol 16:15-36

Krulwich TA, Guffanti AA, Seto-Young D (1990) pH homeostasis and bioenergetic work in alkalophiles. FEMS Microbiol Rev 75:271-278

Krulwich TA, Ito M, Gilmour R, Guffanti AA (1997) Mechanisms of cytoplasmatic pH regulation in alkaliphilic strains of *Bacillus*. Extremophiles 1:163-169

Krulwich TA, Ito M, Gilmour R, Hicks DB, Guffanti AA (1998) Energetics of alkaliphilic *Bacillus* species: physiology and molecules. Adv Microb Physiol 40:402-438

Lara C, Rodriguez R, Guerrero MG (1993) Nitrate transport in the cyanobacterium *Anacystis nidulans*. Physiol Plant 89:582-587

Lucas WJ, Berry JA (eds) (1984) Inorganic carbon uptake by aquatic photosynthetic organisms. American Soc Plant Physiol, Rockville, Maryland

Markarova EN, Kirikova NN, Tambiev AK, Gusev MV (1997) Two questions about physiological role of sodium in cyanobacteria. Vestn Mosk Univ Ser XVI Biol 0(1) :3-7

Merz MUE (1992) The biology of carbonate precipitation by cyanobacteria. Facies 26:81-202

Merz M, Schlue WR, Zankl H (1995) pH-measurements in the sheath of calcifying cyanobacteria. Bull Inst Oceanogr 14(2):281-289

Miller AG, Canvin DT (1987) Na^+ stimulation of photosynthesis in the cyanobacterium *Synechococcus* UTEX 625 grown on high levels of inorganic carbon. Plant Physiol 84:118-124

Nausch M (1998) Alkaline phosphatase activities and the relationship to inorganic phosphate in the Pomeranian Bight (southern Baltic Sea). Aquat Microb Ecol 16:87-94

Nitschmann WH, Peschek GA (1985) Modes of proton translocation across the cell membrane of respiring cyanobacteria. Arch Microbiol 14:330-336

Packer L, Spath S, Martin JB, Roby C, Bligny R (1987) ^{23}Na and ^{31}P-NMR studies of the effect of salt on the fresh water cyanobacterium *Synechococcus* 6311. Arch Biochem Biophys 256:354-361

Park C, Moon JY, Cokie P, Webster DA (1996) Na^+ translocating cytochrome bo terminal oxidase from *Vitreoscilla:* some parameters of its Na^+ pumping and orientation in synthetic vesicles. Biochemistry 35:11895-11900

Paschinger H (1977) DCCD-induced sodium uptake by *Anacystis nidulans*. Arch Microbiol 113:285-292

Peschek GA (1987) Respiratory electron transport. In: Fay P, Van Baalen C (eds) The cyanobacteria. Elsevier, New York, pp 119-161

Pfenninger-Li XD, Albracht SPJ, van Belzen R, Dimroth P (1996) NADH: ubiquinone oxidoreductase of *Vibrio alginolyticus*: purification, properties and reconstitution of the Na^+ pump. Biochemistry 35:6233-6242

Pick U (1992) ATPases and ion transport in *Dunaliella*. In: Avron M, Ben-Amotz A (eds) *Dunaliella*: physiology, biochemistry and biotechnology. CRC Press, Boca Raton, pp 63-97

Popova LG, Balnokin YV (1992) H^+ tranclocating ATPase and antiport activities in the plasma membrane of the marine alga *Platymonas viridis*. FEBS Lett 309:333-336

Popova L, Balnokin Y, Dietz KJ, Gimmler H (1998) Na^+-ATPase from the plasma membrane of the marine alga *Tretraselmis (Platymonas) viridis* forms a phosphorylated intermediate. FEBS Lett 426:161-164

Popova L, Balnokin Y, Dietz KJ, Gimmler H (1999) Characterization of phosphorylated intermediates synthesized during the catalytic cycle of the sodium adenosine triphosphatase in the plasma membrane of the marine unicellular alga *Tetraselmis (Platymonas) viridis*. J Plant Physiol 155:302-309

Price GD, Badger MR (1989) Expression of human carbonic anhydrase in the cyanobacterium *Synechococcus* PCC7942 creates a high CO_2-requiring phenotype. Plant Physiol 91:505-513

Quisel JD, Wykoff DD, Grossman AR (1996) Biochemical characterization of the extracellular phosphatases produced by phosphorus-deprived *Chlamydomonas reinhardtii*. Plant Physiol 111:839-848

Raven JA (1984) The CO_2 concentrating mechanism. In: Lucas WJ, Berry JA (eds) Inorganic carbon uptake by aquatic photosynthetic organisms. American Soc Plant Physiol, Rockville, Maryland

Raven JA (1990) Sensing pH. Plant Cell Environ 13:721-730

Raven JA, Falkowski PG (1999) Oceanic sink for atmospheric CO_2. Plant Cell Environ 22:741-755

Remis D, Simonis W, Gimmler H (1992) Measurement of the transmembrane electrical potential of *Dunaliella acidophila* by microelectrodes. Arch Microbiol 158: 350-355

Richmond A (1986) Microalgae of economic potential. In: Richmond A (ed) Handbook of microalgae mass culture. CRC Press, Boca Raton, pp 199-243

Richmond A (1988) *Spirulina*. In: Borowitzka MA, Borowitzka LJ (eds) Micro-algal biotechnology. Cambridge University Press, Cambridge, pp 85-121

Richmond A, Karg S, Boussiba S (1982) Effects of bicarbonate and carbon dioxide on the competition between *Chlorella vulgaris* and *Spirulina platensis*. Plant Cell Physiol 23:1411-1418

Ritchie R (1991) Membrane potential and pH control in the cyanobacterium *Synechococcus* R-2 (*Anacystis nidulans*) PCC 7942. J Plant Physiol 137:409-418

Ritchie R (1992) Sodium transport and the origin of the membrane potential in the cyanobacterium *Synechococcus* R-2 (*Anacystis nidulans*) PCC 7942. J Plant Physiol 139:320-330

Rivkin RB, Swift E (1980) Characterization of alkaline phosphatase and organic phosphorus utilization in the oceanic dinoflagellate *Pyrocystis noctiluca*. Mar Biol 61:1-8

Schiller P, Heilmeier H, Hartung W (1997) Abscisic acid (ABA) relations in the aquatic resurrection plant *Chamaegigas intrepidus* under naturally fluctuating environmental conditions. New Phytol 136:603-611

Schlesinger WH (1991) Biogeochemistry. An analysis of global change. Academic Press, San Diego

Schlesinger P, Belkin S, Boussiba S (1996) Sodium deprivation under alkaline conditions causes a rapid death of the filamentous cyanobacterium *Spirulina platensis*. J Phycol 32:608-613

Sethi V, Kaushik A (1993) Tolerance of *Anabaena oryzae* to salt stress. Geobios (Jodhpur) 20:10-13

Shiraiwa Y, Goyal A, Tolbert NE (1993) Alkalization of the medium by unicellular green algae during uptake of dissolved inorganic carbon. Plant Cell Physiol 34:649-657

Shono M, Wada M, Fujii T (1995) Partial purification of a Na^+-ATPase from the plasma membrane of the marine alga *Heterosigma akashiwo*. Plant Physiol 108:1615-1621

Shono M, Hara Y, Wada M, Fujii T (1996) A sodium pump in the plasma membrane of the marine alga *Heterosigma akashiwo*. Plant Cell Physiol 37:385-388

Shukla SP, Singh SH, Mishra AK (1997) Energetics of alkaline phosphatase (monoesterase) activity in antarctic and tropical isolates of a cyanobacterium, *Anabaena*. Cytobios 92:29-33

Skulachev VP (1985) Membrane-linked energy transductions. Bioenergetic functions of sodium. Proton is not unique as a coupling ion. Eur J Biochem 151:199-208

Skulachev VP (1988) Membrane energetics. Springer, Berlin Heidelberg New York

Skulachev VP (1994) The latest views from the sodium world. Biochim Biophys Acta 1187:216-221

Stein WD (1986) Transport and diffusion across cell membranes. Academic Press, San Diego, pp 477-571

Talling JF, Wood RB, Prosser MV, Baxter RM (1973) The upper limit of photosynthetic productivity by phytoplakton: evidence from Ethiopian soda lakes. Freshwater Biol 3:53-76

Thielmann J, Tolbert NE, Goyal A, Senger H (1990) Two systems for concentrating carbon dioxide and bicarbonate during photosynthesis by *Scenedesmus*. Plant Physiol 92:622-629

Thompson JB, Ferris FG (1990) Cyanobacterial precipitation of gypsum, calcite, and magnesite from natural alkaline lake water. Geology 18:995-998

Tokuda H, Unemoto T (1981) A respiration-dependent primary sodium extrusion system functioning at alkaline pH in the marine bacterium *Vibrio alginolyticus*. Biochem Biophys Res Commun 102:265-271

Tokuda H, Unemoto T (1984) Na^+ is translocated at NADP: quinone oxidoreductase segment in the respiratory chain of *Vibrio alginolyticus*. J Biol Chem 259:7785-7790

Tokuda H, Ugadawa T, Unemoto T (1985) Generation of the electrical potential of Na^+ by the Na^+ motive NADH oxidase in inverted membrane vesicles of *Vibrio alginolyticus*. FEBS Lett 183:95-98

Ullrich WR, Glaser E (1982) Sodium-phosphate cotransport in the green alga *Ankistrodesmus braunii*. Plant Sci Lett 27:155-161

Ullrich-Eberius CJ, Yingchol Y (1974) Phosphate uptake and its dependence in halophytic and glycophytic algae and higher plants. Oecologica 17:17-26

Unemoto T, Hayashi M (1993) Na^+-translocating NADH-quinone reductase of marine and halophilic bacteria. J Bioenerg Biomembr 25:385-391

Vonshak A, Boussiba S, Abeliovich A, Richmond A (1983) Production of *Spirulina platensis* biomass: maintenance of monoalgal cultures outdoors. Biotechnol Bioeng 25:341-350

Vonshak A, Guy R, Guy M (1988) The response of the filamentous cyanobacterium *Spirulina platensis* to salt stress. Arch Microbiol 150:414-420

Wada M, Satoh S, Kasamo K, Fujii T (1989) Presence of a sodium-activated ATPase in the plasma membrane of the marine raphidophycean *Heterosigma akashiwo*. Plant Cell Physiol 30:923-928

Wada M, Urayama O, Satoh S, Hara Y, Ikawa Y, Fujii T (1992) A marine algal Na^+-activated ATPase possesses an immunologically identical epitope to Na^+, K^+-ATPase. FEBS Lett 309:272-274

Walther K, Fries L (1976) Extracellular alkaline phosphatase in multicellular marine algae and their utilization of glycero-phosphate. Physiol Plant 36:118-122

Weiss M, Pick U (1990) Transient Na^+ flux following hyperosmotic shock in the halotolerant alga *Dunaliella salina*. J Plant Physiol 136:429-438

Wood RB, Talling JF (1988) Chemical and algal relationships in a salinity series of Ethiopian inland waters. Hydrobiologia 158:29-67

Yates KK, Robbins LL (1995) Experimental evidence for a $CaCO_3$ precipitation mechanism for marine *Synechocystis*. Bull Inst Oceanogr14:51-59

11 Salt Acclimation of Algae and Cyanobacteria: A Comparison

Norbert Erdmann and Martin Hagemann*

Contents

---

* Department of Biosciences, University of Rostock, 18051 Rostock, Germany

11.1 Introduction

Almost all cells are able to live within a certain range of enhanced salt concentrations or changing salinities, since it is generally believed that all life originated in the oceans, a highly saline environment. However, during evolution, the degree of salt resistance became very divergent among present-day organisms. Salt acclimation has generated great interest for two main reasons: (1) salt represents one of the environmental master factors of aquatics and, increasingly, of terrestrial habitats, and (2) salt resistance constitutes a real biotechnological challenge in the field of crop plants. Thus, algae and cyanobacteria have attracted considerable attention since they are inhabitants of these biotopes characterized by high and also changing salinities and can serve as model organisms for higher plants. In the second half of the 20th century, the fundamental aspects of salt adaptation were unravelled and subjected to several chapters: Kirst (1990) chaptered salinity tolerance of marine macroalgae and phytoplankton species; Galinski (1993), Da Costa et al. (1998), Kempf and Bremer (1998) as well as Poolman and Glaasker (1998) discussed compatible solutes of bacteria; Serrano et al. (1999) stressed the mechanism of ion homeostasis; Oren (1999) revised the energetic costs of salt adaptation; Wood (1999) focused on bacterial osmosensing; and Bohnert and Sheveleva (1998), Yeo (1998) and McNeil et al. (1999) reviewed the metabolic engineering of salt stress resistance. Salt acclimation of cyanobacteria was reviewed by Reed and Stewart (1988), as well as Hagemann and Erdmann (1997). Comprehensive definitions of terms are to be found in the chapter of Kirst (1990) and Wood (1999). This chapter is a comparative update of recent progress in salt-adaptation research on algae and cyanobacteria.

Enhanced salt concentrations change the growth conditions in a manner unfavourable for most organisms. The increase in external concentrations of inorganic ions (dominated by Na^+ and Cl^-) impairs the well-balanced osmotic relation between the cell interior and the surrounding medium and forces water efflux from the cell. In parallel, the increased exogenous ion concentration tightens the influx of these ions into the cell according to their electrochemical gradients. The loss of water and the invading ions activate a concerted acclimation process that leads to acclimated cells with a new steady state of growth. This acclimation process includes three basal processes (Fig. 11.1): (1) Water loss affects cell turgor and volume and can cause critical changes. Therefore, restoration of the turgor and volume are the main goals of the acclimation process and involve water fluxes which are controlled by water channels and the concentrations of those cellular solutes that predominantly contribute to the osmotic potential (osmolytes). (2) The permanent influx of ions disturbs the cellular ion homeostasis. Acclimation to the altered internal ion situation is managed by a regulated uptake and export of ions (K^+ and Na^+, respectively) and by changes in the permeability of the cell membrane. (3) Elevated ion concentrations and the diminished water activity caused by the water loss impair the function of biopolymers and induce protective measures

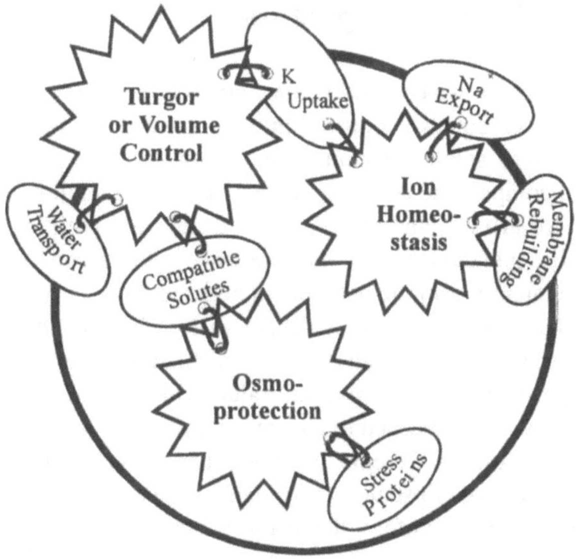

Fig. 11.1. Main targets (*stars*) and basal elements (*ovals*) of cellular salt acclimation processes

which culminate in the accumulation of osmoprotecting compatible solutes and stress proteins. These molecules ensure the protection and renaturation of damaged proteins, nucleic acids as well as membrane lipids. Besides these main processes, several secondary responses are needed to ensure a successful salt acclimation, e.g. the scavenging of liberated free radicals, increase in energy-supplying reactions and, finally, the adjustment of the whole metabolism to the new situation. All these efforts are realized by an altered gene expression and changes in the activity of the enzymes involved.

The main difference in the salt adaptation of cyanobacteria and microalgae seems to be related to the more complex structure of the algal cells. This led Ginzburg et al. (1995) to propose a more complex model of salt acclimation in which an outer (cytoskeleton-characterised) and an inner (organelle-based) compartment are specifically considered. The outer compartment is controlled by the mechanochemical alteration of a macromolecular network and is dominated by Na^+ and Cl^-, whereas the inner compartment executes volume changes according to the osmotic pressure maintained by the accumulation of compatible solutes. Furthermore, many algal cells, like plant cells, harbour a vacuole as a special compartment which is less active in cellular metabolism but plays an important role in the osmotic state of a given cell. Here, the concentrations of K^+ and Cl^- represent the central part of the acclimation process.

11.2 Turgor or Volume Regulation

The osmotically forced water flux is uncontrolled and lasts until the water potential inside the cell equals the outer value. This water flux lasts for seconds to a few minutes, alters solute concentration, turgor (walled cells) and volume (naked cells) and compels the cells into a harmful stress situation since sufficient turgor is needed for growth, and the volume is genetically prefixed. For survival, restoration of the prestress values is necessary and occurs mainly by the adjustment of the internal osmotic potential in hours to days.

For years, it has been a matter of debate as to which cellular parameter is measured to understand the regulation of the acclimation process of salt-treated cells. Since turgor represents a universal value which is negatively influenced after salt stress and which is needed to ensure cell growth, many authors prefer the idea that turgor pressure represents the basic parameter regulated during salt acclimation (Csonka and Epstein 1998). Direct measurements of turgor are difficult to carry out and, therefore, the turgor values reported for cells vary over quite a large range (i.e. for *E. coli* very high values around 20 bar have been reported). Generally, the degree to which cells regain turgor is unclear even for bacteria (Wood 1999). For algae and cyanobacteria, some evidence exists that the prestress values are not always achieved (Reed 1983; Bisson and Bartholomew 1984; Reed and Stewart 1988; Karsten and Kirst 1989; Kirst 1990). Interestingly, estimations of the turgor threshold and salt-induced volume changes which were traditionally measured by light scattering or impedance changes could also be assessed by the chlorophyll a fluorescence as proved for two cyanobacterial freshwater strains (Papageorgiou et al. 1999; Stamatakis et al. 1999). This method represents a relatively easy way to directly obtain turgor values. Unfortunately, until now it has not been applied to compare turgor pressure in cells before and after salt acclimation.

Wall-less algal cells respond to changes in the external salt concentration with changes in their cell volume. Cells of *Dunalliela* spp., *Chlorella emersonii* and *Poterioochromonas malhamensis*, for instance, behaved as nearly ideal osmometers at different salt concentrations (see Reed 1984, and references therein). For these species it was shown that a successful salt acclimation results in almost the same cell volume as was found before stress application.

Four different mechanisms are involved in retaining cell turgor or volume: uptake/export of ions, accumulation/deprivation of organic osmolytes, control of opening of water and solute channels and activation/inactivation of contractile vacuoles. In the case of an enhanced external salt concentration (upshock), the energetically most favourable mode to increase the internal osmotic value would be the uptake of ions from the surrounding. However, this salt-in strategy is confined to some halophilic archea of the order Halobacteriales, and bacteria of the order Haloanaerobiales, which cope with salt stress by the accumulation of K^+ and other ions (see Oren 1999). All other organisms, including algae and cyanobacteria, restrict the balancing of water fluxes by the accumulation of ions in the initial phase immediately after shock treatments (Reed 1984; Nitschmann and Packer 1992; Ginzburg et al. 1995). During the further progress of acclimation, they pass over to the cytoplasmic accumulation of special osmolytes called compatible solutes (Brown 1976). Only giant algae cells containing a large vacuole use inorganic ions to balance the osmotic potential of this less metabolic active compartment to save energy since, in the case of de novo synthesis of the compatible solutes, this strategy of salt acclimation requires about 100 times more energetic efforts than the salt-in strategy using ion accumulation (Oren 1999). However, accumulation of compatible solutes has the advantage that the same cell interior could be used independently of the external salt concentration, and no special adaptation of all intracellular components to a high salt environment has to take place (e.g. evolution of salt-resistant enzymes). Obviously, this advantage was the basis for the general success of the compatible-solute strategy and seems to be a prerequisite for survival in a wide range of salinity (from 0.05 to 5.5 M NaCl) as known in *Dunaliella* species (Ginzburg et al. 1995).

11.3 Compatible Solutes: Universal and Successful

Compatible solutes are low-molecular weight, highly hydrophilic organic compounds without net charge and, contrary to ions, can be accumulated in high concentrations without interfering with the cellular metabolism. Due to the increase of the number of dissolved molecules, the internal osmolality is raised, forcing water to reenter the cell and restore turgor and volume. This is the inherent osmolytic function of any compatible solute. About 20 different compatible solutes have been known to occur in algae and cyanobacteria for a long time (Fig. 11.2) though reliable corroboration remains outstanding in some cases (Reed and Stewart 1988; Kirst 1990; Bisson and Kirst 1995; McNeil et al. 1999;). No new structure has been discovered in the past decade except for an isomer of cyclohexanetetrol which increased with increasing salinity in the diatom *Chaetoceros muelleri* (Fujii et al. 1995). Algae show considerable biochemical diversity and make use of all compounds mentioned except glucosylglycerol and glutamate betaine. In cyanobacteria, seven different compounds have been found so far (glycine betaine, glutamate betaine, DMSP (3-dimethylsulfonium propionate), proline, glucosylglycerol, sucrose and trehalose). The hydrophilic nature of compatible solutes (Fig. 11.2) arises from polyhydroxy groups, quarternary N and tertiary S atoms. Highest salt concentrations are tolerated by those algae and cyanobacteria which accumulate glycerol or glycine betaine. The least salt concentration is tolerated by organisms that accumulate disaccharides. This ranking correlates more or less with the energetic budget necessary for their biosyntheses and with the degree of solubility (Fig. 11.2).

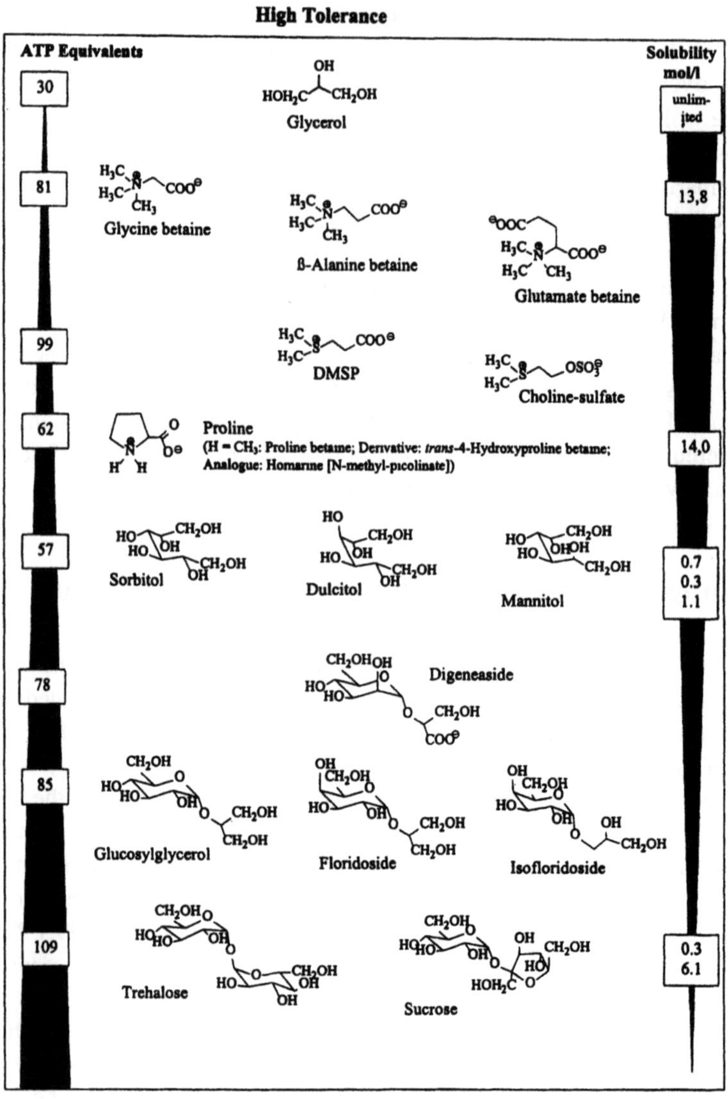

Fig. 11.2. Compatible solutes of algae and cyanobacteria: structures, energy demand and solubility. The compatible solutes are displayed according to their osmoprotective quality (top–down). This corresponds, except for some exceptions, with energy demand and solubility in water. The former was calculated using autotrophic pathways from CO_2, NO_3^- and SO_4^{2-} and recycling of all metabolites. Glycine betaine biosynthesis was assumed via serine and choline (Nuccio et al. 1999), DMSP via methionine (McNeil et al. 1999), proline via glutamate (Yoshiba et al. 1997), polyols, disaccharide and heteroside along known pathways (Hagemann and Erdmann 1994; Nuccio et al. 1999), and transmethylation via S-adenosylmethionine. ATP equivalents mean the sum of ATP and NAD(P)H (NAD(P)H = 3 ATP). DMSP 3-dimethylsulfonium propionate

Glycerol represents the champion among all compatible solutes since it demands the least energetic cost (30 ATP equivalents) and is the most hydrophilic compound (enormous solubility in water). Accumulation of glycerol allows survival at all Na^+ concentrations. A representative example is the genus *Dunaliella*, which can live in habitats saturated with NaCl. However, the use of glycerol is restricted to the domain of Eukarya. This was explained by the occurrence of sterols in eukaryotic membranes and their ability to make the membrane less permeable and prevent leakage of glycerol (see Oren 1999). Nevertheless, some yeasts have evolved an additional active uptake system to regain leaked glycerol (Lages and Lucas 1995).

Cyanobacteria acquire highest salt tolerance by the accumulation of glycine betaine which is also ubiquitously found in plants (see Rhodes and Hanson 1993), algae (Mason and Blunden 1989), bacteria (DaCosta et al. 1998) and archaea (Lai et al. 1999). However, the autotrophic accumulation is rather expensive (81 ATP equivalents per molecule in plants).

Related quarternary ammonium structures, such as ß-alanine betaine, proline betaine, *trans*-4-hydroxyproline betaine and homarine are equally effective, but have been detected in only a few plant families and marine algae (Blunden and Gordon 1986; Dickson and Kirst 1986; Rhodes and Hanson 1993; Karsten et al. 1994; DaCosta et al. 1998). This is also true for another betaine, glutamate betaine, which was found only in two cyanobacteria isolates (*Calothrix* spp.) (Mackay et al. 1984).

Among the known compatible solutes, proline is probably the most widely distributed member (Delauney and Verma 1993; Yoshiba et al. 1997). Its concentration increases rapidly after salt stress in bacteria (Csonka 1989; Csonka and Epstein 1998), cyanobacteria (Riccardi et al. 1983; Reed and Stewart 1988; Fulda et al. 1999), protozoa, marine invertebrates (Schoffeniels and Gilles 1970), algae (Schobert 1977; Greenway and Setter 1979; Wegmann 1986) and higher plants (Yoshiba et al. 1997; Nanjo et al. 1999). Most reports assign proline a role as a compatible solute, though its cellular concentrations are often found to be too low to contribute significantly to the internal osmotic potential. However, osmolyte-relevant concentrations were also amply reported. For instance, in the Gram-positive bacterium *Bacillus subtilis,* proline is the main compatible solute synthesised de novo or imported from the environment in response to high osmolalities (Kempf and Bremer 1998).

Due to the onium structure, tertiary sulfonium compounds are as favourable as quarternary ammonium compounds. However, in accordance with the additional ATP equivalents for the assimilation of S in comparison to N, tertiary sulfonium compounds are not found in organisms living at highest salinities. DMSP is a dominant osmolyte of marine green algae (Malin and Kirst 1997). For instance, in field studies along the east coast of the Columbia District, a close correlation was found between DMSP (and also DMS) and salinity (Iverson et al. 1989). Besides its osmoregulatory function, it affects, via its metabolite DMS, the global climate (Malin and Kirst 1997). Cyanobacteria are considered to be minor producers of

DMSP. A survey of 29 marine strains showed 23 to be DMSP producers, but the DMSP concentrations were unimportant (Vogt et al. 1998). A relative of DMSP, choline sulfate was found in only a few organisms (Csonka and Hanson 1991).

Except for glycerol, polyhydric alcohols belong to the medium energy-demanding compatible solutes (57 ATP equivalents) and attract attention with respect to their specialised, but widespread occurrence in higher plants, algae and fungi, but not in cyanobacteria and bacteria. D-Mannitol represents the principal osmolyte of brown algae (Karsten et al. 1996), but has also recently been detected in the mangrove red alga *Caloglossa leprieurii* and has been considered responsible for its salt tolerance (Karsten et al. 1997). *Bostrychia* spp. of mangrove as well as marine ecosystems use both D-sorbitol and D-dulcitol as osmolytes with a preference for sorbitol in low-temperature waters (Karsten et al. 1995).

The energy-expensive class of compatible solutes is opened by O-α-D-mannopyranosyl-(1→2)-glycerate (digeneaside) which needs 78 ATP equivalents per molecule. Red algae of the order Ceramiales accumulate digeneaside as one of their major osmolytes (Bisson and Kirst 1995; Karsten et al. 1995). Though a direct relationship between digeneaside and increasing salinities was often reported, an osmotic function remains uncertain since the concentrations were too low to contribute significantly to the internal osmotic potential and, at least in *Caloglossa leprieurii,* they hardly varied with the external salinity (Karsten et al. 1994,1995). Furthermore, digeneaside is one of the exceptions in the field of compatible solutes since it carries a net charge.

O-α-D-glucopyranosyl-(1→2)-glycerol (glucosylglycerol) was the first compatible solute found in cyanobacteria (Borowitzka et al. 1980) and was subsequently detected in the majority of moderately halotolerant cyanobacteria (Erdmann 1983; Mackay et al. 1983, 1984; Richardson et al. 1983; Reed et al. 1984,1986; Reed and Stewart 1988), but it was also observed in some bacteria (Galinski and Truper 1994; Pocard et al. 1994). Today, it is regarded as one of the best-studied cyanobacterial compatible solutes (Reed and Stewart 1988; Joset et al. 1996; Hagemann and Erdmann 1997; DaCosta et al. 1998).

Glucosylglycerol is the only osmotically relevant heteroside found in cyanobacteria. Algae are much more diverse. All red algae, except those of the order Ceramiales, generally accumulate the heteroside O-α-D-galactopyranosyl-(1→2)-glycerol (floridoside; Kirst 1990), structurally closely related to glucosylglycerol. Recently, it was found in the mangrove ecosystem species *Caloglossa* (Karsten et al. 1996a,b). Members of the order Bangiales additionally accumulate O-α-galactopyranosyl-(1→1)-glycerol (isofloridoside) as D and L enantiomer (Karsten et al. 1993). From the detailed studies on the biochemical regulation of its biosynthesis, isofloridoside is best known as the compatible solute of the unicellular flagellate *Poterioochromonas malhamensis* (Kauss and Thomson 1982; Kauss 1987; Brunner and Kauss 1988).

Disaccharide accumulation requires the highest energetic cost (109 ATP equivalents per molecule) and confers the lowest tolerance level. Trehalose is common in cyanobacteria (Reed et al. 1984; Reed and Stewart 1988), yeast (Hohmann 1997), bacteria and archaea (DaCosta et al. 1998). Trehalose is less distributed in higher plants and algae (Müller et al. 1995). Recently, its occurrence and strict dependence on the osmolality of the medium has been reported for *Euglena gracilis* (Takenaka et al. 1997) and *Scytonema* species (Page-Sharp et al. 1999). In the latter case, sucrose was also detected, but it reached only half the concentration of trehalose. Besides its osmotic role, trehalose has been found to be involved in the desiccation tolerance of several cyanobacterial strains (Potts 1994).

Sucrose is common in photosynthetic organisms and its role as compatible solute in plants, algae and cyanobacteria has long been known (Warr et al. 1988; Kirst 1990; Potts 1994; Huber and Huber 1996). Despite its poor tolerance-mediating ability, half of all cyanobacterial strains studied so far accumulate sucrose (Hagemann and Erdmann 1997). Often, sucrose is produced as a short-term response to hypersalinity (Reed 1989) and becomes the dominating compatible solute when the temperature rises (Müller and Wegmann 1978; Reed and Stewart 1988; Page-Sharp et al. 1999).

11.4 From Osmolytes to Osmoprotectants

For most, if not all compatible solutes, it is assumed that they exert a direct protective effect on biomolecules against the damage of low water activity and/or high concentrations of inorganic ions (Csonka and Hanson 1991; Winzor et al. 1992; Galinski and Truper 1994; Potts 1994; Bisson and Kirst 1995; DaCosta et al. 1998; McNeil et al. 1999). The protective effect has been investigated by in vitro and in vivo studies and, recently, by genetic engineering.

In vitro studies probe the mitigating effect of a compatible solute on the stress-dependent activity of an enzyme in the presence of different salt concentrations. Many compatible solutes, especially glycine betaine, and many enzymes (e.g. lactate dehydrogenase, phosphofructokinase, glutamine synthetase, glucose-6 phosphate dehydrogenase, Rubisco, PSII proteins) have been examined in this way and stress protection has always been documented (Warr et al. 1988; Galinski 1993; Ishitani et al. 1993; Xing and Rajashekar 1999). A recent example of the protective effect of glycerol on the stability and extraction of PSII protein complexes from membranes was reported for *Dunaliella salina* (Khristin and Simonova 1998) and *Synechococccus* sp. PCC 7942 (Allakhverdiev et al. 2000). Similarly, the protective effect of mannitol, sorbitol and floridoside on malate dehydrogenase and glucose-6 phosphate dehydrogenase has been confirmed recently (Karsten et al. 1996).

The same holds true for the in vivo approaches which use whole cells and measure the NaCl-affected growth rate in the presence of a compatible solute. For instance, of 14 compounds of algal origin, glycine betaine, proline betaine and DMSP were most and equally effective to alleviate osmotic stress exercised on cells of *Klebsiella pneumoniae* (Mason and Blunden 1989). Whole plants behaved in a similar manner on glycine betaine supply (Xing and Rajashekar 1999).

In the third approach, many experiments were carried out to transfer genes involved in the synthesis of compatible solutes into less salt-resistant organisms to produce or overproduce an osmoprotective compound. Main targets have been the higher plants displaying the protective effects to all compatible solutes produced through genetic engineering (see Table 11.4). In another way, mutants which had lost the ability to synthesize a compatible solute and became salt-sensitive have been restored by the intra- or extracellular supply of their own or a foreign compatible solute. For instance, a glycerol-defective mutant of *Saccharomyces cerevisiae* was supplemented with the *mtlD* gene of *E. coli* which encodes for the key enzymes of the mannitol biosynthesis. Mannitol accumulation restored completely the lost salt tolerance of the glycerol-defective mutant strain (up to 1.6 M NaCl) and also its tolerance to oxidative stress (Chaturvedi et al. 1997). Similarly, a mutant of *Synechocystis* PCC 6803 impaired in the synthesis of glucosylglycerol retained growth by the exogenous supply of three compatible solutes in the order glucosylglycerol > trehalose > sucrose (Mikkat et al. 1997).

Most of these studies documented the superior role of glycine betaine. In most eubacteria glycine betaine is preferred over all other compatible solutes (Csonka and Hanson 1991). Plants effectively accumulate glycine betaine to cope with salt, drought and cold stress (for chapter see Sakamoto and Murata 2000). Similarly, proline accumulates in many eukaryotes exposed to environmental stresses such as high salinity, heat, UV radiation, heavy metals, chemicals, freezing and drought (Liu and Zhu 1997; Nanjo et al. 1999), indicating its roles in stress adaptation; but the role of proline still remains controversial. For instance, among three macrophytes tested, the species with the highest level of proline (*Hydrilla verticillata)* behaved as the most sensitive to NaCl treatment (Rout et al. 1997). Similarly, the *sos1* mutant of *Arabidopsis thaliana*, 20 times more sensitive to NaCl than the wild type, accumulated more proline (Liu and Zhu 1997). Thus, proline accumulation is often interpreted as a result of stress injury, rather than an inherent element of stress tolerance (Delauney and Verma 1993; Lutts et al. 1996; Liu and Zhu 1997; Rout et al. 1997). However, the majority of reports claim a protective role of proline (see Yoshiba et al. 1997) or suggest even additional effects (Nanjo et al. 1999). Moreover, it has been demonstrated that overproduction of proline results in increased tolerance to cold stress of *Dunaliella salina* (Helliot and MortainBertrand 1999).

A similar situation was reported for trehalose. It is assumed to be more a general stress protectant for both proteins and membrane lipids than an osmolyte. For instance, endogenous contents of trehalose in NaCl-stressed rice did not reach the necessary molarity in the cells to balance the osmotic potential, but led still to a significant degree of protection (Garcia et al. 1997). The wealth of information reveals that trehalose, and in a similar manner sucrose, are effective in particular against drought and thermal stress. This has led to the pronounced biotechnological application of trehalose in such cases where water is withdrawn from enzymes, vaccines, membranes and whole cells (Potts 1994; DaCosta et al. 1998). DMSP is as effective in osmoprotection as glycine betaine (Mason and Blunden 1989; Cosquer et al. 1999), it also protects algae from grazing stress (Wolfe et al. 1997).

Different suitability for protection compared with different environmental constraints may be the reason for the concomitant accumulation of more than one compatible solute. A recent example has been reported for the trehalose and sucrose accumulation of *Euglena gracilis* during salt stress. Sucrose dominates in the late exponential, trehalose in the stationary phase and both seem to be responsible for salt acclimation (Porchia et al. 1999). A similar change was observed in the cyanobacterium *Synechococcus* sp. PCC 7418. After the 6th hour of a salt shock (1.2 M NaCl), glucosylglycerol held more than the half of the osmolyte pool, but after 24 h it was completely replaced by glycine betaine. Surprisingly, this strain accumulated a third compatible solute, proline, which increased consistently up to 15% with time (Fulda et al. 1999).

11.5 Enigmatic Function of Osmoprotectants

The effects of osmoprotectants have been poorly understood until now. Glycerol was thought to mimic water molecules since, like cyclic polyols with equatorial OH-groups, it fits very well into the hexameric arrangement of water molecules (Schobert 1980). Proline forms supramolecular assemblies at concentrations far above 0.1 mM generating amphiphilic aggregates which are able to function as minidetergents and solubilize proteins (Samuel et al. 1997). Moreover, some compatible solutes (mannitol, sorbitol, proline) were considered to be radical scavengers (Smirnoff and Cumbes 1989). However, the most promising view which is applicable to all compatible solutes was introduced by Thimasheff and his colleagues and has gathered attention and experimental evidence during the past 15 years (see Liu and Bolen 1995, and references therein). According to their hypothesis, molecules of the compatible solutes are thought to favour less dense (structured) water which is characteristic of the bulk solution and are less distributed in dense (weakly bound) water in the hydration shell of proteins (more counterions). Therefore, compatible solutes are preferentially excluded from the vicinity of the proteins, yielding a better hydratation sphere of the protein surface. Thermodynamically, this is an unfavourable event and requires more energy when increased surface of the protein becomes exposed to the bulk solution. This happens when a protein denatures (unfolds). Therefore, compatible solutes favour the native form of the majority of proteins, even under conditions of a reduced water activity (DaCosta et al. 1998). By transferring amino acids from pure water into osmolyte-containing solutions and calculating the changes of the free energy, Liu and Bolen (1995) could show that the peptide backbone of the protein provides the main impetus to the protein packing since it overrides the unfolding effect of the side chains.

11.6 De Novo Biosynthesis Versus Uptake

Phototrophic organisms usually synthesize compatible solutes de novo and are, contrary to heterotrophic bacteria, less equipped with transport proteins to take up them from the medium. Nevertheless, uptake systems have been detected in several cyanobacteria for the solutes glycine betaine, glucosylglycerol, trehalose and sucrose (Moore et al. 1987; Mikkat et al. 1996, 1997). For the latter three compatible solutes, a common transporter was found in *Synechocystis* sp. PCC 6803 which transported the solutes in the order glucosylglycerol > sucrose > trehalose (Mikkat et al.1996). The transporter has been physiologically and genetically characterized. It belongs to the class of ABC transporters and consists of four subunits. The encoding genes are not organized in an operon and were identified recently (Hagemann et al. 1997; Mikkat and Hagemann 2000). The glucosylglycerol transporter has no direct importance in high salt tolerance since a transporter-deficient mutant has shown the same growth and regulation behaviour as the wild type (Hagemann et al. 1997). Its main function seems to be related to reuptake of glucosylglycerol leaked through the cytoplasmic membrane thus minimizing losses of energy and organic carbon (Hagemann et al. 1997). It seems possible that under the restricted conditions of natural microbial mats, where compatible solutes belong to the most abundant low-molecular mass compounds, the uptake of glucosylglycerol from the environment relieves the cells of the costly biosynthesis and will favour cells with this uptake ability against competing organisms.

Biosyntheses of all the polyols mentioned in Fig. 11.2 proceed in a two-step way: the first step leads to the polyol phosphate and is the rate-limiting reaction. In the second step, the polyol phosphate is hydrolyzed by a specific phosphatase to release the osmolyte. This mechanism is well understood for glucosylglycerol of *Synechocystis* sp. PCC 6803 since all the enzymes and genes involved are well known (Hagemann and Erdmann 1994; Hagemann et al. 1997; Marin et al. 1998):

ADP-glucose + glycerol-3-phosphate —1→ glucosylglycerol-3-phosphate (+ ADP) —2→ glucosylglycerol (+ P_i)

Enzymes and genes:

1. ADP-glucose:glycerol-3-phosphate 2-glucosyltransferase (glucosylglycerol-phosphate synthase, GGPS, EC 2.4.1.), *ggpS* (sll 1566).
2. Glucosylglycerol-3-phosphate phosphohydrolase (glucosylglycerol-phosphate phosphatase, GGPP, EC 3.1.3.), *stpA* (slr 0746).

Analogous pathways have been described for sucrose and trehalose in *Euglena gracilis* (Porchia et al. 1999b) and *Scytonema* sp. (Page-Sharp et al. 1999), sucrose in *Synechocystis* sp. PCC 6803 (Curatti et al. 1998; Hagemann and Marin 1999; Lunn et al. 1999) and in *Anabaena* sp. (Porchia and Salerno 1996; Porchia et al. 1999a), floridoside in *Porphyra perforata* (Meng and Srivastava 1991), isofloridoside in *Poterioochromonas malhamensis* (Brunner and Kauss 1988) and mannosylglycerate in *Rhodothermus marinus* (Martins et al. 1999).

Polyhydric alcohols are integrated in low-step metabolic cycles which enable fast building and degradation. This was demonstrated for glycerol and mannitol. The glycerol cycle consists of two anabolic and two catabolic reactions which are well characterized for algae (see Haus and Wegmann 1984) and yeast (see Hohmann 1997; Nevoigt and Stahl 1997). However, active reuptake of external glycerol is hardly documented (Lages and Lucas 1997; Nevoigt and Stahl 1997).

A similar four-step cycle has been described for mannitol and recently verified in *Caloglossa leprieurii*. The anabolic pathway starts from fructose-6 phosphate and is catalyzed by mannitol-1 phosphate dehydrogenase. In the second step, mannitol-1 phosphate is hydrolysed by mannitol-1 phosphate phosphatase. The catabolic mannitol pathway via mannitol dehydrogenase to fructose and an unspecific hexokinase to fructose-6 phosphate is two orders of magnitude less active (Karsten et al. 1997).

DMSP-specific transporters have not been identified in any alga or bacterium, but competition studies indicate that glycine betaine transporters of bacteria recognize DMSP as substrate at concentrations typical for marine ecosystems (Kempf and Bremer 1998; Cosquer et al. 1999). DMSP is synthesized from methionine, but three independent pathways have evolved, two in higher plants and one in algae. They differ in the mode of oxidative decarboxylation and order of methylation (McNeil et al. 1999).

Biosynthesis of glycine betaine was studied in plants (see Rhodes and Hanson 1993), in the cyanobacterium *Aphanothece halophytica* (Sibley and Yopp 1987) and bacteria (see Kempf and Bremer 1998). From the two different pathways (glycine or choline as source), plants seem to favour the two-step oxidation of choline by a ferredoxin-dependent monooxygenase and betaine aldehyde dehydrogenase (McNeil et al. 1999). Glycine betaine transporters are widespread in bacteria (DaCosta et al. 1998), archaebacteria and cyanobacteria (Moore et al. 1987), and are studied in depth in *E. coli* (see Csonka and Epstein 1998).

Due to its pivotal role in stress tolerance, proline synthesis was especially studied in higher plants (Delauney and Verma 1993) and also in algae (Laliberte and Hellebust 1989). It was found to proceed mainly via glutamate. All genes involved in the biosynthesis of proline have been cloned (see Peng et al. 1996; Yoshiba et al. 1997). The rate-limiting bifunctional enzyme, Δ^1-pyrroline-5-carboxylate synthetase (P5CS), is induced by salt stress, dehydration and ABA.

Usually, compatible solutes are thought to be final metabolites. Indeed, turnover of compatible solutes was often found to be low, e.g. for glycerol (see Shariati and Lilley 1994) and glucosylglycerol (Hagemann et al. 1987; Mackay and Norton 1987; Mikkat et al.1996). A clear exception to this rule is proline, which is actively metabolized like other amino acids (e.g. Yoshiba et al. 1997).

11.7 Sensing of Osmotic Stress, Gene Induction and Enzyme Activation

Sensing and signal transduction of osmostress in algae was chaptered by Bisson and Kirst (1995). Different mechanisms were thought to facilitate osmotic sensing such as subcellular structures (cytoskeleton), anisotropic changes in the membrane due to tension or stretch, and electromechanical components of the membranes (Kirst 1990); but no osmosensor could be identified in algae and cyanobacteria as yet. A negative feedback signal from turgor (walled cells) and volume (wall-less cells), respectively, is usually preferred to control the biosynthetic pathways. Investigations of the varying glycerol content of *Dunaliella* cells, while maintaining a constant osmotic pressure, have shown that it was not the external osmotic pressure per se, but a parameter directly linked to the cell volume that was sensed to set the intracellular glycerol content (Shariati and Lilley 1994). This way is quite different from that realized in the yeast *Saccharomyces cerevisiae* and, perhaps, in *Arabidopsis thaliana*, which involves a mitogen-activated type protein kinase cascade (Hohmann 1997; Urao et al. 1999).

Progress of knowledge and a comprehensive discussion of the situation in bacteria are given by Kempf and Bremer (1998) and Wood (1999). At least two low-turgor sensors, Trk and KdpD, one high-turgor sensor (Msc) and one true Na^+ sensor, NhaR, exist in *E. coli.* The low-turgor sensors induce K^+ uptake, which is the initial, but transient response to high osmolarity. It triggers the accumulation of compatible solutes, which replace the K^+ ions progressively. The high-turgor sensor provides an immediate efflux of compatible solutes after a downshock (for recent chapters see Kempf and Bremer 1998; Poolman and Glaasker 1998; Wood 1999); but pH, P_i, different substrate affinities and ion dependencies have been postulated alike to trigger the signal transduction chain (Haus and Wegmann; 1984; Bental et al. 1990; Weiss and Pick 1990; Bisson and Kirst 1995). The Na^+ sensor NhaR is a regulatory protein which controls the expression of the Na^+/H^+ antiporter. NhaR binds to DNA and induces or suppresses the *nhaA* promotor depending on the Na^+ concentration (Carmel et al. 1997).

After sensing, a signal transduction chain should amplify and direct the osmosignal to the response sites. The signal transduction chain is not known in detail for any alga or cyanobacterium. Usually, Ca^{2+} is assumed to act as a link between the membrane and biochemical pathways. The role of calcium in the osmoregulation under salt stress in *Dunaliella salina* and the characterization of a calcium-binding protein including the cloning of its cDNA were studied in detail (Ko and Lee 1995). An indication of the involvement of a two-component system in the osmotic signal transduction of cyanobacteria comes from *Anabaena* sp. PCC 7120. One of its salt-induced genes (*orrA*) shows sequence similarities to response regulators of two-component systems (Schwartz et al. 1998).

Target of the signal transduction chain is either the DNA, leading to the transcriptional induction of stress-relevant genes, or proteins, the activity of which are thereby modulated (posttranslational regulation). Posttranslational enzyme activation should represent the superior strategy during salt acclimation since the activation/deactivation of existing/synthesized proteins ensures a fast accumulation/degradation of the osmolytes. Moreover, it has been shown that the increased cytoplasmic ion concentration leads to a transient inhibition of protein synthesis (e.g. Hagemann et al. 1990). It recovers in parallel to the immediate and massive accumulation of the compatible solutes.

Four different posttranslational mechanisms are described for salt acclimation processes (Fig. 11.3). The most effective way, (1) in Fig. 11.3. seems to be the direct change in the conformation of a key enzyme due to the altered endogenous concentration of Na^+. As shown by in vitro studies, a variety of osmolyte-synthesizing enzymes altered their activities in a close correlation to the Na^+ concentration (Table 11.1). Salt-induced increase of the activity was also reported for a

Table 11.1. Direct modulation of enzyme activities by enhanced NaCl concentration in vitro

Enzyme	Compatible solute	Increase of enzyme activity by salts	Organism	References
Trehalose-6--phosphate synthase	Trehalose	KCl, NaCl and LiCl, optimal at 100 – 300 mM	*Escherichia coli*	Giaever et al. (1988); Lippert et al. (1993)
Mannitol-1-phosphate DH	Mannitol	Maximal activity at 400 – 800 mM NaCl	*Platymonas subcordiformis*	Richter and Kirst (1987)
Glycerol-3-phosphate DH	Glycerol	K^+ and Na^+ ions, optimum at ca. 100 mM	*Dunaliella tertiolecta*	Haus and Wegmann (1984)
Floridoside--phosphate synthase	Floridoside	Different salts, optimum at 50 – 200 mM	*Porphyra perforata*	Meng and Srivastava (1990)
Glucosylglycerol-phosphate synthase	Glucosylglycerol	Different salts, optimum at ca. 100 mM	*Synechocystis* sp. PCC 6803	Hagemann and Erdmann (1994); Hagemann et al. (1996); Schoor et al. (1999)
Sucrose-phosphate synthase	Sucrose	Three- to fourfold increase by NaCl	*Synechococcus* sp. PCC 6301, *Euglena gracilis*	Hagemann and Marin (1999); Porchia et al. (1999)
Mannosyl-3-phosphoglycerate synthase	Mannosylglycerate	NaCl or KCl are required	*Rhodothermus marinus*	Martins et al. (1999)
Glycine betaine aldehyde DH	Glycine betaine	150% by K^+ or Na^+ ions up to 100 mM	*Bacillus subtilis*	Boch et al. (1997)

DH = dehydrogenase

number of other enzymes such as carbonic anhydrase of *Dunaliella salina* (Fisher et al. 1996), cytochrome oxidase of *Spirulina subsalsa* (Gabbay-Azaria et al. 1992) and a number of osmolyte transporters such as GgtA of *Synechocystis* sp. PCC 6803 (Hagemann et al. 1997), but a direct effect of Na^+ has been hardly documented. Nevertheless, if this mechanism is working in vivo too, a direct and, therefore, highly effective way will be ready for the fast recovery after a salt shock: a preexisting, inactive key enzyme is activated by the enhanced Na^+ concentration after the shock and ensures the immediate and massive accumulation of a compatible solute which is obligatory for the survival.

The most detailed studies confirming this mechanism were done with GgpS, the key enzyme of the biosynthesis of glucosylglycerol. Irrespective of the enzyme source (non-stressed, shocked or salt-adapted cells of *Synechocystis* sp. PCC 6803), the enzyme was only active when NaCl was added to the assay buffer (Hagemann and Erdmann 1994; Hagemann et al. 1996; Schoor et al. 1999). Furthermore, the NaCl-caused activation was reversible since the subsequent removal of the Na^+ ions from the buffer restored the inactive state of the enzyme. However, this model of activation is obviously too simple and can only partly reflect the true cellular situation. This is deduced from the behaviour of a GgpS protein which was produced transgenically in *E. coli* by overexpression of the cyanobacterial gene *ggpS*. The recombinant GgpS is independent of the Na^+ concentration (Marin et al. 1998). This means that not the native but an altered GgpS protein is responsive to Na^+. Alteration may be generated by covalent modification like phosphorylation or by aggregation with a regulatory protein.

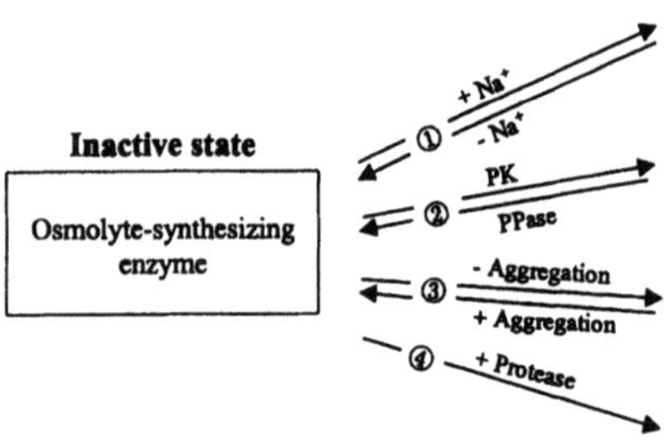

Active State	Example	References
Altered protein conformation	As yet 8 key enzymes of osmotic acclimation	See Table 11.1
Phosphorylated protein (and vice versa)	Sucrosephosphate synthase	Huber and Huber (1996), Toroser and Huber (1997)
Free Protein	Sucrosephosphate synthase	Toroser et al. (1998)
Processed protein	Isofloridoside-phosphate synthase	Brunner and Kauss (1988), Kausch et al. (1991)

Fig. 11.3. Different kinds of posttranslational activation of enzymes involved in the biosyntheses of compatible solutes. (1) Direct enzyme activation by ions, (2) stress-dependent covalent modification by protein phosphorylation/dephosphorylation: salt stimulates phosphorylation of serine-424, light induces dephosphorylation of serine-158-phosphate (PK = proteinkinase, PPase = proteinphosphate phosphatase, (3) salt-dependent activity control by a regulatory protein. (4) salt-stimulated protein processing by partial proteolysis

Way (2) in Fig. 11.3 indicates protein phosphorylation/dephosphorylation in both osmotic signal transduction and osmotic enzyme activation which have been proved several times in bacteria (Xing and Rajashekar 1999), yeast (Varela and Mager 1996; Li et al. 1998) and plants (Toroser and Huber 1997; Jonak et al. 1999).

In the halotolerant green alga *Dunaliella tertiolecta* peptides of 28-32 kDa are transiently phosphorylated in response to a hypoosmotic shock (Yuasa and Muto 1992). Furthermore, two Ca^{2+}-dependent protein kinases (Lap and Hap) were activated when *Dunaliella* cells were exposed to a hyper- and hypoosmotic shock, respectively. K-252 inhibited both kinases and, additionally, the recovery of the cell volume after a hypoosmotic shock. It was concluded that both protein kinases act in signalling cascades of salt acclimation (Yuasa et al. 1995; Yuasa and Muto 1996). Furthermore, enzyme activation by protein phosphorylation/ dephosphorylation was reported for the ATP-dependent Na^+ pump of *Heterosigma akashiwo* and *Tetraselmis (Platymonas) viridis* (Wada et al. 1989; Shono et al.1996; Popova et al. 1999). The involvement of protein phosphorylation/dephosphorylation in the cyanobacterial acclimation to salt stress was principally documented for *Synechocystis* sp. PCC 6803. Hypo- and hyperosmotic shifts led to the appearance and disappearance of ^{32}P-labelled proteins (Hagemann et al. 1993).

With regard to compatible solutes, the most convincing example is represented by the sucrose-phosphate synthase of spinach leaves. At least a part of osmotic stress activation of the sucrose-phosphate synthase results from serine phosphorylation at position 424 catalyzed by a Ca^{2+}-dependent, 150-kDa protein kinase (Toroser and Huber 1997). The sucrose-phosphate synthase from *Synechocystis* sp. strain PCC 6803 behaves differently from the plant enzyme, since it is not stress-dependent and does not contain the regulatory phosphorylation motif (Curatti et al. 1998; Porchia et al. 1999a).

Way (3) (Fig. 11.3) of salinity-induced enzyme activation is only represented by one example. Leaf sucrose-phosphate synthase from spinach is, in addition to the activation by phosphorylation, controlled by the interaction with a regulatory protein. It could be shown that sucrose-phosphate synthase binds to a 14-3-3 protein and that the activity of the enzyme was inversely related to the amount of the attached 14-3-3 protein (Toroser et al. 1998).

Way (4) (Fig. 11.3) is characterized by a proteolytic processing of the native enzyme and was described for isofloridosidephosphate synthase of the unicellular alga *Poterioochromonas malhamensis.* A salt shock leads to a proportional release of Ca^{2+} which activates a special protease and thereby the subsequent processing of the enzyme (Kauss and Thomson 1982; Kauss 1987; Brunner and Kauss 1988; Rausch et al. 1991). The concept of the proteolytic activation was represented two decades ago and has been neither updated for *P. malhamensis* nor confirmed for other osmolyte-synthesizing enzymes.

In comparison to the posttranslational regulation, transcriptional gene induction was thought to be of less importance for osmolyte-synthesizing enzymes (Potts 1994). This was really found for GgpS since the amount of the enzyme increased only three fold during salt acclimation (Hagemann et al. 1996). However, there are several evidence that osmotic acclimation is associated with gene induction via a number of two-component systems and phosphorylation cascades, especially in bacteria and yeast (Hohmann 1997; Csonka and Epstein 1998). A precisely tuned example of plant transcriptional control is obvious for the proline accumulation in *Arabidopsis thaliana* in response of environmental stresses (drought, salinity, low temperature). The two anabolic genes (*p5cs* and *p5cr*) and the first catabolic gene (*proDH*) are induced and repressed, respectively, by water stress. On the contrary, during rehydration the two catabolic genes (*proDH* and *p5cDH*) are induced while the anabolic gene *p5cs* is repressed (Yoshiba et al. 1997). The close correlation between *p5cs* expression and accumulation of proline has made this gene important for genetic manipulation (see Table 11.4).

11.8 Membrane Alterations, Water and Solute Channels

Altered membrane composition in response to salt stress relates to phospholipid composition, sterols and unsaturated fatty acids.

Bacteria respond to increasing salinity by a decrease in the proportions of zwitterionic phospholipids (Russell 1995). Changes in the sterol content were documented by a comparison of the mesophilic *Saccharomyces cerevisiae* with the halophilic *Hortea werneckii* which also showed differences in regulation of the sterol synthesis (Petrovic et al. 1999).

The degree of unsaturation of the acyl chains of membrane lipids determines the acclimation ability of organisms against changing temperature, and also salinity. Whereas the former was extensively studied, only little is known for the latter. Most information is available for cyanobacteria and their mutants defective in desaturases. The most worked-out example is the double mutant *desA⁻/desD⁻* of *Synechocystis* sp. PCC 6803 which is unable to synthesize polyunsaturated fatty acids. This mutant could withstand chilling, but was affected by NaCl stress by inactivation of the oxygen-evolving machinery. The effect has been explained by the lesser fluidity of the membrane which leads to a less active Na^+/H^+ antiporter (Allakhverdiev et al. 1999).

Originally, it was assumed that the permeability of the membrane allows a sufficient water exchange during osmotic shocks. During the past several years, it has become clear that water crosses membranes additionally via water channels (aquaporins) and integral membrane proteins intended for other functions (e.g. solute or ion transport).

Water channels facilitate water molecules to pass across membranes depending on osmotic or hydrostatic pressure. Studies on expression of the *E. coli* aquaporin gene, *aqpZ*, demonstrated a 15-fold increase in the rate of water exchange. Expression of the gene is repressed at increasing osmolality and, to a lesser extent, during the entry into the stationary phase, but continues to function at the highest osmolalities. Furthermore, *aqpZ*-deficient mutants demonstrated that the non-promoted water transport is sufficient for most purposes, but the rapid movement of water across the cytoplasmic membrane is required for optimal growth (Booth and Louis 1999). Water channels have been observed in higher plants (e.g. several members in *Arabidopsis*: Weig et al. 1997) and indirectly detected in algae. In *Chara*, hydraulic conductivity of the cytoplasmic membrane could be reduced and restored by Hg^{2+} and dithiothreitol, respectively, documenting the presence of water channels (Henzler and Steudle 1995). In cyanobacteria, a water channel contributes to the inactivation of PSI and II in response to an osmotic stress in *Synechococcus* sp. PCC 7942 (Allakhverdiev et al. 2000). It has been suggested that engineering manipulation of water channels might be a promising way to improve salt tolerance of higher plants (Bohnert and Jensen 1996).

Mechano-sensitive channels allow the flow of water and small molecules (ions and organic solutes) through the membrane in both directions along their gradients. They prevent a deleterious increase of the turgor by over-accumulation of solutes or influx of water during a transition from high to low osmolarity. It seems that all unicellular organisms make use of this mechanism to rapidly dispose osmolytes after a hypoosmotic shock. In *E. coli*, three mechano-sensitive channels (MscL, MscM and MscS) have been characterised by their different pressure thresholds and conductances, and are able to substitute for each other, at least partly. The largest channel, MscL, has been well characterized at the protein and gene level (see Poolman and Glaasker 1998) and offers the way for studies on similar proteins in algae and cyanobacteria. The fast efflux of glucosylglycerol out of the cells of *Synechocystis* sp. PCC 6803 (Fulda et al. 1990) after an osmotic downshock points to the participation of such a mechano-sensitive channel. Furthermore, a homologous gene of *mscL* is found in the genome of *Synechocystis* (Kaneko et al. 1996).

In the yeast *Saccharomyces cerevisiae*, an aquaporin-, but not MscL-like protein (Fps1p) enables the rapid disposal of glycerol after a hypoosmotic shock and constitutes an important element in the cellular osmoadaptation (Luyten et al. 1995; Tamas et al. 1999).

11.9 Ion Homeostasis

Enhanced concentrations of NaCl exhibit toxic effects on cells and it is this toxic component that differentiates a salt from an osmotic stress and can cause different responses. For instance, two related strains of *Anabaena* displayed different responses in nitrogenase activity, NH_4^+ protection and protein pattern after ion or osmotic shocks (Fernandes et al. 1993). Recent expositions on the toxicity of Na^+- and Cl^--ion are given by Kinraide (1999) and Serrano et al. (1999). Firstly, enhanced external Na^+ displaces Ca^{2+} from the cell surface. Secondly, Na^+ leaks into cells when the exogenous salt concentration is increased. In the initial situation after a salt shock, this is probably due to a sudden breakdown of the permeability of the cell membrane (Reed 1984; Reed et al. 1985). During steady state, leakage seems to be weak (Oren 1999) and is mediated by ion passage through non-specific channels rather than by permeation. Transporter for glucose, amino acids, choline, divalent cations and others is supposed to transport Na^+, K^+, Rb^+ and Li^+ also with K_m values of the order of 100 mM (Serrano et al. 1999). In each case, the additional Na^+ ions affect the metabolism and interfere with balanced ion relations.

True concentrations of internal Na^+ in salt-acclimated cells are still a matter of debate (see Hagemann and Erdmann 1997)). It seems that the most reliable values for bacteria, cyanobacteria and microalgae were obtained by the non-invasive ^{23}Na NMR method. As shown in Table 11.2, the internal Na^+ concentration does not severely change during salt acclimation. Similar results have been obtained by another reliable technique, X-ray microanalysis. Freeze-prepared sections of *Dunaliella tertiolecta* cells subjected to a salt stress of 3 M NaCl indicated an internal Na^+ concentration of 65 mM (Wegmann 1986).

Table 11.2. Intracellular Na^+ concentration as related to the external Na^+ concentration in the cells of two autotrophic and two heterotrophic unicellular organisms. Concentration was determined by the use of ^{23}Na NMR. Visibility (that means free Na^+) of the intracellular Na^+ was on an average 40%

Organism	Extracellular NaCl concentration (M)	Intracellular NaCl concentration (mM)	Reference
Dunaliella salina	0.1	66	Bental et al. (1988)
	1.0	65	
	2.0	114	
	4.0	93	
Synechococcus sp. PCC 6311	0.05	50	Nitschmann and Packer (1992)
	0.5	120	
Vibrio costicola	0.6	140 ± 34	Gilboa et al. (1991)
	1.0	81 ± 12	
	2.0	68 ± 9	
Brevibacterium sp.	0.5	35 ± 4.7	Nagata et al. (1995)
	1.0	43	
	2.0 (24 h)	51 ± 9.3	
	0.5	80 nmol mg $protein^{-1}$	Nagata et al. (1998)
	2.0 (30 min)	Unchanged	

These almost unchanged internal salt concentrations clearly indicate that organisms living under elevated salt concentrations actively export Na^+ ions out of the cytoplasmic space against the electrochemical Na^+ gradient. This is chiefly carried out by Na^+/H^+ antiporter secondarily energized by the protonmotive force. The Na^+/H^+ antiporter has been characterized in all types of organisms including microalgae (Katz et al. 1989) and cyanobacteria (see Hagemann and Erdmann 1997; Stamatakis et al. 1999, and references therein). Importance of the Na^+/H^+

antiporter was demonstrated by inhibitors interfering directly or indirectly with its activity (Stamatakis et al. 1999). The protonmotive force is generally generated by a H^+- ATPase of the plasma membrane. In *Synechocystis* sp. PCC 6803, two genes encoding unidirectional H^+-ATPase were identified and cloned. Mutagenesis of these genes led to salt- and osmosensitive derivatives of the wild-type strain (Kanamaru et al. 1993), but at least cyanobacteria exert a second mechanism: the translocation of protons by the cytochrome oxidase. It showed enhanced activity in isolated plasma membranes of *Spirulina subsalsa* when salt-stressed (Gabbay-Azaria et al. 1992, 1994). Similarly, salt-adjusted *Synechocystis* sp. PCC 6803 cells showed two fold activity, but not the amount of cytochrome oxidase (Jeanjean et al. 1990). Besides its function in salt acclimation, the Na^+/H^+ antiporter is responsible for the pH homeostasis and becomes activated when the pH rises over its set point. The inflow of external H^+ ions reverses the pH rise (Serrano et al. 1999).

A second type of Na^+ transporter was characterized as an ATP-driven primary Na^+ pump in the aquatic macrophyte *Chara australis* (Smith and Walker 1989) and, recently, has been studied in detail in two marine algae, *Tetraselmis (Platymonas) viridis* (Balnokin and Popova 1994; Balnokin et al. 1997; Popova et al. 1999) and *Heterosigma akashiwo* (Shono et al. 1996; Wada et al. 1989). Plasma membrane vesicles isolated from *T. viridis* transferred ATP-dependent $^{22}Na^+$ in the presence of the protonophore CICCP, indicating that Na^+ translocation was performed by a mode which is independent of the protonmotive force (pmf) (Balnokin et al. 1997). Organisms living in a more alkaline environment depend on this pmf-independent mechanism (Ritchie 1992). This type of Na^+ transporter has not been detected in plants as yet and, vice versa, a third type of Na^+ transporter (ABC-ATPase) not in algae and cyanobacteria (Serrano et al. 1999).

Dominance of K^+ against Na^+ in the cytoplasm and fast uptake of K^+ after an upshock are fundamentals of bacterial salt acclimation. Some of the three K^+ transporters in *E. coli* (Trk, Kdp and Kup; Kempf and Bremer 1998) and *S. cerevisiae* (Trk1, Trk2, Nhx1; Schachtman and Liu 1999) for uptake as well as a K^+ channel for efflux (Kef, as yet only for *E. coli*; Kempf and Bremer 1998) are osmotically responsive and guarantee a turgor-relevant ion homeostasis. Such a detailed knowledge is rare for algae and cyanobacteria. A stress-dependent K^+ transferring protein of the Kdp type has been detected in the nitrogen-fixing *Anabaena torulosa* (Apte et al. 1998). A Ca^{2+}-dependent K^+ channel has been described in the unicellular green alga *Eremosphaera viridis*, but not with respect of salt tolerance (see Bauer et al. 1998).

11.10 Supply of Energy for Successful Acclimation

All these acclimation efforts need extra energy, which is supplied by an enhancement of respiratory and/or photosynthetic activities. There are many examples that respiratory activity is increased in salt-stressed cyanobacterial cells (Jeanjean et al.

1993; Molitor et al. 1990). However, initial photosynthesis is often impeded by an increase in salinity, as shown recently for the cyanobacterium *Spirulina platensis* (Liu et al. 1998; Lu and Zhang 1999; Zeng and Vonshak 1998), for earlier reports see Hagemann and Erdmann (1997). PS II and its oxygen-evolving complex seems to be especially sensitive towards increased ion concentrations while the PS I shows a higher tolerance.

In salt-stressed cyanobacteria, an increase in the activity, content of PSI and the PSI/ PSII ratio has been found. Further studies showed that these alterations mainly improve the cyclic photophosphorylation by an optimized electron flux around PSI (Hibino et al. 1996; Jeanjean et al. 1998; Hagemann et al. 1999). In cyanobacteria, alternative routes exist at the acceptor site of PSI for the cyclic electron flow. Electrons seem to flow mainly via NAD(P)H using the NAD(P)H dehydrogenase complex 1. This was shown in a NAD(P)H dehydrogenase-inactivated mutant of *Synechocystis* sp. PCC 6803, which grew like the wild-type at low salinity, but showed impaired oxygen evolution when the NaCl concentration increased above 0.6 M (Tanaka et al. 1997). Furthermore, flavodoxin, which can functionally replace ferredoxin in the photosynthetic electron flow, was markedly enhanced in salt-shocked and salt-adapted cells of *Synechocystis* sp. PCC 6803 (Fulda and Hagemann 1995). A flavodoxin-deficient mutant displayed a clear decrease in the salt-stimulated cyclic electron flow around PSI. This led to an enhanced lag phase in salt-shocked cells, but it did not change the upper resistance level (Hagemann et al. 1999).

11.11 Salt-Inducible Genes and (Stress) Proteins

Enhanced salinity leads to transient or lasting changes of the protein pattern. The appearance of stress proteins represents the most remarkable alteration and is accompanied by suppression of other proteins. The exploration of the nature of these salt-stress proteins represents a main target of acclimation research and could proceed via both the encoding genes and their products.

The search for salt-inducible genes was performed by different strategies. Some success was obtained by screening spontaneously or chemically generated mutants for their gene defect (Chitlaru and Pick 1989; Jeanjean et al. 1990). The most successful way that was executed in cyanobacteria consists in the generation of salt-sensitive mutants by random cartridge (Hagemann and Zuther 1992) and Tn5-based transposon (Schwartz et al. 1998) mutagenesis and looking for the gene-molecular background. Using the first approach, three key genes of salt acclimation of *Synechocystis* sp. PCC 6803 (*ggpS, stpA, ggtA*) could be detected and identified. They encode for the enzymes involved in biosynthesis and tansport of the compatible solute glucosylglycerol (Table 11.3). The mutation approach could only lead to the discovery of genes which are essential for the growth at enhanced salt concentrations, but are not essential for normal growth. This restriction can be

Table 11.3. Cloned salt stress-inducible genes and proteins of *Synechocystis* sp. PCC 6803

Gene	Cyano-base[a]	**Protein**	**Function**	**Reference**
GgpS	sll1566	Glucosylglycerol-phosphate synthase	Glucosylglycerol biosynthesis	Engelbrecht et al. (1999); Marin et al. (1998)
StpA	slr0746	Glucosylglycerol-phosphate phosphatase	Glucosylglycerol biosynthesis	Hagemann et al. (1997a)
GgtA,B,C, D	slr0747 slr0529 slr0530 slr0531	Glucosylglycerol transporter	Transport of glucosylglycerol, trehalose and sucrose SU: A = ATP binding, B = substrate binding, C,D = membrane protein	Hagemann et al. (1997b); Mikkat and Hagemann (2000)
IsiA	sll0247	Iron stress-induced protein	Chlorophyll a binding protein	Vinnemeier and Hagemann (1999); Vinnemeier et al. (1998)
IsiB	sll0248	Flavodoxin	Carrier in cyclic electron transport	Fulda and Hagemann (1995); Hagemann et al. (1999); Vinnemeier and Hagemann (1999)
Cpn60	sll0416	GroEL (HS60)	Chaperon	Vinnemeier et al. (1998)
DnaK	several	DnaK (HS70)	Chaperon	Fulda et al. (1999)
Gcp	slr0807	Hypothetical	Putative sialoglycoprotease	Zuther et al. (1998)
-	slr0082	Hypothetical	-	Vinnemeier et al.
-	slr0083	Hypothetical	Putative RNA helicase	(1998)
GlpD	sll1085	Glyceraldehyde phosphate DH	Biosynthesis of glycerol	Marin et al. (1998)
CtaCDE	several	Cytochrome oxidase SU I-III	Energizing the Na^+ export	Alge and Peschek (1993)
-	sll0273	Na^+/H^+ antiporter	Na^+ exclusion, pH homeostasis	Mikkat et al. (2000)
Pp2	slr1485	Phosphatidyl inositol phosphate kinase	Lipid biosynthesis	Fulda et al. (2000)
-	sll1549	Phosphoglycerate mutase	Polysaccharide synthesis	
-	slr0513	Iron-binding protein	Iron uptake	

DH = dehydrogenase, SU = subunit.

[a] *Synechocystis* sp. PCC 6803 genomic database (Kaneko et al. 1996)

overcome by using methods to screen genes according to their enhanced transcription during salt stress. In a pioneering work, Apte and Haselkorn (1990) used subtractive RNA hybridization and came to the conclusion that approximately 100 genes of *Anabaena torulosa* are involved in salt acclimation. The same method was applied to *Synechocystis* sp. PCC 6803 and enabled the identification of four additional salt-inducible genes (*isiA, cpn60,* slr0082, slr0083) for this strain (Table 11.3). Among them, the gene of the chaperon GroEL (HS60) could be identified (Vinnemeier and Hagemann 1999) which was also found in the pool of salt-inducible genes of *Anabaena torulosa* (Apte et al. 1998). Salt-sensible mutants and salt-inducible genes, respectively, are little known for algae at present.

Salt-stress proteins are usually identified by labelling them with ^{35}S-methionine during the stress application and their subsequent separation by polyacrylamide gel electrophoresis (PAGE). Using one- and two-dimensional PAGE, some stress proteins were detected in cyanobacteria (Apte and Bhagwat 1989; Apte et al. 1998; Fulda et al. 1999, 2000; Hagemann et al. 1990, 1991) and microalgae, mainly *Dunaliella* species (Fisher et al. 1994; Golldack et al. 1995a,b). To differentiate between unique and universal stress proteins, the protein pattern of three cyanobacteria of different tolerance levels (*Synechococcus* sp. PCC 7942 (tolerance limit 0.5 M NaCl), *Synechocystis* sp. PCC 6803 (1.2 M) and *Synechococcus* sp. PCC 7418 (2.6 M), were compared in response to salt, temperature and light stress. About ten stress proteins were seen in each strain. Half of the proteins were more or less common for the two strains and two stressors, whereas the other half represent unique stress proteins (Fulda et al. 1999). Only one protein was induced in all strains, but not at all stressors. It belongs to the HS70 family (DnaK), as proved by a monoclonal antibody. DnaK was characterized in depth in the cyanobacterium *Aphanothece halophytica* (Hibino et al. 1999). Its stress-relevant function has been verified by coexpression of the cyanobacterial *dnaK* gene with a plant plastocyanin precursor in *E. coli* and the finding that solubility of the recombinant plant protein increased dramatically (Lee et al. 1997).

Earlier, the dominating salt-stress protein of *Synechocystis* sp. PCC 6803 could be identified as flavodoxin (Fulda and Hagemann 1995), which contributes to an enhanced cyclic electron transport around PS I under salt, though the quantitative dominance of this stress protein, a mutant deficient in the *isiB* gene, encoding flavodoxin, displayed no altered phenotype (Hagemann et al. 1999). The gene *isiB* comprises an operon with *isiA* which was identified as a salt-inducible gene by the subtractive RNA hybridization (Vinnemeier and Hagemann 1999). It encodes for a chlorophyll-binding protein and is known to be induced by iron depletion (Straus 1994). Its enhanced accumulation under salt stress suggests that iron uptake or use are impeded. This is confirmed by the analysis of the salt-stress proteins of *Dunaliella salina*. A salt-inducible protein, p150, was identified as a transferrin-like protein which is involved in iron uptake and, so far, found only in animals (Fisher et al. 1997,1998).

Two other stress proteins were enhanced in the plasma membrane of *D. salina* during acclimation to high salinity. cDNA cloning and enzymatic analysis led to the identification of carbonic anhydrase, which optimises CO_2 uptake by cells under hypersaline conditions (Fisher et al. 1996).

The actual establishment of protein projects has brought about a breakthrough in the number of proteins thought to be integrated in stress response. By the combination of modern techniques such as mass spectroscopy with two-dimensional PAGE and genome analysis, the knowledge of stress proteins broadens for such organisms the genome of which has been completely sequenced. For example, Fulda et al. (2000) studied the salt-dependent accumulation of proteins in the periplasm of *Synechocystis* sp. PCC 6803. Of about 100 proteins separated by two-dimensional PAGE, 80 were selected and 57 identified. Nine proteins were amplified or newly induced by salt stress and are mainly involved in the biosynthesis of external cell layers, but the most obvious stress protein was identified to be a periplasmatic iron-binding protein. This is another indication for iron depletion under salt stress.

11.12 Engineering of Salt Tolerance

Conventional breeding has brought about only limited progress in the long-standing goal to improve salt resistance in crop plants. This may be understandable with respect to the high interconnection and complexity of the traits involved. For instance, in the halotolerant Indica rice cultivar Bura Rata, 35 salinity stress-induced and 17 repressed proteins were detected (Ramani and Apte 1997). With the advent of genetic transformation technologies, manipulation of plants has become more easy and less time- and labour-consuming. From all fields of the salt-adaptation scenario (Fig. 11.1), genes have been transformed into crop plants, especially those encoding for biosyntheses of compatible solutes, stress proteins, lipid desaturation and Na^+/H^+ exchange (Table 11.4).

Regarding compatible solutes, genes encoding the key enzymes of nine different osmoprotectants have been transferred into other, mainly vegetable crops until now. Five of them (Table 11.4) are ubiquitous in algae and cyanobacteria, but none of the genes was taken from these organisms and only in one case (glycine betaine) was a cyanobacterium (*Synechococcus* sp. PCC 7942) the target of the gene transfer. A comparison between Table 11.4 and Fig. 11.2 reveals that many of the presently known osmolytes of algae and cyanobacteria have not been exploited for the improvement of plant salt tolerance. Therefore, the genes encoding enzymes in those biosynthetic pathways in algae and cyanobacteria represent a rich source of future applications.

In almost all the cases, the introduction of the transgene improved the stress tolerance of the target organism. Glycine betaine emerged as the most versatile protective compound, ameliorating the resistance against salt, cold and light stress. All but one (trehalose) are successful in salt adaptation though the concentrations of the compatible solutes were always too low to contribute to the osmotic

Table 11.4. Engineering of salt tolerance by gene transfer genes encoding key proteins involved in stress response (mainly enzymes of the biosyntheses of compatible solutes) were cloned from donor organisms, fused with a constitutive promotor (as a rule) and brought to expression in a non-resistant organism to increase stress tolerance.

Target	Resistance gene/protein	Source	Target organism	Scores of tolerance increase against	*Reference
Glycine betaine	*CodA* Choline oxidase *BetB* Betaine-aldehyde DH *betA,B,I,T* Complete set of ezymes	*Arthrobacter globiformis* *E. coli*	*A. thaliana*, *Synechococcus*, *O. sativa* *N. tabacum*	Salt^{+++++} Cold^{+++} Light^{+++}	[1, 2, 3, 4, 5, 6, 7]
Mannitol	*mtlD* Man-1-P DH	*E. coli*	*N. tabacum*, *S. cerevisiae*, *A. thaliana*	Salt^{+++++} ROS^{+++}	[8, 9, 10, 11, 12]
Sorbitol	*stpd1* Sor-6-P DH	*Malus domestica*	*N. tabacum*, *S. cerevisiae*	Salt^{++} ROS^{+}	[13, 14]
Proline	*p5cs* Δ^1-Pyrroline-5-carboxylate synthetase *atprodh* Antisense-proline DH	*E. coli*, *Vigna aconitifolia* *A. thaliana*	*N. tabacum* *O. sativa* *A. thaliana*	Salt^{+++} Drought^{++} Cold^{+}	[15, 16, 17]
Trehalose	*tps1*, *otsA* (+ *otsB*) Tre-P synthase (+ Tre-P phosphatase)	*S. cerevisiae* *E. coli*	*N. tabacum* *Solanum tuberosum*	Drought^{++++} ROS^{+}	[18, 19, 20, 21]
Na^+ export	sod2, atnhx1 *Na^+/H^+ antiporter*	*S. cerevisiae* *A. thaliana*	*A. thaliana*	Salt^{+}	[22, 23]
Unsaturated FA	des9 Δ^9-Desaturase	*Anacystis nidulans*	*N. tabacum*	Salt$^{?}$ Cold^{+}	[24]
Stress protein	dnaK1 Heat shock protein	*Aphanothece halophytica*	*N. tabacum*	Salt^{+}	[25]

Abbreviations: *A.* = *Arabidopsis*, DH = dehydrogenase, *E.* = *Escherichia*, FA = fatty acid, Man = mannitol, *N.* = *Nicotiana*, *O.* = *Oryza*, P = phosphate, ROS = reactive oxygen species, *S.* = *Saccharomyces*, Sor = sorbitol.

*[1] Alia et al. (1999), [2] Deshnium et al. (1995, 1997), [3] Hayashi et al. (1997), [4] Holmström et al. (1994), [5] Nomura et al. (1995, 1998), [6] Sakamoto et al. (1998), [7] Sakamoto and Murata (2000), [8] Chaturvedi et al. (1997), [9] Karakas et al. (1997), [10] Shen et al. (1997, 1999), [11] Tarczynski et al. (1992), [12] Thomas et al. (1995), [13] Shen et al. (1999), [14] Sheveleva et al. (1998), [15] Kishor et al. (1995), [16] Nanjo et al. (1999), [17] Zhu et al. (1998), [18] Goddijn et al. (1997), [19] Holmström et al. (1996), [20] PilonSmits et al. (1998), [21] Romero et al. (1997), [22] Apse et al. (1999), [23] Dibrov et al. (1998), [24] Ishizaki-Nishizawa et al. (1996), [25] Sugino et al. (1999)

potential of the cells. This may be an indication that the protective function of compatible solutes is more important than their task of being an osmolyte (Nuccio et al. 1999; Zhang et al. 1999). The almost low concentrations in the transformed cells may result from the poor capacity of the one-gene transfer approach executed until now. It is argued that the introduction of only one gene of the biosynthetic pathway of a compatible solute leads to such a rigid disturbance of the metabolic network that transgenic cells with a massive accumulation of the compatible solute do not survive the selection process. Drought tolerance was especially introduced by proline and trehalose. Sugar alcohols and trehalose proved to be radical scavengers. Proline overexpression was not only achieved by the stimulation of the biosynthesis (expression of the key enzyme gene *p5cs*), but also by suppression of the degradation using the antisense technique *antiprodh* (Nanjo et al. 1999).

Gene transfer regarding the other fields of the adaptation scenario is less extensive. Transformation of the *nhaAv* gene encoding an Na^+/H^+ antiporter from *Vibrio alginolyticus* into the freshwater cyanobacterium *Synechocystis* sp. PCC 7942 led to a functional expression of the antiporter, but only tolerance to LiCl and not to NaCl appeared (Kaku et al. 1999). On the contrary, a similar approach was still quite successful with higher plants and bacteria: overexpression of the Na^+/H^+ antiporter in *Arabidopsis thaliana* conferred enhanced salt tolerance (Apse et al. 1999) and overexpression of the Na^+/H^+ antiporter of *S. cerevisiae* in *E. coli* raised the NaCl tolerance of recombinant cells, but led to complete Li^+ sensitivity (Dibrov et al. 1998).

The potential of cyanobacteria has been exploited in two important cases. The gene *des9* from *Anacystis nidulans,* encoding for a Δ^9 desaturase, was introduced into *Nicotiana tabacum* and improved its chilling resistance. Nothing was reported about the salt tolerance which also should have been increased (Allakhverdiev et al. 1999). Furthermore, *dnaK1* taken from *Aphanothece halophytica* and overexpressed in *N. tabacum* alleviates the decrease in CO_2 fixation and increase of intracellular Na^+ in response to a salt stress in 0.6 M (Sugino et al. 1999).

11.13 Outlook

Modern knowledge of salt acclimation has come mainly from unicellular organisms: *E. coli* for bacteria, *Saccharomyces cerevisiae* for eukaryotes and *Synechocystis* sp. PCC 6803 for cyanobacteria. None of these model organisms is able to withstand the highest level of salinity. This is the domain of the genus *Dunaliella.* However, knowledge of the molecular mechanisms responsible for this incredible acclimation capability is less spectacular, though still leading in comparison to other algae. Therefore, the prediction of Kirst drawn 10 years ago (Kirst 1990) holds until now: "It is unlikely that in the near future the complex responses to osmotic stress may be traced to algal genes analogous to the *osm* genes in *E. coli* that govern the production of proline and betaine to protect the cell against dehydration."

It should, however, be tempting to increase the efforts in the field of algae, since the cheapest way to acquire salt tolerance (glycerol accumulation) as exemplified by *Dunaliella* species should, in principle, be applicable to other organisms.

The comparison between algae and cyanobacteria corroborates the dominant role of compatible solutes, chaperones and proteases in the process of salt acclimation, but points also to the view that iron becomes depleted under salt stress. Obviously, the corresponding indications are underestimated as yet and should attract more attention in the future.

References

Alge D, Peschek GA (1993) Characterization of a cta/CDE operon-like genomic region encoding subunits I-III of the cytochrome-c oxidase of the cyanobacterium *Synechocystis* PCC- 6803. Biochem Mol Biol Int 29:511-525

Alia, Kondo Y, Sakamoto A, Nonaka H, Hayashi H, Saradhi PP, Chen TH, Murata N (1999) Enhanced tolerance to light stress of transgenic *Arabidopsis* plants that express the *codA* gene for a bacterial choline oxidase. Plant Mol Biol 40:279-288

Allakhverdiev SI, Nishiyama Y, Suzuki I, Tasaka Y, Murata N (1999) Genetic engineering of the unsaturation of fatty acids in membrane lipids alters the tolerance of *Synechocystis* to salt stress. Proc Natl Acad Sci USA 96:5862-5867

Allakhverdiev SI, Sakamoto A, Nishiyama Y, Murata N (2000) Inactivation of photosystem I and II in response to osmotic stress in *Synechococcus*. Contribution of water channels. Plant Physiol 122:1201-1208

Apse MP, Aharon GS, Snedden WA, Blumwald E (1999) Salt tolerance conferred by overexpression of a vacuolar Na^+/H^+ antiporter in *Arabidopsis*. Science 285:1256-1258

Apte SK, Bhagwat AA (1989) Salinity-stress-induced proteins in two nitrogen-fixing *Anabaena* strains differentially tolerant to salt. J Bacteriol 171:909-915

Apte SK, Haselkorn R (1990) Cloning of salinity stress-induced genes from the salt-tolerant nitrogen-fixing cyanobacterium *Anabaena torulosa*. Plant Mol Biol 15:723-733

Apte SK, Fernandes T, Badran H, Ballal A (1998) Expression and possible role of stress-responsive proteins in *Anabaena*. J Biosci 23:399-406

Balnokin Y, Popova L, Gimmler H (1997) Further evidence for an ATP-driven sodium pump in the marine alga *Tetraselmis (Platymonas) viridis*. J Plant Physiol 150:264-270

Balnokin YV, Popova LG (1994) ATP-driven Na^+-pump in the plasma membrane of the marine unicellular alga *Platymonas viridis*. FEBS Lett 342:61-64

Bauer CS, Plieth C, Hansen UP, Simonis W, Schonknecht G (1998) A steep Ca^{2+} dependence of a K^+ channel in a unicellular green alga. J Exp Bot 49:1761-1765

Bental M, Degani H, Avron M (1988) ^{23}Na-NMR studies of the intracellular sodium ion concentration in the halotolerant alga *Dunaliella salina*. Plant Physiol 87:813-817

Bental M, Pick U, Avron M, Degani H (1990) The role of intracellular orthophosphate in triggering osmoregulation in the alga *Dunaliella salina*. Eur J Biochem 188:117-122

Bisson MA, Bartholomew D (1984) Osmoregulation or turgor regulation in *Chara*? Plant Physiol 74:252-255

Bisson MA, Kirst GO (1995) Osmotic acclimation and turgor pressure regulation in algae. Naturwissenschaften 82:461-471

Blunden G, Gordon SM (1986) Betaines and their sulphonio analogues in marine algae. In: Round FE, Chapman DJ (eds) Progress in phycological research. Biopress, Bristol, pp 39-80

Boch J, Nau-Wagner G, Kneip S, Bremer E (1997) Glycine betaine aldehyde dehydrogenase from *Bacillus subtilis*: characterization of an enzyme required for the synthesis of the osmoprotectant glycine betaine. Arch Microbiol 168:282-289

Bohnert HJ, Jensen RG (1996) Metabolic engineering for increased salt tolerance —the next step. Aust J Plant Physiol 23:661-667

Bohnert HJ, Sheveleva E (1998) Plant stress adaptations — making metabolism move. Curr Opin Plant Biol 1:267-274

Booth IR, Louis P (1999) Managing hypoosmotic stress: aquaporins and mechanosensitive channels in *Escherichia coli*. Curr Opin Microbiol 2:166-169

Borowitzka LJ, Demmerle S, Mackay MA, Norton RS (1980) Carbon-13 nuclear magnetic resonance study of osmoregulation in a blue-green alga. Science 210:650-651

Brown AD (1976) Microbial water stress. Bacteriol Rev 40:803-846

Brunner G, Kauss H (1988) Deaggregation and proteolytic modification of a galactosyltransferase of *Poterioochromonas malhamensis*. Physiol Plant 74:708-714

Carmel O, Rahav-Manor O, Dover N, Shaanan B, Padan E (1997) The Na^+-specific interaction between the LysR-type regulator, NhaR, and the *nhaA* gene encoding the Na^+/H^+ antiporter of *Escherichia coli*. EMBO J 16:5922-5929

Chaturvedi V, Bartiss A, Wong B (1997) Expression of bacterial *mtlD* in *Saccharomyces cerevisiae* results in mannitol synthesis and protects a glycerol-defective mutant from high-salt and oxidative stress. J Bacteriol 179:157-162

Chitlaru E, Pick U (1989) Selection and characterization of *Dunaliella salina* mutants defective in haloadaptation. Plant Physiol 91:788-794

Cosquer A, Pichereau V, Pocard JA, Minet J, Cormier M, Bernard T (1999) Nanomolar levels of dimethylsulfoniopropionate, dimethylsulfonioacetate, and glycine betaine are sufficient to confer osmoprotection to *Escherichia coli*. Appl Environ Microbiol 65:3304-3311

Csonka LN (1989) Physiological and genetic responses of bacteria to osmotic stress. Microbiol Rev 53:121-147

Csonka LN, Epstein W (1998) Osmoregulation. In: Neidhardt FC (ed) *Escherichia coli* and *Salmonella typhimurium*. ASM Press, pp 1210-1223

Csonka LN, Hanson AD (1991) Prokaryotic osmoregulation: genetics and physiology. Annu Rev Microbiol 45:569-606

Curatti L, Folco E, Desplats P, Abratti G, Limones V, HerreraEstrella L, Salerno G (1998) Sucrose-phosphate synthase from *Synechocystis* sp. strain PCC 6803: identification of the *spsA* gene and characterization of the enzyme expressed in *Escherichia coli*. J Bacteriol 180:6776-6779

DaCosta MS, Santos H, Galinski EA (1998) An overview of the role and diversity of compatible solutes in bacteria and archea. Adv Biochem Eng Biotechnol 61:117-153

Delauney AJ, Verma DPS (1993) Proline biosynthesis and osmoregulation in plants. Plant J 4:215-223

Deshnium P, Los DA, Hayashi H, Mustardy L, Murata N (1995) Transformation of *Synechococcus* with a gene for choline oxidase enhances tolerance to salt stress. Plant Mol Biol 29:897-907

Deshnium P, Gombos Z, Nishiyama Y, Murata N (1997) The action in vivo of glycine betaine in enhancement of tolerance of *Synechococcus* sp. strain PCC 7942 to low temperature. J Bacteriol 179:339-344

Dibrov P, Young PG, Fliegel L (1998) Physiological consequences of expression of the Na^+/H^+ antiporter *sod2* in *Escherichia coli*. Mol Cell Biochem 183:125-132

Dickson DMJ, Kirst GO (1986) The role of dimethylsulphoniopropionate, glycine betaine and homarine in the osmoacclimation of *Platymonas subcordiformis*. Planta 167:536-543

Engelbrecht F, Marin K, Hagemann M (1999) Expression of the *ggpS* gene, involved in osmolyte synthesis in the marine cyanobacterium *Synechococcus* sp. strain PCC 7002, revealed regulatory differences between this strain and the freshwater strain *Synechocystis* sp. strain PCC 6803. Appl Environ Microbiol 65:4822-4829

Erdmann N (1983) Organic osmoregulatory solutes in blue-green algae. Z Pflanzenphysiol 110:147-155

Fernandes TA, Iyer V, Apte SK (1993) Differential responses of nitrogen-fixing cyanobacteria to salinity and osmotic stresses. Appl Environ Microbiol 59:899-904

Fisher M, Pick U, Zamir A (1994) A salt-induced 60-kilodalton plasma membrane protein plays a potential role in the extreme halotolerance of the alga *Dunaliella*. Plant Physiol 106:1359-1365

Fisher M, Gokhman I, Pick U, Zamir A (1996) A salt-resistant plasma membrane carbonic anhydrase is induced by salt in *Dunaliella salina*. J Biol Chem 271:17718-17723

Fisher M, Gokhman I, Pick U, Zamir A (1997) A structurally novel transferrin-like protein accumulates in the plasma membrane of the unicellular green alga *Dunaliella salina* grown in high salinities. J Biol Chem 272:1565-1570

Fisher M, Zamir A, Pick U (1998) Iron uptake by the halotolerant alga *Dunaliella* is mediated by a plasma membrane transferrin. J Biol Chem 273:17553-17558

Fujii S, Nishimoto N, Notoya N, Hellebust JA (1995) Growth and osmoregulation of *Chaetoceros muelleri* in relation to salinity. Plant Cell Physiol 36:759-764

Fulda S, Hagemann M (1995) Salt treatment induces accumulation of flavodoxin in the cyanobacterium *Synechocystis* sp. PCC 6803. J Plant Physiol 146:520-526

Fulda S, Hagemann M, Libbert E (1990) Release of glucosylglycerol from the cyanobacterium *Synechocystis* sp. SAG 92.79 by hypoosmotic shock. Arch Microbiol 153:405-408

Fulda S, Huckauf J, Schoor A, Hagemann M (1999) Analyses of stress response in the cyanobacterial strains *Synechococcus* sp. PCC 7942, *Synechocystis* sp. PCC 6803, and *Synechococcus* sp. PCC 7418: osmolyte accumulation and stress protein synthesis. J Plant Physiol 154:240-249

Fulda S, Huang F, Nilsson F, Hagemann M, Norling B (2000) Proteomics of *Synechocystis* sp. strain PCC 6803: identification of periplasmatic proteins in cells grown at low and high salt concentrations. FEBS Lett (submitted)

Gabbay-Azaria R, Schonfeld M, Tel-Or S, Messinger R, Tel-Or E (1992) Respiratory activity in the marine cyanobacterium *Spirulina subsalsa* and its role in salt tolerance. Arch Microbiol 157:183-190

Gabbay-Azaria R, Pick U, Benhayyim G, Tel-Or E (1994) The involvement of a vanadate-sensitive ATPase in plasma membranes of a salt-tolerant cyanobacterium. Physiol Plant 90:692-698

Galinski EA (1993) Compatible solutes of halophilic eubacteria: molecular principles, water-solute interaction, stress protection. Experientia 49:487-496

Galinski EA, Truper HG (1994) Microbial behaviour in salt-stressed ecosystems. FEMS Microbiol Rev 15:95-108

Garcia AB, Engler JD, Iyer S, Gerats T, VanMontagu M, Caplan A (1997) Effects of osmoprotectants upon NaCl stress in rice. Plant Physiol 115:159-169

Giaever HM, Styrvold OB, Kaasen I, Strøm AR (1988) Biochemical and genetic characterization of osmoregulatory trehalose synthesis in *Escherichia coli*. J Bacteriol 170:2841-2849

Gilboa H, Kogut M, Chalamish S, Regev R, Avi-Dor Y, Russell NJ (1991) Use of ^{23}Na nuclear magnetic resonance spectroscopy to determine the true intracellular concentration of free sodium in a halophilic eubacterium. J Bacteriol 173:7021-7023

Ginzburg BZ, Ginzburg M, Cohen M (1995) A salt-sensitive *Dunaliella* mutant I. Growth and osmoregulatory responses to increase in salt concentration. J Exp Bot 46:1505-1513

Goddijn OJ, Verwoerd TC, Voogd E, Krutwagen RWHH, de Graaf PTHM, Poels J, van Dun K, Ponstein AS, Damm B, Pen J (1997) Inhibition of trehalase activity enhances trehalose accumulation in transgenic plants. Plant Physiol 113:181-190

Golldack D, Dietz K-J, Gimmler H (1995a) Salt-dependent protein composition in the halotolerant green alga, *Dunaliella parva*. Bot Acta 108:227-232

Golldack D, Dietz K-J, Gimmler H (1995b) The effect of sudden salt stress on protein synthesis in the green alga *Dunaliella parva*. J Plant Physiol 146:508-514

Greenway H, Setter TL (1979) Accumulation of proline and sucrose during the first hours after transfer of *Chlorella emersonii* to high NaCl. Aust J Plant Physiol 69-79

Hagemann M, Erdmann N (1994) Activation and pathway of glucosylglycerol synthesis in the cyanobacterium *Synechocystis* sp. PCC 6803. Microbiology 140:1427-1431

Hagemann M, Erdmann N (1997) Environmental stresses. In: Rai AK (ed) Cyanobacterial nitrogen metabolism and environmental biotechnology. Springer, Berlin Heidelberg New York & Narosa, New Dehli, pp 156-221

Hagemann M, Marin K (1999) Salt-induced sucrose accumulation is mediated by sucrose-phosphate-synthase in cyanobacteria. J Plant Physiol 155:424-430

Hagemann M, Zuther E (1992) Selection and characterization of mutants of the cyanobacterium *Synechocystis* sp PCC-6803 unable to tolerate high salt concentrations. Arch Microbiol 158:429-434

Hagemann M, Erdmann N, Wittenburg E (1987) Synthesis of glucosylglycerol in salt-stressed cells of the cyanobacterium *Microcystis firma*. Arch Microbiol 148:275-279

Hagemann M, Woelfel L, Krueger B (1990) Alterations of protein synthesis in the cyanobacterium *Synechocystis* sp. PCC 6803 after a salt shock. J Gen Microbiol 136:1393-1399

Hagemann M, Techel D, Rensing L (1991) Comparison of salt- and heat- induced alterations of protein synthesis in the cyanobacterium *Synechocystis* sp. PCC 6803. Arch Microbiol 155:587-592

Hagemann M, Golldack D, Biggins J, Erdmann N (1993) Salt-dependent protein phosphorylation in the cyanobacterium *Synechocystis* PCC 6803. FEMS Microbiol Lett 113:205-210

Hagemann M, Schoor A, Erdmann N (1996) NaCl acts as a direct modulator in the salt adaptive response: salt-dependent activation of glucosylglycerol synthesis in vivo and in vitro. J Plant Physiol 149:746-752

Hagemann M, Richter S, Mikkat S (1997a) The *ggtA* gene encodes a subunit of the transport system for the osmoprotective compound glucosylglycerol in *Synechocystis* sp. strain PCC 6803. J Bacteriol 179/3:714-720

Hagemann M, Schoor A, Jeanjean R, Zuther E, Joset F (1997b) The *stpA* gene from *Synechocystis* sp. strain PCC 6803 encodes the glucosylglycerol-phosphate phosphatase involved in cyanobacterial osmotic response to salt shock. J Bacteriol 179:1727-1733

Hagemann M, Jeanjean R, Fulda S, Havaux M, Joset F, Erdmann N (1999) Flavodoxin accumulation contributes to enhanced cyclic electron flow around photosystem I in salt-stressed cells of *Synechocystis* sp. strain PCC 6803. Physiol Plant 105:670-678

Haus M, Wegmann K (1984) Glycerol-3-phosphate dehydrogenase (EC 1.1.1.8) from *Dunaliella tertiolecta*. II. Influence of phosphate esters and different salts on the enzymatic activity with respect to osmoregulation. Physiol Plant 60:289-293

Hayashi H, Alia, Mustardy L, Deshnium P, Ida M, Murata N (1997) Transformation of *Arabidopsis thaliana* with the *codA* gene for choline oxidase; accumulation of glycinebetaine and enhanced tolerance to salt and cold stress. Plant J 12:133-142

Helliot B, MortainBertrand A (1999) Accumulation of proline in *Dunaliella salina* (Chlorophyceae) in response to light transition and cold adaptation. Effect on cryopreservation. Cryo Lett 20:287-296

Henzler T, Steudle E (1995) Reversible closing of water channels in *Chara* internodes provides evidence for a composite transport model of the plasma membrane. J Exp Bot 46:199-209

Hibino T, Lee BH, Rai AK, Ishikawa H, Kojima H, Tawada M, Shimoyama H, Takabe T (1996) Salt enhances photosystem I content and cyclic electron flow via NAD(P)H dehydrogenase in the halotolerant cyanobacterium *Aphanothece halophytica.* Aust J Plant Physiol 23:321-330

Hibino T, Kaku N, Yoshikawa H, Takabe T (1999) Molecular characterization of DnaK from the halotolerant cyanobacterium *Aphanothece halophytica* for ATPase, protein folding, and copper binding under various salinity conditions. Plant Mol Biol 40:409-418

Hohmann S (1997) Shaping up: the response of yeast to osmotic stress. In: Hohmann S, Mager WH (eds) Yeast stress responses. R G Landes, Georgetown, pp 101-145

Holmström K-O, Welin B, Mandal A, Kristiansdottir I, Teeri TH, Lamark T, Strom AR, Palva ET (1994) Production of the *Escherichia coli* betaine-aldehyde dehydrogenase, an enzyme required for the synthesis of the osmoprotectant glycine betaine, in transgenic plants. Plant J 6:749-758

Holmström K-O, Mäntylä E, Welin B, Mandal A, Palva ET, Tunnela OE, Londesborough J (1996) Drought tolerance in tobacco. Nature 379:683-684

Huber SC, Huber JL (1996) Role and regulation of sucrose-phosphate synthase in higher plants. Annu Rev Plant Physiol Plant Mol Biol 47:431-444

Ishitani M, Takabe T, Kojima K (1993) Regulation of glycinebetaine accumulation in the halotolerant cyanobacterium *Aphanothece halophytica.* Aust J Plant Physiol 20:693-703

Ishizaki-Nishizawa O, Fujii T, Azuma M, Sekiguchi K, Murata N, Ohtani T, Toguri T (1996) Low-temperature resistance of higher plants is significantly enhanced by a nonspecific cyanobacterial desaturase. Nat Biotechnol 14:1003-1006

Iverson RL, Nearhoof FL, Andreae MO (1989) Production of dimethylsulfonium propionate and dimethylsulfide by phytoplankton in estuarine and coastal waters. Limnol Oceanogr 34:53-67

Jeanjean R, Onana B, Peschek GA, Joset F (1990) Mutants of the cyanobacterium *Synechocystis* PCC 6803 impaired in respiration and unable to tolerate high salt concentrations. FEMS Microbiol Lett 68:125-130

Jeanjean R, Matthijs HP, Onana B, Havaux M, Joset F (1993) Exposure of the cyanobacterium *Synechocystis* PCC 6803 to salt stress induces concerted changes in respiration and photosynthesis. Plant Cell Physiol 34:1073-1079

Jeanjean R, Bedu S, Havaux M, Matthijs HC, Joset F (1998) Salt-induced photosystem I cyclic electron transfer restores growth on low inorganic carbon in a type 1 NAD(P)H dehydrogenase-deficient mutant of *Synechocystis* PCC6803. FEMS Microbiol Lett 167:131-137

Jonak C, Ligterink W, Hirt H (1999) MAP kinases in plant signal transduction. Cell Mol Life Sci 55:204-213

Joset F, Jeanjean R, Hagemann M (1996) Dynamics of the response of cyanobacteria to salt stress: deciphering the molecular events. Physiol Plant 96:738-744

Kaku N, Hibino T, Tanaka Y, Takabe T, Nakamura T (1999) Expression of *nhaAv* gene encoding Na^+/H^+ antiporter from *Vibrio alginolyticus* in a freshwater cyanobacterium *Synechococcus* sp. PCC 7942 confers lithium tolerance, but not sodium tolerance. Plant Cell Physiol 40:557-564

Kanamaru K, Kashiwagi S, Mizuno T (1993) The cyanobacterium, *Synechococcus* sp. PCC 7942, possesses two distinct genes encoding cation-transporting P-type ATPases. FEBS Lett 330:99-104

Kaneko T, Sato S, Kotani H, Tanaka A, Asamizu E, Nakamura Y, Miyajima N, Hirosawa M, Sugiura M, Sasamoto S, Kimura T, Hosouchi T, Matsuno A, Muraki A, Nakazaki N, Naruo K, Okumura S, Shimpo S, Takeuchi C, Wada T, Watanabe A, Yamada M, Yasuda M, Tabata S (1996) Sequence analysis of the genome of the unicellular cyanobacterium *Synechocystis* sp. strain PCC 6803. II. Sequence determination of the entire genome and assignment of potential protein-coding regions. DNA Res 3:109-136

Karakas B, OziasAkins P, Stushnoff C, Suefferheld M, Rieger M (1997) Salinity and drought tolerance of mannitol-accumulating transgenic tobacco. Plant Cell Environ 20:609-616

Karsten U, Kirst GO (1989) Incomplete turgor pressure regulation in the "terrestrial" red alga, *Bostrychia scorpioides* (Huds.) Mont. Plant Sci 61:29-36

Karsten U, Barrow KD, King RJ (1993) Floridoside, L-isofloridoside, and D-isofloridoside in the red alga *Porphyra columbina*. Plant Physiol 103:485-491

Karsten U, Barrow KD, Mostaert AS, King RJ, West JA (1994) ^{13}C- and ^{1}H-NMR studies on digeneaside in the red alga *Caloglossa leprieurii*. A re-evaluation of its osmotic significance. Plant Physiol Biochem 32:669-676

Karsten U, Bock C, West JA (1995) Low molecular weight carbohydrate patterns in geographically different isolates of the eulittoral red alga *Bostrychia tenuissima* from Australia. Bot Acta 108:321-326

Karsten U, Barrow KD, Nixdorf B, King RJ (1996a) The compatibility with enzyme activity of unusual organic osmolytes from mangrove red algae. Aust J Plant Physiol 21:577-582

Karsten U, Mostaert AS, King RJ, Kamiya M, Hara Y (1996b) Osmoprotectors in some species of Japanese mangrove. Phycol Res 44:109-112

Karsten U, Barrow KD, Nixdorf O, West JA, King RJ (1997) Characterization of mannitol metabolism in the mangrove red alga *Caloglossa leprieurii* (Montagne) J Agardh. Planta 201:173-178

Katz A, Pick U, Avron M (1989) Characterization and reconstitution of the Na^{+}/H^{+} antiporter from the plasma membrane of the halotolerant alga *Dunaliella*. Biochim Biophys Acta 983:9-14

Kauss H (1987) Some experiments of calcium-dependent regulation in plant metabolism. Annu Rev Plant Physiol 38:47-72

Kauss H, Thomson K-S (1982) Biochemistry of volume control in *Poterioochromonas*. In: Marme D, Marre E, Hertel R (eds) Plasmalemma and tonoplast: their functions in the plant cell. Elsevier, Amsterdam, pp 255-262

Kempf B, Bremer E (1998) Uptake and synthesis of compatible solutes as microbial stress responses to high-osmolality environments. Arch Microbiol 170:319-330

Khristin MS, Simonova NB (1998) Effect of glycerol on stability and extraction of photosystem II pigment-protein complexes from membranes of the unicellular green alga *Dunaliella salina*. Biol Membr 15:48-55

Kinraide TB (1999) Interactions among Ca^{2+}, Na^{+} and K^{+} in salinity toxicity: quantitative resolution of multiple toxic and ameliorative effects. J Exp Bot 50:1495-1505

Kirst GO (1990) Salinity tolerance of eukaryotic marine algae. Annu Rev Plant Physiol Plant Mol Biol 41:21-53

Kishor PBK, Hong Z, Miao G-H, Hu C-A, Verma DPS (1995) Over-expression of 1-pyrroline-5-carboxylate synthetase increases proline production and confers osmotolerance in transgenic plants. Plant Physiol 108:1387-1394

Ko JH, Lee SH (1995) Role of calcium in the osmoregulation under salt stress in *Dunaliella salina*. J Plant Biol 38:243-250

Lages F, Lucas C (1995) Characterization of a glycerol/H^{+} symport in the halotolerant yeast *Pichia solbitophila*. Yeast 11:111-119

Lages F, Lucas C (1997) Contribution to the physiological characterization of glycerol active uptake in *Saccharomyces cerevisiae*. Biochim Biophys Acta 1322:8-18

Lai MC, Yang DR, Chuang MJ (1999) Regulatory factors associated with synthesis of the osmolyte glycine betaine in the halophilic methanoarchaeon *Methanohalophilus portucalensis*. Appl Environ Microbiol 65:828-833

Laliberte G, Hellebust JA (1989) Pyrroline-5-carboxylate reductase in *Chlorella autotrophica* and *Chlorella saccharophila* in relation to osmoregulation. Plant Physiol 91:917-923

Lee BH, Hibino T, Jo J, Viale AM, Takabe T (1997) Isolation and characterization of a dnaK genomic locus in a halotolerant cyanobacterium *Aphanothece halophytica*. Plant Mol Biol 35:763-775

Li S, Ault A, Malone CL, Raitt D, Dean S, Johnston LH, Deschenes RJ, Fassler JS (1998) The yeast histidine protein kinase, Sln1p, mediates phosphotransfer to two response regulators, Ssk1p and Skn7p. EMBO J 17:6952-6962

Lippert K, Galinski EA, Truper HG (1993) Biosynthesis and function of trehalose in *Ectothiorhodospira halochloris*. Antonie Leeuwenhoek Int J Gen M 63:85-91

Liu J, Zhu JK (1997) Proline accumulation and salt-stress-induced gene expression in a salt-hypersensitive mutant of *Arabidopsis*. Plant Physiol 114:591-596

Liu Y, Bolen DW (1995) The peptide backbone plays a dominant role in protein stabilization by naturally occurring osmolytes. Biochemistry 34:12884-12891

Lu C, Zhang JH, Vonshak A (1998) Inhibition of quantum yield of PS II electron transport in *Spirulina platensis* by osmotic stress may be explained mainly by an increase in the proportion of the Q(B)-non-reducing PS II reaction centres. Aust J Plant Physiol 25:689-694

Lu CM, Zhang JH (1999) Effects of salt stress on PSII function and photoinhibition in the cyanobacterium *Spirulina platensis*. J Plant Physiol 155:740-745

Lunn JE, Price GD, Furbank RT (1999) Cloning and expression of a prokaryotic sucrose-phosphate synthase gene from the cyanobacterium *Synechocystis* sp PCC 6803. Plant Mol Biol 40:297-305

Lutts S, Kinet JM, Bouharmont J (1996) Effects of various salts and of mannitol on ion and proline accumulation in relation to osmotic adjustment in rice (*Oryza sativa* L.) callus cultures. J Plant Physiol 149:186-195

Luyten K, Albertyn J, Skibbe WF, Prior BA, Ramos J, Thevelein JM, Hohmann S (1995) Fps1, a yeast member of the MIP family of channel proteins, is a facilitator for glycerol uptake and efflux and is inactive under osmotic stress. EMBO J 14:1360-1371

Mackay AM, Norton RS (1987) ^{13}C nuclear magnetic resonance study of biosynthesis of glycosylglycerol by a cyanobacterium under osmotic stress. J Gen Microbiol 133:1535-1542

Mackay AM, Norton RS, Borowitzka LJ (1983) Marine blue-green algae have a unique osmoregulatory system. Mar Biol 73:301-307

Mackay AM, Norton RS, Borowitzka LJ (1984) Organic osmoregulatory solutes in cyanobacteria. J Gen Microbiol 30:2177-2191

Malin G, Kirst GO (1997) Algal production of dimethyl sulphide and its atmospheric role. J Phycol 33:889-896

Marin K, Zuther E, Kerstan T, Kunert A, Hagemann M (1998) The *ggpS* gene from *Synechocystis* sp. strain PCC 6803 encoding glucosyl-glycerol-phosphate synthase is involved in osmolyte synthesis. J Bacteriol 180:4843-4849

Martins LO, Empadinhas N, Marugg JD, Miguel C, Ferreira C, DaCosta MS, Santos H (1999) Biosynthesis of mannosylglycerate in the thermophilic bacterium *Rhodothermus marinus*—biochemical and genetic characterization of a mannosylglycerate synthase. J Biol Chem 274:35407-35414

Mason TG, Blunden G (1989) Quarternary ammonium and tertiary sulphonium compounds of algal origin as alleviators of osmotic stress. Bot Mar 32:313-316

McNeil SD, Nuccio ML, Hanson AD (1999) Betaines and related osmoprotectants. Targets for metabolic engineering of stress resistance. Plant Physiol 120:945-949

Meng J, Srivastava LM (1990) Extraction, assay and some properties of floridoside phosphate synthase from *Porphyra perforata*, Rhodophyta. J Phycol 26:683-688

Meng J, Srivastava LM (1991) Partial purification and characterization of floridoside phosphate synthase from *Porphyra perforata*. Phytochemistry 30:1763-1766

Mikkat S, Hagemann M (2000) Molecular analysis of the *ggtBCD* gene cluster of *Synechocystis* sp. PCC 6803 encoding subunits of an ABC transporter for osmoprotective compounds. Arch Microbiol (submitted)

Mikkat S, Hagemann M, Schoor A (1996) Active transport of glucosylglycerol is involved in salt adaptation of the cyanobacterium *Synechocystis* sp. strain PCC 6803. Microbiology 142:1725-1732

Mikkat S, Effmert U, Hagemann M (1997) Uptake and use of the osmoprotective compounds trehalose, glucosylglycerol, and sucrose by the cyanobacterium *Synechocystis* sp. PCC6803. Arch Microbiol 167:112-118

Mikkat S, Milkowski C, Hagemann M (2000) The gene *sll*0273 of the cyanobacterium *Synechocystis* sp. strain PCC 6803 encodes a protein essential for growth at low Na^{+}/K^{+} ratios. Plant Cell Environ 23(in press)

Molitor V, Trnka M, Erber W, Steffan I, Riviere ME, Arrio B, Springer-Lederer H, Peschek G (1990) Impact of salt adaptation on esterified fatty acids and cytochromeoxidase in plasma and thylakoid membranes from the cyanobacterium *Anacystis nidulans*. Arch Microbiol 154:112-119

Moore DJ, Reed RH, Stewart WP (1987) A glycine betaine transport system in *Aphanothece halophytica* and other glycine betaine-synthesising cyanobacteria. Arch Microbiol 147:399-405

Müller J, Boller T, Wiemken A (1995) Trehalose and trehalase in plants: recent developments. Plant Sci 112:1-9

Müller W, Wegmann K (1978) Sucrose biosynthesis in *Dunaliella*. I. Thermic and osmotic regulation. Planta 141:155-158

Nagata S, Adachi K, Shirai K, Sano H (1995) ^{23}Na NMR spectroscopy of free Na^{+} in the halotolerant bacterium *Brevibacterium* sp. and *Escherichia coli*. Microbiology 140:729-736

Nagata S, Adachi K, Sano H (1998) Intracellular changes in ions and organic solutes in halotolerant *Brevibacterium* sp. strain JCM 6894 after exposure to hyperosmotic shock. Appl Environ Microbiol 64:3641-3647

Nanjo T, Kobayashi M, Yoshiba Y, Kakubari Y, Yamaguchi-Shinozaki K, Shinozaki K (1999a) Antisense suppression of proline degradation improves tolerance to freezing and salinity in *Arabidopsis thaliana*. FEBS Lett 461:205-210

Nanjo T, Kobayashi M, Yoshiba Y, Sanada Y, Wada K, Tsukaya H, Kakubari Y, YamaguchiShinozaki K, Shinozaki K (1999b) Biological functions of proline in morphogenesis and osmotolerance revealed in antisense transgenic *Arabidopsis thaliana*. Plant J 18:185-193

Nevoigt E, Stahl U (1997) Osmoregulation and glycerol metabolism in the yeast *Saccharomyces cerevisiae*. FEMS Microbiol Rev 21:231-241

Nitschmann WH, Packer L (1992) NMR studies on Na^{+} transport in *Synechococcus* PCC 6311. Arch Biochem Biophys 294:347-352

Nomura M, Ishitani M, Takabe T, Rai AK (1995) *Synechococcus* sp. PCC 7942 transformed with *Escherichia coli bet* genes produces glycine betaine from choline and acquires resistance to salt stress. Plant Physiol 107:703-708

Nomura M, Hibino T, Takabe T, Sugiyama T, Yokota A, Miyake H (1998) Transgenically produced glycinebetaine protects ribulose 1,5-bisphosphate carboxylase/oxygenase from inactivation in *Synechococcus* sp. PCC 7942 under salt stress. Plant Cell Physiol 39:425-432

Nuccio ML, Rhodes D, McNeil SD, Hanson AD (1999) Metabolic engineering of plants for osmotic stress resistance. Curr Opin Plant Biol 2:128-134

Oren A (1999) Bioenergetic aspects of halophilism. Microbiol Mol Biol Rev 63:334

Page-Sharp M, Behm CA, Smith GD (1999) Involvement of the compatible solutes trehalose and sucrose in the response to salt stress of a cyanobacterium *Scytonema* species isolated from desert soils. Biochim Biophys Acta 1472:519-528

Papageorgiou GC, Govindjee, Govindjee R, Mimuro M, Stamatakis K, AlygizakiZorba A, Marata N (1999) Light-induced and osmotically induced changes in chlorophyll a fluorescence in two *Synechocystis* sp. PCC 6803 strains that differ in membrane lipid unsaturation. Photosynth Res 59:125-136

Peng Z, Lu Q, Verma DPS (1996) Reciprocal regulation of *delta*^{1}-pyrroline-5-carboxylate synthase and proline dehydrogenase genes controls proline levels during and after osmotic stress in plants. Mol Gen Genet 253:334-341

Petrovic U, GundeCimerman N, Plemenitas A (1999) Salt stress affects sterol biosynthesis in the halophilic black yeast *Hortaea werneckii*. FEMS Microbiol Lett 180:325-330

PilonSmits EA, Terry N, Sears T, Kim H, Zayed A, Hwang SB, vanDun K, Voogd E, Verwoerd TC, Krutwagen RW, Goddijn OJ (1998) Trehalose-producing transgenic tobacco plants show improved growth performance under drought stress. J Plant Physiol 152:525-532

Pocard JA, Smith LT, Smith GM, LeRudulier D (1994) A prominent role for glucosylglycerol in the adaptation of *Pseudomonas mendocina* SKB70 to osmotic stress. J Bacteriol 176:6877-6884

Poolman B, Glaasker E (1998) Regulation of compatible solute accumulation in bacteria. Mol Microbiol 29:397-407

Popova L, Balnokin Y, Dietz KJ, Gimmler H (1999) Characterization of phosphorylated intermediates synthesised during the catalytic cycle of the sodium adenosine triphosphatase in the plasma membrane of the marine unicellular alga *Tetraselmis (Platymonas) viridis*. J Plant Physiol 155:302-309

Porchia AC, Salerno GL (1996) Sucrose biosynthesis in a prokaryotic organism: presence of two sucrose-phosphate synthases in *Anabaena* with remarkable differences compared with the plant enzymes. Proc Natl Acad Sci USA 93:13600-13604

Porchia AC, Curatti L, Salerno GL (1999a) Sucrose metabolism in cyanobacteria: sucrose synthase from *Anabaena* sp. strain PCC 7119 is remarkably different from the plant enzymes with respect to substrate affinity and amino-terminal sequence. Planta 210:34-40

Porchia AC, Fiol DF, Salerno GL (1999b) Differential synthesis of sucrose and trehalose in *Euglena gracilis* cells during growth and salt stress. Plant Sci 149:43-49

Potts M (1994) Desiccation tolerance of prokaryotes. Microbiol Rev 58:755-805

Ramani S, Apte SK (1997) Transient expression of multiple genes in salinity-stressed young seedlings of rice (*Oryza sativa* L) cv. Bura Rata. Biochem Biophys Res Commun 233:663-667

Rausch U, Ahmad H, Maier G, Kauss H (1991) Proteinase nature of an enzyme capable of activating the galactosyltransferase involved in volume regulation of *Poterioochromonas malhamensis*. Physiol Plant 82:93-98

Reed RH (1983) The osmotic response of *Polysiphonia lanosa* (L.) Tandy from marine and estuarine sites: evidence for incomplete recovery of turgor. J Exp Mar Biol Ecol 68:169-193

Reed RH (1984) Transient breakdown in the selective permeability of the plasma membrane of *Chlorella emersonii* in response to hyperosmotic shock: implications for cell water relations and osmotic adjustment. J Membr Biol 82:83-88

Reed RH (1989) Osmotic adjustment and organic solute accumulation in *Chaetomorpha capillaris*. Br Phycol J 24:21-37

Reed RH, Stewart WP (1988) The response of cyanobacteria to salt stress. In: Rogers LJ GR (ed) Biochemistry of the algae and cyanobacteria. Clarendon Press, Oxford, pp 217-231

Reed RH, Richardson DL, Warr, SRC, Stewart WDP (1984) Carbohydrate accumulation and osmotic stress in cyanobacteria. J Gen Microbiol 130:1-4

Reed RH, Richardson DL, Stewart WDP (1985) Na^+ uptake and extrusion in the cyanobacterium *Synechocystis* PCC 6714 in response to hypersaline treatment. Evidence for transient changes in plasmalemma Na^+ permeability. Biochim Biophys Acta 814:347-355

Reed RH, Borowitzka LJ, Mackay MA, Chudek JA, Foster R, Warr SRC, Moore DJ, Stewart WDP (1986) Organic solute accumulation in osmotically stressed cyanobacteria. FEMS Microbiol Rev 39:51-56

Rhodes D, Hanson AD (1993) Quarternary ammonium and teriary sulfonium compounds in higher plants. Annu Rev Plant Physiol Plant Mol Biol 44:357-384

Riccardi G, Cella R, Camerino G, Cifferi O (1983) Resistance to azetidine-2-carboxylic acid and sodium chloride tolerance in carrot cell cultures and *Spirulina platensis*. Plant Cell Physiol 24:1078

Richardson DL, Reed RH, Stewart WDP (1983) *Synechocystis* PCC 6803: a euryhaline cyanobacterium. FEMS Microbiol Lett 18:99-102

Richter DFE, Kirst GO (1987) D-Mannitol dehydrogenase and D-mannitol-1-phosphate dehydrogenase in *Platymonas subcordiformis*: some characteristics and their role in osmotic adaptation. Planta 170:528-534

Ritchie RJ (1992) Sodium transport and the origin of the membrane potential in the cyanobacterium *Synechococcus* R-2 (*Anacystis nidulans*) PCC 7942. J Plant Physiol 139:320-330

Romero C, Belles JM, Vaya JL, Serrano R, CulianezMacia FA (1997) Expression of the yeast trehalose-6-phosphate synthase gene in transgenic tobacco plants: pleiotropic phenotypes include drought tolerance. Planta 201:293-297

Rout NP, Tripathi SB, Shaw BP (1997) Effect of salinity on chlorophyll and proline contents in three aquatic macrophytes. Biol Plant 40:453-458

Russell NJ (1995) Membranes as a target for stress adaptation. Int J Food Microbiol 28:255-261

Sakamoto A, Murata N (2000) Genetic engineering of glycinebetaine synthesis in plants: current status and implications for enhancement of stress tolerance. J Exp Bot 51:81-88

Sakamoto A, Murata A, Murata N (1998) Metabolic engineering of rice leading to biosynthesis of glycinebetaine and tolerance to salt and cold. Plant Mol Biol 38:1011-1019

Samuel D, Kumar TK, Jayaraman G, Yang PW, Yu C (1997) Proline is a protein solubilizing solute. Biochem Mol Biol Int 41:235-242

Schachtman D, Liu WH (1999) Molecular pieces to the puzzle of the interaction between potassium and sodium uptake in plants. Trends Plant Sci 4:281-287

Schobert B (1977) The influence of water stress on the metabolism of diatoms. I. Osmotic resistance and proline accumulation under different conditions of stress and light. Z Pflanzenphysiol 85:451-461

Schobert B (1980) The importance of water activity and water structure during hyperosmotic stress in algae and higher plants. Biochem Physiol Pflanz 175:91-103

Schoffeniels E, Gilles R (1970) Osmoregulation in marine mollusca. In: Florkin M, Scheer B (eds) Chemical zoology. Academic Press, London, pp 343-320

Schoor A, Hagemann M, Erdmann N (1999) Glucosylglycerol-phosphate synthase: target for ion-mediated regulation of osmolyte synthesis in the cyanobacterium *Synechocystis* sp. strain PCC 6803. Arch Microbiol 171:101-106

Schwartz SH, Black TA, Jager K, Panoff JM, Wolk CP (1998) Regulation of an osmoticum-responsive gene in *Anabaena* sp. strain PCC 7120. J Bacteriol 180:6332-6337

Serrano R, Mulet JM, Rios G, Marquez JA, deLarrinoa IF, Leube MP, Mendizabal I, PascualAhuir A, Proft M, Ros R, Montesinos C (1999) A glimpse of the mechanisms of ion homeostasis during salt stress. J Exp Bot 50:1023-1036

Shariati M, Lilley RM (1994) Loss of intracellular glycerol from *Dunaliella* by electroporation at constant osmotic pressure: subsequent restoration of glycerol content and associated volume changes. Plant Cell Environ 17:1295-1304

Shen B, Jensen RG, Bohnert HJ (1997) Increased resistance to oxidative stress in transgenic plants by targeting mannitol biosynthesis to chloroplasts. Plant Physiol 113:1177-1183

Shen B, Hohmann S, Jensen RG, Bohnert HJ (1999) Roles of sugar alcohols in osmotic stress adaptation. Replacement of glycerol by mannitol and sorbitol in yeast. Plant Physiol 121:45-52

Sheveleva EV, Marquez S, Chmara W, Zegeer A, Jensen RG, Bohnert HJ (1998) Sorbitol-6-phosphate dehydrogenase expression in transgenic tobacco — high amounts of sorbitol lead to necrotic lesions. Plant Physiol 117:831-839

Shono M, Hara Y, Wada M, Fujii T (1996) A sodium pump in the plasma membrane of the marine alga *Heterosigma akashiwo*. Plant Cell Physiol 37:385-388

Sibley MH, Yopp JH (1987) Regulation of S-adenosylhomocysteine hydrolase in the halophilic cyanobacterium *Aphanothece halophytica*: a possible role in glycinebetaine biosynthesis. Arch Microbiol 149:43-46

Smirnoff N, Cumbes QJ (1989) Hydroxyl radical scavenging activity of compatible solutes. Phytochemistry 28:1057-1060

Smith FA, Walker NA (1989) Transport of potassium in *Chara australis*: I. A symport with sodium. J Membr Biol 108:125-137

Stamatakis K, Ladas NP, AlygizakiZorba A, Papageorgiou GC (1999) Sodium chloride-induced volume changes of freshwater cyanobacterium *Synechococcus* sp PCC 7942 cells can be probed by chlorophyll a fluorescence. Arch Biochem Biophys 370:240-249

Straus NA (1994) Iron deprivation: physiology and gene regulation. In: Bryant DA (ed) The molecular biology of cyanobacteria. Kluwer, Dordrecht, pp 731-750

Sugino M, Hibino T, Tanaka Y, Nii N, Takabe T (1999) Overexpression of DnaK from a halotolerant cyanobacterium *Aphanothece halophytica* acquires resistance to salt stress in transgenic tobacco plants. Plant Sci 146:81-88

Takenaka S, Kondo T, Nazeri S, Tamura Y, Tokunaga M, Tsuyama S, Miyatake S, Nakano Y (1997) Accumulation of trehalose as a compatible solute under osmotic stress in *Euglena gracilis* Z. J Eukaryot Microbiol 44:609-613

Tamas MJ, Luyten K, Sutherland FC, Hernandez A, Albertyn J, Valadi H, Li H, Prior BA, Killan SG, Ramos J, Gustafsson L, Thevelein JM, Hohmann S (1999) Fps1p controls the accumulation and release of the compatible solute glycerol in yeast osmoregulation. Mol Microbiol 31:1087-1104

Tanaka Y, Katada S, Ishikawa H, Ogawa T, Takabe T (1997) Electron flow from NAD(P)H dehydrogenase to photosystem I is required for adaptation to salt shock in the cyanobacterium *Synechocystis* sp. PCC 6803. Plant Cell Physiol 38:1311-1318

Tarczynski MC, Jensen RG, Bohnert HJ (1992) Expression of a bacterial *mtlD* gene in transgenic tobacco leads to production and accumulation of mannitol. Proc Natl Acad Sci USA 89:2600-2604

Tarczynski MC, Jensen RG, Bohnert HJ (1993) Stress protection of transgenic tobacco by production of the osmolyte mannitol. Science 259:508-510

Thomas JC, Sepahi M, Arendall M, Bohnert HJ (1995) Enhancement of seed germination in high salinity by engineering mannitol expression in *Arabidopsis thaliana*. Plant Cell Environ 18:801-806

Toroser D, Huber SC (1997) Protein phosphorylation as a mechanism for osmotic-stress activation of sucrose-phosphate synthase in spinach leaves. Plant Physiol 114:947-955

Toroser D, Athwal GS, Huber SC (1998) Site-specific regulatory interaction between spinach leaf sucrose-phosphate synthase and 14-3-3 proteins. FEBS Lett 435:110-114

Urao T, Yakubov B, Satoh R, YamaguchiShinozaki K, Seki M, Hirayama T, Shinozaki K (1999) A transmembrane hybrid-type histidine kinase in *Arabidopsis* functions as an osmosensor. Plant Cell 11:1743-1754

Varela JCS, Mager WH (1996) Response of *Saccharomyces cerevisiae* to changes in external osmolarity. Microbiology 142:721-731

Vinnemeier J, Hagemann M (1999) Identification of salt-regulated genes in the genome of the cyanobacterium *Synechocystis* sp. strain PCC 6803 by subtractive RNA hybridization. Arch Microbiol 172:377-386

Vinnemeier J, Kunert A, Hagemann M (1998) Transcriptional analysis of the *isiAB* operon in salt-stressed cells of the cyanobacterium *Synechocystis* sp. PCC 6803. FEMS Microbiol Lett 169:323-330

Vogt C, Rabenstein A, Rethmeier J, Fischer U (1998) Alkali-labile precursors of dimethyl sulfide in marine benthic cyanobacteria. Arch Microbiol 169:263-266

Wada M, Satoh S, Kasamo K, Fujii T (1989) Presence of a Na^+-activated ATPase in the plasma membrane of the marine raphidophycean *Heterosigma akashiwo*. Plant Cell Physiol 30:923-928

Warr SRC, Reed RH, Stewart WDP (1988) The compatibility of osmotica in cyanobacteria. Plant Cell Environ 11:137-142

Wegmann K (1986) Osmoregulation in eucaryotic algae. FEMS Microbiol Rev 39:37-43

Weig A, Deswarte C, Chrispeels MJ (1997) The major intrinsic protein family of *Arabidopsis* has 23 members of three distinct groups with functional aquaporins in each group. Plant Physiol 114:1347-1357

Weiss M, Pick U (1990) Transient Na^+ flux following hyperosmotic shock in the halotolerant alga *Dunaliella salina:* a response to intracellular pH changes. J Plant Physiol 136:429-438

Winzor CL, Winzor DJ, Paleg LG, Jones GP, Naidu BP (1992) Rationalization of the effects of compatible solutes on protein stability in terms of thermodynamic nonideality. Arch Biochem Biophys 296:102-107

Wolfe GV, Steinke M, Kirst GO (1997) Grazing-activated chemical defence in a unicellular marine alga. Nature 387:894-897

Wood JM (1999) Osmosensing by bacteria: signals and membrane-based sensors. Microbiol Mol Biol Rev 63:230

Xing WB, Rajashekar CB (1999) Alleviation of water stress in beans by exogenous glycine betaine. Plant Sci 148:185-192

Yeo A (1998) Molecular biology of salt tolerance in the context of whole-plant physiology. J Exp Bot 49:915-929

Yoshiba Y, Kiyosue T, Nakashima K, Yamaguchi-Shinozaki K, Shinozaki K (1997) Regulation of levels of proline as an osmolyte in plants under water stress. Plant Cell Physiol 38:1095-1102

Yuasa T, Muto S (1992) Change of protein phosphorylation of *Dunaliella tertiolecta* cells in osmotic shock. In: Murata N (ed) Research in photosynthesis. Kluwer, Dordrecht, pp 263-266

Yuasa T, Muto S (1996) Activation of 40-kDa protein kinases in response to hypo- and hyperosmotic shock in the halotolerant green alga *Dunaliella tertiolecta.* Plant Cell Physiol 37:35-42

Yuasa T, Takahashi K, Muto S (1995) Purification and characterization of a Ca^{2+}-dependent protein kinase from the halotolerant green alga *Dunaliella tertiolecta.* Plant Cell Physiol 36:699-708

Zeng MT, Vonshak A (1998) Adaptation of *Spirulina platensis* to salinity-stress. Comp Biochem Physiol (A) 120:113-118

Zhang JX, Nguyen HT, Blum A (1999) Genetic analysis of osmotic adjustment in crop plants. J Exp Bot 50:291-302

Zhu B, Su J, Chan M, Verma DPS, Fan YL, Wu R (1998) Overexpression of a Delta(1)-pyrroline-5-carboxylate synthetase gene and analysis of tolerance to water- and salt-stress in transgenic rice. Plant Sci 139:41-48

Zuther E, Schubert H, Hagemann M (1998) Mutation of a gene encoding a putative glycoprotease leads to reduced salt tolerance, altered pigmentation, and cyanophycin accumulation in the cyanobacterium *Synechocystis* sp. strain PCC 6803. J Bacteriol 180:1715-1722

12 Heavy Metal Tolerance in Algae

J.P. Gaur and L.C Rai*

Contents

*Department of Botany, Banaras Hindu University, Varanasi 221 005, India

12.1 Introduction

Algae are often exposed to heavy metal pollution due to the disposal of industrial and domestic wastes into waterways. Many algae growing in metal-polluted environments display an ability to tolerate high concentrations of toxic metals (De Filippis and Pallaghy 1994). In fact, even laboratory cultures of some algae can be acclimated to elevated concentrations of toxic metals (Shehata and Whitton 1982; Kuwabara and Leland 1986; Rai et al. 1991; Twiss et al. 1993). Metal tolerance in algae may be genetic or physiological, and the available reports often do not distinguish between these two different sets of mechanisms. The phrases tolerance and resistance have often been used interchangeably, although the latter phrase may be preferred if ability to withstand high concentrations of metals is genetically fixed.

Different groups of algae exhibit varying levels of tolerance to heavy metals. Whitton (1970) and Foster (1982b) found that the members of Ulotrichales (Chlorophyceae) are generally tolerant to metals. Takamura et al. (1989) also observed greater Cu^{2+}, Cd^{2+} and Zn^{2+} tolerance among the members of the Chlorophyceae than of the Cyanophyceae (Cyanobacteria). According to them, blue-greens were sensitive to these three metals whether or not they were isolated from polluted waters. On the other hand, Harding and Whitton (1976), Say et al. (1977) and Whitton (1980) showed greater Zn^{2+} tolerance of filamentous chlorophytes from high Zn sites than those from low Zn sites. Likewise, Agrawal and Kumar (1975) also demonstrated greater tolerance to Hg^{2+} in *Chlorella* sp. isolated from a mercury-contaminated habitat than the laboratory strain. Genter et al. (1987) noticed a distinct change in composition of an algal community from diatoms to filamentous green algae and finally to unicellular green algae with an increase in the Zn concentration of an experimental stream.

Foster (1982a) reasoned that there is a natural diversity in heavy metal susceptibility at the species level. Kessler (1985) found that one strain of *Chlorella saccharophila* was 100 times more sensitive than 14 other strains belonging to five species of *Chlorella*. Despite the fact that several algal species can tolerate high concentrations of heavy metals, there is no single species that can be said to invariably indicate metal pollution (Foster 1982b) because species occurring at high concentrations of heavy metals have been reported from unpolluted environments as well.

An alga may at times tolerate more than one toxic metal. Multiple metal resistance may be of two possible types: (1) in multiple tolerance an alga concurrently shows tolerance to several metals, although mechanisms of tolerance to the various metals may differ, whereas (2) in cotolerance one metal resistance automatically confers tolerance to another metal (see Rai et al. 1981a). Hall (1980) demonstrated cotolerance of a Cu-tolerant population of *Ectocarpus siliculosus* to Co^{2+} and Zn^{2+}. Takamura et al. (1989) also reported cotolerance to Zn^{2+} and Cd^{2+} in some freshwater benthic algae. Bariaud et al. (1985) similarly showed that a Cd-tolerant strain of *Euglena gracilis* was tolerant to Co^{2+} and Zn^{2+}

also. Many algae display cotolerance to Cu^{2+} and Pb^{2+}, therefore suggesting that in some circumstances, non-specific metal tolerance may be more advantageous or more easy to induce than the metal-specific tolerance (Foster 1982b).

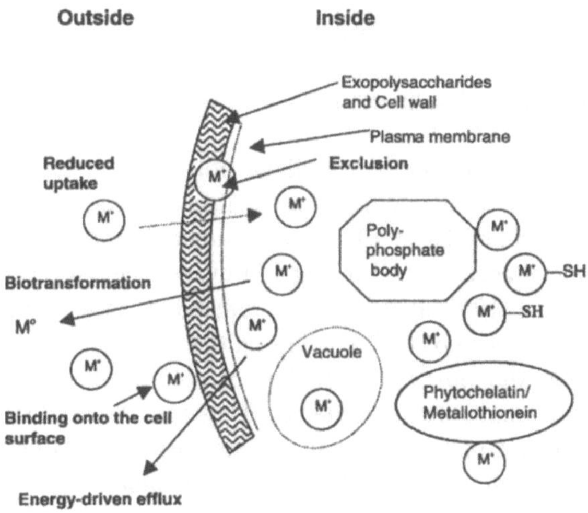

Fig. 12.1. A schematic presentation of some common mechanisms of metal tolerance in algae

Figure 12.1 presents some of the mechanisms for tolerating high concentrations of toxic metals. These can be put under three broad categories. The first comprises mechanisms that sequester metals on extracellular components that reduce metal bioavailability, entry into the cell and intracellular concentration. The second increases the efflux of metals from the cell. The third includes various types of intracellular detoxification. Intracellular detoxification mechanisms include sequestration in polyphosphate bodies, phytochelatins and metallothioneins, or binding to strong chelators. Detoxification can also involve the oxidation or reduction of transition metals to yield states that are less toxic or less soluble. In addition, altered responses to the effects of metals may enhance resistance. For example, some metals cause oxidative stress in algae and cyanobacteria, with the result that metal toxicity can by altered by the synthesis of appropriate enzymes or metabolites that counteract metal-induced oxidative stress. Tolerance to one metal might depend on the concurrent deployment of several strategies.

Several reviews have appeared earlier on metal tolerance in algae. Some are of a general nature describing all the mechanisms of tolerance (Reed and Gadd 1990), while others concentrate on phytochelatins and metallothioneins (Robinson 1989). Silver and Phung (1996) reviewed new developments in metal tolerance in bacteria, whereas Fiore and Trevors (1994) briefly discussed metal tolerance in cyanobacteria. This chapter focuses on the numerous significant advances that have been made recently.

12.2 Extracellular Complexation

12.2.1 Binding of Metal Ions on the Cell Surface

The binding of metal ions onto extracellular polymeric materials and cell walls has been often invoked as an important mechanism of metal tolerance in algae and other microbes (Reed and Gadd 1990). A considerable fraction, up to 80-90% in some cases, of metal accumulated by algal or cyanobacterial cells lies on the cell surface (Knauer et al. 1997; Parker et al. 1998; Mehta et al. 2000). Winter et al. (1994) presented evidence of Cd^{2+} binding on the outer surface layers of the cell wall of *Ectocarpus siliculosus*. The binding of gold on the surface of *Chlorella vulgaris* was subsequently demonstrated by Ting et al. (1995). Similarly, binding of appreciable amount of heavy metals on the surface of *Microcystis* has been demonstrated by Singh et al. (2000) using differential pulse anodic stripping voltammetry. Many cyanobacteria and algae secrete copious amounts of mucilaginous materials of various types —often primarily polysaccharide— that can be arbitrarily classified as sheath, capsule or slime (De Philippis and Vincenzini 1998). These materials frequently have a substantial ability to bind various metal ions (Tease and Walker 1987; Weckesser et al. 1988; Parker et al. 1996). De Philippis and Vincenzini (1998) expressed the possibility that exopolysaccharides play an important role in safeguarding cyanobacteria and algae from extreme environments and other harmful conditions. For example, the alga *Mesotaenium* (Zygnemales) produced a water-soluble slime consisting of 30-40% uronic acid residues capable of binding Cu^{2+} (Eichenberger 1986). The binding of Cu^{2+} by this slime lowered intracellular accumulation of Cu^{2+} and decreased its toxicity to this organism. Similarly, the metal-binding capsules (Singh et al. 1998) of the cyanobacteria *Microcystis* spp. C3-40 (Plude et al. 1991) and K3-A (Nakagawa et al. 1986) both contain polysaccharide with a large percentage of uronic acid. Some researchers believe that an abundance of mucilage-producing algae is a general feature of metal-polluted waterbodies (Whitton and Say 1975; Foster 1982a).

The cell wall carries a net negative charge due to the presence of carboxyl, phosphatic and other groups (Crist et al. 1988; Xue and Sigg 1990). These groups are believed to be involved in binding of metals through ion exchange and other mechanisms, as reviewed by Stokes (1983). An indication that the cell wall can sometimes influence metal susceptibility comes from a cell wall-deficient mutant of *Chlamydomonas reinhardtii* that was more sensitive to Cd^{2+}, Co^{2+} and Ni^{2+} than the wild type (Genter 1996). Rijstenbil et al. (1998b) report that *Enteromorpha lanza* with a thicker cell wall (1-9 μm) can better modulate Cu^{2+} toxicity in comparison to *E. prolifera* with a thinner cell wall (0.5-5 μm).

That the cell wall provides protection against metal toxicity has been a subject of considerable dispute because many researchers believe that the surface adsorption constitutes the first step of metal uptake by an algal cell. Knauer et al. (1997) found that Cu^{2+} adsorbed by *Scenedesmus subspicatus* was transported into the cell, thereby leading to an increase in intracellular Cu and a concomitant decrease in adsorbed Cu. Because the calculated maximum concentration of binding sites on the surface of *S. subspicatus* was initially three times higher than under steady state conditions, these authors suggested that the initial binding to the algal surface comprised the binding to specific transport ligands as well as to the inert adsorption sites.

Whether cell wall and capsular materials act to increase or decrease intracellular metal uptake may vary depending on the organism and the adsorbing material. A critical factor is whether the adsorption acts to increase or decrease the bioavailability of the metal. Strong adsorbers decrease the concentration of free metal ion and thus usually tend to decrease bioavailability. However, some adsorbers may be part of transport systems and thus would have the opposite effect.

12.2.2 Exudation of Metal-Complexing Ligands

The concentration of free metal ion, instead of the total metal concentration, determines the response of any aquatic organism to these toxic substances (Rai et al. 1981a; De Filippis and Pallaghy 1994). The ionic species of metals are freely and easily available to aquatic organisms, and measurement of their concentrations is therefore considered a prerequisite for interpreting toxicological data. The concentration of free metal ion can be determined using a computer programme, such as MINEQL, or experimentally through voltammetry or other suitable techniques. Natural waters contain a variety of ligands to which metal ions can be adsorbed and thus rendered inaccessible to living organisms. Algae may themselves excrete such substances.

Fogg and Westlake (1955) first demonstrated complexation and detoxification of Cu^{2+} by extracellular material produced by *Anabaena cylindrica*. Eukaryotic algae and some of the cyanobacterial species tested by McKnight and Morel (1979) produced organic acids, weak Cu-complexing agents.

McKnight and Morel (1979) also reported the production of hydroxamates, strong metal complexing agents, by four cyanobacterial species. Hydroxamates, a type of siderophore, were produced at a concentration of about 5×10^{-7} M with conditional stability constants in the range 10^{8} – 10^{12} for iron. Siderophore production is an important strategy of microorganisms to overcome iron deficiency. Iron can be scavenged from the siderophore complex through the use of specific membrane transport systems. Clarke et al. (1987) demonstrated chelation of Cu^{2+} by schizokinen, a siderophore, released by *Anabaena* strain 7120 as a mechanism to escape Cu toxicity. This strain could efficiently distinguish and transport ferric schizokinen, but was unable to recognise cupric schizokinen. Thus, the cyanobacterium could be protected from Cu toxicity under conditions favourable for the production of schizokinen. On the contrary, the ferric schizokinen uptake system of *Bacillus megaterium* could recognize cupric schizokinen, and Cu^{2+} toxicity was greatly enhanced in the presence of schizokinen (Arceneaux et al. 1984). Clarke et al. (1987) suggested that the siderophore system of *Anabaena* sp. had developed primarily to overcome iron starvation and that the alleviation of Cu^{2+} toxicity was a secondary benefit.

The production of strong metal-complexing agents by eukaryotes has also been reported (Xue and Sigg 1990). The conditional stability constants for ligands excreted by marine algae range from $10^7 - 10^{13}$ for Cu (Florence and Stauber 1986; Morelli et al. 1989; Robinson and Brown 1991; Gerringa et al. 1995; Gledhill et al. 1999). Gledhill et al. (1999) were not able to ascertain whether extracellular ligands were exuded already complexed to Cu or were released to complex Cu^{2+} in the external environment. Nevertheless, they favoured the former possibility for detoxification of Cu^{2+} in *Fucus vesiculosus.*

The available evidence points towards the secretion of both strong and weak complexing ligands by algae. A particular organism usually produces a specific kind of ligand, but sometimes more than one type of ligand may be produced by one organism (Morelli et al. 1989). Both strong and weak ligands could be important in protecting algae from metal toxicity. Initially, strongly complexing ligands might be used for the binding of each metal, but after they had become saturated, the weaker ligands could assume significance, especially if present in high concentrations. Regardless of affinity of ligands for metals, the most important factor is the total concentration of ligands in natural waters. A bloom-like situation with large populations of various algal species could lead to a plethora of metal-complexing ligands and greatly reduced metal toxicity. A similar situation would arise if algal ligands were persistent, i.e. not prone to rapid degradation in the natural environment. Research on metal-complexing ligands of algae has so far focused on Cu^{2+} because it can be most strongly complexed to certain ligands (Starodub et al. 1987). There is a pressing need to assess the binding by algal exudates of other metals also.

12.3 Exclusion and Efflux of Metal Ions

Although considered to be rare in higher plants (Woolhouse 1983), metal exclusion may be an important mechanism of metal tolerance in algae and cyanobacteria. Consistent with this idea, several reports indicate that metal-tolerant strains of these organisms accumulate less intracellular metal in comparison to non-tolerant strains (Foster 1977; Hall et al. 1979; Bariaud et al. 1985; Singh and Yadava 1986; Rai et al. 1991; Jin et al. 1996). A metal-tolerant organism could conceivably decrease metal influx in various ways, including: (1)

alterations in membrane function, particularly those leading to reduced permeability for a metal ion (Rai et al. 1991; DeFilippis and Pallaghy 1994; Silver and Phung 1996), and (2) perhaps modifications of the cell wall (Twiss and Nalewajko 1992). Rai et al. (1991) isolated a Cu-tolerant strain of *Anabaena doliolum*, and compared its physiological and biochemical characteristics with the Cu-sensitive strain. They observed reduced transport of Cu^{2+} and, more importantly, greater lipid content and improved membrane integrity (as indicated by reduced efflux of K^{+}) in Cu-treated cells of the tolerant strain. Asthana et al. (1993) mentioned alterations in membrane permeability as the possible explanation for Ni^{2+} tolerance in *Nostoc muscorum*. Wagatsuma and Akiba (1989) suggested that strong surface negativity could result in greater binding of the positively charged Al species to the membrane, rendering membrane structure and functions more sensitive to it. Tillberg et al. (1984) reported an increase in cell wall thickness in *Scenedesmus obtusiusculus* under phosphorus-deficient conditions, and this observation led Twiss and Nalewajko (1992) to speculate that a thick cell wall may partially reduce metal uptake by algal cells. Their speculation, however, needs convincing experimental support.

Many metal-resistant bacteria have a mechanism for exporting toxic metals from the cells (Silver and Phung 1996). Metal efflux is considered to be an active mechanism requiring energy expenditure. Nonetheless, Jin et al. (1996) suggested that Ni^{2+} efflux in some strains of *Scenedesmus acutus* occurred by passive diffusion in the dark after they were transferred to Ni-free medium. Verma and Singh (1991) showed the presence of an energy-dependent Cu-efflux system in a Cu-resistant mutant of *Nostoc calcicola*. An active efflux mechanism has also been suggested for Al tolerance in *Cyanidum caldarium* (Yoshimura et al. 1999). It is yet to be established how widely the phenomenon of metal efflux occurs in algae and cyanobacteria.

De Filippis (1978) presented evidence that *Chlorella* sp. has a general enzymatic system for reducing mercury compounds (mercuric chloride and phenyl mercuric acetate) to metallic mercury (Hg^{0}), which is rapidly volatilised and lost from the cell. This system utilizes NADPH or to a lesser extent NADH. It requires –SH groups that can be replaced by $-SCH_3$ groups at reduced efficiency, but cannot be replaced by disulphides.

12.4 Sequestration by Phytochelatins and Metallothioneins

A super family of metal-binding proteins, called metallothioneins, has recently been shown to regulate the cytosolic concentration of free metal ions in algae, plants and other organisms (Rauser 1990). A similar role is performed by certain enzymatically synthesized peptides called phytochelatins (Steffens 1990). A brief description of these compounds is presented below.

12.4.1 Nomenclature and Structure

Metallothionein (MT) was discovered by Margoshes and Vallee in 1957, in equine renal cortex. The term MT may be confusing to readers, as different names have been used to describe the same compounds, and in some cases, even the same name to describe different compounds. MT was first used to define Cd-, Zn- and Cu-containing, S-rich proteins (Kagi and Vallee 1960). In the second International Meeting on Metallothionein and other low Molecular Weight Metal-Binding Proteins held in 1985, however, the scope of MT was expanded to include metal-binding polypeptides reported in seven classes of eukaryotic algae investigated (Weber et al. 1987; Gekeler et al. 1988; Shaw et al. 1989). Metal-binding polypeptides can be designated metallothioneins when they share features common to horse kidney metallothionein, especially: (1) the presence of Cys-Xaa-Cys sequences, where Xaa is an amino acid other than Cys (Kagi and Kojima 1987), and (2) increased synthesis following exposure to elevated levels of certain heavy metals. These proteins are characterised by low molecular weight, high metal and cysteine (Cys) content, absence of aromatic amino acids and histidine, and abundance of Cys-Xaa-Cys.

Algae, plants and some fungi also produce a class of metal-binding peptides different from the animal metallothioneins (Steffens et al. 1986; Gekeler et al. 1989; Robinson 1989). These organisms synthesise a family of thiol-containing metal-binding peptides, designated phytochelatins (Steffens 1990). A cadmium-binding peptide from algae was first described by Stokes et al. (1977). Grill et al. (1985) discovered phytochelatins from *Rauvolfia surpentina* growing luxuriantly on a mine spoil. It was subsequently reported that the major Cd^{2+}-binding ligands in metal-stressed plants are composed of poly(γ-glutamylcysteinyl)-glycine (Grill et al. 1985; Bernard and Kagi 1985; Robinson et al. 1988). These polypeptides, as well as the other γ-glutamylisopeptides in which Gly is either absent or substituted with β-alanine, were given the name class III metallothionein or phytochelatin. These metal-binding peptides have been referred to by various names: those chemically related to glutathione (γ-Glu-Cys-Gly) have been called cadystin (Murasugi et al. 1981; Kondo et al. 1985), phytochelatin (Grill et al. 1985), γ-glutamyl-peptides (Reese et al. 1988) and poly(γ-glutamylcysteinyl)-glycine or $(\gamma\text{-EC})_n$-G (Robinson and Jackson 1986; Robinson et al. 1988). Likewise, the polypeptides related to homoglutathione (γ-Glu-Cys-β-Ala) have been given the name homophytochelatins (Grill et al. 1986). The term phytochelatin, as proposed by Grill et al. (1988), does not encompass structures containing Glu-Cys and Ser that are found in Cd-binding peptides from plants.

Taking recourse to the above and other characteristics, these metal-binding proteins and peptides have been classified as follows (Robinson 1989; De Filippis and Pallaghy 1994). We have slightly expanded the system to accommodate the following data:

Class I. (MT I) Polypeptides with locations of cysteine closely resembling those of horse kidney metallothionein, found in most vertebrates, observed in *Neurospora*, and *Agaricus biosporus*. (not reported in algae).
Class II. (MT II) Polypeptides with locations of cysteine only distantly related to those of horse kidney metallothionein (identified in the cyanobacteria *Anacystis nidulans*, *Synechococcus* PCC 7942, and *Synechococcus* TX-20 by Olafson et al. 1988).
Class III. (MT III or phytochelatins and homophytochelatins) Non-translationally synthesized peptides from higher plants, eukaryotic algae and certain fungi. Phytochelatins (PC) are chemically related to glutathione (γ-glu-cys-gly). Homophytochelatins (HPC) are related to homoglutathione (γ-glu-cys-βala).

MT I and MT II are proteins encoded by structural genes, whereas MT III is an aggregate of peptides not encoded by structural genes.

Biochemical analysis of polypeptides from phylogenetically divergent plants from Orchidales (the most advanced of higher plants) to the algae have elegantly demonstrated that phytochelatins contain three amino acids: Glu, Cys and Gly (Rauser 1990). These inducible metal-binding polypeptides, in sharp contrast to animal metallothionein, are non-protein polypeptides. They differ in structure and biosynthesis but depict analogy in function. The building block of metal-binding polypeptides is glutathione (GSH) with structure as γ-Glu-Cys-Gly. In some plants, such as soybean (*Glycine max*) the basic glutathione structure is replaced by homoglutathione (γ-Glu-Cys-β-Ala), but no such polypeptide has so far been recorded in algae (Rauser 1990). In glutathione, the peptide bond is formed between γ- or side-chain carboxylate of glutamic acid. In polypeptide, the α-carboxylate of the peptide bond is utilised and its synthesis is ribosomal in nature. The general structure for polypeptide may be given as $(\gamma\text{-Glu-Cys})_n$-Gly where n = 2-11. The molecular weight ranges from 2000 to 10 000 Da. Thus, the discovery of γ-carboxyamide linkage in polypeptides was the turning point for the classification of these polypeptides as class III MTs. Amino acid analysis of Cd-binding polypeptides from *Chlorella fusca*, *Euglena gracilis* and other eukaryotic algae (Gekeler et al. 1988; Shaw et al. 1989) precisely demonstrated a $(\gamma\text{-Glu-Cys})_n$-Gly structure for this peptide. Notwithstanding, Murasugi et al. (1981) were the first to characterise and assign the name cadystin to Cd-inducible peptides from *Saccharomyces pombe*. The major difference between phytochelatins and mammalian MT is the recurrence of γ-carboxyamide linkage in the peptide that does not require ribosomes for its synthesis. This compilation does not intend to provide a detailed survey of polypeptide structure from plants (for details readers may refer to Steffens 1990; Rauser 1990). Our effort is to provide current opinion on the physiological, biochemical and molecular strategies deployed by this protein in offering tolerance to heavy metals by algae. However, attempts have also been made to use the pertinent information from organisms belonging to other classes.

12.4.2 Occurrence of Metallothioneins and Phytochelatins in Algae and Cyanobacteria

During the past three decades, metal-binding proteins have been isolated and characterised from different organisms belonging to diverse groups (Table 12.2). All of these proteins have been found to sequester several metal ions, but depict a high affinity for Zn^{2+} and Cu^{2+}.

Table 12.2. A tentative list of algae showing occurrence of MT/PC. (Data from Rauser 1990; De Filippis and Pallaghy 1994; Ahner et al. 1995a,b; Rijstenbil and Wijnholds 1996; Mallick and Rai 1998; Rijstenbil et al. 1998a,b; Satoh et al. 1999)

Division	Species	Peptide	Structure
Chlorophyta	*Chlorella fusca*	PC	p ps
	Chlorella ellipsoida	PC	p
	Scenedesmus acutiformis	PC	p
	S. quadricauda	PC	p
	Enteromorpha	PC	p
	Chlamydomonas reinhardii	PC	p ps
	Monoraphidium minutum	PC	p
	Stichococcus bacillaris	PC	p
	Dunaliella bioculata	PC	p
	D. tertiolecta	PC	p
Prasinophyta	*Tetraselmis tetrathele*	PC	p
	T. maculata	PC	p
Euglenophyta	*Euglena gracilis*	PC	p ps
Bacillariophyta	*Fragilaria crotonensis*	PC	p
	Navicula pelliculosa	PC	p
	Thalassiosira weissflogii	PC	p
	T. pseudonana	PC	p
	T. oceanica	PC	p
	Ditylum brightwellii	PC	p
Chrysophyta	*Ochromonas danica*	PC	P
Phaeophyta	*Sargassum muticum*	PC	P
	Phaeodactylum tricornutum	PC	P
Xanthophyta	*Bumilleriopsis filiformis*	PC	P
	Vaucheria compacta	PC	P
	V. debaryana	PC	P
Rhodophyta	*Porphyrdium cruentum*	PC	
Prymnesiophyta	*Pleurochrysis carterae*	PC	
	Emiliania huxleyi	PC	
	Pavlova lutheri	PC	
Dinoflagellate	*Heterocapsa pygmaea*	PC	
Cyanophyta	*Synechococcus* TX 20	MT II	p ps
	Anacystis nidulans	MT II	p ps
	Anabaena doliolum	PC*	
	Synechococcus PCC 7942	MT II	p ps

p= primary. s= secondary. * Not characterised.

12.4.3 Phytochelatins in Metal Sequestration

Binding to different metals is the fundamental property of phytochelatins. The metal-binding polypeptides (γ-Glu-Cys)$_n$-Gly from a contingent of organisms have been isolated, purified and studied (Robinson et al. 1993), and are still being investigated. Biochemical analyses revealed that only Cd and Cu participate in the formation of the peptide-metal complex in vivo in higher plants and algae (Grill et al. 1987; Robinson 1989), although indirect evidence does suggest binding with Hg and Ag (Grill et al. 1987) and weak association with Zn (Robinson 1989).

Cd- and Cu-phytochelatin complexes absorb maximally in the range of 250 and 275 nm (Rauser 1990). Following acidification to pH 2, the absorption peaks are either reduced or eliminated. However, on back titration to alkaline pH the absorption peaks recover. The loss of absorbance correlates with the breaking of Cd-mercaptide bonds (Weber et al. 1987; Reese et al. 1988). The metal-binding of the polypeptides is determined by the pH and the number of peptides. The pH at which 50% of the metal is displaced from the complex provides an estimate of the binding affinity of metals. The Cd-binding complex from *E. gracilis* showed 50% displacement at pH 5.7 and Cu-binding complex from *Saccharomyces pombe* at pH 1.3. Further, the polypeptides with high polymerization (*n*) values bound Cd more intensely than those with low *n* values.

Cd-Phytochelatin contains acid-labile S as sulphide S^{2-}. Such an aggregate shows high affinity and capacity for Cd binding (Reese et al. 1988). With isolated phytochelatin from *E. gracilis*, Robinson (1989) further demonstrated that high Cd binding by the polypeptide was due to the presence of S^{2-} and not due to the high *n* value of the complex, because removal of S^{2-} from the dissociated Cd-phytochelatin decreased the Cd-binding capacity of the complex. Isolated phytochelatin from *E. gracilis* grown in high Zn-containing medium showed an extremely weak binding of the thiol complex with Zn.

12.4.4 Induction and Biosynthesis of Phytochelatins

Generally speaking, phytochelatins have never been isolated from cells without prior exposure to high concentrations of heavy metals. The occurrence of extremely low levels (1 pmol μg^{-1} protein) of (γ-Glu-Cys)$_2$–Gly and (γ-Glu-Cys)$_3$–Gly in tomato cells following transfer to fresh medium (Steffens et al. 1986) has been explained by the fact that the culture medium always contained Cu and Zn in traces. Phytochelatin biosynthesis can be induced by a variety of heavy metals such as Cd^{2+}, Ag^{+}, Bi^{3+}, Pb^{2+}, Zn^{2+}, Cu^{2+}, $Hg^{2+}$and Au^{2+} both in vivo and in vitro (Robinson 1989). Additionally, induction of phytochelatin synthesis has been observed in the marine diatom *Thalassiosira weissflogii* following exposure to Cd, Pb, Ni, Cu, Zn, Co, Ag and Hg at concentrations occuring in natural waters (Ahner and Morel 1995a,b). Amazingly, only Cd^{2+} and Cu^{2+} have been reported to activate phytochelatin production in vivo in higher plants (Grill et al. 1986,

1989). Analysis of polypeptides from *S. pombe* and the green algae *Chlorella fusca* and *Scenedesmus acutiformis* demonstrated that $(\gamma\text{-Glu-Cys})_2$-Gly and $(\gamma\text{-Glu-Cys})_3$-Gly were induced at high concentrations of Cd, Cu, Pb, Zn and Ag (Grill et al. 1986, 1987). All the above studies demonstrated that Cd is the most potent inducer of phytochelatin.

The structural similarity between glutathione (GSH) and phytochelatin prompted scientists to believe that biosynthesis of both the compounds might involve common enzymes. Glutathione biosynthesis involves two steps.

1. Glu + Cys ⟶ γ-glu-Cys.

2. γ-Glu-Cys + Gly ⟶ γ-Glu-Cys-Gly.

Step (1) is catalysed by the enzyme γ–glutamylcysteine synthetase, while step (2) is catalysed by glutathione synthetase. The enzyme γ–glutamylcysteine synthetase is inhibited by buthionine sulfoximine (BSO). The inhibition of phytochelatin biosynthesis by BSO suggests the involvement of this enzyme in phytochelatin biosynthesis in higher plants (Steffens et al. 1986; Grill et al. 1987). Cells treated with BSO fail to produce phytochelatin and become extremely sensitive to the metals used to induce phytochelatin synthesis (Rauser 1990). Further support for GSH involvement in phytochelatin biosynthesis came from the work done on *S. pombe*. Mutants of this organism deficient in the activity of either γ-glutamyl-cysteine synthetase or glutathione synthetase failed to produce phytochelatin (Mutoh and Hayashi 1988), indicating thereby that both the enzymes are involved in phytochelatin production. Grill and associates (1987), who worked on *Rauvolfia surpentina* cells in culture, found an inverse relation between glutathione consumption and phytochelatin production. Supplementation of 200 μM of $Cd(NO_3)_2$ to the culture medium not only induced the synthesis of phytochelatin, but also created a rapid decrease in the level of glutathione. The disappearance of glutathione was almost equal to the incorporation of γ-glutamyl-cysteine into phytochelatin and was unaffected by BSO. Similar results have been found with *S. pombe,* and seedlings, roots and shoots of maize (Steffens 1990). Plant cells supplemented with GSH as the sole source of sulphur take the molecule intact. In these cells, the GSH concentration was found to decrease to at least half of its original level within 2-3 h of exposure to Cd. Measurable synthesis of phytochelatin occurred within 5 min of Cd exposure. However, peptides with increasing n (n=2-11) values required respectively more time for biosynthesis and aggregation. Grill et al. (1986, 89) suggested that $(\gamma\text{-Glu-Cys})_n$-Gly biosynthesis occurs through a sequential addition of γ-Glu-Cys residue to glutathione. Experiments with (^{35}S) glutathione revealed maximum incorporation of radio-activity into the smaller polypeptide (n=2) and progressively lesser into the larger ones with $n > 3$.

Howe and Merchant (1992) demonstrated that phytochelatins in *Chlamydomonas reinhardtii* are synthesised via a posttranslationally activated metal-dependent enzymatic pathway from a precursor. This view has been supported by genetic evidence that yeast cells deficient in GSH biosynthesis fail to accumulate phytochelatins following exposure to metals (Mutoh and Hayashi 1988).

Analysis of the pathway of phytochelatin biosynthesis in *Silene cucubalus* led to the identification of an enzyme, phytochelatin synthase (Grill et al. 1989). This enzyme is known to possess metal-dependent γ-glutamyl-cysteine dipeptidyl transpeptidase activity responsible for in vivo phytochelatin biosynthesis. Phytochelatin synthase catalyses the transfer of γ-glytamyl-cysteine from GSH to another molecule of GSH, forming $(\gamma\text{-Glu-Cys})_2$-Gly. It also catalyses the stepwise addition of γ-glutamyl-cysteine from GSH to phytochelatin oligomers ranging from $(\gamma\text{-Glu-Cys})_2$-Gly to $(\gamma\text{-Glu-Cys})_5$-Gly. This enzyme has been purified and studied in detail (Steffens 1990). Available reports suggest that the biosynthesis of phytochelatins is insensitive to cyclohexamide, an inhibitor of de novo protein synthesis, in Cd-treated plants. This observation implies that the enzymes responsible for the synthesis of phytochelatins and their precursors are constitutive in the absence of high concentration of metals.

Environmental factors affecting phytochelatin production have not received due attention from researchers. Skowronski et al. (1998) demonstrated the increased production of PC in the light as compared to the dark and also at high (16 ^{0}C) as compared to low (4 ^{0}C) temperatures in *Vacucheria compacta* and *V. debaryana* (Xanthophyceae). Increased production at high temperature was related with the temperature optimum (35 ^{0}C) of phytochelatin synthase responsible for phyto-chelatin biosynthesis. However, decreased production in the dark, as well as at low temperature (4 ^{0}C), could also be due to decreased rates of other cellular reactions including the reduced transport of cadmium leading to its decreased availability for PC induction.

12.4.5 Role of Phytochelatins in Metal Detoxification and Tolerance

High concentrations of Cd, Cu, Zn, Ag, Au, Co, Ni and Bi can induce the synthesis of phytochelatin in plants, fungi, cyanobacteria, algae and animals (Robinson et al. 1993). Metal-induced biosynthesis of phytochelatins vis-à-vis its binding with metals argues in favour of its role in metal sequestration. Moreover trans-boundary results from algae (*Scenedesmus acutiformis* and *Chlorella fusa* for Cd, Pb, Zn, Ag, Cu and Hg), yeast (*Saccharomyces pombe* for Cu) and plants (*Datura innoxia* for Cd) provide an excellent correlation between increased production and accumulation of $(\gamma\text{-Glu-Cys})_n$-Gly and metal resistance by cells (Clements et al. 1999). Also, the synergistic inhibition of growth by Cd and BSO has been found to directly correlate with decreased production of $(\gamma\text{-Glu-Cys})_n$-Gly. *Anabaena doliolum* cells pretreated with BSO failed to survive at high concentrations of Cd, while untreated cells could grow and produce a low molecular weight Cd-binding protein (Mallick et al. 1994). The amounts of metal complexed by phytochelatin can be quite large. For example, Howe and

Merchant (1992) reported sequestration of approximately 70% of cytosolic Cd from *Chlamydomonas reinhardtii* by the polypeptide aggregates.

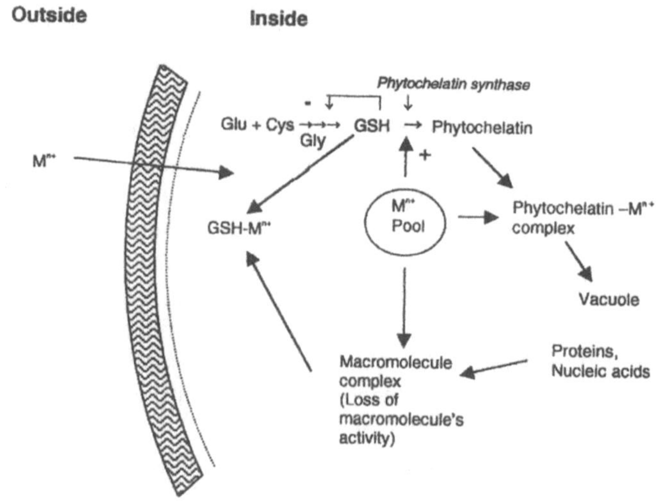

Fig. 12.2. Proposed model for phytochelatin synthesis, metal detoxification and homeostasis. Metals when spiked into cell suspension enter the cells and accumulate to form a metal pool (M^{n+}). Glutathione (GSH), present in the cell is utilised for the biosynthesis of PC. GSH concentration in the cell is regulated by feedback inhibition. If overproduced, GSH may directly bind with metals like Hg, Ag more efficiently and with Cd less efficiently (see Howe and Merchant 1992). Cytosolic metal catalyses formation of phytochelatin which, in turn, binds metals to from phytochelatin-metal complex. The metal-phytochelatin complex produced sequesters metals from the cytosol and detoxifies it, or safely transfers into the vacuole. A part of metals can also be compartmentalised by proteins, nucleic acids and enzymes. Metals from the pool may be used by apoenzymes and proteins for metabolic processes

Based on available data concerning the possible involvement of phytochelatins in metal detoxification, the entire sequence of events can be summarised (Fig. 12.2) into three steps: (1) induced synthesis of the enzyme phytochelatin synthase, consequent to intracellular accumulation of heavy metals, resulting in production of phytochelatins, (2) a decreased concentration of unbound metal ions in the cytosol through sequestration/chelation by phytochelatins, and (3) reduced metal toxicity because of the lower concentration of unbound metal. This process, in a way, offers tolerance to the cell because this step protects many functional attributes of the cell from the activity of metal ions. Vogeli-Lange and Wagner (1990) assign an additional task to metal chelation in the cytosol, whereby phytochelatins could function as (or interact with) a transporter of metals from the cytoplasm to a vacuole.

12.4.6 Cyanobacterial Metallothionein

An exciting development was the identification of two prokaryotes, viz. the cyanobacterium *Synechococcus* and the bacterium *Pseudomonas putida*, capable of metallothionein synthesis on exposure to heavy metals. Maclean et al. (1972) were the first to demonstrate the production of a MT-like protein and to correlate its role in resistance to heavy metals in *Anacystis nidulans.* Soon thereafter, class II MTs were isolated from *Synechococcus* PRIM NI (a marine cyanobacterium), *Synechococcus* UTEX-625 and *Synechococcus* TX-20 (freshwater strains), purified and biochemically characterised (see Turner and Robinson 1995). The metallothionein from *Synechococcus* is peculiar in the sense that its biosynthesis can be appreciably enhanced by Cd and Zn but not by Cu. Its metal thiolate cluster, although quite similar to the eukaryotic (double domain) MT, differs in having a single domain. The purified MT from *Synechococcus* is known to bind Cd^{2+} very profusely; but no such binding has been reported for Zn and Cu. In sharp contrast to *Synechococcus* MTs, Cu is the preferred inducer of MTs in yeast.

The MT gene from *Synechococcus* (*SmtA*) has been expressed in *E. coli* (Erbe et al. 1995). Recombinant proteins from the transformed *E. coli* cells grown at metal excess have been isolated and characterised (Erbe et al. 1995). These proteins possess a high affinity for Zn^{2+}, Cd^{2+}, Cu^{2+} and Hg^{2+}.

Robinson et al. (1990) used *Synechococcus* DNA as the template in the polymerase chain (PCR) reaction to generate a part of MT genes and to sequence them. A gene thus identified was given the name *smtA*. Huckle et al. (1993) used these fragments as probes to isolate the MT divergon (*smt)* which contains two genes: *smtA* and *smtB;* the latter is a divergently transcribed gene. The *smtA* DNA sequence predicts an amino acid sequence that is largely identical to polypeptides purified by Olafson et al. (1988) from *Synechococcus* sp. In *SmtA,* cysteine was substituted by serine and two additional amino acids (histidine and glycine) are present at the carboxy terminal end. A gene isolated from *Synechococcus vulcanes,* similar to *smtA,* was given the name *mtnA.*

Eukaryotes endowed with increased metal tolerance may possess several copies of MT genes and consequently a corresponding increase in the MT mRNA and protein (Palmiter 1987). In close parallel to metal tolerance in eukaryotes, Cd tolerance in *Synechococcus* PCC 6031 was found to coincide with increased production of MT (Olafson 1986). Olafson (1984, 1986) has proposed that Cd tolerance in *Synechococcus* could be due to amplification of extra chromosomal genes. Experimental support to this hypothesis was provided by Gupta et al. (1993), who successfully isolated *smtA* and reported at least four fold amplification and rearrangement of the *smt* locus in metal-resistant derivatives of *Synechococcus* PCC 6031.

12.4.7. Biosynthesis of Metallothioneins

The understanding of MT biosynthesis in *Synechococcus* has been greatly facilitated by molecular analysis of the genetic architecture and DNA sequence of the *smt* divergon (Robinson 1990; Turner and Robinson 1995). The MT locus consists of *smtA*, which encodes a class II MT, the divergently transcribed gene *smtB* (Huckle et al. 1993), and a 100-bp promoter-operator region between the *smtA* and *smtB* coding regions. *SmtB* encodes a *trans*-acting repressor of *smtA* and is proposed to exert metal ion-inducible negative control over *smtA* transcription. An electrophoretic mobility shift assay using either —10 recombinant *smtB* expressed in *E. coli* or the protein extract from *Synechococcus* PCC 6301 demonstrated concentration-dependent formation of three specific complexes with the *smtB* operator-promoter. *SmtB* was found to bind directly with Zn and to inhibit repressor–DNA complex formation (Erbe et al. 1995). The 100-bp promoter-operator region contains divergent promoters similar to *E. coli* –10 promoter consensus sequences (Turner et al. 1995). The polypeptide sequences deduced for *smtB* resemble a number of bacterial proteins (Morby et al. 1993). Some of the *SmtB*-like polypeptides are known to be involved in the regulation of transcription and metal metabolism.

Turner and Robinson (1995) reported that the *smtA* transcripts increased markedly after exposure of *Synechococcus* PCC 6301 to elevated concentrations of Cd^{2+}, Zn^{2+}, Cu^{2+}, Hg^{2+}, Co^{2+} or Ni^{2+}. Of the different metals used, only Zn showed appreciable induction of expression of the reporter gene (lacZ). Cd and Cu showed a very mild induction. This expression was driven by the operator-promoter region of *smtA*. In addition, Northern analysis of nucleic acid from *Synechococcus* PCC 6301 treated with high concentrations of Cd showed an increased number of *smtA* transcripts. However, it failed to show any detectable Cd effect on the stability of the transcripts.

Some information has also accumulated on the interaction of protein with the operator-promoter region of *smt.* An analysis of 100 bp of the *smt* operator-promoter region using electrophoretic mobility shift assay clearly indicated the existence of three protein-DNA complexes, MAC1, MAC2 and MAC3. MAC1 is situated upstream of the ATG of *smtA* and has been studied to some extent (Turner and Robinson 1995). Not much is known about the other two proteins in cyanobacteria.

Olafson et al. (1980) observed a perceptible lag in the growth of *Synechococcus* sp. exposed to Cd^{2+}. The onset of growth coincided with the accumulation of MT. They put forward a proposal to amplify the MT gene to confirm the role of Cd^{2+} in inducing MT accumulation. Subsequently, Gupta et al. (1992) demonstrated amplification of *smtA* in *Synechococcus* PCC 6301 cells destined to exhibit Cd^{2+} resistance; these Cd^{2+}-tolerant cells showed specific rearrangements within *smt*.

The information generated so far provided several lines of evidence for the involvement of cyanobacterial *Smt* in Zn^{2+} homeostasis. Turner et al. (1993)

demonstrated that mutants lacking a functional metallothionein locus (*smt*) were hypersensitive to Zn^{2+}. Recombinant *E. coli* with *smt* from *Synechococcus* has been found to possess a tremendously increased affinity for Zn^{2+} (Shi et al. 1992). Expression from the *smtA* operator–promoter is maximally induced by Zn^{2+} (Turner and Robinson 1995).

12.4.8 Functions of Metallothioneins

A host of functions have been assigned to animal MT, including: Zn^{2+} homeostasis, regulation of gene expression, programmed expression during development, redistribution of Zn^{2+} from the cytosol to the nucleus, and in vitro Zn^{2+} transfer between transcription factors (Turner and Robinson 1995), biomonitoring of low levels of metals from the environment (Erbe et al. 1996) and offering manifold resistance to Cd (Ren et al. 1998; Chen et al. 1999). However, Zn^{2+}-requiring transcriptional factors have not been characterised in prokaryotes (Turner and Robinson 1995). Since Zn^{2+} is a trace nutrient, *smt* may also play some role in maintaining an optimum concentration of this metal in the cell. Nonetheless, it is quite plausible that under the circumstances where *smt* is absent, some other proteins may substitute for S*mt*. As yet it is not known if S*mt* would be required for the survival of organisms in oligotrophic habitats deficient in heavy metals. It remains to be discovered what mechanisms *Smt* uses for conferring tolerance to Zn^{2+} and Cd^{2+}. The S*mtA* may be speculated to act as metal sink, as is known for eukaryotes. A more dynamic role for *smtA* could be its involvement in the transfer of metals to efflux systems in metal-stressed cells.

12.5 Other Intracellular Sites for Metal Complexation

Under phosphorus surplus conditions, many cyanobacteria and eukaryotic microalgae take up concentrations of phosphorus far exceeding their metabolic requirements and sequester it in polyphosphate bodies. Polyphosphates are polymers of orthophosphate that can be mobilised by the organism for its metabolic needs if it is exposed to phosphorus deficiency. Increased phosphorus availability has often been found to mitigate toxicity of metal ions to algae (Rai et al. 1981b). Such a response has been related to the polyanionic character of polyphosphates that allows attraction of electrophilic cations to the electron-rich oxygen atoms of polyphosphates (Twiss and Nalewajko 1992). Polyphosphates also serve as an osmotically inert sink for metallic cations, such as K^+, Na^+, Ca^{2+} and Mg^{2+}, which are required for metabolism (Pick et al. 1986). Many of these essential cations that are complexed to polyphosphate bodies can be lost upon exposure of algae to high concentrations of heavy metals (Daniel and Chamberlain 1981; Jensen et al. 1982a; Rachlin et al. 1984). X-ray energy-dispersive analysis has shown that polyphosphates can accumulate several heavy metals: Pb in *Diatoma tenue* (Sicko-Goad and Stoermer 1979) and *Navicula inerta* (Jensen et al. 1982b); Cd, Cu, Co, Hg, Ni, Pb and Zn in *Plectonema boryanum* (Jensen et al. 1982a); Al in *Anabaena cylindrica* (Petersson et al. 1985); Ti in *Anacystis nidulans* (Crang and Jensen 1975). Using the same technique for intracellular localization of metals on polyphosphate, Wong et al. (1994) demonstrated elevated concentrations of Al, Fe, Cu and Zn in *Chlorella* exposed to a municipal effluent containing Al, Mn, Fe, Ni, Cu and Zn. Some exceptions to these observations are available in the literature. For instance, the Cu-tolerant *Anabaena variabilis* did not show Cu accumulation in polyphosphate bodies, although the number of polyphosphate bodies was increased after exposure to high concentration of Cu (Hashemi et al. 1994). Twiss and Nalewajo (1992) reasoned that polyphosphate plays a passive role in protecting cells of a Cu-sensitive strain of *Scenedesmus acutus* from Cu stress. However, Cu sequestering by polyphosphate in two tolerant strains of this green alga was relatively unimportant.

Sequestration of metals in polyphosphate bodies does not end the problem of metal toxicity for the organism accumulating it. If the organism experiences phosphorus deficiency, the metal-containing polyphosphate bodies can be mobilised to release orthophosphate for metabolic demands. Concomitantly, metal ions would also be released from the polyphosphate body, thereby stressing the organism.

The presence of intranuclear inclusions in a Cu-tolerant *Scenedesmus* led Silverberg et al. (1976) to conclude that the formation of these inclusions caused lowering of cytoplasmic concentrations of Cu^{2+} eventually leading to reduced toxicity of Cu^{2+} to the organism. Organic acids, like malic acid and citric acid, may also chelate heavy metals inside the cell; however, these strategies of metal detoxification have been reported in higher plants only (see Ernst et al. 1992).

12.6 Role of Enzymes and Metabolites in Counteracting Metal-Induced Damage

Algae have several mechanisms for counterbalancing the damage caused by metals. However, the damage once caused by a metal cannot be completely rectified, as Mallick and Rai (1999) recently pointed out. Most heavy metals, particularly transitition metals like Cu and Fe, instigate oxidative stress in algae (Vangronsveld and Clijsters 1994). This may be induced indirectly after depletion of thiol pools due to binding of metals (Zn, Cd, Pb, Ni, Hg) to them; the scarcity of thiols lowers the antioxidant level in the cell, thus making it vulnerable to oxidative stress. Glutathione (GSH) is presumably the major intracellular metal-complexing thiol in algae at low concentrations of metals, but phytochelatins become important at concentrations of metals causing acute toxicity (Rijstenbil et al. 1994b; Rijstenbil and Wijnholds 1996).

Heavy metals can also generate harmful oxyradicals, thereby directly provoking oxidative stress in the cell (Vangronsveld and Clijsters 1994). High concentrations of metal-induced reactive species of oxygen disturb the pro- vs. antioxidant balance and enhance lipid peroxidation followed by membrane disruption which shortens the lifetime of algal cells (Rijstenbil et al. 1994a). Various reductants and oxygen radical scavengers may be crucial in protecting cells from metal-induced oxidative stress (Vangronsveld and Clijsters 1994). Rijstenbil et al. (1994b) noticed a higher total SH and reduced+oxidized glutathione (GSH+GSSG) contents in *Thalassiosira pseudonana* than in *Ditylum brighwellii*. Consequently, the former diatom was more tolerant to Cu than the latter. Many enzymes (superoxide dismutase, catalase and peroxidase) can function as efficient quenchers of reactive intermediary forms of oxygen. Several workers observed an increased activity of SOD in algal species exposed to heavy metals (Rijstenbil et al. 1994a; Mallick and Rai 1999). Sauser et al. (1999) noticed increased peroxidase activity in *Selenastrum capricornutum* by some metals. The relative sensitivity to various metals at their 100-μM concentrations followed the following hierarchy Cu>Co>Pb>Ni. Twenty μM of Cu^{2+} could bring about a significant increase in peroxidase activity; other metals could not do so at this concentration. It is quite obvious that antioxidant systems of algae play a crucial role in counteracting metal toxicity, however, more efforts should be made to show that metal-tolerant strains have a more efficient antioxidant system compared to metal-sensitive strains. All kinds of environmental stresses cause oxidative damage in algae (see Mehta and Gaur 1999), and it would be worthwhile to thoroughly examine if tolerance mechanisms against oxidative damage are common for all kinds of stresses, including high concentrations of toxic metals.

Heavy metal-stressed algae and cyanobacteria tend to rapidly accumulate high concentrations of free proline in their cells (Wu et al. 1995, 1998; Mehta and Gaur 1999). Mehta and Gaur (1999) noticed that the greater the toxicity of a metal, the greater was the intracellular concentration of proline in *Chlorella vulgaris*. They further showed a positive relationship between intracellular levels of proline and heavy metal concentrations, and revealed a definite role of proline in protecting this green alga from metal toxicity. Wu et al. (1995) showed that Cu-induced loss of K^+ was reduced by proline, and thus suggested that proline perhaps protected the plasma membrane from metal toxicity. Subsequently, Wu et al. (1998) demonstrated a reduced Cu uptake by cells containing high concentrations of proline. Mehta and Gaur (1999), however, could not detect any reduced metal uptake due to buildup of proline concentration in algal cells. These latter authors suggested that intracellular proline counteracted metal-induced lipid peroxidation and K^+ efflux. Mehta and Gaur (1999), however, were not sure as to the manner in which intracellular proline displayed a protective effect; complexation of metals and reduction in production or scavenging of harmful oxyradicals by proline may be probable mechanisms.

12.7 Interrelation Between Metal and Acid Tolerance

Increased acidification of aquatic ecosystems due to acid rain and disposal of industrial wastes may simultaneously expose algae to metal and acid stresses, because many metals are more soluble at low pH. Laboratory studies on acidophiles and acid-tolerant strains of algae have demonstrated their tolerance to toxic metals as welt. A number of physiological and biochemical features enable these organisms to concomitantly tolerate acid and metal stresses (see Chap. 9.9 for the details of coupling between acid and metal tolerance).

Acknowledgements. We are grateful to Dr. Nirupma Mallick, ICG-6 Forschungszentrum, Juelich, Germany (presently at the Indian Institute of Technology, Kharagpur), for providing valuable literature on metal-binding proteins, metal toxicity and tolerance, and to Professor Dorothy L. Parker, Department of Biology, Scripps Institution of Oceanography, La Jolla, San Diego, California, USA, for thoughtful comments on the manuscript. L. C. Rai is thankful to the Ministry of Environment and Forest, Govt. of India, for the Pitamber Pant National Environmental Fellowship Award, and to NSF, USA and DST, India, for financial support through an Indo-US Project.

References

Agrawal M, Kumar HD (1975) Response of *Chlorella* to mercury pollution. Ind J Ecol 21:94-98

Ahner BA, Kong S, Morel FMM (1995a) Phytochelatin production in marine algae. 1. An interspecies comparison. Limnol Oceanogr 40:649-657

Ahner BA, Kong S, Morel FMM (1995b) Phytochelatin production in marine algae. 2. Induction by various metals. Limnol Oceanogr 40:658-665

Arceneaux JE, Boutwell ME, Byers BR (1984) Enhancement of copper toxicity by siderophores in *Bacillus megaterium*. Antimicrob Agents Chemother 25:650-652

Asthana RK, Singh AL, Singh SP (1993) Comparison of Ni-sensitive and N-resistant strains of *Nostoc muscorum*. World J Microbiol Biotechnol 9:323-327

Bariaud A, Bury M, Mestre JC (1985) Mechanism of Cd^{2+} resistance in *Euglena gracilis*. Physiol Plant 63:382-386

Bernhard W, Kagi JHR (1985) Purification and partial characterisation of atypical cadmium-binding polypeptides from *Zea mays*. In: Abstr 2nd Int Meet on Metallothionein and Other Low Molecular Weight Metal-Binding Protein. Zurich, Switzerland, 25 pp

Chen Z, Ren Q, Shi D, Ru B (1999) Expression of mammalian metallothionein–I gene in cyanobacteria to enhance heavy metal resistance. Mar Pollut Bull 39:155-158

Clarke SE, Stuart J, Sanders-Loehr J (1987) Induction of siderophore activity in *Anabaena* spp. and its moderation of copper toxicity. Appl Environ Microbiol 53:917-922

Clemens S, Kim EJ, Neumann D, Schroeder JI (1999) Tolerance to toxic metals by a gene family of phytochelatin synthases from plants and yeast. EMBO J 18:3325-3333

Crang RE, Jensen TE (1975) Incorporation of titanium in polyphosphate bodies of *Anacystis nidulans*. J Cell Biol 67:80a

Crist RH, Oberholser K, Schwartz D, Marzoff J, Ruder D, Crist DR (1988) Interaction of metals and protons with algae. Environ Sci Technol 22:755-760

Daniel GF, Chamberlain AHL (1981) Copper immobilization in fouling diatoms. Bot Mar 24:229-243

De Filippis LF (1978) The effect of sublethal concentrations of mercury and zinc on *Chlorella*. IV. Characteristics of a general reducing enzyme system for metallic ions. Z Pflanzenphysiol 86:339-352

De Filippis LF, Pallaghy CK (1994) Heavy metals: Sources and biological effects. In: Rai LC, Gaur JP, Soeder CJ (eds) Algae and water pollution. E. Schweizerbart'sche Verlagsbuchhandlung, Stuttgart, pp 31-77

De Philippis RD, Vincenzini M (1998) Exocellular polysaccharides from cyanobacteria and their possible applications. FEMS Microbiol Rev 22:151-175

Eichenberger E (1986) The interaction between essentiality and toxicity of metals in the aquatic ecosystem. In: Metal ions in biological systems, vol 20. Concepts on metal ion toxicity. Marcel Dekker, New York, pp 67-100

Erbe JL, Taylor KB, Hall LM (1995) Metalloregulation of the cyanobacterial *smt* locus: identification of Smt B binding sites and direct interaction with metals. Nucleic Acids Res 23: 2472-2478

Erbe JL, Adams AC, Taylor KB, Hall LM (1996a) Cyanobacteria carrying a *smt-lux* transcriptional fusion as biosensors for the detection of heavy metal cations. J Ind Microbiol 17:80-83

Erbe JL, Taylor KB, Hall LM (1996b) Expression of mouse metallothionein in the cyanobacterium *Synechococcus* PCC7942. J Ind Microbiol 17:41-46

Ernst WHO, Verkleij JAC, Schat H (1992) Metal tolerance in plants. Acta Bot Neerl 41:229-249

Fiore MF, Trevors JT (1994) Cell composition and metal tolerance in cyanobacteria. BioMetals 7:83-103

Florence TM, Stauber JL (1986) Toxicity of copper complexes to the marine diatom *Nitzschia closterium*. Aquat Toxicol 8:11-26

Fogg GE, Westlake DF (1955) The importance of extracellular products of algae in freshwater. Proc Int Assoc Theor Appl Limnol 12:2119-2132

Foster PL (1977) Copper exclusion as a mechanism of heavy metal tolerance in a green alga. Nature 269:322-323

Foster PL (1982a) Species associations and metal contents of algae from rivers polluted by heavy metals. Freshwater Biol 12:17-39

Foster PL (1982b) Metal resistance of Chlorophyta from rivers polluted by heavy metals. Freshwater Biol 12:41-61

Gekeler W, Grill E, Winnacker E-L, Zenk MH (1988) Algae sequester heavy metals via synthesis of phytochelatin complexes. Arch Microbiol 150:197-202

Gekeler W, Grill E, Winnacker E-L, Zenk MH (1989) Survey of the plant kingdom for their ability to bind heavy metals through phytochelatins. Z Naturforsch 44:361-369

Genter RB (1996) Ecotoxicology of inorganic chemical stress on algae. In: Stevenson RJ, Bothwell ML, Lowe RL (eds) Algal ecology — freshwater benthic ecosystems. Academic Press, California, pp 403-468

Genter RB, Cherry DS, Smith EP, Cairns J Jr (1987) Algal-periphyton population and community changes from zinc stress in stream mesocosms. Hydrobiologia 153:261-275

Gerringa LJA, Rijstenbil JW, Poorvleit TCW, van Drie J, Schot MC (1995) Speciation of copper and responses of the marine diatom *Ditylum brightwellii* upon increasing copper concentrations. Aquat Toxicol 31:77-90

Gledhill M, Nimmo M, Hill SJ, Brown MT (1999) The release of copper-compexing ligands by the brown alga *Fucus vesiculosus* (Phaeophyceae) in response to increasing total copper levels. J Phycol 35:501-509

Grill E, Winnacker E-L, Zenk MH (1985) Phytochelatins: the principal heavy-metal complexing peptides of higher plants. Science 230:674-676

Grill E, Gekeler W, Winnacker E-L, Zenk MH (1986) Homophytochelatins are heavy metal-binding peptides of homo-glutathione containing Fabales. FEBS Lett 205:47-50

Grill E, Winnacker E-L, Zenk MH (1987) Phytochelatins: a class of heavy-metal-binding peptides from plants, are functionally analogous to metallothioneins. Proc Natl Acad Sci USA 84:439-443

Grill E, Thumann J, Winnacker E-L, Zenk MH (1988) Induction of heavy-metal-binding phytochelatins by inoculation of cell cultures in standard media. Plant Cell Rep 7:375-378

Grill E, Löffler S, Winnacker E-L, Zenk MH (1989) Phytochelatins, the heavy-metal-binding peptides of plants, are synthesised from glutathione by a specific γ–glutamylcysteine dipeptidyl transpeptidase (phytochelatin synthase). Proc Natl Acad Sci USA 86:6838-6842

Gupta A, Whitton BA, Morby AP, Huckle JW, Robinson NJ (1992) Amplification and rearrangement of a prokaryotic metallothionein locus *smt* in *Synechococcus* PCC 6301 selected for tolerance to cadmium. Proc R Soc Lond B 248:73-281

Gupta A, Morby AP, Turner JS, Whitton BA, Robinson NJ (1993) Deletion within the metallothionein locus of Cd-tolerant *Synechococcus* PCC 6301 involving a highly iterated palindrome (HIP1). Mol Microbiol 7:189-195

Hall A (1980) Heavy metal cotolerance in a copper tolerant population of the marine fouling alga *Ectocarpus siliculosus* (Dillw.) Lyngbye. New Phytol 85:73-78

Hall A, Fielding AH, Butler M (1979) Mechanisms of copper tolerance in the marine fouling alga *Ectocarpus siliculosus* — evidence for an exclusion mechanism. Mar Biol 54:195-199

Harding JPC, Whitton BA (1976) Resistance of *Stigeoclonium tenue* in the field and the laboratory. Br Phycol J 11:417-426

Hashemi F, Leppard GG, Kushner DJ (1994) Copper resistance in *Anabaena variabilis*: Effects of phosphate nutrition and polyphosphate bodies. Microb Ecol 27:1159-1176

Howe G, Merchant S (1992) Heavy metal-activated synthesis of peptides in *Chlamydomonas reinhardtii*. Plant Physiol 98:127-136

Huckle JW, Morbey AP, Turner JS, Robinson NJ (1993) Isolation of a prokaryotic metallothionein locus and analysis of transcriptional control by trace metal ions. Mol Microbiol 7:177-187

Jensen TE, Baxter M, Rachlin JW, Jani V (1982a) Uptake of heavy metals by *Plectonema boryanum* (Cyanophyceae) into cellular components, especially polyphosphate bodies: an X-ray energy dispersive study. Environ Pollut Ser A Ecol Biol 27:119-127

Jensen TE, Rachlin JW, Jani V, Warkentine BE (1982b) An X-ray dispersive study of cellular compartmentalization of lead and zinc in *Chlorella saccharophila* (Chlorophyta), *Navicula incerta* and *Nitzschia closterium* (Bacillariophyta). Environ Exp Bot 22:319-328

Jin X, Kushner DJ, Nalewajko C (1996) Nickel uptake and release in nickel-resistant and -sensitive strains of *Scenedesmus acutus* f. *alternans* (Chlorophyceae). Environ Exp Bot 36:401-411

Kagi JHR, Kojima Y (1987) Chemistry and biochemistry of metallothionein. In: Kagi JHR, Kojima Y (eds) Metallothionein II. Birkhäuser, Basel, pp 25-61

Kagi JHR, Vallee BL (1960) Metallothionein: a cadmium-and zinc containing protein from equine renal cortex. J Biol Chem 235:3460-3465

Kessler E (1985) An extremely cadmium sensitive strain of *Chlorella*. Experientia 41:1621

Knauer K, Behra R, Sigg L (1997) Adsorption and uptake of copper by the green alga *Scenedesmus subspicatus* (Chlorophyta). J Phycol 33:596-601

Kondo N, Isobe M, Imai K, Goto T (1985) Synthesis of metallothionein-like peptides cadystin A and B occurring in fission yeast, and their isomers. Agric Biol Chem 49:71-83

Kuwabara JS, Leland HV (1986) Adaptation of *Selenastrum capricornutum* (Chloropyceae) to copper. Environ Toxicol Chem 5:197-203

Maclean FI, Lucis OJ, Shaikh ZA, Mansz ER (1972) The uptake and subcellular distribution of Cd and Zn in microorganisms. Fed Proc 31:699

Mallick N, Rai LC (1998) Characterisation of Cd-induced low molecular weight protein in a N_2-fixing cyanobacterium *Anabaena doliolum* with special reference to Co-/ multiple tolerance. Biometals 11:55-61

Mallick N, Rai LC (1999) Respose of the antioxidant system of the nitrogen fixing cyanobacterium *Anabaena doliolum* to copper. J Plant Physiol 155:146-149

Mallick N, Pandey S, Rai LC (1994) Involvement of a low molecular weight cadmium binding protein in regulating Cd toxicity in *Anabaena doliolum* . Biometals 7:71-83

Margoshes M, Vallee BL (1957) A cadmium protein from equine kidney cortex. J Am Chem Soc 79:4813-4814

McKnight DM, Morel FMM (1979) Release of weak and strong copper-complexing agents by algae. Limnol Oceanogr 24:823-837

Mehta SK, Gaur JP (1999) Heavy-metal-induced proline accumulation and its role in ameliorating metal toxicity in *Chlorella vulgaris*. New Phytol 143:253-259

Mehta SK, Tripathi BN, Gaur JP (2000) Influence of pH, culture age and cations on adsorption and uptake of Ni by *Chlorella vulgaris*. Eur J Protistol 36:443-450

Morby AP, Turner JS, Huckle JW, Robinson NJ (1993) *Smt*B is a metal regulated repressor of the cyanobacterial metallothionein gene *smt*A: identification of a Zn inhibited DNA-protein complex. Nucleic Acids Res 21:921-925

Morelli E, Scarano G, Ganni M, Nannicini L, Serriti A (1989) Copper binding ability of the extracellular organic material released by *Skeletonema costatum*. Chem Spec Bioavail 1:71-76

Murasugi A, Wada C, Hayashi Y (1981) Purification and unique properties in UV and Cd spectra of Cd-binding peptide 1 from *Schizosaccharomyces pombe*. Biochem Biophys Res Commun 103:1021-1028

Mutoh N, Hayashi Y (1988) Isolation of mutants of *Schizosaccharomyces pombe* unable to synthesize cadystin, small cadmium-binding peptides. Biochem Biophys Res Commun 151:32-39

Nakagawa M, Takamura Y, Yagi O (1986) Isolation and characterization of the slime from a cyanobacterium *Microcystis aeruginosa* R-3A. J Agric Biol Chem 1:329-337

Olafson RW (1984) Prokaryotic metallothionein. Int J Peptide Protein Res 24:303-308

Olafson RW (1986) Physiological and chemical characterization of cyanobacterial metallothionein. Environ Health Perspect 65:71-75

Olafson RW, Loya S, Sim RG (1980) Physiological parameters of prokaryotic metallothionein induction. Biochem Biophys Res Commun 95:1495-1503

Olafson RW, McCubbin WD, Kay CM (1988) Primary and secondary-structural analysis of a unique prokaryotic metallothionein from a *Synechococcus* sp. cyanobacterium. Biochem J 251:691-699

Palmiter RD (1987) Molecular biology of metallothionein gene expression. Experientia Suppl 52:63-80

Parker DL, Schram BR, Plude JR, Moore RE (1996) Effect of metal cations on the viscosity of a pectin-like capsular polysaccharide from the cyanobacterium *Microcystis flos-aquae* C3-40. Appl Environ Microbiol 62:1208-1213

Parker DL, Rai LC, Mallick N, Rai PK, Kumar HD (1998) Effect of cellular metabolism and viability on metal ion accumulation by cultured biomass from a bloom of the cyanobacterium *Microcystis aeruginosa*. Appl Environ Microbiol 64:1545-1547

Pettersson A, Kunst L, Bergman B, Roomans GM (1985) Accumulation of aluminium by *Anabaena cylindrica* into polyphosphate granules and cell walls: an X-ray energy-dispersive microanalysis. J Gen Microbiol 131:2545-2548

Pick U, Ben Amotz A, Karni L, Seebergts CJ, Avron M (1986) Partial characterization of K and Ca uptake systems in the halotolerant alga *Dunaliella salina*. Plant Physiol 81:875-881

Plude JL, Parker DL, Schommer OJ, Timmerman, RJ, Hagstrom A, Joers JM, Hnasko R (1991) Chemical characterization of polysaccharide from the slime layer of the cyanobacterium *Microcystis flos-aquae*. Appl Environ Microbiol 57:1696-1700

Pradhan S, Singh S, Rai LC, Parker DL (1998) Evaluation of metal sorption efficiency of laboratory-grown *Microcystis* under various environmental conditions. J Microbiol Biotechnol 8:53-60

Rachlin JW, Jensen TE, Warkentine BE (1984) The toxicological response of the alga *Anabaena flos-aquae* (Cyanophyceae) to cadmium. Arch Environ Contam Toxicol 13:143-151

Rai LC, Gaur JP, Kumar HD (1981a) Phycology and heavy metal pollution. Biol Rev 56:99-151

Rai LC, Gaur JP, Kumar HD (1981b) Protective effects of certain environmental factors on the toxicity of zinc, mercury and methylmercury to *Chlorella vulgaris* Beij. Environ Res 25:250-259

Rai LC, Mallick N, Singh JB, Kumar HD (1991) Physiological and biochemical characteristics of a copper tolerant and a wild-type strain of *Anabaena doliolum*. J Plant Physiol 138:68-74

Rauser WE (1990) Phytochelatins. Annu Rev Biochem 59: 61-86

Reed RH, Gadd GM (1990) Metal tolerance in eukaryotic and prokaryotic algae. In: Shaw AJ (ed) Heavy metal tolerance in plants: evolutionary aspects. CRC Press, Boca Raton, pp 1105-1118

Reese RN, Mehra RK, Tarbet EB, Winge DR (1988) Studies on the γ-glutamyl Cu-binding peptide from *Schizosaccharomyces pombe*. J Biol Chem 263: 4186-4192

Ren L, Shi D, Dai J, Ru B (1998) Expression of mouse metallthionein-I gene conferring cadmium resistance in a transgenic cyanobacterium.FEMS Microbiol Lett 158: 127-132

Rijstenbil JW, Wijnholds JA (1996) HPLC analysis of non-protein thiols in planktonic diatoms: pool size, redox state and response to copper and cadmium exposure. Mar Biol 127:45-54

Rijstenbil JW, Derksen JWM, Gerringa LJA, Poortvliet TCW, Sandee A, van den Berg M, van Drie J, Wijnholds JA (1994a) Oxidative stress induced by copper:defense and damage in the marine planktonic diatom *Ditylum brightwellii* grown in continuous cultures with high and low zinc levels. Mar Biol 119:583-590

Rijstenbil JW, Sandee A, van Drie J, Wijnholds JA (1994b) Interaction of toxic trace metals and mechanisms of detoxification in the planktonic diatoms *Ditylum brightwellii* and *Thalassiosira pseudonana*. FEMS Microbiol Rev 14:387-396

Rijstenbil JW, Dehairs F, Ehrlich R, Wijnholds JA (1998a) Effect of the nitrogen status on copper accumulation and pools of metal-binding peptides in the planktonic diatom *Thalassiosira pseudonana*. Aquat Toxicol 42:187-209

Rijstenbil JW, Haritonidis S, Malea P, Seferlis M, Wijnholds JA (1998b) Thiol pools and glutathione redox ratio as possible indicators of copper toxicity in the green macroalgae *Enteromorpha* spp. from the Scheldt Estuary (SW) Netherlands, Belgium and Thermaikos Gulf (Greece, N Aegean Sea). Hydrobiologia 385:171-181

Robinson MG, Brown LN (1991) Copper complexation during bloom of *Gymnodinium sanguineum* Hiraska (Dinophyceae) measured by ASV. Mar Chem 33:105-118

Robinson NJ (1989) Algal metallothioneins: secondary metabolites and proteins. J Appl Phycol 1:5-18

Robinson NJ, Jackson PJ (1986) Metallothionein-like metal complexes in angiosperms: their structure and function. Physiol Plant 67:499-506

Robinson NJ, Ratliff RL, Anderson PJ, Delhaize E, Berger JM, Jackson PJ (1988) Biosynthesis of poly(gamma-glutamylcysteinyl)-glycine in cadmium-resistant *Datura innoxia* cells. Plant Sci 56:197-204

Robinson NJ, Gupta A, Fordham-Skelton AP, Croy RRD, Whitton BA, Huckle JW (1990) Prokaryotic metallothionein gene characterisation and expression: chromosome crawling by ligation mediated PCR. Proc R Soc Lond B 242:241-247

Robinson NJ, Tommey AM, Kuske C, Jackson PJ (1993) Plant metallothioneins. Biochem J 295:1-10

Satoh M, Karaki E, Kakehashi M, Okazaki E, Gotoh T, Oyama Y (1999) Heavy-metal induced changes in non-proteinaceous thiol levels and heavy-metal-binding peptide in *Tetraselmis tetrathele* (Prasinophyceae). J Phycol 35:989-994

Sauser KR, Liu JK, Wong T-Y (1999) Identification of a copper-sensitive ascorbate peroxidase in the unicellular green alga *Selenastrum capricornutum*. Biometals 10:163-168

Say PJ, Diaz BM, Whitton BA (1977) Influence of zinc on lotic plants. 1. Tolerance of *Hormidium* sp. to zinc. Freshwater Biol 7:357-376

Shaw CF, Petering DH, Weber DN, Gingrich DJ (1989) Inorganic studies of the cadmium-binding peptides from *Euglena gracilis*. In: Winge D, Hamer D (eds) Metal ion homeostasis. Alan R Liss, New York

Shehata FHA, Whitton BA (1982) Zinc tolerance in strains of the blue-green alga *Anacystis nidulans*. Br Phycol J 17:5-12

Shi J, Lindsay WP, Huckle JW, Morby AP, Robinson NJ (1992) Cyanobacterial metallothionein gene expressed in *Escherichia coli*-metal-binding properties of the expressed protein. FEBS Lett 303:159-163

Sicko-Goad LM, Stoermer EF (1979) A morphometric study of lead and copper effects on *Diatoma tenue* var. *elongatum* (Bacillariophyta). J Phycol 15:316-321

Silver S, Phung LT (1996) Bacterial heavy metal resistance: new surprises. Annu Rev Microbiol 50:753-789

Silverberg BA, Stokes PM, Ferstenberg LB (1976) Intranuclear complexes in copper tolerant algae. J Cell Biol 69:210-214

Singh S, Pradhan S, Rai LC (1998) Comparative assessment of Fe^{3+} and Cu^{2+} biosorption by field and laboratory-grown *Microcystis*. Process Biochem 33:495-504

Singh S, Pradhan S, Rai LC (2000) Metal removal from single and multimetallic systems by different biosorbent materials as evaluated by differential pulse anodic stripping voltammetry. Process Biochem 36:175-182

Singh SP, Yadava V (1986) Cadmium tolerance in the cyanobacterium *Anacystis nidulans*. Biol Zentralbl 105:539-542

Skowronski T, De Knecht JA, Simons J, Verkleij JAC (1998) Phytochelatin synthesis in response to cadmium uptake in *Vaucheria* (Xanthophyceae). Eur J Phycol 33:87-91

Starodub ME, Wong PTS, Mayfield CI, Chau YK (1987) Influence of complexation and pH on individual and combined heavy metal toxicities to a freshwater green alga. Can J Fish Aquat Sci 44:1173-1180

Steffens JC (1990) The heavy metal-binding peptides of plants. Annu Rev Plant Physiol Plant Mol Biol 41:553-575

Steffens JC, Hunt DF, Williams BG (1986) Accumulation of non-protein metal-binding polypeptides (gamma-glutamyl-cysteinyl)$_n$-glycine in selected cadmium-resistant tomato cells. J Biol Chem 261:13879-13882

Stokes PM (1983) Responses on freshwater algae to metals. In: Round FE, Chapman DJ (eds) Advances in phycological research, vol II. Elsevier, Amsterdam, pp 87-112

Stokes PM, Maler T, Riordan JR (1977) A low molecular weight copper-binding protein in a copper tolerant strain of *Scenedesmus acutiformis*. In: Hemphil DD (ed) Trace substances in environmental health. University of Missouri Press, Columbia, pp 146-154

Takamura N, Kasai F, Watanabe MM (1989) Effects of Cu, Cd and Zn on photosysnthesis of freshwater benthic algae. J Appl Phycol 1:39-52

Tease BC, Walker R (1987) Comparative composition of the sheath of the cyanobacterium *Gloeothece* ATCC 27512 cultured with and without combined nitrogen. J Gen Microbiol 133:3331-3339

Tillberg JE, Barnard T, Rowley JR (1984) Phosphorus status and cytoplasmic structure in *Scenedesmus* (Chlorophyceae) under different metabolic regimes. J Phycol 20:124-136

Ting YP, Teo WK, Soh CY (1995) Gold uptake by *Chlorella vulgaris*. J Appl Phycol 7:97-100

Turner JS, Robinson NJ (1995) Cyanobacterial metallothioneins: biochemistry and molecular genetics. J Ind Microbiol 14:119-125

Turner JS, Morby AP, Whitton BA, Gupta A, Robinson NJ (1993) Construction and characterisation of Zn^{2+}/Cd^{2+} hypersensitive cyanobacterial mutants lacking a functional metallothionein locus. J Biol Chem 268:4494-4498

Twiss MR, Nalewajko C (1992) Influence of phosphorus nutrition on copper toxicity to three strains of *Scenedesmus acutus* (Chlorophyceae). J Phycol 28:291-298

Twiss MR, Welbourn PM, Schwärtzel E (1993) Laboratory selection of copper tolerance in *Scenedesmus acutus* (Chlorophyceae). Can J Bot 71:333-338

Vangronsveld J, Clijsters H (1994) Toxic effects of metals. In: Farago ME (ed) Plants and the chemical elements: biochemistry, uptake, tolerance and toxicity. VCH Verlagsgesellschaft, Weinheim, pp 149-177

Verma SK, Singh HN (1991) Evidence for energy-dependent copper efflux as a mechanism of Cu^{2+} resistance in the cyanobacterium *Nostoc calcicola*. FEMS Microbiol Lett 84:291-294

Vogeli-Lange R, Wagner GJ (1990) Subcellular localization of cadmium and cadmium-binding peptides in tobacco leaves. Plant Physiol 92:1086-1093

Wagatsuma T, Akiba R (1989) Low surface negativity of root protoplasts from aluminum-tolerant species. Soil Sci Plant Nutrition 35:443-452

Weber DN, Shaw CF, Petering DH (1987) *Euglena gracilis* cadmium-binding protein-II contains sulfide ion. J Biol Chem 262:6962-6964

Weckesser J, Hoffmann K, Jurgens UJ, Whitton BA (1988) Isolation and chemical analysis of the sheaths of the filamentous cyanobacteria *Calothrix parietina* and *C. scopulorum*. J Gen Microbiol 134:629-634

Whitton BA (1970) Toxicity of heavy metals to Chlorophyta from running waters. Arch Mikrobiol 72:353-360

Whitton BA (1980) Zinc and plants in rivers and streams. In: Nriagu JO (ed) Zinc in the environment, Part II Health effects. John Wiley, New York, pp 364-400

Whitton BA, Say PJ (1975) Heavy metals. In: Whitton BA (ed) River ecology. Blackwell, Oxford, pp 286-311

Winter C, Winter M, Pohl P (1994) Cadmium adsorption by non-living biomass of the semi-macroscopic alga, *Ectocarpus siliculosus*, grown in axenic mass culture and localisation of the adsorbed Cd by transmission electron microscopy. J Appl Phycol 6:479-487

Wong SL, Nakamoto L, Wainwright JF (1994) Identification of toxic metals in affected algal cells in assay of wastewaters. J Appl Phycol 6:405-414

Woolhouse HW (1983) Toxicity and tolerance in the response of plants to metals. In: Lange O, Nobel PS, Osmond CB, Ziegler H (eds) Physiological plant ecology. III Responses to the chemical and biological environment. Springer, Berlin Heidelberg New York, pp 125-300

Wu J-T, Chang S-C, Chen K-S (1995) Enhancement of intracellular proline level in cells of *Anacystis nidulans* (Cyanobacteria) exposed to deleterious concentrations of copper. J Phycol 31:376-379

Wu J-T., Hsieh M-T, Kow L-C (1998) Role of proline accumulation in response to toxic copper in *Chlorella* sp. (Chlorophyceae). J Phycol 34:113-117

Xue H-B, Sigg L (1990) Binding of Cu (II) to algae in a metal buffer. Water Res 24:1129-1136

Yoshimura E, Nagasaka S, Sato Y, Satake K, Mori S (1999) Extraordinary high aluminium tolerance of the acidophilic alga, *Cyanidium caldarium*. Soil Sci Plant Nutr 45:721-724

13 Eukaryotic Algae, Cyanobacteria and Pesticides

F. Koenig[*]

Contents

[*] Fachbereich 2: Biologie/Chemie, Molekulare Pflanzenphysiologie, Universität Bremen, Postfach 33 04 40, 28334, Bremen, Germany

13.1 Introduction

Pesticides are of equal importance for the farmer for controlling weeds, undesirable plants in monocultures of crop plants (cf. Fedtke 1982), and for the plant physiologists trying to understand their mode of action on one hand and, in addition, using pesticides of known specificity and mode of action as tools in the elucidation of physiological mechanisms on the other (cf. Trebst 1980, 1987).

Unwanted side effects of agricultural application are, of course, to be expected on non-target animal and plant life, for example crop plants and algae in the aquatic ecosystems adjacent to areas subject to agricultural activities: toxicity on one hand and the appearance of herbicide-resistant species on the other (Hirschberg and McIntosh 1983; Hirschberg 1984; Berard et al. 1999; Kasai 1999; Nelson et al. 1999).

Typical tests required for environmental acceptability of pesticides make use of some species of fish, of *Daphnia* and of several species of algae (see Cobb 1992; Cerejeira et al. 1998; Nitschke et al. 1999; Girling et al. 2000). One of the very commonly used test organisms is the unicellular green alga *Chlorella*, some species of which have recently been renamed *Scenedesmus* (Kessler et al. 1997). An algal growth and reproduction inhibition test with *Scenedesmus vacuolatus* (formerly *Chlorella fusca* var. *vacuolata* Shihita et Krauss) has been developed by Grimme and coworkers. This organism was employed to estimate the combined effects of pesticides and surfactants in aquatic toxicology (Altenburger et al. 1990), to study herbicide bioconcentration in algae on the basis of lipophilicity – sorption – activity relationship (LSAR) in toxikokinetics (Manthey et al. 1993) and to follow the phase-I biotransformation capacity of green algae for cytochrome P450-interacting plant growth retardents, fungicides and related compounds (Thies and Grimme 1996). Using synchronized algal cell populations under highly standardized conditions this bioassay has revealed its suitability for hazard assessments of multiple chemical mixtures even at no observed effect concentrations (NOEC) of single toxicants (Faust et al. 2000).

Different herbicides are distinguished on the basis of their site and mode of action (cf. Fedtke 1982; Böger 1983; Böger and Sandmann 1998b): for example. inhibitors interfering with electron transport of photosynthesis (Trebst 1980, 1987; Böger and Sandmann 1998b), inhibitors of carotenoid biosynthesis (Feierabend 1983; Lambert and Böger 1983; Böger and Sandmann 1998a,b), growth regulators like auxin-type herbicides, for example 2,4-D (see Fedtke 1982; Cobb 1992), inhibitors of fatty acid formation (Böger and Sandmann 1998b; Schmalfuß et al. 1998) and compounds affecting tetrapyrrole biosynthesis, thus causing peroxidation processes in the light, consequently called peroxidizing herbicides (Wakayabashi et al. 1988; Nicolaus et al. 1989; Böger and Wakabayashi 1995; Wakabayashi and Böger 1995; Böger and Sandmann 1998b). Cell lethality in the light due to massive protoporphyrin IX accumulation upon inhibition of protoporphyrinogen oxidase by the *N*-phenyl heterocyclic herbicide S-23142 has been observed in the green alga *Chlamydomonas* (Randolph-Anderson et al.

1998). Moreover, inhibitors of the biosynthetic pathway of aromatic amino acids are also powerful herbicides (Steinrucken and Amrhein 1980; see Böger and Sandmann 1998a); these are active not only in higher plants (cf. Amrhein and Roy 1993) but also in algae, as shown for *Euglena* (Wong 2000).

Even a natural algicide, cyanobacterin, has been isolated from the filamentous cyanobacterium *Scytonema hofmanni* and characterized. Cyanobacterin is known to disrupt the thylakoid membrane structure in *Euglena* (Gleason 1990) and inhibits photosynthetic electron transport in other cyanobacteria, such as *Synechococcus*. The activity of the algicide was described to be similar to that of the DCMU-type inhibitors (Gleason and Paulson 1984).

The most prominent target site for pesticides is the photosynthetic apparatus of plants, algae and cyanobacteria; more precisely, the thylakoid membrane. Different inhibitor classes are distinguished due to their site and mode of action (Trebst 1980; Trebst and Draber 1986).

Very well investigated are the so-called DCMU-type inhibitors for their mode of action. They include urea and triazine herbicides such as diuron [3-(3,4-dichlorophenyl)-1,1-dimethylurea], fenuron [N,N-dimethyl-N-phenylurea], atrazine [2-chloro-4-(ethylamino)-6-(isopropylamino)-*s*-trazine], a herbicide used to control broadleaf weeds in corn, simazine [4,6-bis(ethylamino)-2-chloro-*s*-triazine], terbutryn [2-(*tert*-butylamino)-4-(ethylamino)-6-methyl-thio-*s*-triazine], metribuzine [4-amino-6-(1,1-isopropyl)-3-methyl-thio-1,2,4-triazin--5-one] and phenisopham [isopropyl-N-[3-(N-ethyl-N-phenyl-carbamoyloxy)-phenyl]-carbamate], a substituted carbamate, all of which bind to the photosystem II reaction center(RCII), thereby blocking electron transport by replacing the secondary plastoquinone electron acceptor Q_B from the binding pocket of the RCII-D1 protein (Trebst 1980; Trebst and Draber 1986). Comparative binding studies were made for photosystem II herbicides with isolated thylakoids as well as with intact green algae (Laasch et al. 1981, 1982). Also the inhibition of photosynthetic electron transport in cyanobacterial thylakoids, isolated from *Oscillatoria chalybea* (Schmid and Lehmann-Kirk 1977) and *Aphanocapsa* PCC 6803 (Allen et al. 1983) as well as in whole cells (Eley et al. 1983; Koenig 1987b), has been reported.

For the cormophytic crop plants ample literature is available on suitable protective herbicides (see Fedtke 1982; Cobb 1992). Much less impressive is the information on pesticide action on the so-called algae. Those organisms are only affected that commonly dwell in the aquatic ecosystems which are integrally connected to adjacent landscapes and subject to agricultural activities and the accompanying negative impacts on water quality. Over half the length of all the rivers in the United States is moderately to severely impacted by agriculture (Nelson et al. 1999; Scribner et al. 2000; Thurman et al. 2000). Selected genera of algae, viz., *Chlamydomonas*, *Scenedesmus*, *Euglena* and *Chlorella*, are used as test organisms in toxicological studies (see Cobb 1992; Girling et al. 2000). Most scientific information comes, however, from researchers trying to understand plant physiology on one hand and the mode of action of the respective pesticides on the other (cf. Cobb 1992). Moreover, designers of new, more specific and more effective herbicides have contributed tremendously to our knowledge (Trebst and Draber 1986).

13.2 Growth in the Presence of Sublethal Concentrations of DCMU-Type Inhibitors

While lethal doses have been determined for many herbicides under standardized growth conditions also for cyanobacterial cultures (Koenig 1987a,b), even sublethal concentrations of these inhibitors were observed to have an effect on the phenotype of cyanobacteria and algae, as well as on the higher plants.

In principle, the observed effect was found to be equal in all oxygenic organisms tested. Growth in the presence of sublethal concentrations of DCMU-type inhibitors was shown to induce the so-called greening effect in higher plants. The regulatory responses to partial inhibition of photosynthesis led to shade-type chloroplasts in the absence of shade, chloroplasts with a low ratio of chlorophyll a to chlorophyll b and huge grana stacks. Moreover, growth stimulation and delayed senescence have been observed in higher plants under these conditions, herbicides thus simulating a hormone-like effect on the cell division. Nevertheless, hormone-like effects were demonstrated for quite a number of DCMU-type inhibitors (see Fedtke 1982). In 1982 Fedtke summarized the then available data on regulatory responses of plants after partial inhibition of photosynthesis. The so-called greening effect was described as a superposition of growth stimulation, also in non-photosynthetic tissues, and of the induction of shade-type chloroplasts in green tissue. An adaptive reorganization of protein and lipid components in chloroplast membranes as associated with herbicide binding in the aquatic angiosperm *Spirodela* was described by Mattoo et al. (1984). It could be shown that prolonged partial inhibition of photosystem II in the cyanobacterium *Synechococcus* led to alterations in the excitation energy distribution (Srivastava et al. 1994). Tanaka et al. (1977) observed filaments of the green alga *Spirogyra*, treated with diuron, to behave, concerning their mobility, in the same way as if exposed to low light or shade. Growth of cyanobacteria in the presence of sublethal concentrations of DCMU-type inhibitors was observed to induce shade phenotype in strong light (cf. Fig. 13.2, Sect. 13.3; Koenig 1987a,b; 1990c, 1992; Hatfield et al. 1989).

13.3 Atrazine Action on Cyanobacteria

It appears from the literature that the best-investigated example of herbicide action is the exposure of cyanobacteria to DCMU-type inhibitors like the triazine herbicide atrazine, resulting in physiological alterations. Early reports on the inhibition of cyanobacterial photosystem II by this group of inhibitors does not appear exciting any longer, as their mode of action is thoroughly understood. Further, taking into account the background of the similarity of the photosystem II reaction centres (Ohad and Hirschberg 1990), it is not astonishing at all (Schmid and Lehmann-Kirk 1977; Codd and Cossar 1978; Allen et al. 1983).

13.3.1 Cell Division

Astonishing is, however, the fact that with increasing atrazine concentration during growth under otherwise standardized growth conditions, the cell density of *Anacystis nidulans* (*Synechococcus* sp. PCC 6301, SAG 1402-1, UTEX 625) cultures does not gradually decrease as expected, but instead transiently increases (Fig. 13.1), before the cells finally die in the presence of 10^{-5} M atrazine. The highest cell density was observed after 6 days' growth in the presence of atrazine concentrations between 10^{-7} and 10^{-6} M (cf. Koenig 1995). This striking result is most likely the explanation for an earlier observation, not so well understood at the time of the publication: a transient increase in the light-induced O_2 evolution of the cultures depending on the atrazine concentration present during growth. From the very beginning, the evolution of oxygen was observed to increase compared to the higher, as well as lower concentrations of atrazine present during growth (Koenig 1987a). The apparent enhancement of cell division (Fig. 13.1) may be regarded as a part of the greening effect described for higher plants upon treatment with sublethal concentrations of DCMU-type inhibitors (Fedtke 1982). A corresponding observation has recently been made with the green alga *Scenedesmus quadricauda* for two other herbicides at sublethal concentrations, viz., for glyphosate, an inhibitor of aromatic amino acid biosynthesis (cf. Böger and Sandmann 1998a), and for 2,4-D which has for a long time been known to be a growth regulator (Wong 2000).

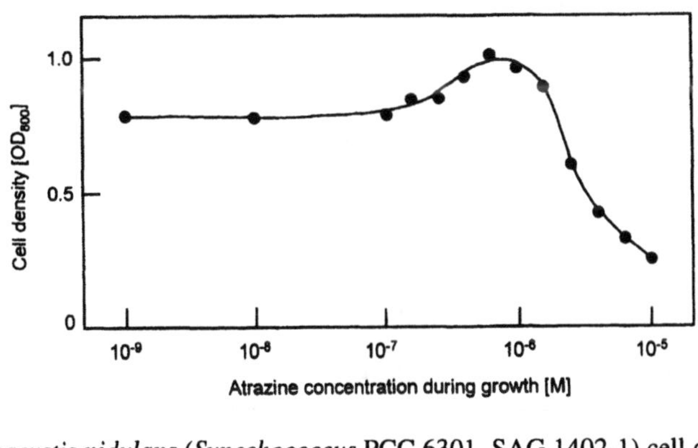

Fig. 13.1. *Anacystis nidulans* (*Synechococcus* PCC 6301, SAG 1402-1) cell density after 6 days' growth in continuous white light of 110 µE $m^{-2}s^{-1}$ at 32 °C. Atrazine was added at the onset of the experiment. (Koenig 1995)

13.3.2 Pigmentation

The second aspect of the greening effect could also be observed in cyanobacteria. It could be shown that besides death upon growth in the presence of lethal concentrations of DCMU-type inhibitors (see above) and enhancement of cell division as described in Section 13.3.1 (Fig. 13.1), some characteristic physiological responses related to growth in the presence of sublethal concentrations occurred in some cyanobacteria, viz., in *Anacystis nidulans* (*Synechococcus* PCC 6301, SAG 1402-1, UTEX 625) (Koenig 1987a,b, 1990c, 1992; Hatfield et al. 1989) and in a green alga, *Chlamydomonas* y-1 (Koenig 1987a). Growth in the presence of sublethal concentrations of all DCMU-type inhibitors tested resulted in the induction of shade phenotype, even under light conditions which, in the absence of herbicide, allowed for the development of sun phenotype (Koenig 1987a,b; 1990c, 1992; Hatfield et al. 1989). Just as in the naturally induced shade phenotype, the molar ratio of phycocyanin to chlorophyll was increased relative to the sun phenotype (Fig. 13.2). The shade character induced by growth in the presence of sublethal concentrations of DCMU-type inhibitors, here 10^{-6} M atrazine, was shown to increase with increase in light intensity for a given temperature and with increasing temperature for a given light intensity (Koenig 1987b). The data are summarized in Table 13.1.

Table 13.1. Molar ratio of phycocyanin to chlorophyll in *Anacystis* (*Synechococcus*) wild type grown in the presence of 10^{-6} M atrazine as a function of quantum flux density and temperature during growth

Quantum flux density (μmol $m^{-2}s^{-1}$)	22 °C	32 °C
5	0.181	0.275
10	0.193	0.281
20	0.220	0.285
24	0.230	0.303
42	0.270	0.315
56	0.285	Not determined
73	0.307	Not determined
110	0.323	Not determined

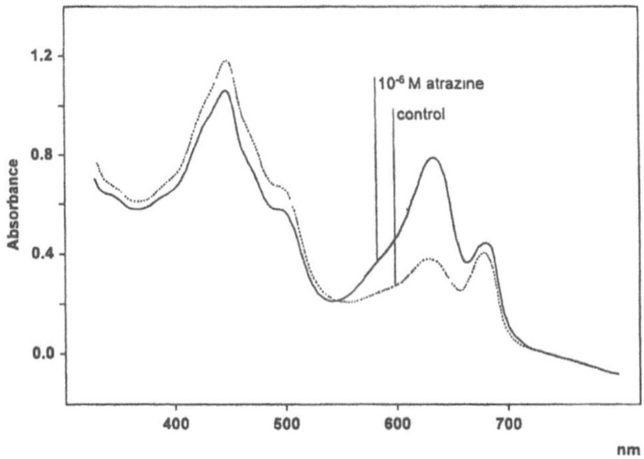

Fig. 13.2. Comparison of absorption spectra of *Anacystis nidulans* (*Synechococcus* PCC 6301, SAG 1402-1) wild type grown in the absence of inhibitor (*dashed line*) and in the presence of 10^{-6} M atrazine (*solid line*) at 22 °C in bright white light of 110 µmol m^{-2} s^{-1}. (Koenig 1987a,b)

13.3.3 Electron Transport

Light-saturated O_2 evolution was also similar for the naturally and "artificially" shade-adapted *Anacystis* (Koenig 1990c). Photosystem I electron transport, measured as Mehler reaction with methylviologen as the electron acceptor, was likewise comparable in naturally and artificially shade-adapted *Synechococcus* (Koenig 1987a).

13.3.4 RC II-D1 Turnover

The turnover of the photosystem (PS) II reaction centre (RC) protein D1 (RC II-D1) was found to be retarded in the artificially shade-adapted cells relative to that in the sun phenotype (Koenig 1992), just as in naturally shade-adapted cells.

There are early reports on DCMU-type inhibitors slowing down the RC II-D1 degradation not only in higher plants but also in cyanobacteria (Goloubinoff et al. 1988; Gong and Ohad 1991; Koenig 1992; Jansen et al. 1993), and similar observations have been made ever since (Komenda and Masojídek 1998; Sippola 2000). It was also shown that DCMU-type inhibitors had a protective role against photoinhibition (Critchley and Russell 1994) by slowing down RC II-D1 turnover (Aro et al. 1993; Komenda and Masojídek 1998). For *Synechococcus* PCC 6301 the half lifetime of the RC II-D1 protein in strong white light increased from 60 to 129 min upon addition of atrazine to the cultures (cf. Koenig 1992).

The level of total mRNA for the RC II-D1 protein was found to double in the presence of DCMU concentrations which reduced the growth rate by 60% in *Synechococcus* (Brusslan and Haselkorn 1989). Of the two functionally different forms of the RC II-D1 protein in *Synechococcus* (Clarke et al. 1993; Campbell et al. 1995, 1996), D1:1 and D1:2, DCMU was recently shown to prevent the upregulation of the mRNA for D1:2 and the downregulation of the mRNA for D1:1 in strong light (Sippola and Aro 1999, 2000), this being the normal adjustment to high light conditions. At the protein level, no exchange of D1 protein forms has been observed so far. RC II-D1:1 was earlier shown to prevail in shade-adapted cells while protein D1:2 was expressed preferentially at high light intensity (Clarke et al. 1993; Schaefer and Golden 1989a,b; Golden 1994; Kulkarni and Golden 1994). A recent article reports a general effect of DCMU-type inhibitors on the translation machinery in *Synechococcus* (Schmitz et al. 1999), thus indicating that no conclusions should be drawn from the respective experiments; but this general recommendation appears to be an overstatement. It has been unequivocally shown (Koenig 1995) that the presence of sublethal concentrations of DCMU-type inhibitors during growth of *Synechococcus* had a selective effect on the synthesis of the RC II-D1 protein, relative to total protein synthesis (Koenig and Samson 2001). In vivo labelling experiments with ^{35}S sulfate clearly demonstrate that translation continues, even at high inhibitor concentrations, and, in addition, they show that the RCII-D1 protein, formerly also called the DCMU-binding protein, is preferentially labelled. The incorporation of ^{35}S into this protein increases with increasing atrazine concentration in the medium. In the presence of 10^{-5} M atrazine 30% of the total ^{35}S incorporated into proteins was detected in the RC II-D1 protein (Koenig 1995).

13.4 Phenolic Inhibitors of Photosynthetic Electron Transport

The phenolic inhibitors of photosynthetic electron transport like dinoseb [2-(1-methylpropyl)-4,6-dinitrophenol] or bromoxynil (4-hydroxy-3,5-dibromobenzonitrile) were found to be of a lower efficiency than DCMU-type inhibitors with respect to their influence on the turnover of the RCII-D1 protein in *Spirodela* (Jansen et al. 1989). This result is consistent with the observation that, contrary to DCMU-type inhibitors, sublethal concentrations of phenolic inhibitors did not induce shade phenotype in *Synechococcus* (Koenig 1987 a,b). A wide spectrum of inhibitors was tested: bromonitrothymol (2-bromo-4-nitrothymol), dinoseb (2,4-dinitro-6-isobutylphenol), DNOC (dinitro-*o*-cresol), DNP-INT (2,4-dinitrophenylether of 2-iodo-4-nitrothymol and iodonitrothymol (2-iodo-4-nitrothymol). Komenda and Masojídek (1998) could not reproduce the results presented by Jansen and coworkers (1989) for *Synechococcus*. In their experiments, the phenolic inhibitor BNT (2-bromo-3-methyl-6-isopropyl-4-nitrophenol) blocked the turnover of the "low-light" form of the RC II-D1 protein D1:1 just as DCMU did (Komenda and Masojídek 1998).

13.5 Herbicide-Resistant Mutants

After the first appearance of higher plants tolerant to atrazine, numerous mutants resistant or tolerant to DCMU-type inhibitors have been generated or selected, as well as characterized in higher plants (see Gasquez et al. 1982; Hirschberg et al. 1984; Howat 1987; Botterman and Leemans 1988; Sato et al. 1988; Havaux 1989a, b; Shigematsu et al. 1989), algae and cyanobacteria (Astier et al. 1979, 1986, 1993; Erickson et al. 1984, 1985, 1989; Pucheu et al. 1984; Golden and Haselkorn 1985; Grizeau et al. 1985; Troton et al. 1986; Hirschberg et al. 1987a,b; Johanningmeier and Hallick 1987; Ohad et al. 1987; Robinson et al. 1987; Brusslan and Haselkorn 1988; Kirilovsky et al. 1988, 1989; Gibson and Barrett 1989; Etienne et al. 1990; Koenig 1990a, b, 1992; Ohad et al. 1990; Trebst 1991; Aiach 1992; Heiss and Johanningmeier 1992; Ohad and Hirschberg 1992; Prášil et al. 1992; Srivastava et al. 1994; Thuillier-Bruston et al. 1996; Oettmeier 1999; Johanningmeier et al. 2000). As in higher plants (Hirschberg and McIntosh 1983; Hirschberg 1984) and in green algae (Erickson et al. 1984, 1985, 1989; Aiach 1992; Thuillier-Bruston et al. 1996), the mutation conferring DCMU-type herbicide resistance was shown to map within the *psbA* gene; this was also found in cyanobacteria (Golden and Haselkorn 1985; Brusslan and Haselkorn 1988). The amino acid exchanges leading to DCMU-type inhibitor tolerance are located in the binding pocket for the secondary plastoquinone electron acceptor Q_B of photosystem II on the RC II-D1 protein in positions 255 and 264, position 255 carrying a phenylalanine and position 264 a serine in *Synechococcus* wild type (see Haworth and Steinback 1987; Ohad and Hirschberg 1992; Oettmeier 1999). Resistance to the photosystem II herbicide was shown to be dominant to sensitivity in *Synechococcus* (Brusslan and Haselkorn 1989). The influence of these mutations on the binding affinities of different classes of herbicides is discussed in a recent review by Oettmeier (1999). This review gives an unsurpassed summary of herbicide-resistant mutants in higher plants, algae and cyanobacteria.

The mutants tolerant to the phenolic inhibitors were also obtained in green algae (Trebst et al. 1993) and in cyanobacteria (Creuzet et al. 1989; Perewoska et al. 1992; Elanskaya et al. 1998, for further references see Oettmeier 1999). Moreover, mutants resistant to the bleaching herbicide norflurazon were described for *Synechococcus* (Chamovitz et al. 1991) and others tolerating porphyric herbicides in *Chlamydomonas reinhardtii* (Randolph-Anderson et al. 1998). Even an *Anacystis* mutant tolerant to the natural algicide cyanobacterin was characterized and shown to be still sensitive to DCMU (Mallipudi and Gleason 1989), while a DCMU-resistant mutant was found sensitive to cyanobacterin (Gleason et al. 1986). Mutants tolerant to glyphosate have been described for higher plants (Shah et al. 1986).

Just as chloroplasts of triazine-resistant weeds and those from leaves grown in shade were characterized by ultrastructural alterations (Vaughn and Duke 1984), cyanobacteria with different degrees of tolerance to DCMU-type inhibitors showed differences in pigmentation, mainly in the molar ratio of phycocyanin to chlorophyll (Fig. 13.3). Cyanobacteria not only depicted changes in pigment but alterations in thylakoid lipid composition were also associated with tolerance to DCMU-type herbicides, as shown for *Euglena* (Troton et al. 1986) and also for higher plants (Chapman et al. 1985).

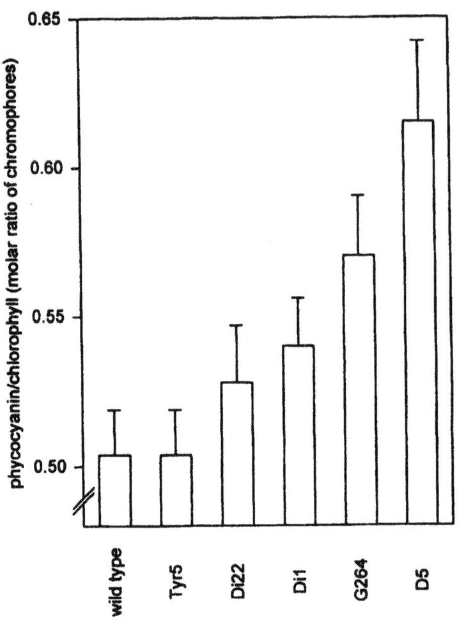

Fig. 13.3. Comparison of *Anacystis nidulans* R2 (*Synechococcus* sp PCC 7942) wild-type and herbicide-tolerant mutants Tyr 5, Di22, Di1, G264 and D5 (cf. Table 13.2) with respect to the ratio of phycocyanin to chlorophyll (Koenig 1990a,b). The molar ratio of chromophores is given on the ordinate. The organism was grown at 32 °C and white light (110 µmol $m^{-2}s^{-1}$)

All *Synechococcus* mutants investigated had higher phycocyanin to chlorophyll ratios, a longer antenna (Fig. 13.3) than the wild type when grown under standardized conditions (Koenig 1990a,b) and also depending on site and mode of the amino acid exchanges in the RC II-D1 protein. The characteristics of investigated mutants are summarised in Table 13.2 with respect to their amino acid exchanges and their relative tolerances to DCMU-type inhibitors (cf. Koenig 1990b). A method predicting the effect of a specific mutation in the RC II-D1 protein was developed by Horovitz et al. (1989).

Table 13.2. Comparison of wild-type *Anacystis* R2 (PCC 7942) with herbicide-tolerant mutants obtained via exchange of amino acids in positions 255 and 264 of the RC II-D1 protein. (Hirschberg et al. 1987a,b; Ohad et al. 1987; Horovitz et al. 1989)

Parameters		Herbicide-tolerant mutants					Wild type
		D5	G264	Di1	Di22	Tyr	
Amino acid in position	255	Tyr	Phe	Phe	Leu	Tyr	Phe
	264	Ala	Gly	Ala	Ala	Ser	Ser
Relative resistance to							
Diuron		167.0	10.0	100.0	1500.0	1.5	
Metribuzine		2000.0		5000.0	175.0	0.8	
Atrazine		200.0	2000.0	30.0	1.3	25.0	
Terbutryn		670.0	450.0	17.0	1.3	13.0	

In close resemblance with this, photosynthetic electron transport was shown to decrease (Prášil et al. 1992) with increasing shade characteristic of the mutants, the latter concluded from increased antenna size, measured as the molar ratio of phycocyanin to chlorophyll (Koenig 1990a,b). Apparently, the photosystem II reaction centre altered by site-directed mutagenesis, point mutations in the Q_B - binding pocket (cf. Table 13.2) at positions 255 and 264 in the RC II-D1 protein, is not so stable as is the wild-type reaction centre (Ohad et al. 1990).

As demonstrated in Table 13.3, the RC II-D1 protein was found to have an accelerated turnover in most of the mutants compared to the wild type (255-Phe, 264-Ser) (Ohad et al. 1990, 1992; Koenig 1992). While the half lifetime in the wild type was 74 min under standardized growth conditions, it was only 49 min for the mutant D (255-Tyr, 264 Ala) and as short as 41 min for the mutant G264 (255-Phe, 264 Gly). The half lifetime of the RC II-D1 protein in the mutant Tyr5 (255-Tyr, 264 Ser) was determined to be 66 min, that of mutant Di22 (255 Leu, 264-Ala) 67 min, very similar to the wild type (Koenig 1992).

The rate of RC II-D1 protein synthesis was determined as ^{35}S incorporation into the respective protein in percent of the total incorporated label into cell proteins (Koenig 1990b). Cells were allowed to take up ^{35}S methionine. The same sequence of mutants was found for the RC II-D1 protein resynthesis as for its degradation (cf. Table 13.3): these mutants showed the highest resynthesis rates of the RC II-

D1 protein (Koenig 1990b) which had the most unstable PS II reaction centre as concluded from the half liftime of the RC II-D1 protein (Ohad et al. 1990, 1992; Koenig 1992).

Table 13.3. Comparison of *Anacystis* R2 (*Synechococcus* PCC 7942) wild-type and herbicide-tolerant mutants with respect to the rate of the RC II-D1 protein synthesis, the rate of the RC II-D1 protein breakdown and with respect to the phycocyanin (PC) to chlorophyll (Chl) ratio, when grown in air without additional CO_2 at 32 °C in strong white light (110 µmol m^{-2} s^{-1} PAR). Cultures carefully normalized to equal cell density were compared

	PC/Chl (molar ratio)	RCII-D1 half life-time (min)	RCII-D1 synthesis in 18 min (arbitrary units)
Wild type	0.504	74	16.1
Tyr5	0.504	66	11.8
Di22	0.528	67	17.5
Di1	0.540	56	20.7
G264	0.570	41	28.6
D5	0.615	49	38.7

For some of the mutants, bicarbonate binding by photosystem II was also found altered (Govindjee et al. 1991; Cao et al. 1992).

Acknowledgements. F. Koenig would like to thank L.H. Grimme for sharing information on his toxicological test system and D. Mangels, as well as G. Lützenkirchen, for technical help with the figures.

References

Aiach A (1992) A herbicide-resistant *Euglena* mutant carrying a Ser to Thr substitution at position 265 in the D1 protein in photosystem II. Z Naturforsch 47c:245-248

Allen MM, Turnburke AC, Lagace EA, Steinback KE (1983) Effects of photosystem II herbicides on the photosynthetic membranes of the cyanobacterium *Aphanocapsa* 6308. Plant Physiol 71:388-392

Altenburger R, Boedeker W, Faust M, Grimme LH (1990) Evaluation of the isobologram method for the assessment of mixtures of chemicals: combination effect studies with pesticides in algal biotests. Ecotoxicol Environ Saf 20:98-114

Amrhein N, Roy P (1993) The buckwheat assay for glyphosate and other inhibitors of aromatic biosynthesis. In: Böger P, Sandmann G (eds) Target assays for modern herbicides and related compounds. Lewis, Boca Raton, Florida, pp109-121

Aro E-M, Virgin I, Andersson B (1993) Photoinhibition of photosystem II. Inactivation, protein damage and turnover. Biochim Biophys Acta 1143:113-134

Astier C, Vernotte C, Der-Vartanian M, Joset-Espardellier F (1979) Isolation and characterization of two DCMU-resistant mutants of the blue-green alga *Aphanocapsa* 6714. Plant Cell Physiol 20:1501-1510

Astier C, Meyer I, Vernotte C, Etienne AL (1986) Photosystem II electron transfer in highly herbicide resistant mutants of *Synechocystis* 6714. FEBS Lett 207:234-238

Astier C, Perewoska I, Picaud M, Kirilovsky D, Vernotte C (1993) Structural analysis of the Q_B pocket of the D1 subunit of photosystem II in *Synechocystis* PCC 6714 and 6803. Z Naturforsch 48c:199-204

Berard A, Leboulanger C, Pelte T (1999) Tolerance of *Oscillatoria limnetica* Lemmermann to atrazine in natural phytoplankton populations and in pure culture: influence of season and temperature. Arch Environ Contam Toxicol 37:472-479

Böger P (1983) Investigating the mode of action of herbicides affecting the chloroplast. J Pestic Sci 8:261-264

Böger P, Sandmann G (1998a) Carotenoid biosynthesis inhibitor herbicides — mode of action and resistance mechanisms. Pestic Outlook 9:29-35

Böger P, Sandmann G (1998b) Action of modern herbicides. In: Raghavendra AS (ed) Photosynthesis. Cambridge University Press, Cambridge, pp 337-351

Böger P, Wakabayashi K (1995) Peroxidizing herbicides (I): mechanism of action. Z Naturforsch 50c:159-166

Botterman J, Leemans (1988) Engineering of herbicide resistance in plants. Biotechnol Genet Engin Rev 6:321-340

Brusslan J, Haselkorn R (1988) Molecular genetics of herbicide resistance in cyanobacteria. Photosynth Res 17:115-124

Brusslan J, Haselkorn R (1989) Resistance to photosystem II herbicide diuron is dominant to sensitivity in the cyanobacterium *Synechococcus* sp. PCC 7942. EMBO J 8:1237-1245

Campbell D, Zhou G, Gustafsson P, Öquist G, Clarke AK (1995) Electron transport regulates exchange of two forms of photosystem II D1 protein in the cyanobacterium *Synechococcus*. EMBO J 14:101-110

Campbell D, Bruce D, Carpenter C, Gustafsson P, Öquist G (1996) Two forms of the photosystem II D1 protein alter energy dissipation and state transitions in the cyanobacterium *Synechococcus* sp. PCC 7942. Photosynth Res 47:131-144

Cao J, Ohad N, Hirschberg J, Xiong J, Govindjee (1992) Binding affinity of bicarbonate and formate in herbicide-resistant D1 mutants of *Synechococcus* sp. PCC 7942. Photosynth Res 34:397-408

Cerejeira MJ, Pereira T, Silva-Fernandez A (1998) Use of new microbiotests with *Daphnia magna* and *Selenastrum capricornutm* immobilized forms. Chemosphere 37:2949-2955

Chamovitz D, Pecker I, Hirschberg J (1991) The molecular basis of resistance to the herbicide norflurazon. Plant Mol Biol 16:967-974

Chapman DJ, De-Felice J, Barber J (1985) Characteristics of chloroplast thylakoid lipid composition associated with resistance to triazine herbicides. Planta 166:280-285

Clarke AK, Hurry VM, Gustafsson P, Öquist G (1993) Two functionally distinct forms of the photosystem II reaction center protein D1 in the cyanobacterium *Synechococcus* sp. PCC 7942. Proc Natl Acad Sci USA 90:11985-11989

Cobb A (1992) Herbicides and plant physiology. Chapman & Hall, London

Codd GA, Cossar JD (1978) The site of inhibition of photosystem II by 3-(3,4-dichlorophenyl)-N,N′-dimethylurea in thylakoids of the cyanobacterium *Anabaena cylindrica*. Biochem Biophys Res Commun 83:342-346

Creuzet S, Ajlani G, Vernotte C, Astier C (1989) A new ioxynil-resistant mutant in *Synechocystis* PCC 6714: hypothesis on the interaction of ioxynil with the D1 protein. Z Naturforsch 45c:436-440

Critchley C, Russell AW (1994) Photoinhibition of photosynthesis in vivo: the role of protein turnover in photosystem II. Physiol Plant 92:188-196

Elanskaya IV, Chesnavichene EA, Vernotte C, Astier C (1998) Resistance to nitrophenolic herbicides and metronidazole in the cyanobacterium *Synechocystis* sp. PCC 6803 as a result of the inactivation of a nitroreductase-like protein encoded by drgA gene. FEBS Lett 428:188-192

Eley JH, McConnell JF, Catlett RH (1983) Inhibition of metribuzin on growth and photosynthesis of the blue-green alga *Anacystis nidulans*. Environ Exp Bot 23:365-368

Erickson JM, Rahire M, Bennoun P, Delepelaire P, Diner B, Rochaix J-D (1984) Herbicide resistance in *Chlamydomonas reinhardtii* results from a mutation in the chloroplast gene for the 32-kilodalton protein of photosystem II. Proc Natl Acad Sci USA 81:3617-3621

Erickson JM, Rahire M., Rochaix JD, Mets L (1985) Herbicide resistance and cross- resistance: changes at three distinct sites in the herbicide-binding protein. Science 228:204-207

Erickson JM, Pfister K, Rahire M, Togasaki RK, Mets L, Rochaix JD (1989) Molecular and biophysical analysis of herbicide-resistant mutants of *Chlamydomonas reinhardtii*: structure-function relationship of the photosystem II D 1 polypeptide. Plant Cell 1:361-371

Etienne A-L, Ducruet J-M, Ajlani G, Vernotte C (1990) Comparative studies on electron transfer in photosystem II of herbicide-resistant mutants from different organisms. Biochim Biophys Acta 1015:435-440

Faust M, Altenburger R, Backhaus T, Boedeker W, Scholz M, Grimme LH (2000) Predicted assessment of the aquatic toxicity of multiple chemical mixtures. J Environ Qual 29:1063-1068

Fedtke C (1982) Biochemistry and physiology of herbicide action. Springer, Berlin

Feierabend J (1983) Comparison of the action of bleaching herbicides. Z Naturforsch 39c:450-454

Gasquez J, Barralis G, Aigle N (1982) Distrubution et extension de l résistence chloroplastique aux triazines chez les adventices annuelles en France. Agronomie 2:119-124

Gibson MT, Barrett PRF (1989) Terbutryn resistance in *Vaucheria dichotoma*. J Appl Phycol 1:167-170

Girling AF, Pascoe D, Janssen CR, Peither A, Wenzel A, Schafer H, Neumeier B, Mitchell GC, Taylor EJ, Maund SJ, Lay JP, Juttner I, Crossland NO, Stephenson RR, Persoone G (2000) Development of methods for evaluating toxicity to freshwater ecosystems. Ecotoxicol Environ Saf 45:148-176

Gleason FK (1990) The natural herbicide, cyanobacterin, specifically disrupts thylakoid membrane structure in *Euglena gracilis* strain Z. FEMS Microbiol Lett 68:77-82

Gleason FK, Paulson JL (1984) Site of action of the natural algicide, cyanobacterin, in the blue-green alga, *Synechococcus* sp. Arch Microbiol 138:273-277

Gleason FK, Case DE, Sipprell KD, Magnuson TS (1986) Effect of the natural algicide, cyanobacterin, on a herbicide-resistant mutant of *Anacystis nidulans* R2. Plant Sci 46:5-10

Golden SS (1994) Light-responsive gene expression and the biochemistry of the photosystem II reaction center. In: Bryant D (ed) The molecular biology of cyanobacteria. Kluwer, Dordrecht, pp 693-714

Golden SS, Haselkorn R (1985) Mutation to herbicide resistance maps within the psbA gene of *Anacystis nidulans* R2. Science 185:1104-1107

Goloubinoff P, Brusslan J, Golden SS, Haselkorn R, Edelman M (1988) Characterization of the photosystem II 32 kDa protein in *Synechococcus* PCC7942. Plant Mol Biol 11:441-447

Gong H, Ohad I (1991) The PQ/PQH2 ratio and occupancy of photosystem II-QB site by plastoquinone control the degradation of D1 protein during photoinhibition in vivo. J Biol Chem 266:21293-21299

Govindjee, Schwarz B, Rochaix J-D, Strasser RJ (1991) The herbicide-resistant D1 mutant L275F of *Chlamydomonas reinhardtii* fails to show the bicarbonate-reversible formate effect on chlorophyll a fluorescence transients. Photosynth Res 27:199-208

Grizeau D, Jeanne N, Calvayrac R (1985) Isolation and characterization of a DCMU (3-(3,4-dichlorophenyl)-1,1-dimethylurea) resistant strain of a marine alga, *Dunaliella vacuolata*. Physiol Plant 65:189-195

Hatfield PM, Guikema JA, St John JB, Gendel SM (1989) Characterization of the adaptation response of *Anacystis nidulans* to growth in the presence of sublethal doses of herbicide. Curr Microbiol 18:369-374

Havaux M (1989a) Photoacoustic characteristics of leaves of atrazine-resistant weed mutants. Photosynth Res 21:51-59

Havaux M (1989b) Comparison of atrazine-resistant and -susceptible biotypes of *Senecio vulgaris* L.: effects of high and low temperatures on the in vivo photosynthetic electron transfer in intact leaves. J Exp Bot 40:849-854

Haworth P, Steinback KE (1987) Interaction of herbicides and quinone with the Q_B- protein of the diuron-resistant *Chlamydomonas reinhardtii* mutant Dr2. Plant Physiol 83:1027-1031

Heiss S, Johanningmeier U (1992) Analysis of a herbicide resistant mutant obtained from transformation of the *Chlamydomonas* chloroplast. Photosynth Res 34:311-317

Hirschberg J (1984) The molecular basis of triazine-herbicide resistance in higher plant chloroplasts. Z Naturforsch 39c:412-420

Hirschberg J, McIntosh L (1983) Molecular basis of herbicide resistance in *Amaranthus hybridus*. Science 222:1346-1349

Hirschberg J, Bleeker A, Kyle DJ, McIntosh L, Arntzen CJ (1984) The molecular basis of triazine-herbicide resistance in higher plant chloroplasts. Z Naturforsch 39c:412-420

Hirschberg J, Ohad N, Rahat A (1987a) Isolation and characterization of herbicide-resistant mutants in the cyanobacterium *Synechococcus* R2. Z Naturforsch 42c:758-761

Hirschberg J, Ben Yehuda A, Pecker I, Ohad N (1987b) Mutants resistant to photosystem II herbicides. In: von Wettstein D, Chua N-H (eds) Plant molecular biology. Plenum, New York, pp 357-366

Horovitz A, Ohad N, Hirschberg J (1989) Predicted effects on herbicide binding of amino acid substitutions in the D1 protein of photosystem II. FEBS Lett 243:161-164

Howat PD (1987) Weeds resistant to herbicides in Australia and contributing factors leading to their appearance. Plant Protect Q 2:82-85

Jansen MAK, Malkin S, Edelman M (1989) Differential sensitivity of 32 kDa-D1 protein degradation and photosynthetic electron flow to photosystem II herbicides. Z Naturforsch 45c:408-411

Jansen MAK, Depka B, Trebst A, Edelman M (1993) Engagement of specific sites in the plastoquinone niche regulates degradation of the D1 protein in photosystem II. J Biol Chem 268:21246-21252

Johanningmeier U, Hallick RB (1987) The psbA gene of DCMU-resistant *Euglena gracilis* has an amino acid substitution at serine codon 265. Curr Genet 12:465-470

Johanningmeier U, Bodner U, Wildner GF (1987) A new mutation in the gene coding for the herbicide-binding protein in *Chlamydomonas*. FEBS Lett 211:221-224

Johanningmeier U, Sopp G, Brauner M, Altenfeld U, Orawski G, Oettmeier W (2000) Herbicide resistance and supersensitivity in Ala 250 or Ala251 mutants of the D1 protein in *Chlamydomonas reinhardtii*. Pestic Biochem Physiol 66:9-12

Kasai F (1999) Shifts in herbicide tolerance in paddy field periphyton following herbicide application. Chemosphere 38:919-931

Kasai F, Arts MT (1998) The interactive effects of UV-B radiation and a herbicide on uptake and allocation of carbon in two strains of the green alga *Scenedesmus*. Aquat Ecol 31:261-272

Kessler E, Schäfer M, Hümmer C, Kloboucek A, Huss VAR (1997) Physiological, biochemical, and molecular characters for the taxonomy of the subgenera of *Scenedesmus* (*Chlorococcales, Chlorophyta*). Bot Acta 110:244-250

Kirilovsky DL, Vernotte C, Astier C, Etienne A-L (1988) Reversible and irreversible photoinhibition in herbicide-resistant mutants of *Synechocystis* 6714. Biochim Biophys Acta 933:124-131

Kirilovsky DL, Ajlani G, Picaud M, Etienne A-L (1989) Mutations responsible for high light sensitivity in an atrazine-resistant mutant of *Synechocystis*. Plant Mol Biol 13:355-363

Koenig F (1987a) Involvement of the Q_B binding protein (M_r 32000) in the adaptation of the photosynthetic apparatus to light intensity. In: Biggins J (ed) Progress in photosynthesis research. Martinus Nijhoff, Dordrecht, IV, pp 95-98

Koenig F (1987b) A role of the Q_B binding protein in the mechanism of cyanobacterial adaptation to light intensity? Z Naturforsch 42c:727-732

Koenig F (1990a) Different amino acid exchanges in the D1 protein cause different degrees of shade type appearance in herbicide-tolerant *Anacystis*. In: Baltscheffsky M (ed) Current research in photosynthesis. Kluwer, Dordrecht, IV, pp 373-376

Koenig F (1990b) *Anacystis* mutants with different tolerances to DCMU-type herbicides show differences in architecture and dynamics of the photosynthetic apparatus, depending on site and mode of the amino acid exchanges in the D1 protein. Z Naturforsch 45c:446-451

Koenig F (1990c) Shade adaptation in cyanobacteria. Photosynth Res 26:29-37

Koenig F (1992) Development of shade-type appearance — light intensity adaptation — and regulation of the D1 protein in *Synechococcus*. In: Argyroudi-Akoyunoglou JH (ed) Regulation of chloroplast biogenesis. Plenum Press, New York, pp 545-550

Koenig F (1995) Effect of different concentrations of atrazine on RC II-D1 protein synthesis and cell division in *Anacystis* in strong white light. In: Mathis P (ed) Photosynthesis: from light to biosphere. Kluwer, Dordrecht, IV pp 653-656

Koenig F, Samson B (2001) Reaction center II-D1 protein and psbA message levels in dependence on quantum flux density and atrazine concentration during growth of *Synechococcus* in strong white light. (in preparation)

Komenda J, Masojídek J (1998) The effect of photosystem II inhibitors DCMU and BNT on the high-light induced D1 turnover in two cyanobacterial strains *Synechocystis* PCC 6803 and *Synechococcus* PCC 7942. Photosynth Res 57:193-202

Kulkarni RB, Golden SS (1994) Adaptation to high light intensity in *Synechococcus* sp. strain PCC 7942: regulation of the three psbA genes and two forms of the D1 protein. J Bacteriol 176:959-965

Laasch H, Pfister K, Urbach W (1981) Comparative binding of photosystem II-herbicides to isolated thylakoid membranes and intact green algae. Z Naturforsch 36c:1041-1049

Laasch H, Pfister K, Urbach W (1982) High- and low-affinity binding of photosystem II herbicides to isolated thylakoid membranes and intact algal cells. Z Naturforsch 37c:620-631

Lambert R, Böger P (1983) Inhibition of carotenogenesis by substituted diphenyl ethers of the m-phenoxybenzamide type. Pestic Biochem Physiol 20:183-187

Mallipudi LR, Gleason FK (1989) Characterization of a mutant of *Anacystis nidulans* R2 resistant to the natural herbicide, cyanobacterin. Plant Sci 60:149-154

Manthey M, Faust M, Smolka S, Grimme LH (1993) Herbicide bioconcentration in algae: studies on lipophilicity-sorption-activity relationships (LSAR) with *Chlorella fusca*. Sci Total Environ Suppl 453-459

Mattoo AK, St John JB, Wergin WP (1984) Adaptive reorganization of protein and lipid components in chloroplast membranes as associated with herbicide binding. J Cell Biochem 24:163-175

Nelson KJ, Hoagland KD, Siegfried BD (1999) Chronic effects of atrazine on tolerance of a benthic diatom. Environ Toxicol Chem 18:1038-1045

Nicolaus B, Sandmann G, Watanabe H, Wakabayashi K, Böger P (1989) Herbicide-induced peroxidation: influence of light and diuron on protoporphyrin IX formation. Pestic Biochem Physiol 35:192-201

Nitschke L, Wilk A, Schussler W, Metzner G, Lind G (1999) Biodegradation in laboratory-activated sludge plants and aquatic toxicity of herbicides. Chemosphere 39:2313-2323

Oettmeier W (1999) Herbicide resistance and supersensitivity in photosystem II. Cell Mol Life Sci 55:1255-1277

Ohad N, Hirschberg J (1990) A similar structure of the herbicide-binding site in photosystem II of plants and cyanobacteria is demonstrated by site-specific mutagenesis of the psbA gene. Photosynth Res 23:73-79

Ohad N, Hirschberg J (1992) Mutations in the D1 subunit of photosystem II distinguish between quinone- and herbicide-binding sites. Plant Cell 4:273-282

Ohad N, Pecker I, Hirschberg J (1987) Biochemical and molecular analyses of herbicide resistant mutants in cyanobacteria. In: Biggins J (ed) Progress in photosynthesis. Martinus Nijhoff, Dordrecht, III pp 807-810

Ohad N, Amir-Shapira D, Koike H, Inoue Y, Ohad I, Hirschberg J (1990) Amino acid substitutions in the D1 protein of photosystem II affect Q_B^- stabilization and accelerate turnover of D1. Z Naturforsch 45c:402-408

Ohad N, Inoue Y, Hirschberg J (1992) Accelerated rate of turnover of the D1 subunit of photosystem II is correlated with inhibition of electron transfer from Q_A to Q_B in cyanobacterial mutants. In: Argyroudi-Akoyunoglou JH (ed) Regulation of chloroplast biogenesis. Plenum Press, New York, pp 589-595

Perewoska I, Vernotte C, Picaud M, Astier C (1992) Mutations in the D_1 subunit of photosystem II and resistance to the phenol-type herbicide ioxynil in *Synechocystis* PCC 6714 and 6803. Z Naturforsch 47c:580-584

Prášil O, Ohad N, Hirschberg J (1992) Mutations in the Q_B-binding niche in the D1 subunit of photosystem II impair electron transport from Q_A to Q_B. In: Argyroudi-Akoyunoglou JH (ed) Regulation of chloroplast biogenesis. Plenum Press, New York, pp 597-602

Pucheu N, Oettmeier W, Heisterkamp U, Masson K, Wildner GF (1984) Metribuzin-resistant mutants of *Chlamydomonas reinhardii*. Z Naturforsch 39c:437-439

Randolph-Anderson BL, Sato R, Johnson AM, Harris EH, Hauser CR, Oeda K, Ishige F, Nishio S, Gillham NW, Boynton JE (1998) Isolation and characterization of a mutant protoporphyrinogen oxidase gene from *Chlamydomonas reinhardtii* conferring resistance to porphyric herbicides. Plant Mol Biol 38: 839-859

Robinson H, Golden S, Brusslan J, Haselkorn R (1987) Functioning of photosystem II in mutant strains of the cyanobacterium *Anacystis nidulans* R2. In: Biggins J (ed) Progress in photosynthesis research. Martinus Nijhoff, Dordrecht, IV pp 825-828

Sato F, Shigematsu Y, Yamada Y (1988) Selection of an atrazine-resistant tobacco cell line having a mutant psbA gene. Mol Gen Genet 214:358-360

Schaefer MR, Golden SS (1989a) Differential expression of members of a cyanobacterial psbA gene family in response to light. J Bacteriol 171:3973-3981

Schaefer MR, Golden SS (1989b) Light availability influences the ratio of two forms of D1 in cyanobacterial thylakoids. J Biol Chem 264:7412-7417

Schmalfuß J, Matthes B, Mayer P, Böger P (1998) Chloroacetamide mode of action, I: inhibition of very long chain fatty acid synthesis in *Scenedesmus acutus*. Z Naturforsch 53c:995-1003

Schmid GH, Lehmann-Kirk U (1977) Photooxidation reactions of diphenylcarbazide and their DCMU sensitivity in thylakoids of the blue-green alga *Oscillatoria chalybea*. Arch Mikrobiol 115:265-269

Schmitz O, Tsinoremas NF, Schaefer MR, Anandan S, Golden SS (1999) General effect of photosynthetic electron transport inhibitors on translation precludes their use for investigating regulation of D1 biosynthesis in *Synechococcus* sp. strain PCC 7942. Photosynth Res 62:261-271

Schwenger-Erger C, Thiemann J, Barz W, Johanningmeier U, Naber D (1983) Metribuzin resistance in photoautotrophic *Chenopodium rubrum* cell cultures. Characterization of double and triple mutations in the psbA gene. FEBS Lett 329:43-46

Scribner EA, Battaglin WA, Goolsby DA, Thurman EM (2000) Changes in herbicide concentrations in Midwestern streams in relation to changes in use, 1989-1998. Sci Total Environ 248:255-263

Shah DM, Horsch RB, Klee HJ, Kishore GM, Winter JA, Tumer NE, Hironaka CM, Sanders PR, Gasser CS, Aykent S, Siegel NR, Rogers SG, Fraley RT (1986) Engineering herbicide tolerance in transgenic plants. Science 233:478-481

Shigematsu Y, Sato F, Yamada Y (1989) The mechanism of herbicide resistance in tobacco cells with a new mutation in the Q_B protein. Plant Physiol 89:986-992

Sippola K (2000) Regulation of psbA gene expression in the cyanobacterium *Synechococcus* sp. PCC 7942. Ann Univ Turkuensis, Ser AII, tom 132

Sippola K, Aro E-M (1999) Exchange of D1 protein forms in photosystem II complexes in the presence of thiol redox reagents or DCMU in *Synechococcus* sp. PCC 7942. Abstr 4th Eur Workshop on the Molecular Biology of Cyanobacteria, Berlin

Sippola K, Aro E-M (2000) Expression of psbA genes is regulated at multiple levels in the cyanobacterium *Synechococcus* sp. PCC 7942. Photochem Photobiol 71:706-714

Srivastava M, Mohanty P, Bose S (1994) Alterations in the excitation energy distribution in *Synechococcus* PCC 7942 due to prolonged partial inhibition of photosystem II. Comparison between inhibition caused by (a) presence of PS II inhibitor, (b) mutation in the D1 polypeptide of PS II. Biochim Biophys Acta 1186:1-11

Steinrucken HC, Amrhein N (1980) The herbicide glyphosate is a potent inhibitor of 5-enolpyruvyl-shikimic acid-3-phosphate synthase. Biochim Biophys Res Commun 94:1207-1212

Tanaka K, Kato S, Yomo S (1977) Studies on the growth and development in *Spirogyra* III. Three types in the pattern of diurnal movement of *Spirogyra* filaments and transformation of the movement pattern by DCMU. Plant Sci Lett 10:31-39

Thies F, Grimme LH (1996) Effects of cytochrome P450-interacting plant growth retardents, fungicides and related compounds on development and phase-I-biotransformation capacity of unicellular photoautotrophic green algae. Pestic Sci 47:337-346

Thuillier-Bruston F, Calvayrac R, Duval E (1996) Partial molecular analysis of the psbA gene in *Euglena gracilis* mutants exhibiting resistance to DCMU and atrazine. Z Naturforsch 51c:711-720

Thurman EM, Bastian KC, Mollhagen T (2000) Occurrence of cotton herbicides and insecticides in playa lakes of the High Plains of West Texas. Sci Total Environ 248:189-200

Trebst A (1980) Inhibitors in electron flow: tools for the functional and structural localization of carriers and energy conservation sites. Methods Enzymol 69:675-715

Trebst A (1987) The three-dimensional structure of the herbicide binding niche on the reaction center polypeptides of photosystem II. Z Naturforsch 42c:742-750

Trebst A (1991) The molecular basis of resistance of photosystem II herbicides. In: Caseley JC, Cussans GW, Atkin RK (eds) Herbicide resistance in weeds and crops. Butterworth-Heinemann, Oxford, pp 145-163

Trebst A, Draber W (1986) Inhibitors of photosystem II and the topology of the herbicide and Q_B binding polypeptide in the thylakoid membrane. Photosynth Res 10:381-392

Trebst A, Hilp U, Draber W (1993) Response in the inhibitor effeciency of substituted phenols on PS II activity in six mutants of the D1 protein subunit in *Chlamydomonas reinhardtii*. Phytochemistry 5:969-977

Troton D, Calvyrac R, Laval-Martin D (1986) Adaptive plasticity of the thylakoid lipids in diuron-resistant *Euglena gracilis*. Biochim Biophys Acta 878:71-78

Vaughn KC, Duke SO (1984) Ultrastructural alterations to chloroplasts in triazine-resistant weed biotypes. Physiol Plant 62:510-520

Wakabayashi K, Böger P (1995) Peroxidizing herbicides (II): structure-activity relationship and molecular design. Z Naturforsch 50c:591-601

Wakabayashi K, Sandmann G, Ohta H, Böger P (1988) Peroxidizing herbicides: comparison of dark and light effects. J Pestic Sci 13:461-471

Wong PK (2000) Effects of 2,4-D, glyphosate and paraquat on growth, photosynthesis and chlorophyll-a synthesis of *Scenedesmus quadricauda*. Chemosphere 41:177-182

Index

MIX
Papier aus verantwortungsvollen Quellen
Paper from responsible sources
FSC® C105338

If you have any concerns about our products,
you can contact us on
ProductSafety@springernature.com

In case Publisher is established outside the EU,
the EU authorized representative is:
Springer Nature Customer Service Center GmbH
Europaplatz 3, 69115 Heidelberg, Germany

Printed by Libri Plureos GmbH
in Hamburg, Germany